STUDY GUIDE AND WORKBOOK:
AN INTERACTIVE APPROACH

STUDY GUIDE AND WORKBOOK: AN INTERACTIVE APPROACH

for Starr's

BIOLOGY

Concepts and Applications
FOURTH EDITION

JOHN D. JACKSON

North Hennepin Community College

JANE B. TAYLOR

Northern Virginia Community College

Brooks/Cole
Thomson Learning™

Pacific Grove • Albany • Belmont • Boston • Cincinnati • Johannesburg • London • Madrid • Melbourne
Mexico City • New York • Scottsdale • Singapore • Tokyo • Toronto

Biology Publisher: Jack Carey
Project Development Editor: Kristin Milotich
Editorial Assistant: Susan Lussier
Marketing Team: Tami Cueny and Dena Donnelly
Print Buyer: Micky Lawler
Production Editor: Mary Vezilich
Production Assistant: Stephanie Andersen
Manuscript Editor: Denise Cook-Clampert
Art and Permissions Editor: Roberta Broyer
Cover Photo: © 1999 John Warden/AlaskaStock.com
Cover Design: Gary Head, Gary Head Design
Typesetting: Publishers' Design & Production Services, Inc.
Printing and Binding: Globus Printing Co.

For more information, contact:
BROOKS/COLE
511 Forest Lodge Road
Pacific Grove, CA 93950 USA
www.brookscole.com

For permission to use material from this work, contact us by
Web: www.thomsonrights.com
fax: 1-800-730-2215
phone: 1-800-730-2214

Printed in the United States of America

10   9   8   7   6   5   4   3   2   1

ISBN 0-534-37270-8

CONTENTS

Photo Credits:

Chapter 20

p. 215 (35): G. Shih & R. Kessel/Visuals Unlimited.

p. 215 (36): Gary W. Grimes and Steven L´Hernault.

p. 215 (37): Photograph courtesy Robert R. Kay from R. R. Kay, et. al., *Development*, 1989 Supplement, pp. 81–90, © The Company of Biologists Ltd. 1989.

p. 215 (38): Davis M. Phillips/Visuals Unlimited.

p. 218: D. J. Patterson/Seaphot Limited: Planet Earth Pictures.

p. 221 (22): Photo courtesy Jack Jones. *Archives of Microbiology*, Volume 136, 1983 pp. 254–261. Reprinted by permission of Springer-Verlag.

p. 221 (23): Tony Brain/SPL/Photo Researchers.

p. 221 (24): Tony Brain and David Parker/SPL/Photo Researchers.

p. 221 (25): M. Abbey/Visuals Unlimited.

p. 221 (26): Greta Fryxell, University of Texas, Austin.

Chapter 21

p. 226: G. T. Cole, University of Texas, Austin/BPS.

p. 227: (inset left); Ed Reschke.

p. 230 (1): Robert C. Simpson/Nature Stock.

p. 230 (2): John E. Hodgin.

p. 230 (3): Jane Burton/Bruce Coleman Ltd.

p. 230 (4): M. Eichelberger/Visuals Unlimited.

p. 231 (6): Edward S. Ross.

p. 231 (7): Robert C. Simpson/Nature Stock.

p. 231 (8): G. L. Barron, University of Guelph.

Chapter 22

p. 239: Jane Burton/Bruce Coleman Ltd.

p. 241: A. & E. Bomford/Ardea, London.

p. 243: Edward S. Ross.

Chapter 23

p. 260: Photograph Carolina Biological Supply.

p. 272: Hervé Chaumeton/Agence Nature.

p. 273 (a): Ian Took/Biofotos.

p. 273 (b): John Mason/Ardea, London.

p. 273 (c): Chris Huss/The Wildlife Collection.

p. 273 (d): Kjell B. Sandved.

Chapter 24

p. 282: Bill Wood/Bruce Coleman Ltd.

p. 293 (20): Christopher Crowley.

p. 293 (21): Bill Wood/Bruce Coleman Ltd.

p. 294 (23): Reinhard/ZEFA.

p. 294 (24): Peter Scoones/Seaphot Limited: Planet Earth Pictures.

p. 294 (25): Erwin Christian/ZEFA.

p. 294 (26): Peter Scoones/Seaphot Limited: Planet Earth Pictures.

p. 294 (27): Allan Power/Bruce Coleman Ltd.

p. 294 (28): Rick M. Harbo.

p. 294 (29): Hervé Chaumeton/Agence Nature.

Chapter 25

p. 302 (1): D. E. Akin and I. L. Rigsby, Richard B. Russel Agricultural Research Center, Agricultural Research Service, U.S., Department of Agriculture, Athens, GA.

p. 302 (2): Biophoto Associates.

p. 302 (3): Biophoto Associates.

p. 302 (4): Kingsley R. Stern.

p. 302 (5): Biophoto Associates.

p. 302 (6): Jan Robert Factor/Photo Researchers.

p. 305: Robert and Linda Mitchell.

p. 306 (center): Carolina Biological Supply Company.

p. 306 (right): James W. Perry.

p. 307 (center): Ray F. Evert.

p. 307 (right): James W. Perry.

p. 309 (above): C. E. Jeffree et. al., *Planta*, 172 (1):20–37, 1987; reprinted by permission of C. E. Jeffree and Springer-Verlag.

p. 309 (below): Jeremy Burgess/SPL Photos/Photo Researchers.

p. 311: Chuck Brown.

p. 314: H. A. Core, W. A. Coté and A. C. Day, *Wood Structure and Identification*, Second edition, Syracuse University Press, 1979.

Chapter 27

p. 328: Photographs Gary Head.

p. 334 (a): Patricia Schulz.

p. 334 (b): Patricia Schulz.

p. 334 (c): Ray F. Evert.

p. 334 (d): Ray F. Evert.

p. 334 (e): Ripon Microslides.

p. 334 (f): Ripon Microslides.

p. 338 (top): Hervé Chaumeton/Agence Nature.

p. 338 (middle): Barry L. Runk/Grant Heilman Photography.

p. 338 (bottom): James D. Mauseth.

Chapter 29

p. 372: Manfred Kage/Peter Arnold, Inc.

p. 375: C. Yokochi and J. Rohen, *Photographic Anatomy of the Human Body*, Second edition, Igaku-Shoin, Ltd., 1979.

Chapter 30

p. 382: Ed Reschke.

Chapter 32

p. 409: Ed Reschke.

p. 413: D. Fawcett, *The Cell*, Philadelphia: W. B. Saunders Co., 1966.

Chapter 38

p. 493: Ed Reschke.

p. 501 (left & right): Lennart Nilsson, *A Child Is Born*, © 1966, 1977 Dell Publishing Company.

PREFACE

Tell me and I will forget, show me and I might remember, involve me and I will understand.
—Chinese Proverb

The proverb outlines three levels of learning, each successively more effective than the method preceding it. The writer of the proverb understood that humans learn most efficiently when they *involve* themselves in the material to be learned. This study guide is like a tutor; when properly used it increases the efficiency of your study periods. The interactive exercises actively involve you in the most important terms and central ideas of your text. Specific tasks ask you to recall key concepts and terms and apply them to life; they test your understanding of the facts and indicate items to reexamine or clarify. Your performance on these tasks provides an estimate of your next test score based on specific material. Most important, though, this biology study guide and text together help you make informed decisions about matters that affect your own well-being and that of your environment. In the years to come, human survival on planet Earth will demand administrative and managerial decisions based on an informed biological background.

HOW TO USE THIS STUDY GUIDE

Following this preface, you will find an outline that will show you how the study guide is organized and will help you use it efficiently. Each chapter begins with a title and an outline list of the 1- and 2-level headings in that chapter. The Interactive Exercises follow, wherein each chapter is divided into sections of one or more of the main (1-level) headings that are labeled 1.1, 1.2, and so on. *For easy reference to an answer or definition, each question and term in this unique study guide is accompanied by the appropriate text page(s), and appears in the form:* [p.352]. The Interactive Exercises begin with a list of Selected Words (other than boldfaced terms) chosen by the authors as those that are most likely to enhance understanding. In the text chapters, the selected words appear in italics, quotation marks, or roman type. This is followed by a list of Boldfaced, Page-Referenced Terms that appear in the text. These terms are essential to understanding each study guide section of a particular chapter. Space is provided by each term for you to formulate a definition in your own words. Next is a series of different types of exercises that may include completion, short answer, true/false, fill-in-the-blank, matching, choice, dichotomous choice, label and match, problems, labeling, sequencing, multiple choice, and completion of tables.

A Self-Quiz immediately follows the Interactive Exercises. This quiz is composed primarily of multiple-choice questions although sometimes we present another examination device or some combination of devices. Any wrong answers in the Self-Quiz indicate portions of the text you need to reexamine. A series of Chapter Objectives/Review Questions follows each Self-Quiz. These are tasks that you should be able to accomplish if you have understood the assigned reading in the text. Some objectives require you to compose a short answer or long essay while others may require a sketch or supplying correct words.

The final part of each chapter is named Integrating and Applying Key Concepts. It invites you to try your hand at applying major concepts to situations in which there is not necessarily a single pat answer and so none is provided in the chapter answer section (except for a problem in Chapter 10). Your text generally will provide enough clues to get you started on an answer, but this part is intended to stimulate your thought and provoke group discussions.

A person's mind, once stretched by a new idea, can never return to its original dimension.
—Oliver Wendell Holmes

STRUCTURE OF THIS STUDY GUIDE

The outline below shows how each chapter in this study guide is organized.

Chapter Number ⟶

4

Chapter Title ⟶

CELL STUCTURE AND FUNCTION
Animalcules and Cells Fill'd With Juices

Chapter Outline ⟶

4.1 BASIC ASPECTS OF CELL STRUCTURE AND FUNCTION
Structural Organization of Cells
Fluid Mosaic Model of Cell Membranes
Overview of Membrane Proteins

4.2 CELL SIZE AND CELL SHAPE
Focus on Science:
MICROSCOPES: GATEWAYS TO CELLS

4.3 THE DEFINING FEATURES OF EUKARYOTIC CELLS
Major Cellular Components
Which Organelles Are Typical of Plants?
Which Organelles Are Typical of Animals?

4.4 THE NUCLEUS
Nuclear Envelope
Nucleolus
Chromosomes
What Happens to the Proteins Specified by DNA?

4.5 THE CYTOMEMBRANE SYSTEM
Endoplasmic Reticulum
Golgi Bodies
A Variety of Vesicles

4.6 MITOCHONDRIA

4.7 SPECIALIZED PLANT ORGANELLES
Chloroplasts and Other Plastids
Central Vacuole

4.8 THE CYTOSKELETON
The Main Components
The Structural Basis of Cell Movements
Flagella and Cilia
Intermediate Filaments

4.9 CELL SURFACE SPECIALIZATIONS
Eukaryotic Cell Walls
Matrixes Between Animal Cells
Cell-to-Cell Junctions

4.10 PROKARYOTIC CELLS-THE BACTERIA

Interactive Exercises ⟶ The interactive exercises are divided into numbered sections by titles of main headings and page references. *This study guide is unique in that each question or term is accompanied by a reference to the text page(s) where that answer or definition may be found.* Each section begins with a list of author-selected words that appear in the text chapter in italics, quotation marks, or roman type. This is followed by a list of important boldfaced, page-referenced terms from each section of the chapter. Each section ends with interactive exercises that vary in type and require constant interaction with the important chapter information.

Self-Quiz ⟶ Usually a set of multiple-choice questions that sample important blocks of text information.

Chapter Objectives/ ⟶ Combinations of relative objectives to be met and questions to be answered.
Review Questions

Integrating and ⟶ Applications of text material to questions for which there may be more than
Applying Key Concepts one correct answer.

Answers to Interactive ⟶ Answers for all interactive exercises can be found at the end of this study
Exercises and Self-Quiz guide by chapter and title, and the main headings with their page references, followed by answers for the Self-Quiz.

1

CONCEPTS AND METHODS IN BIOLOGY

Interactive Exercises

Note: In the answer sections of this book, a specific molecule is most often indicated by its abbreviation. For example, adenosine triphosphate is ATP.

Biology Revisited [pp.2–3]

1.1. DNA, ENERGY, AND LIFE [pp.4–5]

Selected Words: *DNA to RNA to protein* [p.4], *transfer* of energy [p.5], ATP molecules [p.5], *internal* environment [p.5]

In addition to the boldfaced terms, the text features other important terms essential to understanding the assigned material. "Selected Words" is a list of these terms, which appear in the text in italics, in quotation marks, and occasionally in roman type. Latin binomials found in this section are underlined and in roman type to distinguish them from other italicized words.

Boldfaced, Page-Referenced Terms

The page-referenced terms are important; they were in boldface type in the chapter. Write a definition for each term in your own words without looking at the text. Next, compare your definition with that given in the chapter or in the text glossary. If your definition seems inaccurate, allow some time to pass and repeat this procedure until you can define each term rather quickly (how fast you answer is a gauge of your learning effectiveness).

[p.3] biology _____

[p.4] DNA _____

_____

[p.4] reproduction _____

_____

[p.4] development _____

_____

[p.5] energy _____

_____

[p.5] metabolism _____

_____

[p.5] receptors _____

_____

[p.5] stimulus _____

_____

[p.5] homeostasis _____

_____

Fill-in-the-Blanks

(1) _____ [p.4] are the smallest units of matter having the capacity for life. The signature molecule of cells is a nucleic acid known as (2) _____ [p.4]. Encoded in this molecule's structure are the instructions for assembling a dazzling array of (3) _____ [p.4] from a limited number of smaller building blocks, the (4) _____ [p.4] acids. (5) _____ [p.4] are worker proteins that build, split, and rearrange all of the complex molecules of life when they receive an energy boost. (6) _____ [p.4] carry out DNA's protein-building instructions by working as partners with some enzymes. Think about the information encoded in DNA flowing to (7) _____ [p.4] and then to (8) _____ [p.4]. One of the key defining features of life is (9) _____ [p.4], the process by which parents transmit DNA instructions for duplicating their traits to offspring. For frogs and humans and other large organisms, DNA also guides (10) _____ [p.4], the transformation of a fertilized egg into a multicelled adult with cells, tissues, and organs specialized for particular tasks.

(11) _____ [p.5] is most simply defined as the capacity to do work. Nothing in the universe happens without a complete or partial (12) _____ [p.5] of energy. (13) _____ [p.5] refers to the cell's capacity to obtain and convert energy from its surroundings and use energy to maintain itself, grow, and produce more cells. Leaves contain cells that carry on the process of (14) _____ [p.5] by intercepting energy from the sun and converting it to energy molecules called (15) _____ [p.5]. These molecules transfer energy to metabolic workers—in this case, enzymes that assemble sugar molecules. These energy molecules also form by (16) _____ _____ [p.5]. This process can release energy that cells have tucked away in sugars and other kinds of molecules.

Organisms sense changes in their surroundings, then they make controlled, compensatory (17) _____ [p.5] to them. To accomplish this, each organism has (18) _____ [p.5], which are molecules and structures that detect (19) _____ (plural) [p.5]. A (20) _____ (singular) [p.5] is a specific form of energy detected by receptors. Cells adjust metabolic activities in response to signals from receptors. Following a snack, simple sugars and other molecules leave the gut and enter the blood, which is part of a(n) (21) _____ [p.5] environment. Blood sugar level then rises and stimulates secretion of the hormone insulin by the pancreas. Most of your cells have receptors for this hormone. Insulin stimulates cells to take up sugar molecules from the internal environment and return blood sugar concentration levels to normal. Organisms respond so exquisitely to energy changes that their internal operating conditions remain within tolerable limits. This is called a state of (22) _____ [p.5], one of the key defining features of life.

1.2. ENERGY AND LIFE'S ORGANIZATION [pp.6–7]

Boldfaced, Page-Referenced Terms

[p.6] cell _____

_____

[p.6] multicelled organisms _____

_____

[p.6] population _____

_____

[p.6] community _____

_____

[p.6] ecosystem _____

_____

[p.7] biosphere _____

_____

[p.7] producers _____

_____

[p.7] consumers _____

_____

[p.7] decomposers _____

_____

Matching

Choose the most appropriate answer for each term.

1. ___organ system [p.6]
2. ___cell [p.6]
3. ___community [p.6]
4. ___ecosystem [pp.6–7]
5. ___molecule [p.6]
6. ___organelle [p.6]
7. ___population [p.6]
8. ___subatomic particle [p.6]
9. ___tissue [p.6]
10. ___biosphere [pp.6–7]
11. ___multicelled organism [p.6]
12. ___organ [p.6]
13. ___atom [p.6]

A. One or more tissues interacting as a unit
B. A proton, neutron, or electron
C. A well-defined structure within a cell, performing a particular function
D. All of the regions of Earth where organisms can live
E. The smallest unit of life capable of surviving and reproducing on its own
F. Two or more organs whose separate functions are integrated to perform a specific task
G. Two or more atoms bonded together
H. All of the populations interacting in a given area
I. The smallest unit of a pure substance that has the properties of that substance
J. A community interacting with its nonliving environment
K. An individual composed of cells arranged in tissues, organs, and often organ systems
L. A group of individuals of the same species in a particular place at a particular time
M. A group of cells that work together to carry out a particular function

Sequence

Arrange the following levels of organization in nature in the correct hierarchical order. Write the letter of the most inclusive level next to 14. The letter of the least inclusive level is written next to 26.

14. ___
15. ___
16. ___
17. ___
18. ___
19. ___
20. ___
21. ___
22. ___
23. ___
24. ___
25. ___
26. ___

A. Tissue [p.6]
B. Community [p.6]
C. Molecule [p.6]
D. Biosphere [p.6]
E. Organ system [p.6]
F. Organelle [p.6]
G. Ecosystem [p.6]
H. Atom [p.6]
I. Cell [p.6]
J. Population [p.6]
K. Subatomic particle [p.6]
L. Multicelled organism [p.6]
M. Organ [p.6]

Fill-in-the-Blanks

Plants and other organisms that make their own food are (27) _____ [p.7] and serve as an energy entry point for the world of life. Animals feed directly or indirectly on energy stored in tissues of the producers; they are known as (28) _____ [p.7]. (29) _____ [p.7] are bacteria and fungi that feed on tissues or remains of other organisms and break down biological molecules to simple materials that may be cycled back to producers. Thus, there is a one-way flow of (30) _____ [p.7] through organisms and a (31) _____ [p.7] of materials among them that organizes life in the biosphere.

1.3. IF SO MUCH UNITY, WHY SO MANY SPECIES? [pp.8–9]

Selected Words: *family* [p.8], *order* [p.8], *phylum* [p.8], *kingdom* [p.8], *prokaryotic* [p.8], *eukaryotic* [p.9]

Boldfaced, Page-Referenced Terms

[p.8] species _____

_____

[p.8] genus _____

_____

[p.8] Archaebacteria _____

_____

[p.8] Eubacteria _____

_____

[p.8] Protista _____

_____

[p.8] Fungi _____

_____

[p.8] Plantae _____

_____

[p.8] Animalia _____

_____

Fill-in-the-Blanks

Different "kinds" of organisms are referred to as (1) _____ [p.8]. A (2) _____ [p.8] is the first of a two-part name of each organism which encompasses all the species that seem closely related by way of their recent descent from a common ancestor. For example, the pronghorn antelope is known by the two-part name *Antilocapra americana*; *Antilocapra* is the (3) _____ [p.8] name and *americana* is the (4) _____ [p.8] name. In the classification of organisms, genera that share common ancestry are grouped in the same (5) _____ [p.8]. Related families are grouped in the same (6) _____ [p.8] and related orders are grouped in the same (7) _____ [p.8]. Related classes are grouped into a(n) (8) _____ [p.8] that is then assigned to one of the six kingdoms of life. The adjective (9) _____ [p.8] describes single-celled organisms that lack a nucleus while the adjective (10) _____ [p.9] describes single-celled and multicelled organisms whose DNA is located in a nucleus.

Complete the Table

11. Complete the table below by entering the correct name of each kingdom of life described.

Kingdom	Description
[pp.8–9] a.	Eukaryotic, multicelled, photosynthetic producers
[pp.8–9] b.	Prokaryotic, very successful in distribution so they live nearly everywhere, single cells, producers, consumers, or decomposers
[pp.8–9] c.	Eukaryotic, mostly multicelled, decomposers and consumers, food is digested outside their cells and bodies
[pp.8–9] d.	Prokaryotic, live only in extreme habitats, such as the ones that prevailed when life originated
[pp.8–9] e.	Eukaryotic, single-celled species and some multicelled forms, larger than bacteria but internally more complex
[pp.8–9] f.	Eukaryotic, diverse multicelled consumers, actively move at least during some stage of their life.

Sequence

Arrange the following categories in correct hierarchical order. Write the letter of the largest, most inclusive category next to 12. The letter of the smallest, most exclusive category is written next to 18. This exercise classifies an animal with the common name of "beaver." Refer to p.8 in the text and Appendix I.

12. ___   A. Class: Mammalia [p.8]

13. ___   B. Family: Castoridae [p.8]

14. ___   C. Genus: *Castor* [p.8]

15. ___   D. Kingdom: Animalia [p.8]

16. ___   E. Order: Rodentia [p.8]

17. ___   F. Phylum: Chordata [p.8]

18. ___   G. Species: *canadensis* [p.8]

1.4. AN EVOLUTIONARY VIEW OF DIVERSITY [pp.10–11]

Selected Words: *natural* selection [p.10], *antibiotics* [p.11], *Staphylococcus* [p.11]

Boldfaced, Page-Referenced Terms

[p.10] mutation _____

_____

[p.10] adaptive trait _____

_____

[p.10] evolution _____

_____

[p.10] artificial selection _____

_____

[p.11] natural selection _____

_____

Choice

For questions 1–10, choose from the following:

         a. evolution through artificial selection   b. evolution through natural selection

1. ___ Pigeon breeding [p.10]

2. ___ Antibiotics are powerful agents of this process [p.11]

3. ___ A favoring of adaptive traits in nature [p.10]

4. ___ The selection of one form of a trait over another taking place under contrived, manipulated conditions [p.10]

5. ___ Refers to change that is occurring within a line of descent over time [p.10]

6. ___ A difference in which individuals of a given generation survive and reproduce, the difference being an outcome of which ones have adaptive forms of traits [p.11]

7. ___ Darwin viewed this as a simple model for natural selection [p.10]

8. ___ Breeders are the "selective agents" [p.10]

9. ___ The mechanism whereby antibiotic resistance evolves [p.11]

10. ___ Changing of the relative frequencies of different forms of moths through successive generations [p.10]

1.5. THE NATURE OF BIOLOGICAL INQUIRY [pp.12–13]

1.6. *Focus on Science:* THE POWER OF EXPERIMENTAL TESTS [pp.14–15]

1.7. THE LIMITS OF SCIENCE [p.15]

Selected Words: "if-then" process [p.12], *inductive* logic [p.12], *deductive* logic [p.12], sampling error [p.13], *high or low probability* [p.13], *biological therapy* [p.14], *Escherichia coli* [p.14], *subjective* answers [p.15]

Boldfaced, Page-Referenced Terms

[p.12] hypotheses _____

_____

[p.12] prediction _____

_____

[p.12] models _____

_____

[p.12] logic _____

_____

[p.12] test _____

_____

[p.12] experiments _____

_____

[p.13] control group _____

_____

[p.13] variables _____

_____

[p.13] scientific theory _____

_____

Sequence

Arrange the following steps of the scientific method in correct chronological sequence. Write the letter of the first step next to 1, the letter of the second step next to 2, and so on.

1. ___   A. Develop hypotheses about what the solution or answer to a problem might be. [p.12]

2. ___   B. Devise ways to test the accuracy of predictions drawn from the hypothesis (use of observations, models, and experiments). [p.12]

3. ___   C. Repeat or devise new tests (different tests might support the same hypothesis). [p.12]

4. ___   D. Make a prediction, using hypotheses as a guide; the "if-then" process. [p.12]

5. ___   E. If the tests do not provide the expected results, check to see what might have gone wrong. [p.12]

6. ___   F. Objectively analyze and report the results from tests and the conclusions drawn. [p.12]

7. ___   G. Identify a problem or ask a question about nature. [p.12]

Labeling (thought questions)

Assume that you have to identify what object is hidden inside a sealed, opaque box. Your only tools to test the contents are a bar magnet and a triple-beam balance. Label each of the following with an O (for observation) or a C (for conclusion).

8. ___ The object has two flat surfaces.

9. ___ The object is composed of nonmagnetic metal.

10. ___ The object is not a quarter, a half-dollar, or a silver dollar.

11. ___ The object weighs x grams.

12. ___ The object is a penny.

Complete the Table

13. Complete the following table of concepts important to understanding the scientific method of problem solving. Choose from scientific experiment, variable, prediction, inductive logic, control group, hypothesis, deductive logic, and theory.

Concept	Definition
[p.12] a.	An educated guess about what the answer (or solution) to a scientific problem might be
[p.12] b.	A statement of what one should be able to observe in nature if one looks; the "if-then" process
[p.13] c.	A related set of hypotheses that, taken together, form a broad explanation of a fundamental aspect of the natural world
[p.12] d.	A carefully designed test that manipulates nature into revealing one of its secrets
[p.13] e.	Used in scientific experiments to evaluate possible side effects of a test being performed on an experimental group
[p.13] f.	The control group is identical to the experimental group except for the *key factor* under study
[p.12] g.	An individual makes inferences about specific consequences or specific predictions that must follow from a hypothesis
[p.12] h.	An individual derives a general statement from specific observations

Dichotomous Choice

Circle one of two possible answers given between parentheses in each statement; questions 14–18 deal with spontaneous generation.

An Italian physician, Francisco Redi, published a paper in 1688 in which he challenged the doctrine of spontaneous generation, the proposition that living things could arise from dead material. Although many examples of spontaneous generation were described in his day, Redi's work dealt particularly with disproving the notion that decaying meat could be transformed into flies. He tested his ideas in a laboratory.

14. "Two sets of jars are filled with meat or fish. One set is sealed; the other is left open so that flies can enter the jars." This description deals with a(n) (hypothesis/experiment). [p.12]
15. The description in the last question also includes a (prediction/control). [p.13]
16. Prior to his test, Redi suggested that "Worms are derived directly from the droppings of flies." This statement represents a(n) (theory/hypothesis). [p.12]
17. "Worms (maggots) will appear only in the second set of jars" represents a(n) (prediction/hypothesis). [p.12].
18. The statement, "Mice arise from a dirty shirt and a few grains of wheat placed in a dark corner," is best called a (belief/test). [p.15]
19. From a multitude of individual observations he made of the natural world, Charles Darwin proposed the theory of organic evolution. This was an example of (deductive logic/inductive logic). [p.12]
20. Since the time that Darwin proposed the theory of organic evolution, countless numbers of biologists have discovered evidence of various kinds that conform to the general theory. This is an example of (deductive logic/inductive logic). [p.12]
21. The control group is identical to the experimental group except for the (hypothesis/variable) being studied. [p.13]

22. Systematic observations, model development, and conducting experiments are all methods employed to (make predictions/test predictions). [p.12]
23. Science is distinguished from faith in the supernatural by (cause and effect/experimental design). [p.12]
24. Through their failure to use large-enough samples in their experiments, scientists encounter (bias in reporting results/sampling error). [p.13]
25. Science emphasizes reporting test results in (quantitative/qualitative terms). [p.15]

Completion

26. Questions whose answers are _____ in nature do not readily lend themselves to scientific analysis and experimentation. [p.15]
27. Scientists often stir up controversy when they explain a part of the world that was considered beyond natural explanation—that is, belonging to the "_____." [p.15]
28. The external world, not internal _____, must be the testing ground for scientific beliefs. [p.15]

Self-Quiz

___ 1. About 12 to 24 hours after a meal, a person's blood-sugar level normally varies from about 60 to 90 mg per 100 ml of blood, though it may attain 130 mg/100 ml after meals high in carbohydrates. That the blood-sugar level is maintained within a fairly narrow range despite uneven intake of sugar is due to the body's ability to carry out _____. [p.5]
   a. predictions
   b. inheritance
   c. metabolism
   d. homeostasis

___ 2. Different species of Galapagos Island finches have different beak types to obtain different kinds of food. One species removes tree bark with a sharp beak to forage for insect larvae and pupae while another species has a large, powerful beak capable of crushing and eating large, heavy coated seeds. These statements illustrate _____. [p.10]
   a. adaptation
   b. metabolism
   c. puberty
   d. homeostasis

___ 3. A boy is color-blind just as his grandfather was, even though his mother had normal vision. This situation is the result of _____. [p.4]
   a. adaptation
   b. inheritance
   c. metabolism
   d. homeostasis

___ 4. The digestion of food, the production of ATP by respiration, the construction of the body's proteins, cellular reproduction by cell division, and the contraction of a muscle are all part of _____. [p.5]
   a. adaptation
   b. inheritance
   c. metabolism
   d. homeostasis

___ 5. Which of the following does *not* involve using energy to do work? [p.5]
   a. atoms bonding together to form molecules
   b. the division of one cell into two cells
   c. the digestion of food
   d. none of these

___ 6. The experimental group and control group are identical except for _____. [p.13]
   a. the number of variables studied
   b. the variable under study
   c. the two variables under study
   d. the number of experiments performed on each group

___ 7. A hypothesis should *not* be accepted as valid if _____. [p.12]
   a. the sample studied is determined to be representative of the entire group
   b. a variety of different tools and experimental designs yield similar observations and results
   c. other investigators can obtain similar results when they conduct the experiment under similar conditions
   d. several different experiments, each without a control group, systematically eliminate each of the variables except one

___ 8. The principal point of evolution by natural selection is that _____. [p.11]
   a. it measures the difference in survival and reproduction that has occurred among individuals who differ from one another in one or more traits
   b. even bad mutations can improve survival and reproduction of organisms in a population
   c. evolution does not occur when some forms of traits increase in frequency and others decrease or disappear with time
   d. individuals lacking adaptive traits make up more of the reproductive base for each new generation

___ 9. Which match is incorrect? [pp.8–9]
   a. Kingdom Animalia—multicelled consumers, most move about
   b. Kingdom Plantae—mostly multicelled producers
   c. Kingdom Monera—relatively simple, multicelled organisms
   d. Kingdom Fungi—mostly multicelled decomposers
   e. Kingdom Protista—many complex single cells, some multicellular

___10. The least inclusive of the taxonomic categories listed is _____. [p.8]
   a. family
   b. phylum
   c. class
   d. order
   e. genus

Chapter Objectives/Review Questions

This section lists general and detailed chapter objectives that can be used as review questions. You can make maximum use of these items by writing answers on a separate sheet of paper. Fill in answers where blanks are provided. To check for accuracy, compare your answers with information given in the chapter or glossary.

1. It is at the _____ level that differences between living and nonliving things begin to emerge. [p.4]
2. The signature molecule of cells is a nucleic acid known as _____. [p.4]
3. Describe the nature of the instructions that are encoded in DNA's structure. [p.4]
4. Cells arise only from cells that already exist through the process of _____. [p.4]
5. Briefly describe the role of DNA in the development of organisms. [pp.4–5]
6. Relate the concept of energy to the capacity of cells to metabolize. [p.5]
7. _____ is an energy carrier that helps drive hundreds of metabolic activities. [p.5]
8. By the process of aerobic _____, cells can release stored energy in food molecules and produce ATP molecules. [p.5]
9. _____ are certain molecules and structures that can detect specific kinds of information about the environment. [p.5]
10. List some typical stimuli that are detected by receptors. [p.5]
11. _____ refers to the internal operating conditions of organisms remaining within tolerable limits. [p.5]
12. Arrange in order, from least inclusive to most inclusive, the levels of organization that occur in nature. Define each level as you list it. [pp.6–7]
13. Explain how the actions of producers, consumers, and decomposers create an interdependency among organisms. [p.7]

14. Describe the general pattern of energy flow through Earth's life forms, and explain how Earth's resources are used again and again (cycled). [p.7]
15. Explain the use of genus and species names by considering your Latin name, *Homo sapiens*. [p.8]
16. Arrange in order, from greater to fewer organisms included, the following categories of classification: class, family, genus, kingdom, order, phylum, and species. [p.8]
17. List the six kingdoms of life; briefly describe the general characteristics of the organisms placed in each. [pp.8–9]
18. Distinguish these terms: *prokaryotic, eukaryotic.* [pp.8–9]
19. Explain the origin of trait variations that function in inheritance. [p.10]
20. A(n) _____ trait is any form of a trait that helps an organism survive and reproduce under a given set of environmental conditions. [p.10]
21. _____ means genetically based changes in a line of descent over time. [p.10]
22. Darwin used _____ selection as a model for natural selection. [p.10]
23. Define *natural selection,* and briefly describe what is occurring when a population is said to evolve. [pp.10–11]
24. Explain what is meant by the term *diversity* and speculate about what caused the great diversity of life forms on Earth. [p.11]
25. Be able to list the general steps used in scientific research. [p.12]
26. Be able to distinguish between inductive logic and deductive logic. [p.12]
27. _____ are tests that simplify observation in nature or the laboratory by manipulating and controlling the conditions under which observations are made. [p.12]
28. Generally, members of a control group should be identical to those of the experimental group except for the key factor under study, the _____. [p.13]
29. Define what is meant by scientific theory; cite an actual example. [p.13]
30. Due to possible bias on the part of experimenters, science emphasizes presenting test results in _____ terms. [p.15]
31. Explain how the methods of science differ from answering questions by using subjective thinking and systems of belief. [p.15]

Integrating and Applying Key Concepts

1. Humans have the ability to maintain body temperature very close to 37°C.
   a. What conditions would tend to make body temperature drop?
   b. What measures do you think your body takes to raise body temperature when it drops?
   c. What conditions would cause body temperature to rise?
   d. What measures do you think your body takes to lower body temperature when it rises?
2. Do you think that all humans on Earth today should be grouped in the same species?
3. What sorts of topics are usually regarded by scientists as untestable by the kinds of methods that scientists generally use?

2

CHEMICAL FOUNDATIONS FOR CELLS

Interactive Exercises

Checking Out Leafy Clean-Up Crews [pp.20–21]

2.1. REGARDING THE ATOMS [p.22]

2.2. *Focus on Science:* USING RADIOISOTOPES TO TRACK CHEMICALS AND SAVE LIVES [p.23]

Selected Words: *phytoremediation* [p.20], *trace* element [p.20], *atomic* number [p.22], *mass* number [p.22], *PET* [p.23], *radiation therapy* [p.23]

Boldfaced, Page-Referenced Terms

[p.20] elements _____

_____

[p.22] atoms _____

_____

[p.22] protons _____

_____

[p.22] electrons _____

_____

[p.22] neutrons _____

_____

[p.22] isotopes _____

_____

[p.22] radioisotope _____

_____

[p.23] tracer _____

_____

Matching

Choose the most appropriate answer for each term.

1. ___atoms [p.22]
2. ___protons [p.22]
3. ___trace element [p.20]
4. ___PET [p.23]
5. ___neutrons [p.22]
6. ___electrons [p.22]
7. ___atomic number [p.22]
8. ___mass number [p.22]
9. ___phytoremediation [p.20]
10. ___elements [p.20]
11. ___isotope [p.22]
12. ___radioisotopes [p.22]
13. ___tracer [p.23]
14. ___radiation therapy [p.23]

A. Radioisotope used with scintillation counters to reveal the pathway or destination of a substance
B. Subatomic particles with a negative charge
C. Positively charged subatomic particles within the nucleus
D. Positron Emission Tomography; obtains images of particular body tissues
E. Atoms of a given element that differ in the number of neutrons
F. Refers to the number of protons in an atom
G. Radioactive isotopes
H. Chemical elements representing less than 0.01 percent of body weight
I. Destroys or impairs living cancer cells
J. The use of living plants to withdraw harmful substances from the environment
K. The number of protons and neutrons in the nucleus of one atom nucleus
L. Fundamental forms of matter that occupy space, have mass, and cannot be broken down into something else
M. Smallest units that retain the properties of a given element
N. Subatomic particles within the nucleus carrying no charge

2.3. WHAT HAPPENS WHEN ATOM BONDS WITH ATOM? [pp.24–25]

Selected Words: *orbitals* [p.24], *lowest available energy level* [p.24], *higher energy levels* [p.24], *formulas* [p.24], *chemical equations* [p.24], *reactants* [p.24], *products* [p.24], *mole* [p.24], *inert* atoms [p.25]

Boldfaced, Page-Referenced Terms

[p.24] shell model _____

_____

[p.24] chemical bond _____

_____

[p.25] molecule _____

_____

[p.25] compounds _____

_____

[p.25] mixture _____

Matching

Choose the most appropriate answer for each term.

1. ___mixture [p.25]
2. ___shell [p.24]
3. ___lowest available energy level [p.24]
4. ___inert atoms [p.25]
5. ___orbitals [p.24]
6. ___chemical bond [p.24]
7. ___compounds [p.25]
8. ___molecule [p.25]
9. ___higher energy levels [p.24]

A. Regions of space around an atom's nucleus where electrons are likely to be at any one instant
B. Results when two or more atoms bond together
C. Two or more elements are simply intermingling in proportions that usually vary
D. A union between the electron structures of atoms
E. A series of these enclose all orbitals available to an atom's electrons
F. Types of molecules composed of two or more different elements in proportions that never vary
G. Energy of electrons farther from the nucleus than the first orbital
H. Refers to those with no vacancies in their shells and hence show little tendency to enter chemical reactions
I. Energy of electrons in the orbital closest to the nucleus

Complete the Table

10. Complete the table (text Figure 2.8, p.25) by entering the name of the element and its symbol in the appropriate spaces.

Element	Symbol	Atomic Number	Electron Distribution First Shell	Second Shell	Third Shell	Fourth Shell
[p.25] a.		20	2	8	8	2
[p.25] b.		6	2	4		
[p.25] c.		17	2	8	7	
[p.25] d.		1	1			
[p.25] e.		11	2	8	1	
[p.25] f.		7	2	5		
[p.25] g.		8	2	6		

Fill-in-the-Blanks

The expression 12H$_2$O plus 6 CO$_2$ ⟶ 6O$_2$ plus C$_6$H$_{12}$O$_6$ plus H$_2$O is known as the chemical (11) _____ [p.24] for photosynthesis. H$_2$O is the (12) _____ [p.24] for water. The arrow in the expression above means (13) _____ [p.24]. The (14) _____ [p.24] are to the left of the arrow and the (15) _____ [p.24] are to the right of the arrow. In the expression one can count 12 hydrogen atoms on the left side of the arrow, then one should be able to count (16) _____ [p.24] hydrogen atoms on the right side of the arrow. By reference to text Table 2.1, p.22, one can determine the weight of one mole of H$_2$SO$_4$ to be (17) _____ [p.22].

Identification

18. Following the model (number of protons and neutrons shown in the nucleus), identify the indicated atoms of the elements illustrated by entering appropriate electrons in this form: (2e$^-$). [pp.24–25]

MODEL:

He

H

C

N

O

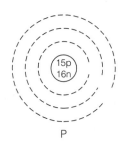

P

S

2.4. IMPORTANT BONDS IN BIOLOGICAL MOLECULES [pp. 26–27]

Selected Words: *sharing* [p.26], *single* covalent bond [p.26], *double* covalent bond [p.26], *triple* covalent bond [p.26], *nonpolar* covalent bond [p.26], *polar* covalent bond [p.27], *no net* charge [p.27]

Boldfaced, Page-Referenced Terms

[p.26] ion _____

_____

[p.26] ionic bond _____

_____

[p.26] covalent bond _____

_____

[p.27] hydrogen bond _____

_____

Identification

1. Following the model, complete the diagram by adding arrows to identify the transfer of electron(s), showing how positive magnesium and negative chlorine ions form ionic bonds to create a molecule of $MgCl_2$ (magnesium chloride). [p.26]

MODEL:

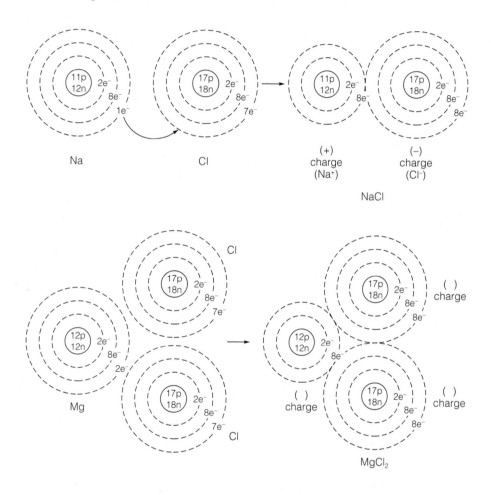

Identification

2. Following the model of hydrogen gas shown, complete the diagram by placing electrons (as dots) in the outer shells to identify the nonpolar covalent bonding that forms oxygen gas; similarly identify polar covalent bonds by completing electron structures to form a water molecule. [p.26]

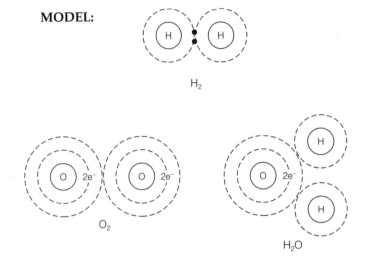

Short Answer

3. Distinguish between a nonpolar covalent bond and a polar covalent bond. Cite an example of each. [pp.26–27] _____

_____

_____

_____

_____

4. Describe one example of a large biological molecule within which hydrogen bonds exist. [p.27]

_____

_____

_____

2.5. PROPERTIES OF WATER [pp.28–29]

Selected Words: dissolved [p.29]

Boldfaced, Page-Referenced Terms

[p.28] hydrophilic substances _____

_____

[p.28] hydrophobic substances _____

_____

[p.29] temperature _____

_____

[p.29] evaporation _____

_____

[p.29] cohesion _____

_____

[p.29] solute _____

_____

Fill-in-the-Blanks

The (1) _____ [p.28] of water molecules allows them to hydrogen-bond with each other. Water molecules hydrogen-bond with polar molecules, which are (2) _____ (water-loving) substances [p.28]. Polarity causes water to repel oil and other nonpolar substances, which are (3) _____ (water-dreading) [p.28]. (4) _____ [p.29] is a measure of the molecular motion of a given substance. Liquid water changes its temperature more slowly than air because of the great amount of heat required to break the high number of (5) _____ [p.29] bonds between water molecules; this property helps stabilize temperature in aquatic habitats and cells. When large energy inputs increase molecular motion to the point where hydrogen bonds stay broken, individual molecules are released from the water surface. This process is known as (6) _____ [p.29].

Below 0°C, water molecules become locked in the less dense latticelike bonding pattern of (7) _____ [p.29], which is less dense than water. Collective hydrogen bonding creates a high tension on surface water molecules resulting in (8) _____ [p.29], the property of water that explains how long, narrow water columns rise to the tops of tall trees. Water is an excellent (9) _____ [p.29] in which ions and polar molecules readily dissolve. Substances dissolved in water are known as (10) _____ [p.29]. A substance is (11) _____ [p.29] in water when spheres of (12) _____ [p.29] form around its individual ions or molecules.

2.6. ACIDS, BASES, AND BUFFERS [pp.30–31]

Selected Words: *donate* H⁺ [p.30], *accept* H⁺ [p.30], *acidic* solutions [p.30], *basic* solutions [p.30], *acid stomach* [p.30], *chemical burns* [p.31], *acid rain* [p.31], *coma* [p.31], *tetany* [p.31], *acidosis* [p.31], *alkalosis* [p.31]

Boldfaced, Page-Referenced Terms

[p.30] hydrogen ions _____

_____

[p.30] hydroxide ions _____

_____

[p.30] pH scale _____

_____

[p.30] acids _____

_____

[p.30] bases _____

_____

[p.31] buffer system _____

[p.31] salts _____

_____

Matching

Choose the most appropriate answer for each term.

1. ___acid stomach [p.30]
2. ___acids [p.30]
3. ___pH scale [p.30]
4. ___chemical burns [p.31]
5. ___H+ [p.30]
6. ___bases [p.30]
7. ___examples of basic solutions [p.30]
8. ___coma [p.31]
9. ___acidosis [p.31]
10. ___OH- [p.30]
11. ___tetany [p.31]
12. ___examples of acid solutions [p.30]
13. ___alkalosis [p.31]
14. ___buffer system [p.31]
15. ___acid rain [p.31]

A. A sometimes irreversible state of unconsciousness
B. CO_2 builds up in the blood, too much H_2CO_3 forms, and blood pH severely decreases
C. Hydroxide ion
D. Substances that accept H^+ when dissolved in water
E. An uncorrected increase in blood pH
F. Used to measure H^+ concentration in various fluids
G. A partnership between a weak acid and the base that forms when it dissolves in water; counters slight pH shifts
H. Hydrogen ion or proton
I. Baking soda, seawater, egg white
J. Can be caused by ammonia, drain cleaner, and sulfuric acid in car batteries
K. Substances that donate H^+ when dissolved in water
L. Lemon juice, gastric fluid, coffee
M. Sulfur dioxide is a major component; some regions are very sensitive to its pH
N. Can be caused by eating too much fried chicken or certain other foods
O. A potentially lethal pH stage in which the body's skeletal muscles enter a state of uncontrollable contraction

Complete the Table

16. Complete the following table by consulting text Figure 2.16, p.30.

Fluid	pH Value	Acid/Base
[p.30] a. Blood		
[p.30] b. Saliva		
[p.30] c. Urine		
[p.30] d. Stomach acid		
[p.30] e. Seawater		
[p.30] f. Some acid rain		

Self-Quiz

___ 1. Each element has a unique _____, which refers to the number of protons present in its atoms. [p.22]
   a. isotope
   b. mass number
   c. atomic number
   d. radioisotope

___ 2. A molecule is _____. [p.25]
   a. a combination of two or more atoms
   b. less stable than its constituent atoms separated
   c. electrically charged
   d. a carrier of one or more extra neutrons

___ 3. If lithium has an atomic number of 3 and an atomic mass of 7, it has _____ neutron(s) in its nucleus. [p.22]
   a. one
   b. two
   c. three
   d. four
   e. seven

___ 4. Substances that are nonpolar and repelled by water are _____. [p.28]
   a. hydrolyzed
   b. nonpolar
   c. hydrophilic
   d. hydrophobic

___ 5. A hydrogen bond is _____. [p.27]
   a. a sharing of a pair of electrons between a hydrogen nucleus and an oxygen nucleus
   b. a sharing of a pair of electrons between a hydrogen nucleus and either an oxygen or a nitrogen nucleus
   c. formed when a small electronegative atom of a molecule weakly interacts with a hydrogen atom that is already participating in a polar covalent bond
   d. none of the above

___ 6. An ionic bond is one in which _____. [p.26]
   a. electrons are shared equally
   b. electrically neutral atoms have a mutual attraction
   c. two charged atoms have a mutual attraction due to electron transfer
   d. electrons are shared unequally

___ 7. A covalent bond is one in which _____. [p.26]
   a. electrons are shared
   b. electrically neutral atoms have a mutual attraction
   c. two charged atoms have a mutual attraction due to electron transfer
   d. electrons are lost

___ 8. A nonpolar covalent bond implies that _____. [pp.26–27]
   a. one negative atom bonds with a hydrogen atom
   b. the bond is double
   c. there is no difference in charge at the ends (the two poles) of the bond
   d. atoms of different elements do not exert the same pull on shared electrons

___ 9. The shapes of large molecules are controlled by _____ bonds. [p.27]
   a. hydrogen
   b. ionic
   c. covalent
   d. inert
   e. single

___ 10. A solution with a pH of 10 is _____ times as basic as one with a pH of 8. [p.30]
   a. 2
   b. 3
   c. 10
   d. 100
   e. 1,000

___ 11. A control that minimizes unsuitable pH shifts is a(n) _____. [p.31]
   a. neutral molecule
   b. salt
   c. base
   d. acid
   e. buffer

Chapter Objectives/Review Questions

1. The use of living plants to withdraw harmful substances from the environment is known as _____. [p.20]
2. Be able to explain what an element is. [p.20]
3. Define *trace* element. [pp.20–21]
4. List and describe the three types of subatomic particles and describe the reason that hydrogen is an exception. [p.22]
5. The number of protons in an atom is referred to as the _____ number of that element; the combined number of protons and neutrons in the atomic nucleus is referred to as the _____ number of that element. [p.22]
6. Distinguish between isotopes and radioisotopes. [p.22]
7. Be able to read a chemical equation. [text Figure 2.7, p.24]
8. Describe what tracers are and how they are used. [p.23]
9. _____ uses radioactively labeled compounds to yield images of metabolically active and inactive tissues in a patient. [p.23]
10. Describe the use of radiation therapy. [p.23]
11. Explain the relationship of orbitals to shells. [p.24]
12. Be able to sketch shell models of the atoms described in the text. [text Figure 2.8, p.25]
13. Explain why helium and neon are known as inert atoms. [p.25]
14. A(n) _____ is formed when two or more atoms bond together. [p.25]
15. A(n) _____ contains atoms of two or more elements whose proportions never vary. [p.25]
16. How does a mixture differ from a compound? [p.25]
17. A(n) _____ is an atom that becomes positively or negatively charged. [p.26]
18. An association of two oppositely charged ions is a(n) _____ bond. [p.26]
19. In a(n) _____ bond, two atoms share electrons. [p.26]
20. Explain why H_2 is an example of a nonpolar covalent bond and H_2O has two polar covalent bonds. [pp.26–27]
21. In a(n) _____ bond, a small, highly electronegative atom of a molecule weakly interacts with a hydrogen atom that is already participating in a polar covalent bond. [p.27]
22. Polar molecules attracted to water are _____; all nonpolar molecules are _____ and are repelled by water. [p.28]
23. _____ is a measure of molecular motion; during _____, an input of heat energy converts liquid water to the gaseous state. [p.29]
24. Describe the formation of ice in terms of hydrogen bonding. [p.29]
25. Completely describe the properties of water that move water from the roots to the tops of the tallest trees. [p.29]
26. Distinguish a solvent from a solute. [p.29]
27. Describe what happens when a substance is dissolved in water. [p.29]
28. The ionization of water is the basis of the _____ scale. [p.30]
29. _____ are substances that donate H^+ ions when they dissolve in water, and _____ are substances that accept H^+ when dissolved in water. [p.30]
30. Black coffee with a pH of 5 is a(n) _____ solution, whereas baking soda with a pH of 9 is a(n) _____ solution. [p.30]
31. Define *buffer system;* cite an example and describe how buffers operate. [p.31]
32. Be able to define the following terms: *acid stomach, chemical burns, coma, tetany, acidosis,* and *alkalosis.* [pp.30–31]
33. A(n) _____ is produced by a chemical reaction between an acid and a base. [p.31]

Integrating and Applying Key Concepts

1. Explain what would happen if water were a nonpolar molecule instead of a polar molecule. Would water be a good solvent for the same kinds of substances? Would the nonpolar molecule's specific heat likely be higher or lower than that of water? Would surface tension be affected? Cohesive nature? Ability to form hydrogen bonds? Is it likely that the nonpolar molecules could form unbroken columns of liquid? What implications would that hold for trees?

3

CARBON COMPOUNDS IN CELLS

Interactive Exercises

Carbon, Carbon, in the Sky—Are You Swinging Low and High? [pp.34–35]

3.1. PROPERTIES OF ORGANIC COMPOUNDS [pp.36–37]

Selected Words: *global warming* [p.35], "inorganic" [p.36], *functional-group transfer* [p.37], *electron transfer* [p.37], *rearrangement* [p.37], *condensation* [p.37], *cleavage* [p.37]

Boldfaced, Page-Referenced Terms

[p.36] organic compounds _____

_____

[p.36] functional groups _____

_____

[p.36] alcohols _____

_____

[p.37] enzymes _____

_____

[p.37] condensation reactions _____

[p.37] hydrolysis _____

Labeling

Study the structural formulas of the organic compounds shown below. Note that each carbon atom can share pairs of electrons with as many as four other atoms. Reference text Figure 3.2, p.36, to identify the circled functional groups (sometimes repeated) by entering the correct name in the blanks with matching numbers.

1. _____ [p.36]

2. _____ [p.36]

3. _____ [p.36]

4. _____ [p.36]

5. _____ [p.36]

6. _____ [p.36]

7. _____ [p.36]

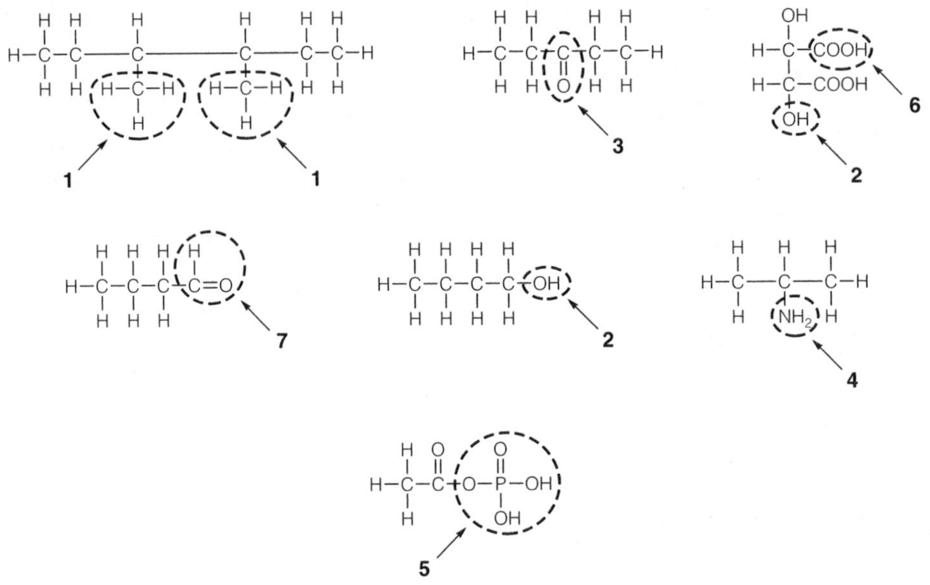

Short Answer

8. State the general role of enzymes as they relate to the chemistry of life. [p.37] _____

_____

_____

_____

Complete the Table

9. Complete the following table summarizing five categories of reactions that are mediated by enzymes.

Reaction Category	Reaction Description
[p.37] a. Rearrangement	
[p.37] b. Cleavage	
[p.37] c. Condensation	
[p.37] d. Electron transfer	
[p.37] e. Functional-group transfer	

Identification

10. Study the structural formulas of the two adjacent amino acids. Identify the enzyme action causing formation of a covalent bond and a water molecule (through a condensation reaction) by circling an H atom from one amino acid and an -OH group from the other amino acid. Also circle the covalent bond that formed the dipeptide [pp.37,43].

amino acid   amino acid   dipeptide

Short Answer

11. Describe hydrolysis through enzyme action for the molecules in exercise 10. [p.37] _____

_____

_____

_____

_____

_____

3.2. CARBOHYDRATES [pp.38–39]

*Selected Words: mono*saccharide [p.38], *oligo*saccharide [p.38], *di*saccharides [p.38], *poly*saccharides [p.38]

Boldfaced, Page-Referenced Terms

[p.38] carbohydrates _____

_____

[p.38] monosaccharides _____

_____

[p.38] oligosaccharides _____

_____

[p.38] polysaccharides _____

_____

Identification

1. In the diagram, identify condensation reaction sites between the two glucose molecules by circling the components of the water removed that allow a covalent bond to form between the glucose molecules (text Figure 3.5, p.38). Note that the reverse reaction is hydrolysis and that both condensation and hydrolysis reactions require enzymes in order to proceed. [p.38]

glucose
(a monosaccharide) + glucose
(a monosaccharide) → enzyme (synthesis) / (hydrolysis) enzyme → maltose
(a disaccharide) + H_2O water

Complete the Table

2. Complete the table below by entering the name of the carbohydrate described by its carbohydrate class and functions.

Carbohydrate	Carbohydrate Class	Function
[p.38] a.		Most plentiful sugar in nature; transport form of carbohydrates in plants; table sugar; formed from glucose and fructose
[p.38] b.		Five-carbon sugar occurring in DNA
[p.38] c.		Main energy source for most organisms; precursor of many organic molecules; serve as building blocks for larger carbohydrates
[p.38] d.		Structural material of plant cell walls; formed from glucose chains
[p.38] e.		Five-carbon sugar occurring in RNA
[p.38] f.		Sugar present in milk; formed from glucose and galactose
[p.39] g.		Muscle cells tap their stores of this compound for a rapid burst of energy
[p.39] h.		Main structural material in some external skeletons and other hard body parts of some animals and fungi
[p.38] i.		Animal starch; stored especially in liver and muscle tissue; formed from glucose chains
[p.38] j.		Storage form for photosynthetically produced sugars

3.3. LIPIDS [pp.40–41]

Selected Words: *unsaturated* tails [p.40], *saturated* tails [p.40], eicosanoids [p.41]

Boldfaced, Page-Referenced Terms

[p.40] lipids _____

_____

[p.40] fats _____

_____

[p.40] fatty acid _____

_____

[p.40] triglycerides _____

_____

[p.41] phospholipid _____

_____

[p.41] sterols _____

_____

[p.41] waxes _____

_____

Labeling

1. In the appropriate blanks, label the molecules shown as saturated or unsaturated.

a   oleic acid

b   stearic acid

a. _____ [p.40]   b. _____ [p.40]

Identification

2. Combine glycerol with three fatty acids to form a triglyceride by circling the participating atoms that will identify three covalent bonds. Also circle the covalent bonds in the triglyceride. [p.40]

glycerol   three fatty acids   triglyceride (a complete fat molecule)

Short Answer

3. Describe the structure and biological functions of phospholipid molecules. [p.41] _____

_____

_____

_____

_____

Choice

For questions 4–18, choose from the following:

    a. fatty acids   b. triglycerides   c. phospholipids   d. waxes   e. sterols

4. ___ richest source of body energy [p.40]

5. ___ honeycomb material [p.41]

6. ___ cholesterol [p.41]

7. ___ saturated tails [p.40]

8. ___ butter and lard [p.40]

9. ___ lack fatty acid tails [p.41]

10. ___ main cell membrane component [p.41]

11. ___ all possess a rigid backbone of four fused carbon rings [p.41]

12. ___ plant cuticles [p.41]

13. ___ precursors of testosterone, estrogen, and bile salts [p.41]

14. ___ unsaturated tails [p.40]

15. ___ vegetable oil [p.40]

16. ___ vertebrate insulation [p.40]

17. ___ furnishes lubrication for skin and feathers [p.41]

18. ___ neutral fats [p.40]

3.4. AMINO ACIDS AND THE PRIMARY STRUCTURE OF PROTEINS [pp.42–43]

3.5. HOW DOES A PROTEIN'S THREE-DIMENSIONAL STRUCTURE EMERGE? [pp.44–45]

3.6. *Focus on the Environment:* FOOD PRODUCTION AND A CHEMICAL ARMS RACE [p.46]

Selected Words: primary structure [p.43], secondary structure [p.44], tertiary structure [p.44], quaternary structure [p.45], *herbicides* [p.46], *insecticides* [p.46], *fungicides* [p.46]

Boldfaced, Page-Referenced Terms

[p.42] proteins _____

_____

[p.42] amino acid _____

_____

[p.42] polypeptide chain _____

_____

[p.44] hemoglobin _____

_____

[p.45] denaturation _____

_____

[p.46] toxin _____

_____

Matching

For exercises 1, 2, and 3, match the major parts of *every* amino acid by entering the letter of the part in the blank corresponding to the number on the molecule.

1. ___[p.42]

2. ___[p.42]

3. ___[p.42]

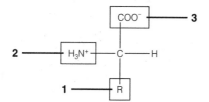

A. R group (a symbol for a characteristic group of atoms that differ in number and arrangement from one amino acid to another)
B. Carboxyl group (ionized)
C. Amino group (ionized)

Identification

4. Four ionized amino acids (in cellular solution) are shown in the upper illustration below; form the proper covalent bonds by circling the acids and ions involved in polypeptide formation. Circle the newly formed peptide bonds of the completed polypeptide on the lower illustration. [p.43]

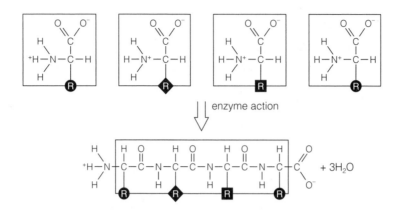

Matching

Choose the most appropriate answer for each term.

5. ___amino acid [p.42]

6. ___peptide bond [p.42]

7. ___polypeptide chain [p.42]

8. ___primary structure [p.43]

9. ___proteins [p.42]

10. ___secondary structure [p.44]

11. ___tertiary structure [p.44]

12. ___dipeptide [p.42]

13. ___quaternary structure [p.45]

14. ___lipoproteins [p.45]

15. ___glycoproteins [p.45]

16. ___denaturation [p.45]

A. A coiled or extended pattern of protein structure caused by regular intervals of H bonds
B. Three or more amino acids joined in a linear chain
C. Proteins with linear or branched oligosaccharides covalently bonded to them; found on animal cell surfaces, in cell secretion, or on blood proteins
D. Folding of a protein through interactions among R groups of a polypeptide chain
E. Form when freely circulating blood proteins encounter and combine with cholesterol, or phospholipids
F. The type of covalent bond linking one amino acid to another
G. Hemoglobin, a globular protein of four chains, is an example
H. Breaking weak bonds in large molecules (such as protein) to change its shape so it no longer functions
I. Formed when two amino acids join together
J. Lowest level of protein structure; has a linear, unique sequence of amino acids, an acid group, a hydrogen atom, and an R group
K. A small organic compound having an amino group, an acid group, a hydrogen atom, and an R group
L. The most diverse of all the large biological molecules; constructed from pools of only twenty kinds of amino acids

3.7. NUCLEOTIDES AND THE NUCLEIC ACIDS [p.47]

Boldfaced, Page-Referenced Terms

[p.47] nucleotides _____

_____

[p.47] ATP _____

_____

[p.47] nucleic acids _____

_____

[p.47] DNA _____

_____

[p.47] RNAs _____

_____

Matching

For exercises 1, 2, and 3, match the following answers to the parts of a nucleotide shown in the diagram.

1. ___[p.47]

2. ___[p.47]

3. ___[p.47]

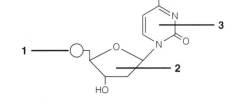

A. A five-carbon sugar (ribose or deoxyribose)
B. Phosphate group
C. A nitrogen-containing base that has either a single-ring or double-ring structure

Identification

4. In the diagram of a single-stranded nucleic acid molecule, encircle as many complete nucleotides as possible. How many complete nucleotides are present? [p.47]

Matching

Choose the most appropriate answer for each term.

5. ___adenosine triphosphate [p.47]

6. ___RNA [p.47]

7. ___DNA [p.47]

A. Single nucleotide strand; function in processes by which genetic instructions are used to build proteins
B. ATP, a cellular energy carrier
C. Double nucleotide strand; encodes genetic instructions with nucleotide sequences

Self-Quiz

Complete the Table

1. Complete the table below by entering the correct name of the major cellular organic compounds suggested in the "types" column (choose from carbohydrates, lipids, proteins, and nucleic acids).

Cellular Organic Compounds   *Types*

[p.41] a.	Phospholipids
[p.47] b.	Nucleotides
[p.42] c.	Weapons against bacteria-causing bacteria
[p.37] d.	Enzymes
[p.47] e.	Genes
[p.38] f.	Glycogen, starch, cellulose, and chitin
[p.47] g.	DNA and RNAs
[p.40] h.	Triglycerides
[p.47] i.	ATP
[p.40] j.	Saturated and unsaturated fats
[pp.40–41] k.	Sterols, oils, and waxes
[p.38] l.	Glucose and sucrose

___ 2. Amino acids are linked by _____ bonds to form the primary structure of a protein. [p.42]
   a. disulfide
   b. hydrogen
   c. ionic
   d. peptide

___ 3. Proteins _____. [p.42]
   a. are weapons against disease-causing bacteria and other invaders
   b. are composed of nucleotide subunits
   c. translate protein-building instructions into actual protein structures
   d. lack diversity of structure and function

___ 4. Lipids _____. [p.40]
   a. include fats that are broken down into one fatty acid molecule and three glycerol molecules
   b. are composed of monosaccharides
   c. include triglycerides that serve as energy sources
   d. include cartilage and chitin

___ 5. DNA _____. [p.47]
   a. is one of the adenosine phosphates
   b. is one of the nucleotide coenzymes
   c. contains protein-building instructions
   d. is composed of monosaccharides

___ 6. Most of the chemical reactions in cells must have _____ present before they proceed. [p.37]
a. RNA
b. salt
c. enzymes
d. fats

___ 7. Carbon is part of so many different substances because _____. [p.36]
a. carbon generally forms two covalent bonds with a variety of other atoms
b. a carbon atom generally forms four covalent bonds with a variety of atoms
c. carbon ionizes easily
d. carbon is a polar compound

___ 8. The information built into a protein's amino acid sequence plus a coiled pattern of that chain and the addition of more folding yields the _____ level of protein structure. [p.44]
a. quaternary
b. primary
c. secondary
d. tertiary

___ 9. _____ are compounds used by cells as transportable packets of quick energy, storage forms of energy, and structural materials. [p.38]
a. Lipids
b. Nucleic acids
c. Carbohydrates
d. Proteins

___10. Hydrolysis could be correctly described as the _____. [p.37]
a. heating of a compound in order to drive off its excess water and concentrate its volume
b. breaking of a long-chain compound into its subunits by adding water molecules to its structure between the subunits
c. linking of two or more molecules by the removal of one or more water molecules
d. constant removal of hydrogen atoms from the surface of a carbohydrate

___11. Genetic instructions are encoded in the base sequence of _____; molecules of _____ function in processes using genetic instructions to construct proteins. [p.47]
a. DNA; DNA
b. DNA; RNA
c. RNA; DNA
d. RNA; RNA

Chapter Objectives/Review Questions

1. The molecules of life are _____ compounds. [p.36]
2. Each carbon atom can share pairs of electrons with as many as _____ other atoms. [p.36]
3. A _____ has only hydrogen atoms attached to a carbon backbone. [p.36]
4. Hydroxyl, amino, and carboxyl are examples of _____ groups. [p.36]
5. Describe the role of enzymes in the metabolism of life. [pp.36–37]
6. _____ use some assortment of the simple sugars, fatty acids, amino acids, and nucleotides for building all the organic compounds they require for their structure and functioning. [p.36]
7. _____ make specific metabolic reactions proceed faster than they would on their own and different _____ mediate different kinds of reactions. [p.37]
8. Be able to define the following: *functional-group transfer, electron transfer, rearrangement, condensation,* and *cleavage.* [p.37]
9. Define and distinguish between *condensation reactions* and *hydrolysis*; cite a general example. [p.37]
10. Define *carbohydrates*; be able to list their two general functions. [p.38]
11. Most carbohydrates consist of _____, _____, and _____ in a 1:2:1 ratio. [p.38]
12. Name and generally define the three classes of carbohydrates. [p.38]

13. A(n) _____ means "one monomer of sugar"; be able to give common examples and their functions. [p.38]
14. A(n) _____ is a short chain of two or more covalently bonded sugar monomers; be able to give examples of well-known disaccharides and their functions. [p.38]
15. A(n) _____ is a straight or branched chain of hundreds or thousands of sugar monomers, of the same or different kinds; be able to give common examples and their functions. [pp.38–39]
16. Define *lipids*; list their general functions. [p.40]
17. Describe a "fatty acid"; a(n) _____ molecule has three fatty acid tails attached to a backbone of glycerol; distinguish a saturated fatty acid from an unsaturated fatty acid. [p.40]
18. Describe the structure of triglycerides; list examples and their functions. [p.40]
19. A(n) _____ has two fatty acid tails and a hydrophilic head attached to a glycerol backbone; list examples and their cellular functions. [p.41]
20. Define *sterols,* and describe their chemical structure; cite the importance of the sterols known as cholesterol and the steroids. [p.41]
21. Lipids called _____ have long-chain fatty acids, tightly packed and linked to long-chain alcohols or to carbon rings; list examples and their functions. [p.41]
22. Describe protein structure and cite their general functions; be able to sketch the three general parts of any amino acid. [p.42]
23. Describe how the primary, secondary, tertiary, and quaternary structure of proteins results in complex three-dimensional structures. [pp.42–45]
24. Distinguish lipoproteins from glycoproteins. [p.45]
25. _____ refers to the loss of a molecule's three-dimensional shape through disruption of the weak bonds responsible for it. [p.45]
26. A(n) _____ is an organic compound, a normal metabolic product of one species, but its chemical effects can harm or kill individuals of a different species that come in contact with it. [p.46]
27. Distinguish between *herbicides, insecticides,* and *fungicides.* [p.46]
28. List the three parts of every nucleotide. [p.47]
29. The nucleic acids _____ and _____, built of nucleotides, are the basis of inheritance and reproduction. [p.47]

Integrating and Applying Key Concepts

1. Humans can obtain energy from many different food sources. Do you think this ability is an advantage or a disadvantage in terms of long-term survival? Why?
2. If the ways that atoms bond affect molecular shapes, do the ways that molecules behave toward one another influence the shapes of organelles? Do the ways that organelles behave toward one another influence the structure and function of the cells?

4

CELL STRUCTURE AND FUNCTION

Interactive Exercises

Animalcules and Cells Fill'd With Juices [pp.50–51]

4.1. BASIC ASPECTS OF CELL STRUCTURE AND FUNCTION [pp.52–53]

4.2. CELL SIZE AND CELL SHAPE [p.54]

4.3. *Focus on Science:* MICROSCOPES—GATEWAYS TO CELLS [pp.54–55]

Selected Words: cellulae [p.50], *transport* proteins [p.53], *receptor* proteins [p.53], *recognition* proteins [p.53], *adhesion* proteins [p.53], *compound light microscope* [p.54], *transmission electron microscope* [p.54], *scanning electron microscope* [p.54], *scanning tunneling microscope* [p.55], phase-contrast process [p.55], Nomarski process [p.55]

Boldfaced, Page-Referenced Terms

[p.51] cell theory _____

[p.52] cell _____

_____

[p.52] plasma membrane _____

_____

[p.52] DNA-containing region _____

_____

[p.52] cytoplasm _____

_____

[p.52] ribosomes _____

_____

[p.52] eukaryotic cells _____

_____

[p.52] prokaryotic cells _____

_____

[p.52] lipid bilayer _____

_____

[p.52] fluid mosaic model _____

_____

[p.54] surface-to-volume ratio _____

_____

[p.54] micrograph _____

_____

[p.54] wavelength _____

_____

Label–Match

Although cells vary in many specific ways, they are all alike in a few basic respects. Identify each part of the illustration. Choose from cytoplasm, DNA-containing region (nucleus in eukaryotes), and plasma membrane. Complete the exercise by matching and entering the letter of the proper description in the parentheses following each label.

1. _____ (   ) [p.52]

2. _____ (   ) [p.52]

3. _____ (   ) [p.52]

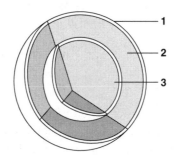

A. Includes everything enclosed by the plasma membrane; a semifluid substance in which particles, filaments, and often compartments are organized

B. Thin, outermost membrane that maintains the cell as a distinct entity with metabolism occurring within

C. Molecules of heredity are here along with molecules that can read or copy hereditary instructions

Labeling

Identify each numbered membrane protein in the illustration below.

4. _____ protein

5. through 8. are types of _____ proteins.

9. _____ protein

10. _____

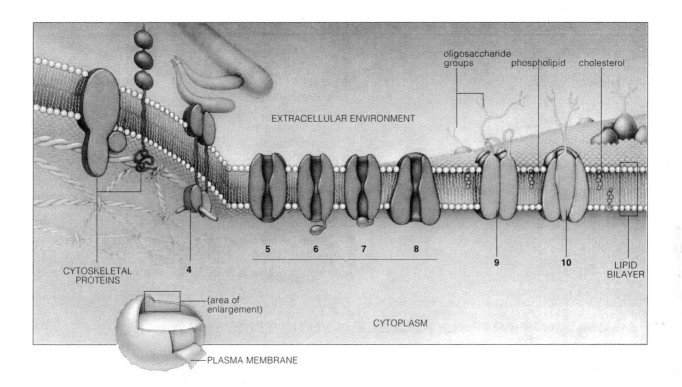

Matching

Choose the most appropriate answer for each term.

11. ___ "fluid" [p.52]

12. ___ phospholipid molecule [p.52]

13. ___ adhesion proteins [p.53]

14. ___ prokaryotic [p.52]

15. ___ transport proteins [p.53]

16. ___ "mosaic" [p.52]

17. ___ recognition proteins [p.53]

18. ___ eukaryotic [p.52]

19. ___ fluid mosaic [p.52]

20. ___ receptor proteins [p.53]

A. Allow water-soluble substances to move through their interior, thus crossing the bilayer
B. Refers to the mixed composition of diverse phospholipids, glycolipids, sterols, and proteins
C. Bacteria cells lacking a nucleus
D. Of multicelled organisms; helps cells of the same type locate and stick to one another and stay positioned in the proper tissues
E. By this model, membranes have a mixed composition of lipids and proteins
F. Bind extracellular substances, such as hormones, that trigger changes in cell activities
G. Their cytoplasm includes organelles, which are tiny sacs and other compartments bounded by membranes
H. Like molecular fingerprints, their oligosaccharide chains identify a cell as being of a specific type
I. Consists of a hydrophilic head and two hydrophobic tails
J. Refers to the motions and interactions of component parts of the membrane

Short Answer

21. Explain why most cells are, of necessity, very small. [p.54] _____

_____

_____

_____

Matching

Choose the most appropriate answer for each term.

22. ___electron wavelength [p.54]

23. ___micrograph [p.54]

24. ___transmission electron microscope [p.54]

25. ___scanning tunneling microscope [p.55]

26. ___wavelength [p.54]

27. ___compound light microscope [p.54]

28. ___scanning electron microscope [p.54]

A. Glass lenses bend incoming light rays to form an enlarged image of a cell or some other specimen
B. The distance from one wave's peak to the peak of the wave behind it
C. A computer analyzes a tunnel formed in electron orbitals, this produced between the tip of a needlelike probe and an atom
D. A narrow beam of electrons moves back and forth across the surface of a specimen coated with a thin metal layer
E. A photograph of an image formed with a microscope
F. About 100,000 times shorter than those of visible light
G. Electrons pass through a thin section of cells to form an image

4.4. THE DEFINING FEATURES OF EUKARYOTIC CELLS [pp.56–59]

Boldfaced, Page-Referenced Terms

[p.56] organelle _____

_____

Complete the Table

1. Complete the following table to identify eukaryotic organelles and their functions.

Organelle or Structure	Main Function
[p.56] a.	Localizing the cell's DNA
[p.56] b.	Synthesis of polypeptide chains
[p.56] c.	Route and modify newly formed polypeptide chains; lipid synthesis
[p.56] d.	Modifying polypeptide chains into mature proteins; sort and ship proteins and lipids for secretion or use inside the cell
[p.56] e.	Different types function in transport or storage of substances; digestion inside cells and other functions
[p.56] f.	Efficient ATP production
[p.56] g.	Overall cell shape and internal organization; moving the cell and its internal structures

4.5. THE NUCLEUS [pp.60–61]

Selected Words: *chromatin [p.61], chromosomes [p.61]*

Boldfaced, Page-Referenced Terms

[p.60] nucleus _____

_____

[p.60] nuclear envelope _____

_____

[p.60] nucleolus (plural, nucleoli) _____

_____

[p.61] chromatin _____

_____

[p.61] chromosome _____

_____

Short Answer

1. List the two major functions of the nucleus. [p.60] _____

_____

_____

_____

_____

_____

Complete the Table

2. Complete this table about the eukaryotic nucleus by entering the name of each nuclear component described.

Nuclear Component	Description
[p.61] a.	Dense cluster of RNA and proteins that will be assembled into subunits of ribosomes
[p.61] b.	Pore-riddled double-membrane system that selectively controls the passage of various substances into and out of the nucleus
[p.61] c.	Total collection of all DNA molecules and their associated proteins in the nucleus
[p.61] d.	Fluid interior portion of the nucleus
[p.61] e.	One DNA molecule and many proteins that are intimately associated with it

Fill-in-the-Blanks

Outside the nucleus, polypeptide chains for proteins are assembled on (3) _____ [p.61]. Many new chains become stockpiled in the (4) _____ [p.61] or get used at once. Many others enter a(n) (5) _____ [p.61] system. This system consists of different organelles, including (6) _____ [p.61] reticulum, (7) _____ [p.61] bodies, and vesicles.

Thanks to (8) _____ [p.61] instructions, many (9) _____ [p.61] take on particular final forms in the cytomembrane system. (10) _____ [p.61] also are packaged and assembled in the system by (11) _____ [p.61] and other proteins constructed according to DNA's instructions. (12) _____ [p.61] deliver the proteins and lipids to specific sites within the cell or to the plasma membrane, for export.

4.6. THE CYTOMEMBRANE SYSTEM [pp.62–63]

Selected Words: *rough* ER [p.62], *smooth* ER [p.62], lysosomes [p.63], peroxisomes [p.63]

Boldfaced, Page-Referenced Terms

[p.62] cytomembrane system _____

_____

[p.62] endoplasmic reticulum, or ER _____

_____

[p.62] Golgi bodies _____

_____

[p.63] vesicles _____

_____

Matching

Study the illustration below and match each component of the cytomembrane system with the most correct description of function. Some components may be used more than once.

1. ___ Assembly of polypeptide chains [p.62]

2. ___ Lipid assembly [p.62]

3. ___ DNA instructions for building polypeptide chains [p.62]

4. ___ Initiate protein modification following assembly [p.62]

5. ___ Proteins and lipids take on final form [p.62]

6. ___ Sort and package lipids and proteins for transport to proper destinations following modification [p.62]

7. ___ Vesicles formed at plasma membrane transport substances into cytoplasm [p.62]

8. ___ Sacs of enzymes that break down fatty acids and amino acids, forming hydrogen peroxide [p.63]

9. ___ Special vesicles budding from Golgi bodies that become organelles of intracellular digestion [p.63]

10. ___ Transport unfinished proteins to a Golgi body [p.62]

11. ___ Transport finished Golgi products to the plasma membrane [p.62]

12. ___ Release Golgi products at the plasma membrane [p.62]

13. ___ Transport unfinished lipids to a Golgi body [p.62]

A. spaces within smooth membranes of ER
B. nucleus
C. Golgi body
D. vesicles from Golgi
E. vesicles budding from rough ER
F. endocytosis with vesicles
G. exocytosis with vesicles
H. spaces within rough ER
I. ribosomes in the cytoplasm
J. vesicles budding from smooth ER
K. lysosomes
L. peroxisomes

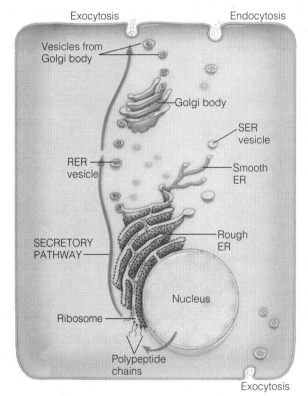

4.7. MITOCHONDRIA [p.64]
4.8. SPECIALIZED PLANT ORGANELLES [p.65]

Selected Words: granum [p.65], chlorophylls [p.65], carotenoids [p.65], chromoplasts [p.65], amyloplasts [p.65]

Boldfaced, Page-Referenced Terms

[p.64] mitochondrion _____

_____

[p.65] chloroplasts _____

_____

[p.65] central vacuole _____

_____

Choice

For questions 1–20, choose from the following:

    a. mitochondria   b. chloroplasts   c. amyloplasts   d. central vacuole   e. chromoplasts

1. ___ Occur only in photosynthetic eukaryotic cells [p.65]

2. ___ ATP molecules form when organic compounds are completely broken down to carbon dioxide and water [p.64]

3. ___ Plastids that lack pigments [p.65]

4. ___ A muscle cell might have thousands [p.64]

5. ___ Plastids that have an abundance of carotenoids but no chlorophylls [p.65]

6. ___ ATP-forming reactions require oxygen [p.64]

7. ___ Causes fluid pressure to build up inside a living plant cell [p.65]

8. ___ Organelles that convert sunlight energy to the chemical energy of ATP [p.65]

9. ___ The source of the red-to-yellow colors of many flowers, autumn leaves, ripening fruits, and carrots or other roots [p.65]

10. ___ Inner folds are called cristae [p.64]

11. ___ May increase so much in volume that it takes up 50 to 90 percent of the cell's interior [p.65]

12. ___ Plastid with internal areas known as *grana* and *stroma* [p.65]

13. ___ Plastids that resemble photosynthetic bacteria [p.65]

14. ___ Two distinct compartments are created by a double-membrane system [p.64]

15. ___ Store starch grains and are abundant in cells of stems, potato tubers, and seeds [p.65]

16. ___ Resemble nonphotosynthetic bacteria in terms of size and biochemistry [p.64]

17. ___ The site of photosynthesis in plant cells [p.65]

18. ___ Fluid-filled, stores amino acids, sugars, ions, and toxic wastes [p.65]

19. ___ All eukaryotic cells have one or more [p.64]

20. ___ The thylakoid membrane [p.65]

4.9. THE CYTOSKELETON [pp.66–67]

4.10. CELL SURFACE SPECIALIZATIONS [pp.68–69]

Selected Words: *accessory* proteins [p.66], crosslinking proteins [p.66], *Amoeba proteus* [p.66], "9 + 2 array" [p.67], dynein or motor proteins [p.67], pectin [p.68], *tight* junctions [p.69], *adhering* junctions [p.69], *gap* junctions [p.69]

Boldfaced, Page-Referenced Terms

[p.66] cytoskeleton _____

_____

[p.66] microtubules _____

_____

[p.66] microfilaments _____

_____

[p.66] intermediate filaments _____

_____

[p.66] pseudopods _____

_____

[p.67] flagellum (plural, flagella) _____

_____

[p.67] cilium (plural, cilia) _____

_____

[p.67] centriole _____

_____

[p.67] basal body _____

_____

[p.68] cell wall _____

_____

[p.68] primary wall _____

_____

[p.68] secondary wall _____

_____

Dichotomous Choice

Circle one of two possible answers given between parentheses in each statement.

1. The cytoskeleton gives (prokaryotic/eukaryotic) cells their shape, internal organization, and movement. [p.66]
2. (Protein/Carbohydrate) subunits form the basic components of microtubules. [p.66]
3. Microtubules, microfilaments, or both take part in most aspects of cell (motility/chemistry). [p.66]
4. Hollow microtubule cylinders consist of (tubulin/actin) protein subunits. [p.66]
5. Microfilaments consist of two chains of (tubulin/actin) subunits twisted together. [p.66]
6. (Intermediate filaments/Microfilaments) are ropelike cytoskeletal elements that impart mechanical strength to cells and tissues. [p.66]
7. The pseudopods of *Amoeba proteus* are extended by controlled (assembly/disassembly) of microfilaments beneath its plasma membrane. [p.66]
8. Within muscle cells, ATP-activated myosin repeatedly bind and release adjacent microfilaments; this action causes the microfilaments to slide over the myosin strands, towards the center of the contractile unit—which (lengthens/shortens) as a result. [p.66]
9. In response to the sun's position overhead, chloroplasts flowing in plant cells due to myosin monomers "walking" over bundles of microfilaments is known as (amoeboid motion/cytoplasmic streaming). [p.67]
10. Sperm and many other free-living cells use (flagella/cilia) as whiplike tails for swimming. [p.67]
11. The cross-sectional array of 9 + 2 microtubules is found in (cilia/centrioles). [p.67]
12. The human respiratory tract is lined with beating (flagella/cilia). [p.67]
13. 9 + 2 arrays of microtubules arise from a (centriole/pseudopod), one type of microtubule-producing center. [p.67]
14. Flagella and cilia beat by a sliding mechanism involving motor proteins known as (crosslinking proteins/dynein). [p.67]

Matching

Choose the most appropriate answer for each term.

15. ____ primary wall [p.68]
16. ____ secondary wall [p.68]
17. ____ plasmodesmata [p.69]
18. ____ tight junctions [p.69]
19. ____ adhering junctions [p.69]
20. ____ gap junctions [p.69]

A. Link the cells of epithelial tissues lining the body's outer surface, inner cavities, and organs
B. Numerous tiny channels crossing the adjacent primary walls of living plant cells and connect their cytoplasm
C. Quite sticky, and they cement adjacent cells together
D. Link the cytoplasm of neighboring animal cells and are open channels for the rapid flow of signals and substances
E. Formed of rigid cellulose and additional deposits; reinforces plant cell shape
F. Join cells in tissues of the skin, heart, and other organs subject to stretching

4.11. PROKARYOTIC CELLS—THE BACTERIA [p.70]

Selected Words: *prokaryotic* [p.70], pathogenic bacteria [p.70], bacterial flagellum [p.70], pili [p.70], *Escherichia coli* [p.70], *Nostoc* [p.70], *Pseudomonas marginalis* [p.70]

Boldfaced, Page-Referenced Terms

[p.70] nucleoid _____

_____

Fill-in-the-Blanks

With one exception, (1) _____ [p.70] are the smallest and most structurally simple cells. The word *prokaryotic* means "before the (2) _____ ," [p.70] which implies that bacteria evolved before cells possessing nuclei existed. Most bacteria have a semirigid or rigid cell (3) _____ [p.70] that wraps around the plasma membrane; this membrane controls movement of substances to and from the (4) _____ [p.70]. Some bacteria have one or more long, threadlike motile structures, the (5) _____ [p.70], that extend from the cell surface; they permit rapid movements through fluid environments. Bacterial cells have many (6) _____ [p.70] upon which polypeptide chains are assembled. Bacterial cell cytoplasm is continuous with an irregularly shaped region of DNA, the (7) _____ [p.70], that lacks surrounding membranes.

Self-Quiz

Label–Match

Identify each indicated part of the accompanying illustrations. Complete the exercise by matching and entering the letter of the proper function description in the parentheses following each label. Some letter choices must be used more than once.

1. _____ _____ ( ) [p.62]

2. _____ ( ) [p.63]

3. _____ (cytoskeletal component) ( ) [p.66]

4. _____ ( ) [p.64]

5. _____ ( ) [p.65]

6. _____ (cytoskeletal component) ( ) [p.66]

7. _____ _____ ( ) [p.65]

8. _____ _____ _____ ( ) [p.62]

9. _____ ( ) [p.62]

10. _____ _____ _____ ( ) [p.62]

11. _____ + _____ ( ) [pp.60–61]

12. _____ ( ) [pp.60–61]

13. _____ _____ ( ) [p.60]

14. _____ ( ) [p.60]

15. _____ _____ ( ) [p.52]

16. _____ _____ ( ) [p.68]

17. _____ (cytoskeletal component) ( ) [p.66]

18. _____ (cytoskeletal component) ( ) [p.66]

19. _____ _____ ( ) [p.52]

20. _____ ( ) [p.64]

21. _____ _____ ( ) [p.60]

22. _____ ( ) [pp.60–61]

23. _____ + _____ ( ) [pp.60–61]

24. _____ ( ) [p.60]

25. _____ ( ) [p.63]

26. _____ ( ) [p.63]

27. _____ _____ _____ ( ) [p.62]

28. _____ ( ) [p.62]

29. _____ _____ _____ ( ) [p.62]

30. _____ ( ) [p.63]

31. _____ _____ ( ) [p.62]

32. _____ ( ) [p.67]

A. Pore-riddled two-membrane structure; a profusion of ribosomes is found on its outer surface

B. Porous, but provides protection and structural support for some cells

C. Present in mature, living plant cells; a fluid-filled organelle that stores amino acids, sugars, ions, and toxic wastes

D. Free of ribosomes; the main site of lipid synthesis in many cells

E. Formed as buds from Golgi membranes of animal cells and some fungal cells; an organelle of intracellular digestion

F. Barrel-shaped structures that serve as a type of microtubule-producing center

G. Tiny membranous sacs that move through the cytoplasm or take up positions in it; a common type functions as an organelle of intracellular digestion

H. Takes part in diverse cell movements; consist of protein subunits, called tubulins, that form a hollow cylinder

I. A membrane-bound compartment that houses DNA in eukaryotic cells

J. Enzymes put the finishing touches on proteins and lipids, sort them out, and package them inside vesicles for shipment to specific locations

K. Site of assembly of protein and RNA molecules into ribosomal subunits

L. Takes part in diverse cell movements; consist of two chains of actin subunits, twisted together

M. Organelles that convert sunlight energy to the chemical energy of ATP, which is used to make sugars and other organic compounds

N. Thin, outermost membrane that maintains the cell as a distinct entity; substances and signals continually move across it in highly controlled ways

O. Site of aerobic respiration; ATP-producing powerhouse of all eukaryotic cells

P. Stacks of flattened sacs with many ribosomes attached; accomplishes initial modification of protein structure after formation on ribosomes

Q. Genetic material and the fluid interior portion of the nucleus

R. Site of the synthesis of every new polypeptide chain

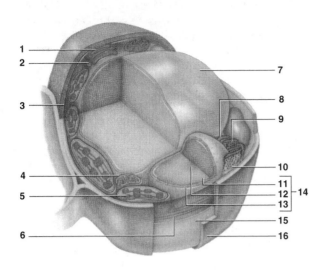

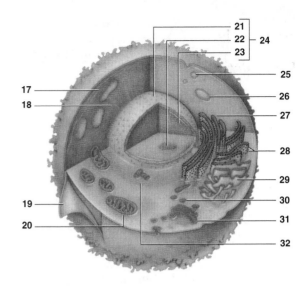

Multiple Choice

____33. Membranes consist of _____. [pp.53–54]
  a. a lipid bilayer
  b. a protein bilayer
  c. phospholipids and proteins
  d. both a and c are correct

____34. Which of the following is *not* found as a part of prokaryotic cells? [p.70]
  a. Ribosomes
  b. DNA
  c. Nucleus
  d. Cytoplasm
  e. Cell wall

____35. Which of the following statements most correctly describes the relationship between cell surface area and cell volume? [p.54]
  a. As a cell expands in volume, its diameter increases at a rate faster than its surface area does.
  b. Volume increases with the square of the diameter, but surface area increases only with the cube.
  c. If a cell were to grow four times in diameter, its volume of cytoplasm increases sixteen times and its surface area increases sixty-four times.
  d. Volume increases with the cube of the diameter, but surface area increases only with the square.

____36. Most cell membrane functions are carried out by _____. [p.53]
  a. carbohydrates
  b. nucleic acids
  c. lipids
  d. proteins

____37. Animal cells dismantle and dispose of waste materials by _____. [p.63]
  a. using centrally located vacuoles
  b. several lysosomes fusing with a vesicle formed at the plasma membrane that encloses the wastes
  c. microvilli packaging and exporting the wastes
  d. mitochondrial breakdown of the wastes

____38. The nucleolus is the site where _____. [pp.60–61]
  a. the protein and RNA subunits of ribosomes are assembled
  b. the chromatin is formed
  c. chromosomes are bound to the inside of the nuclear envelope
  d. chromosomes duplicate themselves

___39. The _____ is free of ribosomes and curves through the cytoplasm like connecting pipes; the main site of lipid synthesis. [p.62]
   a. lysosome
   b. Golgi body
   c. smooth ER
   d. rough ER

___40. Which of the following is *not* present in all cells? [p.52]
   a. Cell wall
   b. Plasma membrane
   c. Ribosomes
   d. DNA molecules

___41. As a part of the cytomembrane system, the _____ put(s) the finishing touches on lipids and proteins to permit sorting and packaging for specific locations. [p.62]
   a. endoplasmic reticulum
   b. Golgi bodies
   c. peroxisomes
   d. lysosomes

___42. Chloroplasts _____. [p.65]
   a. are specialists in oxygen-requiring reactions
   b. function as part of the cytoskeleton
   c. trap sunlight energy and produce organic compounds
   d. assist in carrying out cell membrane functions

___43. Mitochondria convert energy stored in _____ to forms that the cell can use, principally ATP. [p.64]
   a. water
   b. carbon compounds
   c. $NADPH_2$
   d. carbon dioxide

___44. _____ are sacs of enzymes that bud from ER; they produce potentially harmful hydrogen peroxide while breaking down fatty acids and amino acids. [p.63]
   a. Lysosomes
   b. Glyoxysomes
   c. Golgi bodies
   d. Peroxisomes

___45. Two classes of cytoskeletal elements underlie nearly all movements of eukaryotic cells; they are _____. [p.66]
   a. desmins and vimentins
   b. actin and microfilaments
   c. microtubules and microfilaments
   d. microtubules and myosin

Choice

Cells of the organisms in the six kingdoms of life share the following characteristics: plasma membrane, DNA, RNA, and ribosomes. For questions 46–57, choose the kingdom(s) possessing the characteristics listed below. Some questions will require more than one kingdom as the answer.

   a. Archaebacteria, Eubacteria   b. Protistans   c. Fungi   d. Plants   e. Animals

46. _____ cytoskeleton [p.72]

47. _____ cell wall [p.72]

48. _____ nucleus [p.72]

49. _____ nucleolus [p.72]

50. _____ central vacuole [p.72]

51. _____ simple flagellum [p.72]

52. _____ photosynthetic pigment [p.72]

53. _____ chloroplast [p.72]

54. _____ endoplasmic reticulum [p.72]

55. _____ Golgi body [p.72]

56. _____ lysosome [p.72]

57. _____ complex flagella and cilia [p.72]

Chapter Objectives/Review Questions

1. Be able to list the three generalizations that together constitute the cell theory. [p.51]
2. List and describe the three major regions that all cells have in common. [p.52]
3. The cytoplasm of _____ cells includes tiny sacs called organelles; one sac houses the DNA. [p.52]
4. _____ cells, by contrast, have no nuclei. [p.52]
5. Describe the lipid bilayer arrangement for the plasma membrane. [pp.52–53]
6. Describe and list the function(s) of the following membrane proteins: transport, receptor, recognition, and adhesion. [p.53]
7. Cell size is necessarily limited because its volume increases with the _____, but surface area increases only with the _____. [p.54]
8. Briefly describe the operating principles of light microscopes, phase-contrast microscopes, scanning tunneling microscopes, transmission electron microscopes, and scanning electron microscopes. [pp.54–55]
9. Briefly describe the cellular location and function of the organelles typical of most eukaryotic cells: nucleus, ribosomes, endoplasmic reticulum, Golgi body, various vesicles, mitochondria, and the cytoskeleton. [p.56]
10. In eukaryotic cells, _____ separate different incompatible chemical reactions in space and time. [p.56]
11. Describe the nature of the nuclear envelope and relate its function to its structure. [p.60]
12. _____ are sites where the protein and RNA subunits of ribosomes are assembled. [pp.60–61]
13. _____ is the cell's collection of DNA molecules and associated proteins; a(n) _____ is an individual DNA molecule and associated proteins. [p.61]
14. Explain how the endoplasmic reticulum (rough and smooth types), peroxisomes, Golgi bodies, lysosomes, and a variety of vesicles function together as the cytomembrane system. [pp.62–63]
15. Describe the function of exocytic vesicles; also describe the function of endocytic vesicles. [p.62]
16. _____ are organelles of intracellular digestion that bud from Golgi membranes of animal cells and some fungal cells. [p.63]
17. Define and describe the function of peroxisomes. [p.63]
18. Within _____, energy stored in organic molecules is released by enzymes and used to form many ATP molecules in the presence of oxygen. [p.64]
19. Describe the detailed structure of the chloroplast, the site of photosynthesis (include grana and stroma). [p.65]
20. Give the general function of the following plant organelles: chloroplasts, chromoplasts, amyloplasts, and the central vacuole. [p.65]
21. Elements of the _____ give eukaryotic cells their internal organization, overall shape, and capacity to move. [p.66]
22. List the three major structural elements of the cytoskeleton and give the general function of each. [p.66]
23. Describe the general function of accessory proteins. [p.66]
24. *Amoeba proteus*, a soft-bodied protistan, crawls on _____. [p.66]
25. Both cilia and flagella have an internal microtubule arrangement called the "_____" array. [p.67]
26. A centriole remains at the base of a completed microtubule-producing center where it is often called a(n) _____ _____. [p.67]
27. Distinguish a primary cell wall from a secondary cell wall in leafy plants. [p.68]
28. Describe the location and function of plasmodesmata. [p.69]
29. _____ junctions link the cells of epithelial tissues; _____ junctions join cells in tissues of the skin, heart, and other organs subjected to stretching; _____ junctions link the cytoplasm of neighboring cells. [p.69]
30. Describe the structure of a generalized prokaryotic cell. Include the bacterial flagellum, nucleoid, pili, capsule, cell wall, plasma membrane, cytoplasm, and ribosomes. [p.70]

Integrating and Applying Key Concepts

1. Which parts of a cell constitute the minimum necessary for keeping the simplest of living cells alive?
2. How did the existence of a nucleus, compartments, and extensive internal membranes confer selective advantages on cells that developed these features?

5

GROUND RULES OF METABOLISM

Interactive Exercises

You Light Up My Life [pp.74–75]

5.1. ENERGY AND THE UNDERLYING ORGANIZATION OF LIFE [pp.76–77]

5.2. DOING CELLULAR WORK [pp.78–79]

Selected Words: *fluorescent* light [p.74], *bioluminescent gene transfers* [p.74], *Salmonella* [p.74], *chemical* work [p.76], *mechanical* work [p.76], *electrochemical* work [p.76], *energy inputs* [p.78], *"oxidized"* [p.79], *"reduced"* [p.79], *substrates* [p.79], *intermediate* [p.79], *end product* [p.79], *energy carriers* [p.79], *enzymes* [p.79], *cofactors* [p.79], *transport proteins* [p.79]

Boldfaced, Page-Referenced Terms

[p.74] bioluminescence _____

_____

[p.75] metabolism _____

_____

[p.76] potential energy _____

_____

[p.76] kinetic energy _____

_____

[p.76] heat _____

_____

[p.76] chemical energy _____

_____

[p.76] first law of thermodynamics _____

_____

[p.77] second law of thermodynamics _____

_____

[p.77] entropy _____

_____

[p.78] endergonic _____

_____

[p.78] exergonic _____

_____

[p.78] ATP _____

_____

[p.79] phosphorylation _____

_____

[p.79] oxidation-reduction reaction _____

_____

[p.79] metabolic pathways _____

_____

Short Answer

1. The world of life maintains a high degree of organization only because [p.77] _____

_____

True–False

If the statement is true, write a T in the blank. If the statement is false, correct it by changing the underlined word(s) and writing the correct word(s) in the answer blank.

_____2. The first law of thermodynamics states that entropy is constantly increasing in the universe. [p.77]

_____3. Your body steadily gives off heat equal to or greater than that from a 100-watt light bulb. [p.76]

_____ 4. When you eat a potato, some of the stored chemical energy of the food is converted into <u>mechanical</u> energy that moves your muscles. [p.76]

_____ 5. The amount of low-quality energy in the universe is <u>decreasing</u>. [p.77]

Labeling

In the blank preceding each item, indicate whether the first law of thermodynamics (I) or the second law of thermodynamics (II) is best described.

6. ___ Corn plants producing starch molecules [p.76]

7. ___ Evaporation of gasoline into the atmosphere [p.77]

8. ___ A hydroelectric plant at a waterfall producing electricity [p.76]

9. ___ The creation of a snowman by children [p.76]

10. ___ The death and decay of an organism [p.77]

Short Answer

11. Define exergonic reactions; give an example. [p.78] _____

_____

_____

12. Define endergonic reactions; give an example. [p.78] _____

_____

_____

13. A(n) (3) _____ _____ [p.79] is an orderly series of chemical reactions, with each reaction mediated by a specific (14) _____ [p.79].

Labeling

Classify each of the following reactions as *endergonic* or *exergonic*.

15. _____ [p.78]   The product of a chemical reaction has more energy than the reactants.

16. _____ [p.78]   Glucose + oxygen ⟶ carbon dioxide + water + energy

17. _____ [p.78]   The reactants of a chemical reaction have more energy than the product.

Choose the most appropriate answer for each: A or B.

18. ___ a degradative reaction [p.79]

19. ___ an endergonic reaction [p.78]

20. ___ a biosynthetic reaction [p.79]

21. ___ an exergonic reaction [p.78]

Fill-in-the-Blanks

In cells, the release of energy from glucose proceeds in controlled steps, so that lower energy (22) _____ [p.79] molecules are formed along the route from glucose to carbon dioxide and (23) _____ [p.78]. At each step in the pathway, some of the potential (24) _____ [p.78] stored in the chemical bonds of glucose is released as a specific bond is broken; some of this energy can be harnessed to do chemical (25) _____ [p.78]. For example, the released energy can be used to bond a(n) (26) _____ [p.79] group to ADP and form (27) _____ [p.79].

ATP is constructed of the nitrogenous base (28) _____ [p.78], the sugar (29) _____ [p.78], and three (30) _____ [p.78] groups. When ATP is split into (31) _____ [p.79] and a(n) (32) _____ [p.79] group, usable (33) _____ [p.79] is released, which is easily transferred to other molecules in the cell. The splitting apart of ATP provides (34) _____ [p.79] for biosynthesis, active transport across cell membranes, and molecular movements such as those required for muscle contraction. ATP directly or indirectly delivers energy to almost all (35) _____ [p.79] pathways. In the (36) _____ / _____ [p.79] cycle, a phosphate group is linked to adenosine diphosphate, and adenosine triphosphate donates a phosphate group elsewhere and reverts back to adenosine diphosphate. Adding a phosphate group to a molecule is called (37) _____ [p.79]. When this occurs, the molecule increases its store of (38) _____ [p.79] and becomes activated to enter a specific (39) _____ [p.79]. The (40) _____ / _____ [p.79] cycle provides a renewable means of conserving and transferring energy to specific reactions.

Labeling

Identify the molecule at the right and label its parts.

41. _____ _____ _____

42. _____

43. _____

44. The name of this molecule is _____ _____.

Fill-in-the-Blanks

Along some membranes of cellular organelles, the liberated electrons released from the breaking of chemical bonds are sent through (45) _____ _____ [p.79] systems; these systems consist of (46) _____ [p.79] and cofactors bound in a cell membrane that transfer electrons in a highly organized sequence. A molecule that donates electrons in the sequence is being (47) _____ [p.79], while molecules accepting electrons are being (48) _____ [p.79]. As the electrons are transferred from one electron carrier to another, some (49) _____ [p.79] is released that can be harnessed to do chemical (50) _____ [p.79]. One type of electrochemical work occurs when energy released during electron transfers is used to bond a (51) _____ [p.79] group to ADP.

5.3. ENZYME STRUCTURE AND FUNCTION [pp.80–81]

5.4. FACTORS INFLUENCING ENZYME ACTIVITY [pp.82–83]

5.5. REACTANTS, PRODUCTS, AND CELL MEMBRANES [p.83]

Selected Words: *transition* state [p.81], allosteric control [p.83], selective permeability [p.83]

Boldfaced, Page-Referenced Terms

[p.80] chemical equilibrium _____

_____

[p.80] enzymes _____

_____

[p.80] activation energy _____

_____

[p.81] active sites _____

_____

[p.81] induced-fit model _____

_____

[p.82] feedback inhibition _____

_____

Short Answer

1. List four characteristics that enzymes have in common. [p.80] _____

_____

_____

Matching

Match the items on the sketch below with the list of descriptions. Some answers may require more than one letter. [All are from p.81]

2. _____
3. _____
4. _____
5. _____
6. _____
7. _____

A. Transition state, the time of the most precise fit between enzyme and substrate
B. Complementary active site of the enzyme
C. Enzyme, a protein with catalytic power
D. Product or reactant molecules that an enzyme can specifically recognize
E. Product or reactant molecule
F. Bound enzyme-substrate complex

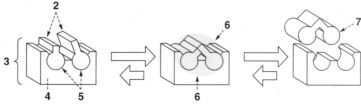

Fill-in-the-Blanks

(8) _____ [p.80] are highly selective proteins that act as catalysts, which means that they greatly

enhance the rate at which specific reactions approach (9) _____ [p.80]. The specific substance on which

a particular enzyme acts is called its (10) _____ [p.81]; this substance fits into the enzyme's crevice,

which is called its (11) _____ _____ [p.81].

The (12) _____ - _____ [p.81] model describes how a substrate contacts the substrate without a

perfect fit. Enzymes increase reaction rates by lowering the required (13) _____ _____ [pp.80–81].

(14) _____ [p.82] and (15) _____ [p.82] are two important factors that influence the rates of enzyme

activity. Extremely high fevers can destroy the three-dimensional shape of an enzyme, which may

adversely affect (16) _____ [p.82] and cause death. (17) _____ [p.82] enzymes have control sites

where specific substances can bind and alter enzyme activity. The situation in which the end product binds

to the first enzyme in a metabolic pathway and prevents product formation is known as (18) _____

_____ [pp.82–83].

Short Answer

19. Cells determine the kinds and concentrations of substances that operate within their boundaries by controlling how their existing enzymes behave. Under what circumstances might feedback inhibition be used by a bacterial cell to shut down the metabolic pathway that produces tryptophan? [pp.82–83] _____

_____

_____

20. What might happen to the bacterial colony if it could not stop producing tryptophan? [p.83] _____

_____

_____

21. Describe the mechanism of feedback inhibition in stopping the synthesis of tryptophan. [p.83] _____

_____

_____

22. Describe how the tryptophan pathway is turned on again. [p.83]

_____

_____

5.6. WORKING WITH AND AGAINST CONCENTRATION GRADIENTS [pp.84–85]
5.7. MOVEMENT OF WATER ACROSS MEMBRANES [pp.86–87]
5.8. EXOCYTOSIS AND ENDOCYTOSIS [p.88]

Selected Words: "concentration" [p.84], "gradient" [p.84], "dynamic equilibrium" [p.84], *electric* gradient [p.84], *net* direction [p.85], *tonicity* [p.87], *osmotic* pressure [p.87], *receptor-mediated* endocytosis [p.88], clathrin [p.88], *bulk-phase* endocytosis [p.88]

Boldfaced, Page-Referenced Terms

[p.84] concentration gradient _____

_____

[p.84] diffusion _____

_____

[p.84] passive transport _____

_____

[p.85] active transport _____

_____

[p.86] bulk flow _____

_____

[p.86] osmosis _____

_____

[p.87] hypotonic solution _____

_____

[p.87] hypertonic solution _____

_____

[p.87] isotonic solution _____

_____

[p.87] hydrostatic pressure _____

_____

[p.88] exocytosis _____

_____

[p.88] endocytosis _____

_____

[p.88] phagocytosis _____

_____

Fill-in-the-Blanks

(1) _____ [p.84] refers to the number of molecules (or ions) of a substance in a specified volume of fluid.
A(n) (2) _____ [p.84] means that one region of the fluid contains more molecules of a particular
substance than a neighboring region. The net movement of like molecules down their concentration
gradient is called (3) _____ [p.84]. A special case of (3) is (4) _____ [p.86]: the movement of water
across membranes in response to a concentration gradients, a pressure gradient, or both.

Complete the Table

Study the illustration below and complete the table by entering the name of the membrane structure(s) involved in the transport mechanism described. Choose from lipid bilayer, "membrane pumps," and transport protein.

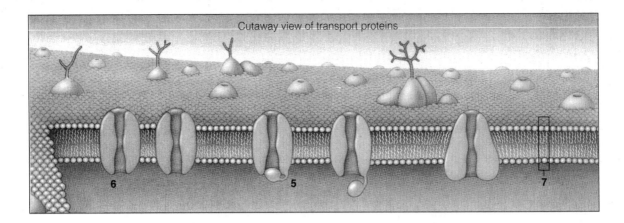

Cutaway view of transport proteins

Membrane Structures(s) Involved	Function
[p.85] 5.	[pp. 85,89,90] Pumps solutes across the membrane against a gradient (energized by ATP)
[pp.84–85] 6.	[p.90] Solute molecules pass through by passive transport (also called facilitated diffusion)
[pp.83–84] 7.	[p.90] Simple diffusion

True–False

If the statement is true, write a T in the blank. If the statement is false, correct it by changing the underlined word(s) and writing the correct word(s) in the answer blank.

_____ 8. Consider a membrane that has different concentrations of water molecules on each side; the side with fewer solute particles has a <u>higher</u> concentration of water molecules than the fluid on the other side of the membrane. [p.86]

_____ 9. An animal cell placed in a <u>hypertonic</u> solution would swell and perhaps burst. [p.86]

_____ 10. Physiological saline is 0.9% NaCl; red blood cells placed in such a solution will not gain or lose water; therefore, one could state that the fluid in red blood cells is <u>hypertonic</u>. [p.86]

_____ 11. See #8. In such a situation, more water molecules will move from the side that has a <u>higher</u> concentration, through the membrane to the side that has a <u>lower</u> concentration of water molecules. [p.86]

_____ 12. Red blood cells shrivel and shrink when placed in a <u>hypotonic</u> solution. [p.86]

_____ 13. Plant cells placed in a <u>hypotonic</u> solution will swell. [p.86]

Labeling

In the blank following each substance listed below, indicate whether the substance moves through the lipid bilayer (LB) or through the interior of transport proteins (ITP) on its path through the membrane. [All answers are from p.83, also recall the meanings of polar (water-soluble) and nonpolar (not soluble in water); additional information about questions 14, 21, and 22 is on the page indicated.]

14. H_2O [p.86] _____

15. CO_2 _____

16. Na^+ _____

17. glucose _____

18. O_2 _____

19. K^+ _____

20. amino acids _____

21. substances pumped through interior of a transport protein; energy input required [p.85] _____

22. substances move through interior of transport protein; no energy input required [p.84] _____

23. lipid soluble substances _____

Short Answer

24. Name the cause that compels molecules or ions of a substance to diffuse from where they are to a different place. [p.85] _____

_____

25. Name the cause that determines the general direction in which the particles diffuse. [p.85] _____

_____

_____

_____

_____

_____

26. Review your understanding of passive transport, in which transport proteins allow substances to "go with the flow" (flow downstream through the protein interiors to the side of the membrane with less concentrated particles of the substance being observed). State what would be required to transport particles (molecules or ions) upstream (against the natural direction of flow). [p.85] _____

_____

_____

27. If you answered #26, you have described the mechanism that drives the process of active transport. Name three kinds of cellular activities that require active transport.

(a). [p.85] _____

(b). [p.85] _____

(c). [p.85] _____

28. Name the process that drives blood through interconnected blood vessels and also causes sap to travel through tubular structures in trees. [p.86] _____

_____

29. Distinguish between hydrostatic pressure and osmotic pressure as each might influence a young plant cell. [p.87] _____

_____

_____

30. Phagocytosis, by which white blood cells consume bacteria and other debris, is an active form of _____. [p.88]

Self-Quiz

___ 1. Essentially, the first law of thermodynamics states that _____. [p.76]
a. one form of energy cannot be converted into another
b. entropy is increasing in the universe
c. energy cannot be created or destroyed
d. energy cannot be converted into matter or matter into energy

___ 2. An important principle of the second law of thermodynamics states that _____. [p.77]
a. energy can be transformed into matter, and because of this we can get something for nothing
b. energy can be destroyed only during nuclear reactions, such as those that occur inside the sun
c. if energy is gained by one region of the universe, another place in the universe also must gain energy in order to maintain the balance of nature
d. matter tends to become increasingly more disorganized

___ 3. In _____ pathways, carbohydrates, lipids, and proteins are broken down in stepwise reactions that lead to products of lower energy. [pp.78–79]
a. phosphorylation
b. biosynthetic
c. endergonic
d. degradative

___ 4. If a phosphate group is transferred to ADP, the bond formed _____. [p.79]
a. absorbs a large amount of free energy when the phosphate group is attached during hydrolysis
b. is oxidized when ATP is hydrolyzed to ADP and one phosphate group
c. is usually found in each glucose molecule; this is why glucose is chosen as the starting point for glycolysis
d. will release a large amount of usable energy later when the phosphate group is split off during an exergonic reaction

___ 5. Any substance (for example, NAD+) that gains electrons is _____. [p.79]
a. oxidized
b. a catalyst
c. reduced
d. a substrate

___ 6. With regard to major function, NAD+, is classified as _____. [p.79]
a. enzymes
b. a phosphate carrier
c. a cofactor that functions as a coenzyme
d. an end product of a metabolic pathway

___ 7. An enzyme is best described as _____. [pp.79,81]
a. an acid
b. a protein
c. a catalyst
d. a fat
e. both (b) and (c)

___ 8. Which is not true of enzyme behavior? [p.80]
   a. Enzyme shape may change during catalysis.
   b. The active site of an enzyme orients its substrate molecules, thereby facilitating interaction of their reactive parts.
   c. All enzymes have an active site where substrates are temporarily bound.
   d. An individual enzyme can catalyze a wide variety of different reactions.

___ 9. An allosteric enzyme _____. [pp.82–83]
   a. has an active site where substrate molecules bind and another site that binds with intermediate or end-product molecules
   b. is an important energy-carrying nucleotide
   c. carries out either oxidation reactions or reduction reactions but not both
   d. raises the activation energy of the chemical reaction it catalyzes

___ 10. Which process would most likely be used to move water from very dry soil into plant roots? [p.85]
   a. diffusion
   b. osmosis
   c. passive trasport
   d. active transport

Chapter Objectives/Review Questions

1. _____ is the controlled capacity to acquire and use energy for stockpiling, breaking apart, building, and eliminating substances in ways that contribute to survival and reproduction. [p.75]
2. Define *energy*; be able to state the first and second laws of thermodynamics. [pp.76–77]
3. _____ is a measure of the degree of randomness or disorder of systems. [p.77]
4. Explain how the world of life maintains a high degree of organization. [pp.77–78]
5. Reactions that show a net loss in energy are said to be _____; reactions that show a net gain in energy are said to be _____. [p.78]
6. ATP is composed of _____, a five-carbon sugar, three _____ groups, and _____, a nitrogen-containing compound. [pp.78–79]
7. Adding a phosphate to a molecule is called _____. [p.79]
8. ATP directly or indirectly delivers _____ to almost all metabolic pathways. [p.79]
9. Explain the functioning of the ATP/ADP cycle. [p.79]
10. Give the function of each of the following participants in metabolic pathways: reactants, intermediates, enzymes, cofactors, and end products. [p.79]
11. Cite one example of a coenzyme. [p.79]
12. What is the function of metabolic pathways in cellular chemistry? [pp.79;82–83]
13. Explain the effects of enzymes on activation energy. [pp.80–81]
14. Describe the induced-fit model. [pp.80–81]
15. Explain what happens to enzymes if temperature and pH continually increase. [p.82]
16. Describe the control mechanism known as feedback inhibition. [p.82]
17. What are "allosteric enzymes," and what is their function? [pp.82–83]
18. Distinguish diffusion from osmosis, and active transport from passive transport. [pp.84–86]
19. Explain what a gradient is in general; then describe (a) concentration gradients, (b) pressure gradients, and (c) electric gradients. [pp.84,87]
20. Water tends to diffuse from (select one) ❏ hypertonic ❏ hypotonic solutions to (select one) ❏ hypertonic ❏ hypotonic solutions. [p.87]
21. Describe and contrast exocytosis and endocytosis. Explain the role of vesicle formation in both processes. [p.88]
22. Describe phagocytosis and describe the kind of phagocytosis that occurs in your body. [p.88]

Integrating and Applying Key Concepts

A piece of dry ice left sitting on a table at room temperature vaporizes. As the dry ice vaporizes into CO_2 gas, does its entropy increase or decrease? Tell why you answered as you did.

6

HOW CELLS ACQUIRE ENERGY

Interactive Exercises

Sunlight and Survival [pp.92–93]

6.1. PHOTOSYNTHESIS—AN OVERVIEW [pp.94–95]

6.2. SUNLIGHT AS AN ENERGY SOURCE [pp.96–97]

Selected Words: *photo*autotrophs [p.92], T. Englemann [p.92], *Spirogyra* [p.92], grana (singular, granum) [p.94], *light-dependent* reactions [p.94], *light-independent* reactions [p.94], *visible* light [p.96], pigments [p.96], accessory *pigments* [p.97]

Boldfaced, Page-Referenced Terms

[p.92] autotrophs _____

_____

[p.92] photosynthesis _____

_____

[p.92] heterotrophs _____

_____

[p.94] chloroplasts _____

_____

[p.94] stroma _____

_____

[p.94] thylakoids _____

_____

[p.96] wavelength _____

_____

[p.96] photons _____

_____

[p.96] chlorophylls _____

_____

[p.97] carotenoids _____

_____

Fill-in-the-Blanks

(1) _____ [p.92] obtain carbon and energy from the physical environment; their carbon source is

(2) _____ _____ [p.92]. (3) _____ [p.92] autotrophs obtain energy from sunlight.

(4) _____ [p.92] feed on autotrophs, each other, and organic wastes; representatives include

(5) _____ [p.92], fungi, many protistans, and most bacteria. Although energy stored in organic

compounds such as glucose may be released by several pathways, the pathway known as (6) _____

_____ [p.93] releases the most energy per molecule of glucose.

7. In the spaces below, supply the missing information to complete the summary equation for photosynthe-sis:

12 _____ + _____CO_2 ⟶ _____ O_2 + $C_6H_{12}O_6$ + 6_____ [p.93]

8. Supply the appropriate information to state the equation (above) for photosynthesis in words:

(a) _____ [p.93] molecules of water plus six molecules of (b) _____ _____ [p.94] (in the

presence of pigments, enzymes, and visible light) yield six molecules of (c) _____ [p.93] plus one

molecule of (d) _____ [p.93] plus (e) _____ [p.93] molecules of water.

The two major sets of reactions of photosynthesis are the (9) _____-_____ [p.94] reactions and

the (10) _____-_____ [p.94] reactions. (11) _____ _____ [p.94] and (12) _____ [p.94]

are the reactants of photosynthesis, and the end product is usually given as (13) _____ [p.94]. The

internal membranes and channels of the chloroplast are the (14) _____ [p.94] membrane system and are

organized into stacks, called (15) _____ [p.94]. Spaces inside the thylakoid disks and channels form a

continuous compartment where (16) _____ [p.94] ions accumulate to be used to produce ATP. The

semifluid interior area surrounding the grana is known as the (17) _____ [p.94] and is the area where

the products of photosynthesis are produced.

The light-capturing phase of photosynthesis takes place on a system of (18) _____ [p.94] membranes.

A(n) (19) _____ [p.96] is a packet of light energy. Thylakoid membranes contain (20) _____ [p.96], which absorb photons of light. The principal pigments are the (21) _____ [p.97], which reflect green wavelengths but absorb (22) _____ [p.97] and (23) _____ [p.97] wavelengths. (24) _____ [p.97] are pigments that absorb violet and blue wavelengths but reflect yellow, orange, and red.

Matching

Choose the most appropriate answer for each term.

25. ___chlorophylls [p.96]

26. ___chlorophyll *b* and carotenoids [p.97]

27. ___carotenoids [p.97]

28. ___violet-blue-green-yellow-red [p.96]

29. ___photons [p.96]

30. ___chlorophyll *a* [p.97]

31. ___chloroplast [p.94]

32. ___phycobilins [p.97]

33. ___grana [p.94]

A. The main pigment of photosynthesis
B. Packets of energy that have an undulating motion through space
C. The two stages of photosynthesis occur here
D. Absorb blue-violet and blue-green wavelengths but reflect red, orange, and yellow
E. Visible light portion of the electromagnetic spectrum
F. Pigments that transfer energy to chlorophyll *a*
G. Absorb violet-to-blue and red wavelengths; the reason leaves appear green
H. The characteristic pigments of red algae and cyanobacteria
I. The site of the first stage of photosynthesis

6.3. THE LIGHT-DEPENDENT REACTIONS [pp.98–99]

6.4. A CLOSER LOOK AT ATP FORMATION IN CHLOROPLASTS [p.100]

Selected Words: "fluorescence" [p.98], *cyclic* pathway [p.98], *type I* photosystem [p.98], *type II* photosystem [p.98], *noncyclic* pathway [p.98], chemiosmotic model [p.100]

Boldfaced, Page-Referenced Terms

[p.98] light-dependent reactions _____

_____

[p.98] photosystems _____

_____

[p.98] electron transport systems _____

_____

[p.98] photolysis _____

_____

Fill-in-the-Blanks

A cluster of 200 to 300 of these pigment proteins is a(n) (1) _____ [p.98]. When pigments absorb

(2) _____ [p.98] energy, a(n) (3) _____ [p.98] is transferred from a photosystem to a(n)

(4) _____ [p.98] molecule. (5) _____ [recall p.79, Ch. 5] refers to the attachment of phosphate

group (P_i) to ADP or other organic molecules. Due to the input of light energy, electrons flow through a

transport system that causes protons (H^+) simultaneously to be pumped into the thylakoid compartments.

Electrons then end up in (6) _____ [p.99] chlorophyll at the end of this transport chain. The flow of protons from the thylakoid compartment through (7) _____ _____ [p.99] drives the enzyme

machine that phosphorylates (8) _____ [p.99], a sequence of events known as the (9) _____ [p.99]

model of ATP formation.

Complete the Table

10. Complete the table by identifying and stating the role of each item given in the cyclic pathway of the light-dependent reactions.

[pp.98–99] a. Type I Photosystem	
[pp.98–99] b. Electrons	
[pp.98–99] c. P700	
[p.98] d. Electron acceptor	
[p.98] e. Electron transport system	
[p.100] f. ADP	

Labeling

The diagram below illustrates noncyclic photophosphorylation. Identify each numbered part of the illustration.

11. _____ _____ [p.99]

12. _____ _____ _____ [p.99]

13. _____ _____ [p.99]

14. _____ [p.99]

15. _____ [p.98]

16. _____ [p.99]

17. _____ [p.98]

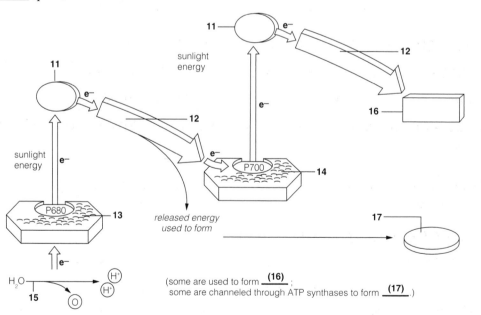

Complete the Table

With a check mark (√) indicate for each phase of the light-dependent reactions all items from the left-hand column that are applicable.

Light-Dependent Reactions	*Cyclic Pathway*	*Noncyclic Pathway*	*Photolysis Alone*
[pp.98–99] Uses H_2O as a reactant	(18)	(30)	(42)
[pp.98–99] Produces H_2O as a product	(19)	(31)	(43)
[pp.98–99] Type I photosystem involved (P700)	(20)	(32)	(44)
[p.98] Type II photosystem involved (P680)	(21)	(33)	(45)
[pp.98–99] ATP produced	(22)	(34)	(46)
[p.99] NADPH produced	(23)	(35)	(47)
[p.99] Uses CO_2 as a reactant	(24)	(36)	(48)
[p.99] Causes H^+ ions to accumulate in the thylakoid compartments	(25)	(37)	(49)
[p.100] Produces O_2 as a product	(26)	(38)	(50)
[pp.98,100] Produces H^+ ions by breaking apart H_2O	(27)	(39)	(51)
[pp.98,100] Uses ADP and P_i as reactants	(28)	(40)	(52)
[p.99] Uses $NADP^+$ as a reactant	(29)	(41)	(53)

6.5. THE LIGHT-INDEPENDENT REACTIONS [p.101]

6.6. FIXING CARBON—SO NEAR, YET SO FAR [pp.102–103]

6.7. *Focus on Science:* LIGHT IN THE DEEP DARK SEA? [p.103]

6.8. *Focus on the Environment:* AUTOTROPHS, HUMANS, AND THE BIOSPHERE [p.104]

Selected Words: photorespiration [p.102], *three*-carbon PGA [p.102], *four*-carbon oxaloacetate [p.102], *chemo*autotrophs [p.104]

Boldfaced, Page-Referenced Terms

[p.101] light-independent reactions _____

_____

[p.101] Calvin-Benson cycle _____

_____

[p.101] RuBP, ribulose bisphosphate _____

_____

[p.101] Rubisco, RuBP carboxylase _____

_____

[p.101] PGA, phosphoglycerate _____

_____

[p.101] carbon fixation _____

_____

[p.101] PGAL, phosphoglyceraldehyde _____

_____

[p.102] stomata (singular, stoma) _____

_____

[p.102] C3 plants _____

_____

[p.102] C4 plants _____

_____

[p.103] CAM plants _____

_____

[p.103] hydrothermal vent _____

_____

Label–Match

Identify each part of the illustration below. Complete the exercise by matching and entering the letter of the proper function description in the parentheses following each label. [All are from p.101]

1. _____ _____ ( )
2. _____ _____ ( )
3. _____ ( )
4. _____ _____ ( )
5. _____ ( )
6. _____ ( )
7. _____ _____ ( )
8. _____ - _____ _____ ( )
9. _____ _____ ( )

A. A three-carbon sugar, the first sugar produced; goes on to form sugar phosphate and RuBP
B. Typically used at once to form carbohydrate end products of photosynthesis
C. A five-carbon compound produced from PGALs; attaches to incoming CO_2
D. A compound that diffuses into leaves; attached to RuBP by enzymes in photosynthetic cells
E. Includes all the chemical reactions that "fix" carbon into an organic compound

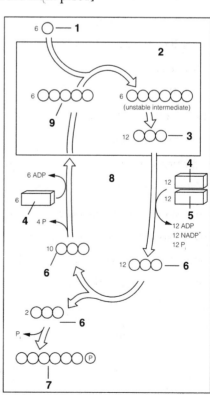

F. Three-carbon compounds formed from the splitting of the six-carbon intermediate compound
G. A molecule that was reduced in the noncyclic pathway; furnishes hydrogen atoms to construct sugar molecules
H. A product of the light-dependent reactions; necessary in the light-independent reactions to energize molecules in metabolic pathways
I. Includes all the chemistry that fixes CO_2, converts PGA to PGAL and PGAL to RuBP and sugar phosphates

Fill-in-the-Blanks

The light-independent reactions can proceed without sunlight as long as (10) _____ [p.101] and (11) _____ [p.101] are available. The reactions begin when an enzyme links (12) _____ _____ [p.101] to (13) _____ _____ [p.101], a five-carbon compound. The resulting six-carbon compound is highly unstable and breaks apart at once into two molecules of a three-carbon compound, (14) _____ [p.101]. This entire reaction sequence is called carbon (15) _____ [p.101]. ATP gives a phosphate group to each (16) _____ [p.101]. This intermediate compound takes on H^+ and electrons from NADPH to form (17) _____ [p.101]. It takes (18) _____ [p.101] carbon dioxide molecules to produce twelve PGAL. Most of the PGAL becomes rearranged into new (19) _____ [p.101] molecules—which can be used to fix more (20) _____ [p.101]. Two (21) _____ [p.101] are joined together to form a (22) _____ _____ [p.101], primed for further reactions. The Calvin-Benson cycle yields enough RuBP to replace those used in carbon (23) _____ [p.101]. ADP, $NADP^+$, and phosphate leftovers are sent back to the (24) _____-_____ [p.101] reaction sites, where they are again converted to (25) _____ [p.101] and (26) _____ [p.101]. (27) _____ _____ [p.101] formed in the cycle serves as a building block for the plant's main carbohydrates. When RuBP attaches to oxygen instead of carbon dioxide, (28) _____ [p.102] results; this is typical of (29) _____ [p.102] plants in hot, dry conditions. If less PGA is available, leaves produce a reduced amount of (30) _____ [p.102]. C4 plants can still construct carbohydrates when the ratio of carbon dioxide to (31) _____ [p.102] is unfavorable because of the attachment of carbon dioxide to produce (32) _____ [p.102] in certain leaf cells.

(33) _____ [p.102] plants open their stomata at night, capture (34) _____ _____ [p.103] as part of a metabolic intermediate stored in central vacuoles and use it in photosynthesis the next day when stomata are closed. (33) plants capture (34) that forms by (35) _____ _____ [p.103] more than once at different times.

Organisms that obtain energy from oxidation of inorganic substances such as ammonium compounds, and iron or sulfur compounds, are known as (36) _____ [p.104] autotrophs. Such organisms use this energy to build (37) _____ [p.104, by inference] compounds. As an example, some hydrothermal vent bacteria use hydrogen sulfide molecules as an energy source, stripping them of (38) _____ [p.104] and (39) _____ [p.104].

Complete the Table

With a check mark (√) indicate for each phase of the light-independent reaction all items from the left-hand column that are applicable. [All are from p.101]

Light-Independent Reactions	CO$_2$ Fixation Alone	Conversion of PGA to PGAL	Regeneration of of RuBP	Formation of Glucose and Other Organic Compounds
Requires RuBP as a reactant	(40)	(51)	(62)	(73)
Requires ATP as a reactant	(41)	(52)	(63)	(74)
Produces ADP as a product	(42)	(53)	(64)	(75)
Requires NADPH as a reactant	(43)	(54)	(65)	(76)
Produces NADP$^+$ as a product	(44)	(55)	(66)	(77)
Produces PGA	(45)	(56)	(67)	(78)
Produces PGAL	(46)	(57)	(68)	(79)
Requires PGAL as a reactant	(47)	(58)	(69)	(80)
Produces P$_i$ as a product	(48)	(59)	(70)	(81)
Produces H$_2$O as a product	(49)	(60)	(71)	(82)
Requires CO$_2$ as a reactant	(50)	(61)	(72)	(83)

Self-Quiz

___ 1. Plants need _____ and _____ to carry on photosynthesis. [pp.93–94]
   a. oxygen; water
   b. oxygen; CO$_2$
   c. CO$_2$; H$_2$O
   d. sugar; water

___ 2. Thylakoid disks are stacked in groups called _____. [p.94]
   a. grana
   b. stroma
   c. lamellae
   d. cristae

___ 3. Chlorophyll is _____. [p.94]
   a. on the outer chloroplast membrane
   b. inside the mitochondria
   c. in the stroma lamellae
   d. part of the thylakoid membrane system

___ 4. The cyclic pathway of the light-dependent reactions functions mainly to _____. [p.98]
   a. fix CO$_2$
   b. make ATP
   c. produce PGAL
   d. regenerate ribulose bisphosphate

___ 5. Plant cells produce O$_2$ during photosynthesis by _____. [p.98]
   a. breaking apart CO$_2$ molecules
   b. breaking apart water molecules
   c. degradation of the stroma
   d. breaking apart sugar molecules

6. The electrons that are passed to NADPH during noncyclic photophosphorylation were obtained from _____. [p.98]
   a. water
   b. CO_2
   c. glucose
   d. sunlight

7. The ultimate electron and hydrogen acceptor in the noncyclic pathway is _____. [p.99]
   a. $NADP^+$
   b. ADP
   c. O_2
   d. H_2O

8. The two products of the light-dependent reactions that are required for the light-independent metabolic pathway are _____ and _____. [p.101]
   a. CO_2; H_2O
   b. O_2; NADPH; inorganic phosphate
   c. O_2; ATP
   d. ATP; NADPH

9. C4 plants have an advantage in hot, dry conditions because _____. [p.102]
   a. their leaves are covered with thicker wax layers than those of C3 plants
   b. their stomates open wider than those of C3 plants, thus cooling their surfaces
   c. CO_2 is fixed in the mesophyll cells, where the C3 pathway occurs, then delivered to the bundle sheath cells
   d. they are also capable of capturing CO_2 by photorespiration

10. Chemosynthetic autotrophs obtain energy by oxidizing such inorganic substances as _____. [p.104]
    a. PGA
    b. PGAL
    c. hydrogen sulfide
    d. water

Chapter Objectives/Review Questions

1. List the major stages of photosynthesis, and state what occurs in those sets of reactions. [p.93]
2. Study the general equation for photosynthesis as shown on text p. 64 until you can remember the reactants and products. Reproduce the equation from memory on another piece of paper. [p.93]
3. Describe the structural details of the green leaf. Begin with the layers of a leaf cross-section and complete your description with the minute structural sites within the chloroplast where the major sets of photosynthetic reactions occur. Explain how each of the reactants needed in various phases of photosynthesis arrive at the place where they are used. Explain what happens to the products of photosynthesis. [pp.94–95]
4. Describe how the pigments found on thylakoid membranes are organized into photosystems and how they relate to photon light energy. [p.96]
5. Describe the role that chlorophylls and the other chloroplast pigments play in the light-dependent reactions. After consulting text Figure 6.5 on p. 97, state which colors of the visible spectrum are absorbed by (a) chlorophyll *a*, (b) chlorophyll *b*, and (c) carotenoids. [pp.96–98]
6. State what T. Englemann's 1882 experiment with *Spirogyra* revealed. [p.92]
7. Two energy-carrying molecules produced in the noncyclic pathways are _____ and _____; explain why these molecules are necessary for the light-independent reactions. [pp.98–99]
8. After evolution of the noncyclic pathway, _____ accumulated in the atmosphere and made _____ respiration possible. [p.99]
9. Explain how the chemiosmotic theory is related to thylakoid compartments and the production of ATP. [p.100]
10. Explain why the light-independent reactions are called by that name. [p.101]
11. Describe the Calvin-Benson cycle as it is related to the four phases shown in the table in 6.8. [p.101]
12. Describe the mechanism by which C4 plants thrive under hot, dry conditions; distinguish this CO_2-capturing mechanism from that of C3 plants. [p.102]
13. Cacti are CAM plants adapted to survive desert conditions by using a mechanism unlike that of C4 plants. Describe the mechanism. [p.103]

14. State some of the observations that support the hypothesis that life on Earth originated near hydrothermal vents. [p.103]
15. Which creatures constitute the "pastures of the seas"? What benefits do they provide for us? How might industrial wastes, fertilizers, and herbicides affect these photoautotrophs? [p.104]

Integrating and Applying Key Concepts

Suppose that humans acquired all the enzymes needed to carry out photosynthesis. Speculate about the attendant changes in human anatomy, physiology, and behavior that would be necessary for those enzymes to actually carry out photosynthetic reactions.

7

HOW CELLS RELEASE STORED ENERGY

Interactive Exercises

The Killers Are Coming! The Killers Are Coming! [pp.108–109]

7.1. HOW DO CELLS MAKE ATP? [pp.110–111]

Selected Words: anaerobic [p.110], *coenzymes* [p.111]

Boldfaced, Page-Referenced Terms

[p.110] aerobic respiration _____

_____

[p.110] glycolysis _____

_____

[p.110] pyruvate _____

_____

[p.111] Krebs cycle _____

_____

[p.111] NAD$^+$ _____

_____

[p.111] FAD _____

_____

[p.111] electron transport phosphorylation _____

_____

Short Answer

1. Although various organisms utilize different energy sources, what is the usual form of chemical energy that will drive metabolic reactions? [p.109] _____

_____

2. Describe the function of oxygen in the main degradative pathway, aerobic respiration. [p.109]

_____

_____

3. List the most common anaerobic pathways, and describe the conditions in which they function. [p.109]

_____

Fill-in-the-Blanks

Virtually all forms of life depend on a molecule known as (4) _____ [p.110] as their primary energy carrier. Plants produce adenosine triphosphate during (5) _____ [p.110], but plants and all other organisms also can produce ATP through chemical pathways that degrade (take apart) food molecules. The main degradative pathway requires free oxygen O_2, and is called (6) _____ _____ [p.110].

There are three stages of aerobic respiration. In the first stage, (7) _____ [p.110], glucose is partially degraded to (8) _____ [p.110]. By the end of the second stage, which includes the (9) _____ [p.111] cycle, glucose has been completely degraded to carbon dioxide. (10) _____ [p.111] is the end product formed during the third stage. Neither of the first two stages produces much (11) _____ [p.111]. During both stages, protons and (12) _____ [p.111] are stripped from intermediate compounds and delivered to a(n) (13) _____ _____ [p.111] system. That system is used in the third stage of reactions, electron transport (14) _____ [p.111]; passage of electrons along the transport system drives the enzymatic "machinery" that phosphorylates ADP to produce a high yield of (15) _____ [p.111]. (16) _____ [p.111] accepts "spent" electrons from the transport system and keeps the pathway clear for repeated ATP production.

Other degradative pathways are (17) _____ [p.110], in that something other than oxygen serves as the final electron acceptor in energy-releasing reactions. (18) _____ [p.110] and anaerobic (19) _____ _____ [p.110] are the most common anaerobic pathways.

Completion

20. Complete the equation below, which summarizes the degradative pathway known as aerobic respiration:

_____ + _____ $O_2 \longrightarrow$ 6 _____ + 6 _____ [p.110]

21. Supply the appropriate information to state the equation (above) for aerobic respiration in words:

One molecule of glucose plus six molecules of _____ [p.110] (in the presence of appropriate enzymes) yield _____ [p.110] molecules of carbon dioxide plus _____ [p.110] molecules of water.

7.2. GLYCOLYSIS: FIRST STAGE OF ENERGY-RELEASING PATHWAYS [pp.112–113]

Selected Words: *energy-requiring* reactions [p.112], "phosphorylation" [p.112], *energy-releasing* reactions [p.112], *net energy* yield [p.112]

Boldfaced, Page-Referenced Terms

[p.112] substrate-level phosphorylation _____

_____

Fill-in-the-Blanks

Glucose is produced by (1) _____ [p.110, recall] organisms as they synthesize and stockpile energy-rich carbohydrates and other food molecules from inorganic raw materials. (2) _____ [p.112] is partially dismantled by the glycolytic pathway; at the end of this process some of its stored energy remains in two (3) _____ [p.112] molecules. Some of the energy of glucose is released during the breakdown reactions and used in forming the nucleotide (4) _____ [recall p.47] and the coenzyme (5) _____ [p.112]. These reactions take place in the cytoplasm. Glycolysis begins with two phophate groups being transferred to (6) _____ [p.113] from two (7) _____ [p.113] molecules. The addition of two phosphate groups to (6) energizes it and causes it to become unstable and split apart, forming two molecules of (8) _____ [p.113]. Each (8) gains one (9) _____ [p.113] group from the cytoplasm, then (10) _____ [p.113] atoms and electrons from each PGAL are transferred to NAD$^+$, changing this coenzyme to NADH. Two (11) _____ [p.113] molecules form by substrate-level phosphorylation; the cell's energy investment is paid off. One (12) _____ [p.113] molecule is released from each 2-PGA as a waste product. The resulting intermediates are rather unstable; each gives up a(n) (13) _____ [p.113] group to ADP. Once again, two (14) _____ [p.113] molecules have formed by (15) _____-_____ [p.113] phosphorylation. For each (16) _____ [p.113] molecule entering glycolysis, the net energy yield is two ATP molecules that the cell can use anytime to do work. The end products of glycolysis are two molecules of (17) _____ [p.113], each with a(n) (18) _____ (number) [p.113]-carbon backbone.

Sequence

Arrange the following events of the glycolysis pathway in correct chronological sequence. Write the letter of the first step next to 19, the letter of the second step next to 20, and so on. [All are from p.113]

19. ___   A. Two ATPs form by substrate-level phosphorylation; the cell's energy debt is paid off

20. ___   B. Diphosphorylated glucose (fructose 1,6-bisphosphate) molecules split to form 2 PGALs; this is the first energy-releasing step

21. ___   C. Two 3-carbon pyruvate molecules form as the end products of glycolysis

22. ___   D. Glucose is present in the cytoplasm

23. ___   E. Two more ATPs form by substrate-level phosphorylation, the cell gains ATP; net yield of ATP from glycolysis is 2 ATPs

24. ___   F. The cell invests two ATPs; one phosphate group is attached to each end of the glucose molecule (fructose 1,6-bisphosphate)

25. ___   G. Two PGALs gain two phosphate groups from the cytoplasm

26. ___   H. Hydrogen atoms and electrons from each PGAL are transferred to NAD$^+$, reducing this carrier to NADH

7.3. SECOND STAGE OF THE AEROBIC PATHWAY [pp.114–115]

Selected Words: ATP synthases [p.114], NADH [p.114], FADH$_2$ [p.114], coenzymes [p.115]

Boldfaced, Page-Referenced Terms

[p.114] mitochondrion (plural, mitochondria) _____

_____

[p.114] acetyl-CoA _____

_____

[p.114] oxaloacetate _____

_____

Fill-in-the-Blanks

If sufficient oxygen is present, the end product of glycolysis enters a preparatory step, (1) _____

_____ [p.114] formation. This step converts pyruvate into acetyl CoA, the molecule that enters the

(2) _____ [p.114] cycle. Acetyl-CoA is then linked to (3) _____ [p.115] to form (4) _____

[p.115] as (5) _____ [p.115] is split off and returns to the preparatory step to be linked with another

acetyl (two-carbon) group. In the preparatory conversions prior to the Krebs cycle and within the Krebs

cycle, the food molecule fragments are further broken down into (6) _____ _____ [p.114]. During

these reactions, hydrogen atoms (with their (7) _____ [p.114]) are stripped from the fragments and

transferred to the energy carriers (8) _____ [p.114] and (9) _____ [p.114].

Labeling

In exercises 10–14, identify the structure or location; in exercises 15–18, identify the chemical substance involved. In exercise 19, name the metabolic pathway. [All are from p.114]

10. _____ _____ of mitochondrion   15. _____

11. _____ _____ of mitochondrion   16. _____

12. _____ _____ of mitochondrion   17. _____

13. _____ _____ of mitochondrion   18. _____

14. _____   19. _____ _____ _____

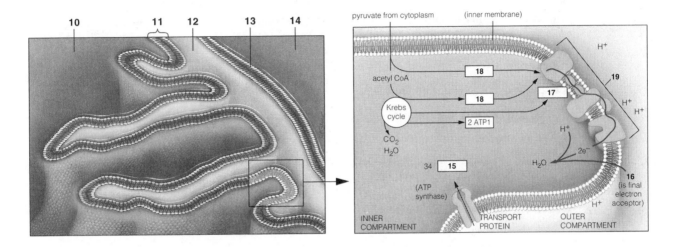

7.4. THIRD STAGE OF THE AEROBIC PATHWAY [pp.116–117]

Selected Words: electron transport phosphorylation [p.116], *chemiosmotic* model [p.116]

Fill-in-the-Blanks

NADH delivers its electrons to the highest possible point of entry into a transport system; from each NADH enough H^+ is pumped to produce (1) _____ [p.117] (number) ATP molecules. $FADH_2$ delivers its electrons at a lower point of entry into the transport system; fewer H^+ are pumped, and (2) _____ [p.117] (number) ATPs are produced. The electrons are then sent down highly organized (3) _____ [p.117] systems located in the inner membrane of the mitochondrion; hydrogen ions are pumped into the outer mitochondrial compartment. According to the (4) _____ [p.116] model, the hydrogen ions accumulate and then follow a gradient to flow through channel proteins, called ATP (5) _____ [pp.116–117], that lead into the inner compartment. The energy of the hydrogen ion flow across the membrane is used to phosphorylate ADP to produce (6) _____ [pp.116–117]. Electrons leaving the electron transport system combine with hydrogen ions and (7) _____ [p.116] to form water. These reactions occur only in (8) _____ [p.116]. From glycolysis (in the cytoplasm) to the final reactions occurring in the mitochondria, the aerobic pathway commonly yields (9) _____ [p.117] (number) ATP or (10) _____ [p.117] (number) ATP for every glucose molecule degraded.

Choice

For questions 11–32, choose from the following. Some correct answers may require more than one letter.

      a. preparatory steps to Krebs cycle   b. Krebs cycle   c. electron transport phosphorylation

11. ___ Chemiosmosis occurs to form ATP molecules [p.116]

12. ___ Three carbon atoms in pyruvate leave as three CO_2 molecules [recall p.115]

13. ___ Chemical reactions occur at transport systems [pp.116–117]

14. ___ Coenzyme A picks up a 2-carbon acetyl group [recall p.115}

15. ___ Makes two turns for each glucose molecule entering glycolysis [recall p.115]

16. ___ Two NADH molecules form for each glucose entering glycolysis [recall p.115]

17. ___ Oxaloacetate forms from intermediate molecules [recall p.115]

18. ___ Named for a scientist who worked out its chemical details [recall p.114]

19. ___ Occurs within the mitochondrion [recall pp.115–116]

20. ___ Two $FADH_2$ and six NADH form from one glucose molecule entering glycolysis [recall p.115]

21. ___ Hydrogens accumulate in the mitochondrion's outer compartment [p.116]

22. ___ Hydrogens and electrons are transferred to NAD^+ and FAD [pp. 116–117]

23. ___ Two ATP molecules form by substrate-level phosphorylation [recall p.115]

24. ___ Free oxygen withdraws electrons from the system and then combines with H^+ to form water molecules [p.116]

25. ___ No ATP is produced [recall p.115]

26. ___ Thirty-two ATPs are produced [p.116]

27. ___ Delivery point of NADH and $FADH_2$ [p.117]

28. ___ Two pyruvates enter for each glucose molecule entering glycolysis [recall p.115]

29. ___ The carbons in the acetyl group leave as CO_2 [recall p.115]

30. ___ One carbon in pyruvate leaves as CO_2 [recall p.115]

31. ___ An electron transport system and channel proteins are involved [pp.116–117]

32. ___ Two $FADH_2$ and ten NADH are sent to this stage [recall p.115]

7.5. ANAEROBIC ROUTES OF ATP FORMATION [pp.118–119]

Selected Words: *pyruvate* [p.118], "lactic acid" [p.118], lactate [p.118], *Lactobacillus* [p.118], *Saccharomyces cerevisiae* [p.118], *S. ellipsoideus* [p.118]

Boldfaced, Page-Referenced Terms

[p.118] lactate fermentation _____

_____

[p.118] alcoholic fermentation _____

_____

[p.119] anaerobic electron transport _____

_____

Fill-in-the-Blanks

If (1) _____ [p.118] is not present in sufficient amounts, the end product of glycolysis enters (2) _____ [p.118] pathways; in some bacteria and muscle cells, pyruvate is converted into such products as (3) _____ [p.118], or in yeast cells it is converted into (4) _____ [p.118] and (5) _____ _____ [p.118].

(6) _____ [p.118] pathways do not use oxygen as the final (7) _____ [p.118] acceptor that ultimately drives the ATP-forming machinery. Anaerobic routes must be used by many bacteria and protistans that live in an oxygen-free environment. (8) _____ [p.118] precedes any of the fermentation pathways. During (8), a glucose molecule is split into two (9) _____ [p.118] molecules, two energy-rich (10) _____ [p.118] intermediate molecules form, and the net energy yield from one glucose molecule is two ATP.

In one kind of fermentation pathway, (11) _____ [p.118] itself accepts hydrogen and electrons from NADH. (11) is then converted to a three-carbon compound, (12) _____ [p.118], during this process in a few bacteria species and some animal cells. Human muscle cells can carry on (12) fermentation in times of oxygen depletion; this provides a low yield of ATP.

In yeast cells, each pyruvate molecule from glycolysis forms an intermediate called (13) _____ [p.118] as a gas (14) _____ _____ [p.118] is detached from pyruvate with the help of an enzyme. This intermediate accepts hydrogen and electrons from NADH and is then converted to (15) _____ [p.119], the end product of alcoholic fermentation.

In both types of fermentation pathways, the net energy yield of two ATPs is formed during (16) _____ [p.119]. The reactions of the fermentation chemistry regenerate the (17) _____ [p.119] needed for glycolysis to occur.

Anaerobic electron transport is an energy-releasing pathway occurring among the (18) _____ [p.119]. For example, sulfate-reducing bacteria living in soil or water produce (19) _____ [p.119] by stripping electrons from a variety of compounds and sending them through membrane transport systems. The inorganic compound (20) _____ (SO_4^-) [p.119] serves as the final electron acceptor and is converted into foul-smelling hydrogen sulfide gas (H_2S). Some sulfate-reducing bacteria live on the deep ocean floor near (21) _____ _____ [p.119]. These bacteria are important because they form the (22) _____ [p.119] production base for this unique ecosystem.

Complete the Table

Include a (√) in each box that correctly links an occurrence (left-hand column) with a process (or processes).

	Glycolysis	Lactate Fermentation	Alcoholic Fermentation	Anaerobic Electron Transport
6-C → 3C / 3C [p.113]	(23)	(36)	(49)	(62)
NADH → NAD+ [pp.118–119]	(24)	(37)	(50)	(63)
NAD+ → NADH [p.113]	(25)	(38)	(51)	(64)
$SO_4^- \rightarrow H_2S$ [p.119]	(26)	(39)	(52)	(65)
H_2O is a waste product	(27)	(40)	(53)	(66)
CO_2 is a waste product [p.119]	(28)	(41)	(54)	(67)
3-C → 3-C [p.118]	(29)	(42)	(55)	(68)
3-C → 2-C [p.119]	(30)	(43)	(56)	(69)
ATP is used as a reactant [p.113]	(31)	(44)	(57)	(70)
ATP is produced [pp.113,119]	(32)	(45)	(58)	(71)
pyruvate → ethanol / CO_2 [p.119]	(33)	(46)	(59)	(72)
Occurs in animal cells [pp.109,118]	(34)	(47)	(60)	(73)
Occurs in yeast cells [p.118]	(35)	(48)	(61)	(74)

7.6. ALTERNATIVE ENERGY SOURCES IN THE HUMAN BODY [pp.120–121]

7.7. *Commentary*: PERSPECTIVE ON LIFE [p.122]

Selected Words: *chemical message* [p.122]

True–False

If the statement is true, write a T in the blank. If the statement is false, explain why in the answer blank.

_____1. Glucose is the only carbon-containing molecule that can be fed into the glycolytic pathway. [p.121]

_____2. If you exceed the glycogen-storage capacity of your liver and muscle cells, the excess is likely to be converted to fats for storage. [p.120]

_____3. In many animals, excess carbon dioxide and water, the products of aerobic respiration, generally diffuse into the blood and are carried to gills or lungs, kidneys, and skin, where they are expelled from the animal's body. [general knowledge]

_____4. Energy is recycled along with materials. [p.122]

_____5. The first forms of life on Earth were most probably photosynthetic eukaryotes. [p.122]

Labeling

Identify the process or substance indicated in the illustration below. [All are from text Figure 7.12, p.121]

6. _____ _____

7. _____

8. _____

9. _____ _____

10. _____ _____

11. _____ _____

12. _____

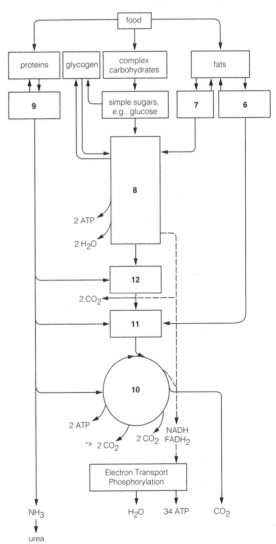

Complete the Table

Across the top of each column are the principal phases of degradative pathways into which food molecules (in various stages of breakdown) enter or in which specific events occur. Put a check (√) in each box that indicates the phase into which a specific food molecule is fed, or in which a specific event occurs. For example: if simple sugars can enter the glycolytic pathway, put a check in the top left-hand box; if not, let the box remain blank.

	Glycolysis (includes pyruvate)	Acetyl CoA Formation	Krebs Cycle	Electron Transport Phosphorylation	Fermentation Alcoholic	Lactate
Complex → Simple → which Carbohydrates Sugars enter [p.121]	(13)	(28)	(43)	(58)	(73)	(88)
Fats ⌐ Fatty acids, which enter [p.121]	(14)	(29)	(44)	(59)	(74)	(89)
└ Glycerol, which enters [p.121]	(15)	(30)	(45)	(60)	(75)	(90)
Proteins → Amino Acids enter [p.121]	(16)	(31)	(46)	(61)	(76)	(91)
Intermediate energy carriers (NADH, $FADH_2$) are produced [pp.115,121]	(17) (18)	(32) (33)	(47) (48)	(62) (63)	(77) (78)	(92) (93)
ATPs produced directly as a result of this process alone [p.121]	(19)	(34)	(49)	(64)	(79)	(94)
NAD^+ produced [pp.117–118]	(20)	(35)	(50)	(65)	(80)	(95)
FAD produced	(21)	(36)	(51)	(66)	(81)	(96)
ADP produced [p.113]	(22)	(37)	(52)	(67)	(82)	(97)
Unbound phosphate (P_i) required [pp.113,115]	(23)	(38)	(53)	(68)	(83)	(98)
CO_2 produced (waste product) [pp.115,118,121]	(24)	(39)	(54)	(69)	(84)	(99)
H_2O produced (waste product) [pp.113,121]	(25)	(40)	(55)	(70)	(85)	(100)
Atoms from O_2 react here [p.121]	(26)	(41)	(56)	(71)	(86)	(101)
NADH required to drive this process [pp.118,121]	(27)	(42)	(57)	(72)	(87)	(102)

Choice

For questions 103–117, refer to the text and Figure 7.12; choose from the following:

> a. glucose   b. glucose-6-phosphate   c. glycogen   d. fatty acids   e. triglycerides
> f. PGAL   g. acetyl-CoA   h. amino acids   i. glycerol   j. proteins

103.___ Fats that are used between meals or during exercise as alternatives to glucose [p.120]

104.___ Used between meals when free glucose supply dwindles; enters glycolysis after conversion [pp.120–121]

105.___ Its breakdown yields much more ATP than glucose [p.120]

106.___ Absorbed in large amounts immediately following a meal [p.120]

107.___ Represents only 1 percent or so of the total stored energy in the body [p.120]

108.___ Following removal of amino groups, the carbon backbones may be converted to fats or carbohydrates or they may enter the Krebs cycle [pp.120–121]

109.___ On the average, represents 78 percent of the body's stored food [p.120]

110.___ Between meals liver cells can convert it back to free glucose and release it [p.120]

111.___ Can be stored in cells but not transported across plasma membranes [p.120]

112.___ Amino groups undergo conversions that produce urea, a nitrogen-containing waste product excreted in urine [pp.120–121]

113.___ Converted to PGAL in the liver, a key intermediate of glycolysis [p.120]

114.___ Accumulate inside the fat cells of adipose tissues, at strategic points under the skin [p.120]

115.___ A storage polysaccharide produced from glucose-6-phosphate following food intake that exceeds cellular energy demand. [p.120]

116.___ Building blocks of the compounds that represent 21 percent of the body's stored food [p.120]

117.___ A product resulting from enzymes cleaving circulating triglycerides; enters the Krebs cycle [pp.120–121]

Self-Quiz

___ 1. Glycolysis would quickly halt if the process ran out of _____, which serves as the hydrogen and electron acceptor. [p.112]
   a. $NADP^+$
   b. ADP
   c. NAD^+
   d. H_2O

___ 2. The ultimate electron acceptor in aerobic respiration is _____. [p.117]
   a. NADH
   b. carbon dioxide (CO_2)
   c. oxygen ($1/2\ O_2$)
   d. ATP

___ 3. When glucose is used as an energy source, the largest amount of ATP is generated by the _____ portion of the entire respiratory process. [p.116]
   a. glycolytic pathway
   b. acetyl-CoA formation
   c. Krebs cycle
   d. electron transport phosphorylation

___ 4. The process by which about 10 percent of the energy stored in a sugar molecule is released as it is converted into two small-organic-acid molecules is _____. [pp.112–113]
   a. photolysis
   b. glycolysis
   c. fermentation
   d. the dark reactions

___ 5. During which of the following phases of aerobic respiration is ATP produced directly by substrate-level phosphorylation? [p.113]
   a. glucose formation
   b. ethyl-alcohol production
   c. acetyl-CoA formation
   d. glycolysis

___ 6. What is the name of the process by which reduced NADH transfers electrons along a chain of acceptors to oxygen so as to form water and in which the energy released along the way is used to generate ATP? [pp.116–117]
   a. glycolysis
   b. acetyl-CoA formation
   c. the Krebs cycle
   d. electron transport phosphorylation

___ 7. Pyruvic acid can be regarded as the end product of _____. [p.113]
   a. glycolysis
   b. acetyl-CoA formation
   c. fermentation
   d. the Krebs cycle

___ 8. Which of the following is not ordinarily capable of being reduced at any time? [p.121]
   a. NAD
   b. FAD
   c. oxygen, O_2
   d. water

___ 9. ATP production by chemiosmosis involves _____. [pp.116–117]
   a. H^+ concentration and electric gradients across a membrane
   b. ATP synthases
   c. formation of ATP in the inner mitochondrial compartment
   d. all of the above

___10. During the fermentation pathways, a net yield of two ATP is produced from _____; the NAD^+ necessary for _____ is regenerated during the fermentation reactions. [pp.113,118]
   a. the Krebs cycle; glycolysis
   b. glycolysis; electron transport phosphorylation
   c. the Krebs cycle; electron transport phosphorylation
   d. glycolysis; glycolysis

Matching

Match the following components of respiration to the list of words below. Some components may have more than one answer.

11. ___lactic acid, lactate [p.118]

12. ___$NAD^+ \rightarrow NADH$ [pp.113,115]

13. ___carbon dioxide is a product [pp.115,118]

14. ___$NADH \rightarrow NAD^+$ [pp.115–118]

15. ___pyruvic acid, pyruvate, used as a reactant [pp.115,118]

16. ___ATP produced by substrate-level phosphorylation [pp.113,115]

17. ___glucose [p.113]

18. ___acetyl CoA is both a reactant and a product [p.115]

19. ___oxygen [pp.117,121]

20. ___water is a product [pp.113,121]

A. Glycolysis
B. Preparatory conversions prior to the Krebs cycle
C. Fermentation
D. Krebs cycle
E. Electron transport

Chapter Objectives/Review Questions

1. What happens to the CO_2 and H_2O produced during acetyl-CoA formation and the Krebs cycle? [pp.109–110]
2. No matter what the source of energy might be, organisms must convert it to _____, a form of chemical energy that can drive metabolic reactions. [pp.109–110]
3. Give the overall equation for the aerobic respiratory route; indicate where energy occurs in the equation. [p.110]
4. In the first of the three stages of aerobic respiration, _____ is partially degraded to pyruvate. [p.112]
5. Glycolysis occurs in the _____ of the cell. [p.112]
6. Explain the purpose served by molecules of ATP reacting first with glucose and then with fructose-6-phosphate in the early part of glycolysis [Consult Figure 7.4 in the text, pp.112–113]
7. Four ATP molecules are produced by _____-_____ phosphorylation for every two used during glycolysis. [Consult Figure 7.4 in the text, pp.112–113]
8. Glycolysis produces _____ (number) NADH, _____ (number) ATP (net), and _____ (number) pyruvate molecules for each glucose molecule entering the reactions. [p.113]
9. State the events that happen during the preparatory steps and explain how the process of acetyl-CoA formation relates glycolysis to the Krebs cycle. [Consult Figures 7.4 and 7.6 in the text, pp.112–113,115]
10. Explain, in general terms, the role of oxygen in aerobic respiration. [p.114]
11. Consult Figure 7.5 in the text and predict what will happen to the NADH produced during acetyl-CoA formation and the Krebs cycle. [p.114]
12. State which factors determine whether the pyruvate (pyruvic acid) produced at the end of glycolysis will enter into the alcoholic fermentation pathway, the lactate fermentation pathway, or the acetyl-CoA formation pathway. [pp.114,118]
13. By the end of the second stage of aerobic respiration, which includes the _____ cycle, _____ has been completely degraded to carbon dioxide and water. [p.115]
14. Be able to account for the total net yield of thirty-six ATP molecules produced through aerobic respiration; that is, state how many ATPs are produced in glycolysis, acetyl-CoA formation, the Krebs cycle, and electron transport phosphorylation. [pp.115,117]
15. Explain how chemiosmotic theory operates in the mitochondrion to account for the production of ATP molecules. [pp.116–117]
16. Briefly describe the process of electron transport phosphorylation by stating what reactants are needed and what the products are. State how many ATP molecules are produced through operation of the transport system. [pp.116–117]
17. List some places where there is very little oxygen present and where anaerobic organisms might be found. [pp.118–119]
18. Describe what happens to pyruvate in anaerobic organisms. Then explain the necessity for pyruvate to be converted to a fermentative product. [pp.118–119]
19. You have been fasting for three days, drinking only water and eating no solid food. Tell which stored molecules your body is using to provide energy, and describe how that is occurring. [pp.120–121]
20. After reading the *Commentary* "Perspective on Life" in the main text, outline the supposed evolutionary sequence of energy-extraction processes. [p.122]
21. Closely scrutinize the diagram of the carbon cycle in the *Commentary*; be able to reproduce the cycle from memory. [p.122]

Integrating and Applying Key Concepts

How is the "oxygen debt" experienced by runners and sprinters related to aerobic and anaerobic respiration in humans?

8

CELL DIVISION AND MITOSIS

Interactive Exercises

Silver in the Stream of Time [pp.126–127]

8.1. DIVIDING CELLS: THE BRIDGE BETWEEN GENERATIONS [p.128]

8.2. THE CELL CYCLE [p.129]

Selected Words: cell division [p.126], *nuclear* division [p.128], 2*n* [p.128], *haploid* chromosome number [p.128], *n* [p.128], G1-S-G2-M [p.129], *Haemanthus* [p.129]

Boldfaced, Page-Referenced Terms

[p.127] reproduction _____

_____

[p.128] mitosis _____

_____

[p.128] meiosis _____

_____

[p.128] somatic cells _____

_____

[p.128] germ cells _____

_____

[p.128] chromosome _____

_____

[p.128] sister chromatids _____

_____

[p.128] centromere _____

_____

[p.128] chromosome number _____

_____

[p.128] diploid _____

_____

[p.129] cell cycle _____

_____

[p.129] interphase _____

_____

Matching

Choose the most appropriate answer for each term.

1. ___centromere [p.128]
2. ___haploid or *n* cell [p.128]
3. ___diploid or *2n* cell [p.128]
4. ___chromosome [p.128]
5. ___nuclear division [p.128]
6. ___germ cells [p.128]
7. ___somatic cells [p.128]
8. ___sister chromatids [p.128]
9. ___mitosis and meiosis [p.128]
10. ___chromosome number [p.128]

A. Cell lineage set aside for forming gametes and sexual reproduction
B. Necessary for the reproduction of eukaryotic cells
C. Any cell having two of each type of chromosome characteristic of the species
D. The number of each type of chromosome present in a cell (*n* or *2n*)
E. Each DNA molecule with attached proteins
F. Sort out and package parent DNA molecules into new cell nuclei
G. A small chromosome region with attachment sites for microtubules
H. The two attached DNA molecules of a duplicated chromosome
I. Body cells that reproduce by mitosis and cytoplasmic division
J. Possess only one of each type of chromosome characteristic of the species

Labeling

Identify the stage in the cell cycle indicated by each number.

11. _____ [p.129]

12. _____ [p.129]

13. _____ [p.129]

14. _____ [p.129]

15. _____ [p.129]

16. _____ [p.129]

17. _____ [p.129]

18. _____ [p.129]

19. _____ [p.129]

20. _____ [p.129]

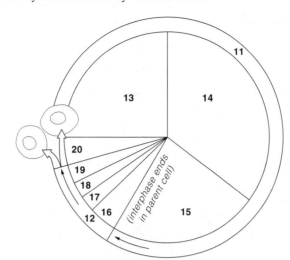

Matching

Link each time span identified below with the most appropriate number in the preceding labeling section.

21. ___Period after duplication of DNA during which the cell prepares for division [p.129]

22. ___The complete period of nuclear division that is followed by cytoplasmic division (a separate event) [p.129]

23. ___DNA duplication occurs now; a time of "synthesis" of DNA and proteins [p.129]

24. ___Period of cell growth before DNA duplication; a "gap" of interphase [p.129]

25. ___Usually the longest part of a cell cycle [p.129]

26. ___Period of cytoplasmic division [p.129]

27. ___Period that includes G1, S, G2 [p.129]

8.3. THE STAGES OF MITOSIS—AN OVERVIEW [pp.130–131]

Selected Words: *mitos* [p.130], prometaphase [p.130], *meta*– [p.131]

Boldfaced, Page-Referenced Terms

[p.130] prophase _____

_____

[p.130] metaphase _____

_____

[p.130] anaphase _____

_____

[p.130] telophase _____

_____

[p.130] spindle apparatus _____

_____

[p.130] centrioles _____

_____

Label–Match

Identify each of the mitotic stages shown by entering the correct stage in the blank beneath the sketch. Select from late prophase, transition to metaphase (prometaphase), cell at interphase, metaphase, early prophase, telophase, interphase-daughter cells, and anaphase. Complete the exercise by matching and entering the letter of the correct phase description in the parentheses following each label.

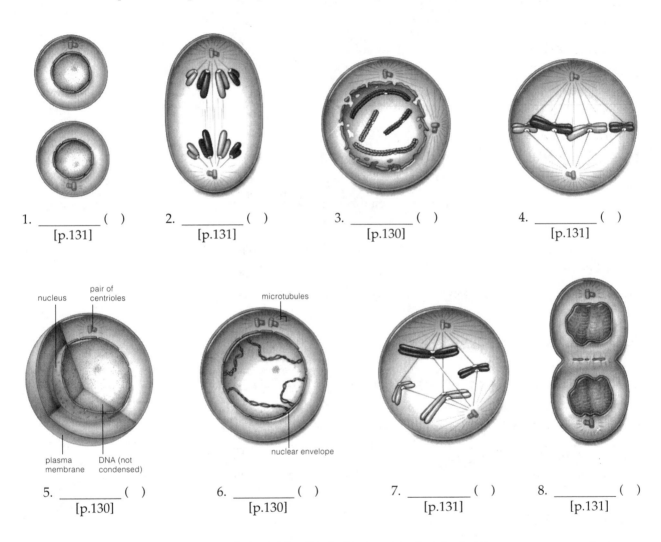

1. _____ ( )
   [p.131]

2. _____ ( )
   [p.131]

3. _____ ( )
   [p.130]

4. _____ ( )
   [p.131]

5. _____ ( )
   [p.130]

6. _____ ( )
   [p.130]

7. _____ ( )
   [p.131]

8. _____ ( )
   [p.131]

A. Attachments between two sister chromatids of each chromosome break; the two are now separate chromosomes that move to opposite spindle poles

B. Microtubules penetrate the nuclear region and collectively form the spindle apparatus; microtubules become attached to the two sister chromatids of each chromosome

C. The DNA and its associated proteins have started to condense

D. All the chromosomes are now fully condensed and lined up at the equator of the spindle

E. DNA is duplicated and the cell prepares for nuclear division

F. Two daughter cells have formed, each diploid with two of each type of chromosome, just like the parent cell's nucleus

G. Chromosomes continue to condense. New microtubules are assembled, and they move one of two centriole pairs toward the opposite end of the cell. The nuclear envelope begins to break up

H. Patches of new membrane fuse to form a new nuclear envelope around the decondensing chromosomes

8.4. DIVISION OF THE CYTOPLASM [pp.132–133]

Selected Words: cytokinesis [p.132]

Boldfaced, Page-Referenced Terms

[p.132] cytoplasmic division _____

_____

[p.132] cell plate formation _____

_____

[p.133] cleavage furrow _____

_____

Choice

For questions 1–10 on cytoplasmic division, choose from the following:

a. plant cells   b. animal cells

1. ___ Formation of a cell plate [p.132]
2. ___ Contractions continue and in time will divide the cell in two [pp.132– 133]
3. ___ Cellulose deposits form a crosswall between the two daughter cells [p.133]
4. ___ Possess rigid walls that cannot be pinched in two [p.132]
5. ___ A cleavage furrow [pp.132–133]
6. ___ Deposits form a cementing middle lamella [p.132]
7. ___ A band of microfilaments beneath the cell's plasma membrane generates the force for cytoplasmic division [p.132–133]
8. ___ Two daughter nuclei are cut off in separate cells, each with its own cytoplasm and plasma membrane [pp.132–133]
9. ___ At the location of a disklike structure, deposits of cellulose accumulate [pp.132–133]
10. ___ The structure grows at its margins until it fuses with the parent cell's plasma membrane [p.132]

8.5. A CLOSER LOOK AT THE CELL CYCLE [pp.134–135]

8.6. *Focus on Science:* HENRIETTA'S IMMORTAL CELLS [p.135]

Selected Words: histone-DNA spool [p.134], *Colchicum* [p.134], Henrietta Lacks [p.135]

Boldfaced, Page-Referenced Terms

[p.134] histones _____

_____

[p.134] nucleosome _____

_____

[p.134] kinetochores _____

_____

[p.134] motor proteins _____

_____

[p.135] HeLa cells _____

_____

Matching

Choose the most appropriate answer for each term.

1. ___G2 [p.135]
2. ___histones [p.134]
3. ___nucleosome [p.134]
4. ___late prophase [p.134]
5. ___kinetochore [p.134]
6. ___centromere [p.134]
7. ___G1 [p.135]
8. ___motor proteins [p.134]
9. ___*Colchicum* [p.134]
10. ___solenoid [p.134]

A. The time in the cell cycle when a chromosome acquires its distinct shape and size
B. A long period when cells assemble most of the carbohydrates, lipids, and proteins they use or export
C. A deeper level of structural organization; chromosomal proteins and DNA are arranged as a cylindrical fiber about thirty nanometers in diameter
D. Small disk-shaped structures at the surface of centromeres that serve as docking sites for spindle microtubules
E. Produces a poison that blocks assembly and promotes disassembly of microtubules
F. Dividing cells produce proteins that will drive mitosis to completion
G. Accessory for cytoskeletal elements; attached to and project from microtubules and microfilaments that take part in cell movements
H. Term for a histone-DNA spool
I. Many act like spools for winding up small stretches of DNA
J. The most prominent constriction of a metaphase chromosome

Short Answer

11. Describe the origin and the significance of HeLa cells. [p.135] _____

_____

_____

_____

_____

_____

_____

Self-Quiz

___ 1. The replication of DNA occurs _____.
[p.129]
a between the growth phases of interphase
b. immediately before prophase of mitosis
c. during prophase of mitosis
d. during prophase of meiosis

___ 2. In the cell life cycle of a particular cell,
_____. [p.129]
a. mitosis occurs immediately prior to G1
b. G2 precedes S
c. G1 precedes S
d. mitosis and S precede G1

___ 3. In eukaryotic cells, which of the following
can occur during mitosis? [p.130]
a. Two mitotic divisions to maintain the
parental chromosome number
b. The replication of DNA
c. A long growth period
d. The disappearance of the nuclear envelope and nucleolus

___ 4. Diploid refers to _____. [p.128]
a. having two chromosomes of each type in
somatic cells
b. twice the parental chromosome number
c. half the parental chromosome number
d. having one chromosome of each type in
somatic cells

___ 5. Somatic cells are _____ cells; germ cells
are _____ cells. [p.128]
a. meiotic; body
b. body; body
c. meiotic; meiotic
d. body; meiotic

___ 6. If a parent cell has sixteen chromosomes
and undergoes mitosis, the resulting cells
will have _____ chromosomes. [p.128]
a. sixty-four
b. thirty-two
c. sixteen
d. eight
e. four

___ 7. The correct order of the stages of mitosis is
_____. [pp.130–131]
a. prophase, metaphase, telophase,
anaphase
b. telophase, anaphase, metaphase,
prophase
c. telophase, prophase, metaphase,
anaphase
d. anaphase, prophase, telophase,
metaphase
e. prophase, metaphase, anaphase,
telophase

___ 8. "The nuclear envelope breaks up completely
into vesicles. Microtubules are now free to
interact with the chromosomes." These sen-
tences describe the _____ of mitosis.
[p.130]
a. prophase
b. metaphase
c. prometaphase
d. anaphase
e. telophase

___ 9. During _____, sister chromatids of
each chromosome are separated from each
other, and those former partners, now chromosomes, are moved toward opposite
poles. [p.131]
a. prophase
b. metaphase
c. anaphase
d. telophase

___10. Each histone-DNA spool is a single struc-
tural unit called a _____. [p.134]
a. kinetochore
b. motor protein
c. centromere
d. nucleosome

___11. In the process of cytokinesis, cleavage fur-
rows are associated with _____ cell division, and cell plate formation is associated
with _____ cell division. [pp.132–133]
a. animal; animal
b. plant; animal
c. plant; plant
d. animal; plant

Chapter Objectives/Review Questions

1. Define the word *reproduction*. [p.126]
2. Mitosis and meiosis refer to the division of the cell's _____. [p.128]
3. Distinguish between somatic cells and germ cells as to their location and function. [p.128]
4. The eukaryotic chromosome is composed of _____ and _____. [p.128]
5. The two attached threads of a duplicated chromosome are known as sister _____. [p.128]
6. The _____ is a small region with attachment sites for the microtubules that move the chromosome during nuclear division. [p.128]
7. Any cell having two of each type of chromosome is a(n) _____ cell; a(n) _____ cell is one set aside for the formation of gametes. [p.128]
8. Be able to list and describe, in order, the various activities occurring in the eukaryotic cell life cycle. [p.129]
9. Interphase of the cell cycle consists of G1, _____, and G2. [p.129]
10. S is the time in the cell cycle when _____ replication occurs. [p.129]
11. Describe the structure and function of the spindle apparatus. [p.130]
12. Describe the number and movements of centrioles in the cell division of some cells. [p.130]
13. The "_____" is a time of transition when the nuclear envelope breaks up into tiny, flattened vesicles prior to metaphase. [p.130]
14. Be able to give a detailed description of the cellular events occurring in the prophase, metaphase, anaphase, and telophase of mitosis. [pp.130–131]
15. Compare and contrast cytokinesis as it occurs in plant and animal cell division; use the following concepts: cleavage furrow, microfilaments at the cell's midsection, and cell plate formation. [pp.132–133]
16. Be able to characterize the organization of metaphase chromosomes using the following terms: histones, nucleosome, kinetochore, and motor proteins. [p.134]
17. Explain the value of the plant *Colchicum* to cellular scientists. [p.134]
18. Describe particular events occurring in G1, S, and G2 of interphase. [p.135]
19. Explain how cells from Henrietta Lacks continue to benefit humans everywhere more than forty years after her death. [p.135]

Integrating and Applying Key Concepts

Runaway cell division is characteristic of cancer. Imagine the various points of the mitotic process that might be sabotaged in cancerous cells in order to halt their multiplication. Then try to imagine how one might discriminate between cancerous and normal cells in order to guide those methods of sabotage most effective in combating cancer.

9

MEIOSIS

Interactive Exercises

Octopus Sex and Other Stories [pp.138–139]

9.1. COMPARING SEXUAL WITH ASEXUAL REPRODUCTION [p.140]

9.2. HOW MEIOSIS HALVES THE CHROMOSOME NUMBER [pp.140–141]

Selected Words: *unfertilized* egg cells [p.139], *sexual* reproduction [p.139], *pairs of genes* [p.140], *hom–* [p.140], *two consecutive divisions* [p.141], *homologue to homologue* [p.141]

Boldfaced, Page-Referenced Terms

[p.139] germ cells _____

_____

[p.139] gametes _____

_____

[p.140] asexual reproduction _____

_____

[p.140] genes _____

_____

[p.140] sexual reproduction _____

_____

[p.140] allele _____

_____

[p.140] meiosis _____

_____

[p.140] chromosome number _____

_____

[p.140] diploid number (2*n*) _____

_____

[p.140] homologous chromosomes _____

_____

[p.141] haploid number (*n*) _____

_____

[p.141] sister chromatids _____

_____

Choice

For questions 1–10, choose from the following:

a. asexual reproduction   b. sexual reproduction

1. ___ Brings together new combinations of alleles in offspring [p.140]
2. ___ One parent alone produces offspring [p.140]
3. ___ Each offspring inherits the same number and kinds of genes as its parent [p.140]
4. ___ Commonly involves two parents [p.140]
5. ___ The production of "clones" [p.140]
6. ___ Involves meiosis, formation of gametes, and fertilization [p.140]
7. ___ Offspring are genetically identical copies of the parent [p.140]
8. ___ Produces the variation in traits that is the foundation of evolutionary change [p.140]
9. ___ Change can only occur by rare mutations [p.140]
10. ___ The first cell of a new individual has pairs of genes on pairs of homologous chromosomes that are of maternal and paternal origin [p.140]

Dichotomous Choice

Circle one of two possible answers given between parentheses in each statement.

11. (Meiosis/Mitosis) divides chromosomes into separate parcels not once but twice prior to cell division. [p.140]
12. Sperms and eggs are sex cells known as (germ/gamete) cells. [p.140]
13. (Haploid/Diploid) germ cells produce haploid gametes. [p.140]
14. (Haploid/Diploid) cells possess pairs of homologous chromosomes. [p.140]
15. (Meiosis/Mitosis) produces cells that have one member of each pair of homologous chromosomes possessed by the species. [p.141]

16. Identical alleles are found on (homologous chromosomes/sister chromatids). [p.141]
17. Two attached DNA molecules are known as (sister chromatids/homologous chromosomes). [p.141]
18. Two attached sister chromatids represent (two/one) chromosome(s). [p.141]
19. One pair of duplicated chromosomes would be composed of (two/four) chromatids. [p.141]
20. With meiosis, chromosomes proceed through (one/two) divisions to yield four haploid nuclei. [p.141]
21. DNA is replicated during (prophase I of meiosis I/interphase preceding meiosis I). [p.141]
22. During meiosis I, each duplicated (chromosome/chromatid) lines up with its partner, homologue to homologue, and then the partners are moved apart from one another. [p.141]
23. Cytoplasmic division following Meiosis I results in two (diploid/haploid) daughter cells. [p.141]
24. During (meiosis I/meiosis II), the two sister chromatids of each chromosome are separated from each other, and each sister chromatid is now a chromosome in its own right. [p.141]
25. If human body cell nuclei contain twenty-three pairs of homologous chromosomes, each resulting gamete will contain (twenty-three/forty-six) chromosomes. [p.141]

Matching

To review the major stages of meiosis, match the following written descriptions with the appropriate sketch. Assume that the cell in this model initially has only one pair of homologous chromosomes (one from a paternal source and one from a maternal source) and crossing over does not occur. Complete the exercise by indicating the diploid ($2n = 2$) or haploid ($n = 1$) chromosome number of the cell chosen in the parentheses following each blank.

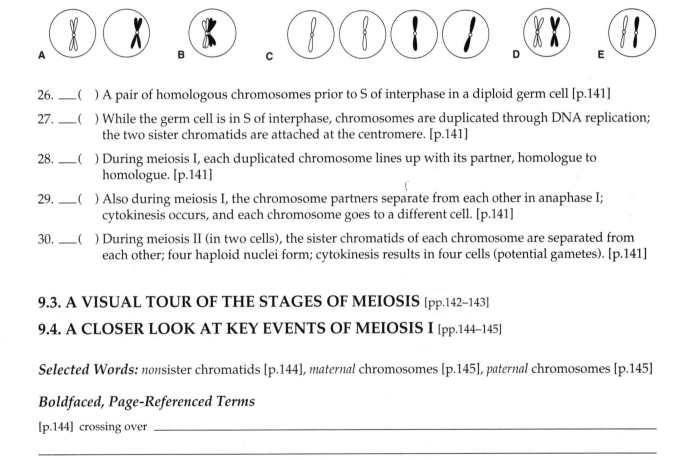

26. ___( ) A pair of homologous chromosomes prior to S of interphase in a diploid germ cell [p.141]

27. ___( ) While the germ cell is in S of interphase, chromosomes are duplicated through DNA replication; the two sister chromatids are attached at the centromere. [p.141]

28. ___( ) During meiosis I, each duplicated chromosome lines up with its partner, homologue to homologue. [p.141]

29. ___( ) Also during meiosis I, the chromosome partners separate from each other in anaphase I; cytokinesis occurs, and each chromosome goes to a different cell. [p.141]

30. ___( ) During meiosis II (in two cells), the sister chromatids of each chromosome are separated from each other; four haploid nuclei form; cytokinesis results in four cells (potential gametes). [p.141]

9.3. A VISUAL TOUR OF THE STAGES OF MEIOSIS [pp.142–143]

9.4. A CLOSER LOOK AT KEY EVENTS OF MEIOSIS I [pp.144–145]

Selected Words: *non*sister chromatids [p.144], *maternal* chromosomes [p.145], *paternal* chromosomes [p.145]

Boldfaced, Page-Referenced Terms

[p.144] crossing over _____

_____

Label–Match

Identify each of the meiotic stages shown below by entering the correct stage of either meiosis I or meiosis II in the blank beneath the sketch. Choose from prophase I, metaphase I, anaphase I, telophase I, prophase II, metaphase II, anaphase II, and telophase II. Complete the exercise by matching and entering the letter of the correct stage description in the parentheses following each label.

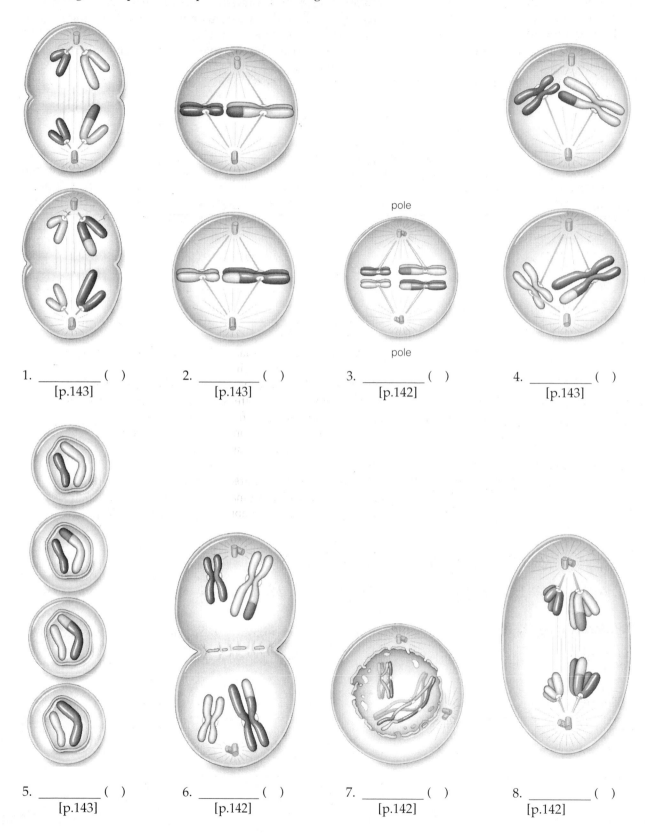

1. _____ ( )
   [p.143]

2. _____ ( )
   [p.143]

3. _____ ( )
   [p.142]

4. _____ ( )
   [p.143]

5. _____ ( )
   [p.143]

6. _____ ( )
   [p.142]

7. _____ ( )
   [p.142]

8. _____ ( )
   [p.142]

A. The spindle is now fully formed; all chromosomes are positioned midway between the poles of one cell
B. In each of two daughter cells, microtubules attach to the kinetochores of chromosomes, and motor proteins drive the movement of chromosomes toward the spindle's equator
C. Four daughter nuclei form; when the cytoplasm divides, each new cell has a haploid chromosome number, all in the unduplicated state; the cells may develop into gametes
D. In one cell, each duplicated chromosome is pulled away from its homologous partner; the partners are moved to opposite spindle poles
E. Each chromosome is drawn up close to its homologous partner; crossing over and genetic recombination (swapping of genes) occurs; each chromosome becomes attached to microtubules of a newly forming spindle
F. Motor proteins and spindle microtubule interactions have moved all of the duplicated chromosomes so that they are positioned at the spindle equator
G. Two haploid cells form, each having one of each type of chromosome that was present in the parent cell; the chromosomes are still in the duplicated state
H. Attachment between the two chromatids of each chromosome breaks; former "sister chromatids" are now chromosomes in their own right and are moved to opposite poles by interactions between motor proteins and kinetochore microtubules

Matching

Choose the most appropriate answer for each term.

9. ___ 2^{23} [p.145]

10. ___ paternal chromosomes [p.145]

11. ___ early prophase I [p.144]

12. ___ nonsister chromatids [p.144]

13. ___ late prophase I [p.144]

14. ___ crossing over [pp.144–145]

15. ___ metaphase I [p.145]

16. ___ each chromosome zippers to its homologue [p.144]

17. ___ function of meiosis [p.144]

18. ___ maternal chromosomes [p.145]

A. Random positioning of maternal and paternal chromosomes at the spindle equator
B. Chromosomes continue to condense and become thicker rodlike forms
C. Reduction of the chromosome number by half for forthcoming gametes
D. Break at the same places along their length and then exchange corresponding segments
E. Twenty-three chromosomes inherited from your mother
F. Combinations of maternal and paternal chromosomes possible in gametes from one germ cell
G. Each duplicated chromosome is in thin threadlike form
H. An intimate parallel array of homologues that favors crossing over
I. Twenty-three chromosomes inherited from your father
J. Breaks up old combinations of alleles and puts new ones together in pairs of homologous chromosomes

9.5. FROM GAMETES TO OFFSPRING [pp.146–147]

Selected Words: *gamete*-producing bodies [p.146], *spore*-producing bodies [p.146]

Boldfaced, Page-Referenced Terms

[p.146] spores _____

_____

[p.146] sperm _____

_____

[p.146] oocyte _____

_____

[p.146] egg _____

_____

[p.146] fertilization _____

_____

Choice

For questions 1–10, choose from the following:

a. animal life cycle   b. plant life cycle   c. both animal and plant life cycles

1. ___ Meiosis results in the production of haploid spores [p.146]
2. ___ A zygote divides by mitosis [p.146]
3. ___ Meiosis results in the production of haploid gametes [p.146]
4. ___ Haploid gametes fuse in fertilization to form a diploid zygote [p.146]
5. ___ A zygote divides by mitosis to form a diploid sporophyte [p.146]
6. ___ A spore divides by mitosis to produce a haploid gametophyte [p.146]
7. ___ A haploid gametophyte divides by mitosis to produce haploid gametes   [p.146]
8. ___ A haploid spore divides by mitosis to produce a gametophyte [p.146]
9. ___ A diploid body forms from mitosis of a zygote [p.146]
10. ___ A gamete-producing body and a spore-producing body develop during the life cycle [p.146]

Sequence

Arrange the following entities in correct order of development, entering a 1 by the stage that appears first and a 5 by the stage that completes the process of spermatogenesis. Complete the exercise by indicating if each cell is *n* or *2n* in the parentheses following each blank.

11. ___   ( ) primary spermatocyte [p.147]
12. ___   ( ) sperm [p.147]
13. ___   ( ) spermatid [p.147]
14. ___   ( ) spermatogonium [p.147]
15. ___   ( ) secondary spermatocyte [p.147]

Matching

Choose the most appropriate answer to match with each oogenesis concept.

16. ___primary oocyte [p.147]

17. ___oogonium [p.147]

18. ___secondary oocyte [p.147]

19. ___ovum and three polar bodies [p.147]

20. ___first polar body [p.147]

A. The cell in which synapsis, crossing over, and recombination occur
B. A cell that is equivalent to a diploid germ cell
C. A haploid cell formed after division of the primary oocyte that does not form an ovum at second division
D. Haploid cells, but only one of which functions as an egg
E. A haploid cell formed after division of the primary oocyte, the division of which forms a functional ovum

Short Answer

21. List the various mechanisms that contribute to the huge number of new gene combinations that may result from fertilization. [pp.146–147] _____

_____

_____

_____

9.6. MEIOSIS AND MITOSIS COMPARED [pp.148–149]

Selected Words: *clones* [p.148], *variation in traits* [p.148], somatic cell [p.148]

Complete the Table

1. Complete the table below by entering the word *mitosis* or *meiosis* in the blank adjacent to the statement describing one of these processes.

Description	Mitosis/Meiosis
[p.149] a. Involves one division cycle	
[p.148] b. Functions in growth and tissue repair	
[p.149] c. Daughter cells are haploid	
[p.148] d. Occurs only in germ cells	
[pp.148–149] e. Involves two division cycles	
[p.149] f. Daughter cells have one chromosome from each homologous pair	
[p.148] g. Asexual reproduction by single-celled eukaryotic species	
[p.149] h. Daughter cells have the diploid chromosome number	
[p.149] i. Completed when four daughter cells are formed	

Matching

The cell model used in this exercise has two pairs of homologous chromosomes, one long pair and one short pair. Match the descriptions to the numbers of chromosomes shown in the sketches below.

2. ___one cell at the beginning of meiosis II [p.148]

3. ___a daughter cell at the end of meiosis II [p.149]

4. ___metaphase I of meiosis [p.148]

5. ___metaphase of mitosis [p.149]

6. ___G1 in a daughter cell following mitosis [p.149]

7. ___prophase of mitosis [p.149]

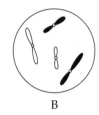

A   B   C   D   E   F

The following thought questions refer to the sketches above; enter answers in the blanks following each question.

8. How many chromosomes are present in cell E? ___

9. How many chromatids are present in cell E? ___

10. How many chromatids are present in cell C? ___

11. How many chromatids are present in cell D? ___

12. How many chromosomes are present in cell F? ___

Self-Quiz

___ 1. Which of the following does not occur in prophase I of meiosis? [pp.144–145]
   a. a cytoplasmic division
   b. a cluster of four chromatids
   c. homologues pairing tightly
   d. crossing over

___ 2. Crossing over is one of the most important events in meiosis because _____. [p.145]
   a. it leads to genetic recombination
   b. homologous chromosomes must be separated into different daughter cells
   c. the number of chromosomes allotted to each daughter cell must be halved
   d. homologous chromatids must be separated into different daughter cells

___ 3. Crossing over _____. [p.144]
   a. generally results in pairing of homologues and binary fission
   b. is accompanied by gene-copying events
   c. involves breakages and exchanges between sister chromatids
   d. is a molecular interaction between two of the nonsister chromatids of a pair of homologous chromosomes

___ 4. The appearance of chromosome ends lapped over each other in meiotic prophase I provides evidence of _____. [p.144]
   a. meiosis
   b. crossing over
   c. chromosomal aberration
   d. fertilization
   e. spindle fiber formation

___ 5. Which of the following does not increase genetic variation? [p.148]
   a. crossing over
   b. random fertilization
   c. prophase of mitosis
   d. random homologue alignments at metaphase I

___ 6. Which of the following is the most correct sequence of events in animal life cycles? [p.146]
   a. meiosis ⟶ fertilization ⟶ gametes ⟶ diploid organism
   b. diploid organism ⟶ meiosis ⟶ gametes ⟶ fertilization
   c. fertilization ⟶ gametes ⟶ diploid organism ⟶ meiosis
   d. diploid organism ⟶ fertilization ⟶ meiosis ⟶ gametes

___ 7. In sexually reproducing organisms, the zygote is _____. [p.146]
   a. an exact genetic copy of the female parent
   b. an exact genetic copy of the male parent
   c. unlike either parent genetically
   d. a genetic mixture of male parent and female parent

___ 8. Which of the following is the most correct sequence of events in plant life cycles? [p.146]
   a. fertilization ⟶ zygote ⟶ sporophyte ⟶ meiosis ⟶ spores ⟶ gametophytes ⟶ gametes
   b. fertilization ⟶ sporophyte ⟶ zygote ⟶ meiosis ⟶ spores ⟶ gametophytes ⟶ gametes

   c. fertilization ⟶ zygote ⟶ sporophyte ⟶ meiosis ⟶ gametes ⟶ gametophyte ⟶ spores
   d. fertilization ⟶ zygote ⟶ gametophyte ⟶ meiosis ⟶ gametes ⟶ sporophyte ⟶ spores

___ 9. The cell in the diagram is a diploid that has three pairs of chromosomes. From the number and pattern of chromosomes, the cell _____. [p.143]
   a. could be in the first division of meiosis
   b. could be in the second division of meiosis
   c. could be in mitosis
   d. could be in neither mitosis nor meiosis, because this stage is not possible in a cell with three pairs of chromosomes

___10. You are looking at a cell from the same organism as in the previous question. Now the cell _____. [p.149]
   a. could be in the first division of meiosis
   b. could be in the second division of meiosis
   c. could be in mitosis
   d. could be in neither mitosis nor meiosis, because this stage is not possible in this organism

---

Chapter Objectives/Review Questions

1. Distinguish between germ cells and gametes. [p.139]
2. "One parent alone produces offspring, and each offspring inherits the same number and kinds of genes as its parent" describes _____ reproduction. [p.140]
3. _____ reproduction involves meiosis, gamete formation, and fertilization. [p.140]
4. _____ divides chromosomes into separate parcels not once but twice prior to cell division. [p.140]
5. Describe the relationship between the following terms: homologous chromosomes, diploid number, and haploid number. [pp.140–141]
6. If the diploid chromosome number for a particular plant species is 18, the haploid gamete number is _____. [p.141]
7. During interphase a germ cell duplicates its DNA; a duplicated chromosome consists of two DNA molecules that remain attached to a constriction called the _____. [p.141]
8. As long as the two DNA molecules remain attached, they are referred to as _____ _____ of the chromosome. [p.141]

9. During meiosis I, homologous chromosomes pair; each homologue consists of _____ chromatids. [p.141]
10. During meiosis II, the two sister _____ of each _____ are separated from each other. [p.141]
11. _____ _____ breaks up old combinations of alleles and puts new ones together during prophase I of meiosis. [p.145]
12. The _____ attachment and subsequent positioning of each pair of maternal and paternal chromosomes at metaphase I lead to different _____ of maternal and paternal traits in each generation of offspring. [p.145]
13. Meiosis in the animal life cycle results in haploid _____; meiosis in the plant life cycle results in haploid _____. [p.146]
14. Using the special terms for the cells at the various stages, describe spermatogenesis in male animals and oogenesis in female animals. [p.147]
15. Crossing over, the distribution of random mixes of homologous chromosomes into gametes, and fertilization contribute to _____ in the traits of offspring. [p.147]
16. Mitotic cell division produces only _____; meiotic cell division, in conjunction with subsequent fertilization, promotes _____ in traits among offspring. [p.148]
17. Be able to list three ways that meiosis promotes variation in offspring. [p.148]

Integrating and Applying Key Concepts

A few years ago, it was claimed that the actual cloning of a human being had been accomplished. Later, this claim was admitted to be fraudulent. In 1997, a mammal was cloned. A sheep named "Dolly" resulted from a cloning experiment in Scotland. If sometime in the future cloning of humans becomes possible, speculate about the effects of reproduction without sex on human populations.

10

OBSERVABLE PATTERNS OF INHERITANCE

Interactive Exercises

A Smorgasbord of Ears and Other Traits [pp.152–153]

10.1. MENDEL'S INSIGHT INTO PATTERNS OF INHERITANCE [pp.154–155]
10.2. MENDEL'S THEORY OF SEGREGATION [pp.156–157]
10.3. INDEPENDENT ASSORTMENT [pp.158–159]

Selected Words: *observable* evidence [p.153], *pairs* of genes [p.153], *Pisum sativum* [p.154], *identical* alleles [p.155], *nonidentical* alleles [p.155], *homozygous* condition [p.155], *heterozygous* condition [p.155], *dominant* allele [p.155], *recessive* allele [p.155], *P, F_1 , F_2* [p.155]

Boldfaced, Page-Referenced Terms

[p.155] genes _____

_____

[p.155] alleles _____

_____

[p.155] true-breeding lineage _____

_____

[p.155] hybrid offspring _____

_____

[p.155] homozygous dominant _____

_____

[p.155] homozygous recessive _____

_____

[p.155] heterozygous _____

_____

[p.155] genotype _____

_____

[p.155] phenotype _____

_____

[p.156] monohybrid crosses _____

_____

[p.157] probability _____

_____

[p.157] Punnett-square method _____

_____

[p.157] testcrosses _____

_____

[p.157] segregation _____

_____

[p.158] dihybrid crosses _____

_____

[p.159] independent assortment _____

_____

Matching

Choose the most appropriate answer for each term.

1. ___genotype [p.155]
2. ___alleles [p.155]
3. ___heterozygous [p.155]
4. ___dominant allele [p.155]
5. ___phenotype [p.155]
6. ___genes [p.155]
7. ___true-breeding lineage [p.155]
8. ___homozygous recessive [p.155]
9. ___recessive allele [p.155]
10. ___homozygous [p.155]
11. ___P, F_1, F_2 [p.155]
12. ___hybrid offspring [p.155]
13. ___diploid cell [p.155]
14. ___gene locus [p.155]
15. ___homozygous dominant [p.155]

A. Parental, first-generation, and second-generation offspring
B. All the different molecular forms of the same gene
C. Particular location of a gene on a chromosome
D. Describes an individual having a pair of nonidentical alleles
E. An individual with a pair of recessive alleles, such as *aa*
F. Allele whose effect is masked by the effect of the dominant allele paired with it
G. Offspring of a genetic cross that inherit a pair of nonidentical alleles for a trait
H. Refers to an individual's observable traits
I. Refers to the particular genes an individual carries
J. When the effect of an allele on a trait masks that of any recessive allele paired with it
K. When both alleles of a pair are identical
L. An individual with a pair of dominant alleles, such as *AA*
M. Units of information about specific traits; passed from parents to offspring
N. Has a pair of genes for each trait, one on each of two homologous chromosomes
O. When offspring of genetic crosses inherit a pair of identical alleles for a trait, generation after generation

Fill-in-the-Blanks

Offspring of (16) _____ [p.156] crosses are heterozygous for the one trait being studied. (17) _____ [p.157] means that the chance that each outcome of a given event will occur is proportional to the number of ways it can be reached. Because (18) _____ [p.157] is a chance event, the rules of probability apply to genetics crosses. The separation of *A* and *a* as members of a pair of homologous chromosomes which move to different gametes during meiosis is known as Mendel's theory of (19) _____ [p.157]. Crossing F_1 hybrids (possibly of unknown genotype) back to a plant known to be a true-breeding recessive plant is known as a(n) (20) _____ [p.157]. From this cross, a ratio of (21) _____ [p.157] is expected. When F_1 offspring inherit two gene pairs, each consisting of two nonidentical alleles, the cross is known as a(n) (22) _____ [p.158] cross. "Gene pairs assorting into gametes independently of other gene pairs located on nonhomologous chromosomes" describes Mendel's theory of (23) _____ _____ [p.159].

Problems

24. In garden pea plants, Tall (*T*) is dominant over dwarf (*t*). In the cross *Tt* × *tt*, the *Tt* parent would produce a gamete carrying *T* (tall) and a gamete carrying *t* (dwarf) through segregation; the *tt* parent could only produce gametes carrying the *t* (dwarf) gene. Use the accompanying Punnett square (refer to text Figures 10.4, 10.6, and 10.7) to determine the genotype and phenotype probabilities of offspring from the above cross, *Tt* × *tt*. [pp.156–157]

Although the Punnett-square (checkerboard) method is a favored method for solving single-factor genetics problems, there is a quicker way. Only six different outcomes are possible from single-factor crosses. Studying the following relationships allows one to obtain the result of any such cross by inspection.

1. *AA* × *AA* = all *AA*
   (Each of the four blocks of the Punnett square would be *AA*.)
2. *aa* × *aa* = all *aa*
3. *AA* × *aa* = all *Aa*
4. *AA* × *Aa* = 1/2 *AA*; 1/2 *Aa*
   or *Aa* × *AA*
   (Two blocks of the Punnett square are *AA*, and two blocks are *Aa*.)
5. *aa* × *Aa* = 1/2 *aa*; 1/2 *Aa*
   or *Aa* × *aa*
6. *Aa* × *Aa* = 1/4 *AA*; 1/2 *Aa*; 1/4 *aa*
   (One block in the Punnett square is *AA*, two blocks are *Aa*, and one block is *aa*.)

Complete the Table

25. Using the gene symbols (tall and dwarf pea plants) in exercise 24, apply the six Mendelian ratios listed above to complete the following table of single-factor crosses by inspection. State results as phenotype and genotype ratios. [pp.156–157]

Cross	Phenotype Ratio	Genotype Ratio
a. *Tt* × *tt*		
b. *TT* × *Tt*		
c. *tt* × *tt*		
d. *Tt* × *Tt*		
e. *tt* × *Tt*		
f. *TT* × *tt*		
g. *TT* × *TT*		
h. *Tt* × *TT*		

When working genetics problems dealing with two gene pairs, one can visualize the independent assortment of gene pairs located on nonhomologous chromosomes into gametes by use of a fork-line device. Assume that in humans, pigmented eyes (*B*) are dominant (an eye color other than blue) over blue (*b*), and right-handedness (*R*) is dominant over left-handedness (*r*). To learn to solve a problem, cross the parents *BbRr* × *BbRr*. A sixteen-block Punnett square is required with gametes from each parent arrayed on two sides of the Punnett. The gametes receive genes through independent assortment using a fork-line method. [pp.158–159]

26. Array the gametes at the right on two sides of the Punnett square; combine these haploid gametes to form diploid zygotes within the squares. In the blank spaces below, enter the probability ratios derived within the Punnett square for the phenotypes listed. [pp.158–159]

$B\ b\ R\ r$ $\times$ $B\ b\ R\ r$

BR, Br, bR, br   BR, Br, bR, br

a. _____ pigmented eyes, right-handed
b. _____ pigmented eyes, left-handed
c. _____ blue-eyed, right-handed
d. _____ blue-eyed, left-handed

27. Albinos cannot form the pigments that normally produce skin, hair, and eye color, so albinos exhibit white hair, pink eyes, and skin (because the blood shows through). To be an albino, one must be homozygous recessive (*aa*) for the pair of genes that code for the key enzyme in pigment production. Suppose a woman of normal pigmentation (*A __*) with an albino mother marries an albino man. State the possible kinds of pigmentation possible for this couple's children, and specify the ratio of each kind of child the couple is likely to have. Show the genotype(s) and state the phenotype(s). [pp.156–157]

_____

_____

28. In horses, black coat color is influenced by the dominant allele (*B*), and chestnut coat color is influenced by the recessive allele (*b*). Trotting gait is due to a dominant gene (*T*), pacing gait to the recessive allele (*t*). A homozygous black trotter is crossed to a chestnut pacer. [pp.158–159]

a. What will be the appearance of the F_1 and F_2 generations?

_____

b. Which phenotype will be most common?_____

c. Which genotype will be most common? _____

d. Which of the potential offspring will be certain to breed true?_____

10.4. DOMINANCE RELATIONS [p.160]

10.5. MULTIPLE EFFECTS OF SINGLE GENES [p.161]

10.6. INTERACTIONS BETWEEN GENE PAIRS [pp.162–163]

Selected Words: ABO blood typing [p.160], *transfusions* [p.160], *sickle-cell anemia* [p.161], *albinism* [p.162], *interaction between two gene pairs* [p.163]

Boldfaced, Page-Referenced Terms

[p.160] incomplete dominance _____

_____

[p.160] codominance _____

_____

[p.160] multiple allele system _____

_____

[p.161] pleiotropy _____

_____

[p.162] epistasis _____

_____

Complete the Table

1. Complete the following table by supplying the type of inheritance illustrated by each example. Choose from these gene interactions: pleiotropy, multiple allele system, incomplete dominance, codominance, and epistasis.

Type of Inheritance	Example
[p.160] a.	Pink-flowered snapdragons produced from red- and white-flowered parents
[p.160] b.	AB type blood from a gene system of three alleles, *A*, *B*, and *O*
[p.160] c.	A gene with three or more alleles, such as the *ABO* blood-typing alleles
[p.162] d.	Black, brown, or yellow fur of Labrador retrievers and comb shape in poultry
[p.161] e.	The multiple phenotypic effects of the gene causing human sickle-cell anemia

Problems

2. Genes that are not always dominant or recessive may blend to produce a phenotype of a different appearance. This is termed *incomplete dominance*. In four o'clock plants, red flower color is determined by gene R and white flower color by R', while the heterozygous condition, RR', is pink. Complete the table below by determining the phenotypes and genotypes of the offspring of the following crosses. [p.160]

Cross	Phenotype	Genotype
a. $RR \times R'R' =$		
b. $R'R' \times R'R' =$		
c. $RR \times RR' =$		
d. $RR \times RR =$		

Sickle-cell anemia is a genetic disease in which children that are homozygous for a defective gene produce defective hemoglobin (Hb^S/Hb^S). The genotype of normal persons is Hb^A/Hb^A. If the level of blood oxygen drops below a certain level in a person with the Hb^S/Hb^S genotype, the hemoglobin chains stiffen and cause the red blood cells to form sickle, or crescent, shapes. These cells clog and rupture capillaries, which results in oxygen-deficient tissues where metabolic wastes collect. Several body functions are badly damaged. Severe anemia and other symptoms develop, and death nearly always occurs before adulthood. The sickle-cell gene is considered *pleiotropic*. Persons who are heterozygous (Hb^A/Hb^S) are said to possess *sickle-cell trait*. They are able to produce enough normal hemoglobin molecules to appear normal, but their red blood cells will sickle if they encounter oxygen tension (such as at high altitudes).

3. A man whose sister died of sickle-cell anemia married a woman whose blood is found to be normal. What advice would you give this couple about the inheritance of this disease as they plan their family? [p.161]

4. If a man and a woman, each with sickle-cell trait, planned to marry, what information could you provide for them regarding the genotypes and phenotypes of their future children? [p.161] _____

In one example of a *multiple allele system* with *codominance*, the three genes $I^A i$, $I^B i$, and i produce proteins found on the surfaces of red blood cells that determine the four blood types in the ABO system—A, B, AB, and O. Genes I^A and I^B are both dominant over i but not over each other. They are codominant. Recognize that blood types A and B may be heterozygous or homozygous ($I^A I^A$, $I^A i$ or $I^B I^B$, $I^B i$) whereas blood type O is homozygous (ii). Indicate the genotypes and phenotypes of the offspring and their probabilities from the parental combinations in questions 5–9. [p.160]

5. $I^A i \times I^A I^B =$ _____

6. $I^B i \times I^A i =$ _____

7. $I^A I^A \times ii =$ _____

8. $ii \times ii =$ _____

9. $I^A I^B \times I^A I^B =$ _____

In one type of gene interaction, two alleles of a gene mask the expression of alleles of another gene, and some expected phenotypes never appear. *Epistasis* is the term given such interactions. Work the following problems on scratch paper to understand epistatic interactions. In sweet peas, genes C and P are necessary for colored flowers. In the absence of either (_ _ pp or cc_ _), or both (ccpp), the flowers are white. What will be the color of the offspring of the following crosses, and in what proportions will they appear? [p.162]

10. *CcPp* x *ccpp* = _____

11. *CcPP* x *Ccpp* = _____

12. *Ccpp* x *ccPp* = _____

In the inheritance of the coat (fur) color of Labrador retrievers, allele B specifies black that is dominant to brown (chocolate), b. Allele E permits full deposition of color pigment, but two recessive alleles, ee, reduce deposition, and a yellow coat results.

13. Predict the phenotypes of the coat color and their proportions resulting from the following cross: *BbEe* x *Bbee* = [p.162] _____

In poultry, an epistatic interaction occurs in which two genes produce a phenotype that neither gene can produce alone. The two interacting genes (*R* and *P*) produce comb shape in chickens. The possible genotypes and phenotypes are [p.162]:

Genotypes	Phenotypes
R_ _P_ _	walnut comb
R_ _pp	rose comb
rrP_ _	pea comb
rrpp	single comb

Hint: where a blank appears in the genotypes above, either the dominant or recessive symbol in that blank yields the same phenotype.

14. What are the genotype and phenotype ratios of the offspring of a heterozygous walnut-combed male and a single-combed female? [p.163]

_____

_____

15. Cross a homozygous rose-combed rooster with a homozygous single-combed hen, and list the genotype and phenotype ratios of their offspring. [p.163]

_____

_____

_____

10.7. HOW CAN WE EXPLAIN LESS PREDICTABLE VARIATIONS? [pp.164–165]
10.8. EXAMPLES OF ENVIRONMENTAL EFFECTS ON PHENOTYPE [p.166]

Selected Words: camptodactyly [p.164], "bell-shaped" curve [p.165], *Hydrangea macrophylla* [p.166]

Boldfaced, Page-Referenced Terms

[p.164] continuous variation _____

_____

Choice

For questions 1–5, choose from the following primary contributing factors:

a. environment   b. a number of genes affecting a trait

1. ___ Height of human beings [p.165]

2. ___ Continuous variation in a trait [p.164]

3. ___ Flower color in *Hydrangea macrophylla* [p.166]

4. ___ The range of eye colors in the human population [p.164]

5. ___ Heat-sensitive version of one of the enzymes required for melanin production in Himalayan rabbits [p.166]

Self-Quiz

___ 1. The best statement of Mendel's principle of independent assortment is that _____. [pp.158–159]
   a. one allele is always dominant to another
   b. hereditary units from the male and female parents are blended in the offspring
   c. the two hereditary units that influence a certain trait separate during gamete formation
   d. each hereditary unit is inherited separately from other hereditary units

___ 2. All the different molecular forms of the same gene are called _____. [p.155]
   a. chiasma
   b. alleles
   c. autosome
   d. locus

___ 3. In the F_2 generation of a monohybrid cross involving complete dominance, the expected phenotypic ratio is _____. [pp.156–157]
   a. 3:1
   b. 1:1:1:1
   c. 1:2:1
   d. 1:1

___ 4. In the F_2 generation of a cross between a red-flowered snapdragon (homozygous) and a white-flowered snapdragon, the expected phenotypic ratio of the offspring is _____. [p.160]
   a. 3/4 red, 1/4 white
   b. 100 percent red
   c. 1/4 red, 1/2 pink, 1/4 white
   d. 100 percent pink

___ 5. In a testcross, F_1 hybrids are crossed to an individual known to be _____ for the trait. [p.157]
   a. heterozygous
   b. homozygous dominant
   c. homozygous
   d. homozygous recessive

___ 6. The tendency for dogs to bark while trailing is determined by a dominant gene, S, whereas silent trailing is due to the recessive gene, s. In addition, erect ears, D, is dominant over drooping ears, d. What combination of offspring would be expected from a cross between two erect-eared bark-

ers who are heterozygous for both genes? [pp.158–159]
   a. 1/4 erect barkers, 1/4 drooping barkers, 1/4 erect silent, 1/4 drooping silent
   b. 9/16 erect barkers, 3/16 drooping barkers, 3/16 erect silent, 1/16 drooping silent
   c. 1/2 erect barkers, 1/2 drooping barkers
   d. 9/16 drooping barkers, 3/16 erect barkers, 3/16 drooping silent, 1/16 erect silent

___ 7. A man with type A blood could be the father of _____. [p.160]
   a. a child with type A blood
   b. a child with type B blood
   c. a child with type O blood
   d. a child with type AB blood
   e. all of the above

___ 8. A single gene that affects several seemingly unrelated aspects of an individual's phenotype is said to be _____. [p.161]
   a. pleiotropic
   b. epistatic
   c. mosaic
   d. continuous

___ 9. Suppose two individuals, each heterozygous for the same characteristic, are crossed. The characteristic involves complete dominance. The expected genotypic ratio of their progeny is _____. [pp.156–157]
   a. 1:2:1
   b. 1:1
   c. 100 percent of one genotype
   d. 3:1

___10. If the two homozygous classes in the F_1 generation of the cross in question 9 are allowed to mate, the observed genotypic ratio of the offspring will be _____. [pp.156–157]
   a. 1:1
   b. 1:2:1
   c. 100 percent of one genotype
   d. 3:1

___11. Applying the types of inheritance learned in this chapter of the text, the skin color trait in humans exhibits _____. [pp.164–165]
   a. pleiotropy
   b. epistasis
   c. environmental effects
   d. continuous variation

Chapter Objectives/Review Questions

1. What was the prevailing method of explaining the inheritance of traits before Mendel's work with pea plants? [p.154]
2. Garden pea plants are naturally _____-fertilizing, but Mendel took steps to _____-fertilize them for his experiments. [p.155]
3. _____ are units of information about specific traits; they are passed from parents to offspring. [p.155]
4. What is the general term applied to the location of a gene on a chromosome? [p.155]
5. Define *allele*; how many types of alleles are present in the genotypes *Tt*? *tt*? *TT*? [p.155]
6. Explain the meaning of a true-breeding lineage. [p.155]
7. When two alleles of a pair are identical, it is a(n) _____ condition; if the two alleles are different, it is a _____ condition. [p.155]
8. Distinguish a dominant allele from a recessive allele. [p.155]
9. _____ refers to the genes present in an individual; _____ refers to an individual's observable traits. [p.155]
10. Offspring of _____ crosses are heterozygous for the one trait being studied. [p.156]
11. Explain why probability is useful to genetics. [p.157]
12. Be able to use the Punnett-square method of solving genetics problems. [p.157]
13. Define *testcross*, and cite an example. [p.157]
14. Mendel's theory of _____ states that during meiosis, the two genes of each pair separate from each other and end up in different gametes. [p.157]
15. Describe the testcross. [p.157]
16. Be able to solve dihybrid genetic crosses. [pp.158–159]
17. Mendel's theory of _____ _____ states that gene pairs on homologous chromosomes tend to be sorted into one gamete or another independently of how gene pairs on other chromosomes are sorted out. [p.159]
18. Distinguish between complete dominance, incomplete dominance, and codominance. [p.160]
19. Define *multiple allele system*, and cite an example. [p.160]
20. Explain why sickle-cell anemia is a good example of pleiotropy. [p.161]
21. Gene interaction involving two alleles of a gene that mask expression of another gene's alleles is called _____. [p.162]
22. List possible explanations for less predictable trait variations that are observed. [pp.164–165]
23. List two human traits that are explained by continuous variation. [pp.164–165]
24. Himalayan rabbits and garden hydrangeas are good examples of environmental effects on _____ expression. [p.166]

Integrating and Applying Key Concepts

Solve the following genetics problem:

In garden peas, one pair of alleles controls the height of the plant, and a second pair of alleles controls flower color. The allele for tall (*D*) is dominant to the allele for dwarf (*d*), and the allele for purple (*P*) is dominant to the allele for white (*p*). A tall plant with purple flowers crossed with a tall plant with white flowers produces 3/8 tall purple, 3/8 tall white, 1/8 dwarf purple, and 1/8 dwarf white. What are the genotypes of the parents?

11

CHROMOSOMES AND HUMAN GENETICS

Interactive Exercises

The Philadelphia Story [pp.170–171]

11.1. THE CHROMOSOMAL BASIS OF INHERITANCE—AN OVERVIEW [p.172]

11.2. *Focus on Science:* KARYOTYPING MADE EASY [p.173]

Selected Words: Philadelphia chromosome [p.170], *leukemias* [p.170], *spectral* karyotyping [p.170], Walther Flemming [p.171], August Weismann [p.171], *wild-type* allele [p.172], *Colchicum* [p.173]

Boldfaced, Page-Referenced Terms

[p.170] karyotype _____

_____

[p.172] genes _____

_____

[p.172] homologous chromosomes _____

_____

[p.172] alleles _____

_____

[p.172] crossing over _____

_____

[p.172] genetic recombination _____

_____

[p.172] X chromosome _____

_____

[p.172] Y chromosome _____

_____

[p.172] sex chromosomes _____

_____

[p.172] autosomes _____

_____

[p.173] in vitro _____

_____

[p.173] centrifugation _____

_____

Fill-in-the-Blanks

The first abnormal chromosome to be associated with cancer was named the (1) _____ [p.170]
chromosome. (2) _____ [p.170] arise when stem cells in bone marrow overproduce the white blood
cells necessary for the body's housekeeping and defense. A preparation of metaphase chromosomes based
on their defining features is a(n) (3) _____ [p.170]. A Philadelphia chromosome is not easy to identify
in standard karyotypes but should be easier using the newer (4) _____ [p.170] karyotyping that
artificially colors chromosomes in an unambiguous way.

Heritable traits arise from units of molecular information (located on chromosomes) that are known as
(5) _____ [p.172]. A pair of chromosomes that have the same length, shape, and gene sequence and
that interact during meiosis are known as (6) _____ [p.172] chromosomes. (7) _____ [p.172] are
different molecular forms of the same gene that arose through mutation. The most common form of an
allele is called a(n) (8) _____-type [p.172] allele. An event known as (9) _____ _____ [p.172]
functions to exchange gene segments between the members of a homologous pair and thus provide genetic
(10) _____ [p.172]. Human X and Y chromosomes that determine gender are examples of
(11) _____ [p.172] chromosomes. Chromosomes that are identical in both sexes of a species are called
(12) _____ [p.172].

The (13) _____ [p.172] of an individual (or species) is a preparation of sorted metaphase
chromosomes. To prepare a karyotype, technicians culture cells (14) _____ _____ [p.173] (in
glass). During culture, a chemical compound named (15) _____ [p.173] is added to the culture
medium. The purpose of this chemical is to block spindle formation and arrest nuclear divisions at the

(16) _____ [p.173] stage of mitosis when the chromosomes are the most condensed and easiest to identify in cells. Cells are then moved to the bottoms of culture tubes by a spinning force called (17) _____ [p.173]. Following saline treatment to separate the chromosomes, cells are fixed, stained, and photographed for study. The photograph is cut apart, one chromosome at a time, and the individual cutouts are arranged according to size, shape, and length of arms. Then all the pairs of homologous chromosomes are horizontally aligned by (18) _____ [p.173].

11.3. SEX DETERMINATION IN HUMANS [pp.174–175]
11.4. EARLY QUESTIONS ABOUT GENE LOCATIONS [pp.176–177]

Selected Words: *nonsexual traits* [p.174], *SRY* gene [p.174], *Drosophila melanogaster* [p.176], *Zea mays* [p.176], Thomas Morgan [p.176]

Boldfaced, Page-Referenced Terms

[p.176] X-linked genes _____

_____

[p.176] Y-linked genes _____

_____

[p.176] linkage group _____

_____

[p.176] reciprocal crosses _____

_____

Matching

Choose the most appropriate answer for each term.

1. ___X-linked genes [p.176]
2. ___linkage group [p.176]
3. ___*SRY* [p.174]
4. ___Y-linked genes [p.176]
5. ___reciprocal crosses [p.176]

A. Male-determining gene found on the Y chromosome; expression leads to testes formation
B. In the first cross, one parent displays the trait of interest, while in the second cross, the other parent displays it
C. The block of genes located on each type of chromosome; these genes tend to travel together in inheritance
D. Found only on the Y chromosome
E. Found only on the X chromosome

Complete the Table

6. Complete the Punnett-square table at the right, which brings Y-bearing and X-bearing sperm together randomly in fertilization. [p.174]

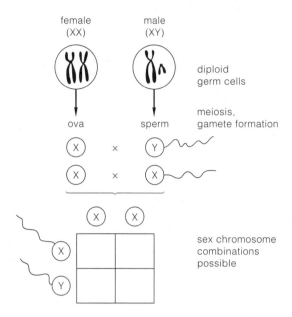

Dichotomous Choice

Answer the following questions related to the Punnett-square table just completed.

7. Male humans transmit their Y chromosome only to their (sons/daughters). [p.174]
8. Male humans receive their X chromosome only from their (mothers/fathers). [p.174]
9. Human mothers and fathers each provide an X chromosome for their (sons/daughters). [p.174]

Problems

10. In Morgan's experiments with *Drosophila*, white eyes are determined by a recessive X-linked gene, and the wild-type or normal brick-red eyes are due to its dominant allele. Use symbols of the following types: X^wY = a white-eyed male; X^WX^W = a homozygous normal red female. [p.176]
    a. What offspring can be expected from a cross of a white-eyed male and a homozygous normal female? [p.176]
    b. In addition, show the genotypes and phenotypes of the F_2 offspring. [p.176]
11. Which of the following represents a chromosome that has undergone crossover and recombination? It is assumed that the organism involved is heterozygous with the genotype:

$$\begin{matrix} A \\ B \end{matrix} \bigg| \ \bigg| \begin{matrix} a \\ b \end{matrix} \cdot \underline{\hspace{3cm}}$$

$$\mathbf{a} \ \begin{matrix} A \\ B \end{matrix} \bigg| \qquad \mathbf{b} \ \begin{matrix} a \\ b \end{matrix} \bigg| \qquad \mathbf{c} \ \begin{matrix} B \\ A \end{matrix} \bigg| \qquad \mathbf{d} \ \begin{matrix} a \\ B \end{matrix} \bigg|$$

[p.177]   [p.177]   [p.177]   [p.177]

11.5. HUMAN GENETIC ANALYSIS [pp.178–179]

Selected Words: polydactyly [p.178], *Huntington disorder* [p.178], *inherited* condition [p.179], *genetic* disease [p.179]

Boldfaced, Page-Referenced Terms

[p.178] pedigree _____

_____

[p.179] genetic abnormality _____

_____

[p.179] genetic disorder _____

_____

[p.179] syndrome _____

_____

[p.179] disease _____

_____

Matching

The following standardized symbols are used in the construction of pedigree charts. Choose the appropriate description for each.

1. ___[p.178] ■ ●
2. ___[p.178] ●
3. ___[p.178] ◆
4. ___[p.178] ■
5. ___[p.178] I, II, III, IV...
6. ___[p.178] ■─●
7. ___[p.178] ●■■●

A. Individual showing the trait being tracked
B. Male
C. Successive generations
D. Offspring with birth order left to right
E. Female
F. Sex unspecified; numerals present indicate number of children
G. Marriage/mating

Short Answer

8. Distinguish between these terms: *genetic abnormality, genetic disorder, syndrome,* and *genetic disease.* [p.179]

_____

_____

_____

Complete the Table

9. Complete the table below by indicating whether the genetic disorder listed is due to inheritance that is autosomal recessive, autosomal dominant, X-linked recessive, or due to changes in chromosome number or changes in chromosome structure.

Genetic Disorder	Inheritance Pattern
[p.179] a. Galactosemia	
[p.179] b. Achondroplasia	
[p.179] c. Hemophilia	
[p.179] d. Anhidrotic dysplasia	
[p.179] e. Huntington disorder	
[p.179] f. Turner and Down syndrome	
[p.179] g. Achoo syndrome	
[p.179] h. Cri-du-chat syndrome	
[p.179] i. XYY condition	
[p.179] j. Color blindness	
[p.179] k. Fragile X syndrome	
[p.179] l. Progeria	
[p.179] m. Klinefelter syndrome	
[p.179] n. Testicular feminization syndrome	
[p.179] o. Phenylketonuria	

11.6. INHERITANCE PATTERNS [pp.180–181]

11.7. *Focus on Health:* TOO YOUNG TO BE OLD [p.182]

Selected Words: galactosemia [p.180], *Huntington disorder* [p.180], *achondroplasia* [p.180], *color blindness* [p.181], *Hemophilia A* [p.181], *Fragile X syndrome* [p.181], *Hutchinson-Gilford progeria syndrome* [p.182]

Choice

For questions 1–18, choose from the following patterns of inheritance; some items may require more than one letter.

a. autosomal recessive   b. autosomal dominant   c. X-linked recessive

1. ___ The trait is expressed in heterozygous females. [pp.180–181]

2. ___ Heterozygotes can remain undetected. [pp.180–181]

3. ___ The trait appears in each generation. [p.180]

4. ___ The recessive phenotype shows up far more often in males than in females. [p.181]

5. ___ Both parents may be heterozygous normal. [p.180]

6. ___ If one parent is heterozygous and the other homozygous recessive, there is a 50 percent chance that any child of theirs will be heterozygous. [p.180]

7. ___ The allele is usually expressed, even in heterozygotes. [p.180]

8. ___ The trait is expressed in heterozygous females. [pp.180–181]

9. ___ Heterozygous normal parents can expect that one-fourth of their children will be affected by the disorder. [p.180]

10. ___ A son cannot inherit the recessive allele from his father, but his daughter can. [p.181]

11. ___ Females can mask this gene; males cannot. [p.181]

12. ___ The trait is expressed in heterozygotes of either sex. [p.180]

13. ___ Heterozygous women transmit the allele to half their offspring, regardless of sex. [p.180]

14. ___ Individuals displaying this type of disorder will always be homozygous for the trait. [p.180]

15. ___ The trait is expressed in both the homozygote and the heterozygote. [p.180]

16. ___ Heterozygous females will transmit the recessive gene to half their sons and half their daughters. [p.181]

17. ___ If both parents are heterozygous, there is a 50 percent chance that each child will be heterozygous. [p.180]

Problems

18. The autosomal allele that causes albinism (*a*) is recessive to the allele for normal pigmentation (*A*). A normally pigmented woman whose father is an albino marries an albino man whose parents are normal. The couple has three children, two normal and one albino. List the genotypes for each person involved. [p.180]

_____

_____

19. Huntington disorder is a rare form of autosomal dominant inheritance, *H*; the normal gene is *h*. The disease causes progressive degeneration of the nervous system, with onset exhibited near middle age. An apparently normal man in his early twenties learns that his father has recently been diagnosed as having Huntington disorder. What are the chances that the son will develop this disorder? [p.180] _____

_____

20. A color-blind man and a woman with normal vision whose father was color-blind have a son. Color blindness, in this case, is caused by an X-linked recessive gene. If only the male offspring are considered, what is the probability that their son is color-blind? [p.181] _____

21. Hemophilia A is caused by an X-linked recessive gene. A woman who is seemingly normal but whose father had hemophilia marries a normal man. What proportion of their sons will have hemophilia? What proportion of their daughters will have hemophilia? What proportion of their daughters will be carriers? [p.181] _____

22. The accompanying pedigree shows the pattern of inheritance of color blindness in a family. (Persons with the trait are indicated by black circles.) What is the chance that the third-generation female indicated by the arrow (below) will have a color-blind son if she marries a normal male? A color-blind male? [p.181]

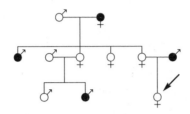

11.8. CHANGES IN CHROMOSOME STRUCTURE [pp.182–183]

11.9. CHANGES IN CHROMOSOME NUMBER [pp.184–185]

*Selected Words: non*homologous chromosome [p.182], *cri-du-chat* [p.183], *miscarriages* [p.184], *tetra*ploid [p.184], "trisomic" [p.184], "monosomic" [p.184], *Down syndrome* [p.184], *Turner syndrome* [p.185], *Klinefelter syndrome* [p.185]

Boldfaced, Page-Referenced Terms

[p.182] duplications _____

_____

[p.182] inversion _____

_____

[p.182] translocation _____

_____

[p.183] deletion _____

_____

[p.184] aneuploidy _____

_____

[p.184] polyploidy _____

_____

[p.184] nondisjunction _____

_____

[p.185] double-blind studies _____

_____

Label-Match

On rare occasions, chromosome structure becomes abnormally rearranged. Such changes may have profound effects on the phenotype of an organism. Label the following diagrams of abnormal chromosome structure as a deletion, a duplication, an inversion, or a translocation. Complete the exercise by matching and entering the letter of the proper description in the parentheses following each label.

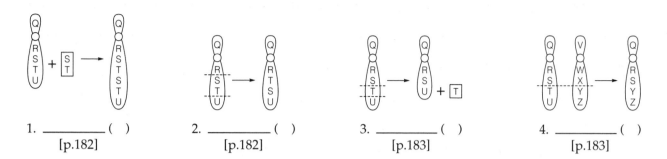

1. _____ ( )
   [p.182]

2. _____ ( )
   [p.182]

3. _____ ( )
   [p.183]

4. _____ ( )
   [p.183]

A. The loss of a chromosome segment; an example is cri-du-chat disorder.
B. A gene sequence in excess of its normal amount in a chromosome; an example is the fragile X syndrome.
C. A chromosome segment that separated from the chromosome and then was inserted at the same place, but in reverse; this alters the position and order of the chromosome's genes; possibly promoted human evolution.
D. The transfer of part of one chromosome to a nonhomologous chromosome; an example is when chromosome 14 ends up with a segment of chromosome 8; the Philadelphia chromosome is also an example.

Complete the Table

5. Complete the table below to summarize the major categories and mechanisms of chromosome number change in organisms.

Category of Change	Description
[p.184] a. Aneuploidy	
[p.184] b. Polyploidy	
[p.184] c. Nondisjunction	

Short Answer

6. If a nondisjunction occurs at anaphase I of the first meiotic division, what will be the proportion of abnormal gametes (for the chromosomes involved in the nondisjunction)? [p.184] _____

7. If a nondisjunction occurs at anaphase II of the second meiotic division, what will be the proportion of abnormal gametes (for the chromosomes involved in the nondisjunction)? [p.184] _____

8. Contrast the effects of polyploidy in plants and humans. [p.184] _____

_____

_____

9. Define the following terms: *tetraploid, trisomic,* and *monosomic.* [p.184] _____

_____

_____

_____

_____

Choice

For questions 10–19, choose from the following:

      a. Down syndrome   b. Turner syndrome   c. Klinefelter syndrome   d. XYY condition

10. ___ XXY male [p.185]

11. ___ Ovaries nonfunctional and secondary sexual traits fail to develop at puberty [p.185]

12. ___ Testes smaller than normal, sparse body hair, and some breast enlargement [p.185]

13. ___ Could only be caused by a nondisjunction in males [p.185]

14. ___ Older children smaller than normal with distinctive facial features; small skin fold over the inner corner of the eyelid [p.184]

15. ___ X0 female; often abort early; distorted female phenotype [p.185]

16. ___ Males that tend to be taller than average; some mildly retarded, but most are phenotypically normal [p.185]

17. ___ Injections of testosterone reverse feminized traits but not the mental retardation [p.185]

18. ___ Trisomy 21; skeleton develops more slowly than normal, with slack muscles [p.184]

19. ___ At one time, these males were thought to be genetically predisposed to become criminals [p.185]

11.10. *Focus on Bioethics:* PROSPECTS IN HUMAN GENETICS [pp.186–187]

Selected Words: PKU [p.186], *cleft lip* [p.186], *genetic counseling* [p.186], *prenatal diagnosis* [p.186], *embryo* [p.186], *fetus* [p.186], *amniocentesis* [p.186], *chorionic villi sampling* [p.187], *fetoscopy* [p.187], *preimplantation diagnosis* [p.187], *pre*-pregnancy stage [p.187], *"test-tube" babies* [p.187]

Boldfaced, Page-Referenced Terms

[p.186] abortion _____

_____

[p.187] in-vitro fertilization _____

_____

Complete the Table

1. Complete the table below, which summarizes methods of dealing with the problems of human genetics. Choose from phenotypic treatments, abortion, genetic screening, preimplantation diagnosis, genetic counseling, and prenatal diagnosis.

Method	Description
[pp.186–187] a.	Detects genetic disorders before birth; may use karyotypes, biochemical tests, amniocentesis, CVS, in-vitro fertilization, and possibly abortion; an example is a pregnancy at risk in a mother forty-five years old
[p.187] b.	A controversial option if prenatal diagnosis reveals a serious problem
[p.187] c.	If a severe heritable problem exists, it includes diagnosis of parental genotypes, detailed pedigrees, and genetic testing for known metabolic disorders; geneticists may be contacted for predictions
[p.187] d.	Relies on in-vitro fertilization; a fertilized egg mitotically divides into a ball of eight cells that provides a cell to be analyzed for genetic defects
[p.186] e.	Suppressing or minimizing symptoms of genetic disorders by surgical intervention, controlling diet or environment, or chemically modifying genes; PKU and cleft lip are examples
[p. 186] f.	Large-scale programs to detect affected persons or carriers in a population; early detection may allow introduction of preventive measures before symptoms develop; PKU screening for newborns is an example

Self-Quiz

___ 1. All the genes located on a given chromosome compose a _____. [p.176]
   a. karyotype
   b. bridging cross
   c. wild-type allele
   d. linkage group

___ 2. Chromosomes other than those involved in sex determination are known as _____. [p.172]
   a. nucleosomes
   b. heterosomes
   c. alleles
   d. autosomes

___ 3. The farther apart two genes are on a chromosome, _____. [p.177]
   a. the less likely that crossing over and recombination will occur between them
   b. the greater will be the frequency of crossing over and recombination between them

   c. the more likely they are to be in two different linkage groups
   d. the more likely they are to be segregated into different gametes when meiosis occurs

___ 4. Karyotype analysis is _____. [pp.172–173]
   a. a means of detecting and reducing mutagenic agents
   b. a surgical technique that separates chromosomes that have failed to segregate properly during meiosis II
   c. used in prenatal diagnosis to detect chromosomal mutations and metabolic disorders in embryos
   d. a process that substitutes defective alleles with normal ones

___ 5. Which of the following did Morgan and his research group not do? [p.176]
   a. They isolated and kept under culture fruit flies with the sex-linked recessive white-eyed trait.
   b. They developed the technique of amniocentesis.
   c. They discovered X-linked genes.
   d. Their work reinforced the concept that each gene is located on a specific chromosome.

___ 6. Red-green color blindness is a sex-linked recessive trait in humans. A color-blind woman and a man with normal vision have a son. What are the chances that the son is color-blind? If the parents ever have a daughter, what is the chance for each birth that the daughter will be color-blind? (Consider only the female offspring.) [p.181]
   a. 100 percent, 0 percent
   b. 50 percent, 0 percent
   c. 100 percent, 100 percent
   d. 50 percent, 100 percent
   e. none of the above

___ 7. Suppose that a hemophilic male (X-linked recessive allele) and a female carrier for the hemophilic trait have a nonhemophilic daughter with Turner syndrome. Nondisjunction could have occurred in _____. [pp.181,184]
   a. both parents
   b. neither parent
   c. the father only
   d. the mother only
   e. the nonhemophilic daughter

___ 8. Nondisjunction involving the X chromosome occurs during oogenesis and produces two kinds of eggs, XX and O (no X chromosome). If normal Y sperm fertilize the two types, which genotypes are possible? [p.184]
   a. XX and XY
   b. XXY and YO
   c. XYY and XO
   d. XYY and YO
   e. YY and XO

___ 9. Of all phenotypically normal males in prisons, the type once thought to be genetically predisposed to becoming criminals was the group with _____. [p.185]
   a. XXY disorder
   b. XYY disorder
   c. Turner syndrome
   d. Down syndrome
   e. Klinefelter syndrome

___10. Amniocentesis is _____. [pp.186–187]
   a. a surgical means of repairing deformities
   b. a form of chemotherapy that modifies or inhibits gene expression or the function of gene products
   c. used in prenatal diagnosis; a small sample of amniotic fluid is drawn to detect chromosomal mutations and metabolic disorders in embryos
   d. a form of gene-replacement therapy
   e. a diagnostic procedure; cells for analysis are withdrawn from the chorion

Chapter Objectives/Review Questions

1. A(n) _____ is a preparation of metaphase chromosomes based on their defining features. [p.170]
2. The units of information about heritable traits are known as _____. [p.172]
3. Diploid (2*n*) cells have pairs of _____ chromosomes. [p.172]
4. _____ are different molecular forms of the same gene that arise through mutation; a(n) _____-type allele is the most common form of a gene. [p.172]
5. State the circumstances required for crossing over, and describe the results. [p.172]
6. Name and describe the sex chromosomes in human males and females. [p.172]
7. Human X and Y chromosomes fall in the general category of _____ chromosomes; all other chromosomes in an individual's cells are the same in both sexes and are called _____. [p.172]
8. Define *karyotype*; briefly describe its preparation and value. [pp.172–173]
9. Explain meiotic segregation of sex chromosomes to gametes and the subsequent random fertilization that determines sex in many organisms. [p.174]
10. A newly identified region of the Y chromosome called _____ appears to be the master gene for sex determination. [p.174]

11. All the genes on a specific chromosome are called a(n) _____ group. [p.176]
12. Define the terms *X-linked* and *Y-linked genes*. [p.176]
13. In whose laboratory was sex linkage in fruit flies discovered? When? [p.176]
14. State the relationship between crossover frequency and the location of genes on a chromosome. [p.177]
15. A(n) _____ chart or diagram is used to study genetic connections between individuals. [p.178]
16. A genetic _____ is a rare, uncommon version of a trait, whereas an inherited genetic _____ causes mild to severe medical problems. [p.179]
17. Describe what is meant by a genetic disease. [p.179]
18. Carefully characterize patterns of autosomal recessive inheritance, autosomal dominant inheritance, and X-linked recessive inheritance. [pp.180–181]
19. Describe the Hutchinson-Gilford progeria syndrome. [p.182]
20. A(n) _____ is a loss of a chromosome segment; a(n) _____ is a gene sequence separated from a chromosome and then inserted at the same place, but in reverse; a(n) _____ is a repeat of several gene sequences on the same chromosome; a(n) _____ is the transfer of part of one chromosome to a non-homologous chromosome. [pp.182–183]
21. When gametes or cells of an affected individual end up with one extra or one less than the parental number of chromosomes, it is known as _____; relate this concept to monosomy and trisomy. [pp.184–185]
22. Having three or more complete sets of chromosomes is called _____. [p.184]
23. _____ is the failure of the chromosomes to separate in either meiosis or mitosis. [p.184]
24. Trisomy 21 is known as _____ syndrome; Turner syndrome has the chromosome constitution _____; XXY chromosome constitution is _____ syndrome; taller than average males with sometimes slightly depressed IQs have the _____ condition. [pp.184–185]
25. Explain what is meant by *double-blind studies*. [p.185]
26. Define *phenotypic treatment*, and describe one example. [p.186]
27. List some benefits of genetic screening and genetic counseling to society. [p.186]
28. Explain the procedures used in three types of prenatal diagnosis: amniocentesis, chorionic villi analysis, and fetoscopy; compare the risks. [p.187]
29. Discuss some of the ethical considerations that might be associated with a decision of induced abortion. [p.187]
30. A procedure known as preimplantation diagnosis relies on _____ fertilization. [p.187]

Integrating and Applying Key Concepts

1. The parents of a young boy bring him to their doctor. They explain that the boy does not seem to be going through the same vocal developmental stages as his older brother. The doctor orders a common cytogenetics test to be done, and it reveals that the young boy's cells contain two X chromosomes and one Y chromosome. Describe the test that the doctor ordered, and explain how and when such a genetic result, XXY, most logically occurred.
2. Solve the following genetics problem. Show rationale, genotypes, and phenotypes. A husband sues his wife for divorce, arguing that she has been unfaithful. His wife gave birth to a girl with a fissure in the iris of her eye, an X-linked recessive trait. Both parents have normal eye structure. Can the genetic facts be used to argue for the husband's suit? Explain your answer.

12

DNA STRUCTURE AND FUNCTION

Interactive Exercises

Cardboard Atoms and Bent-Wire Bonds [pp.190–191]

12.1. DISCOVERY OF DNA FUNCTION [pp.192–193]

Selected Words: J. F. Miescher [p.190], L. Pauling [p.190], J. Watson and F. Crick [p.190], F. Griffith [p.192], *Streptococcus pneumoniae* [p.192], O. Avery [p.192], A. Hershey and M. Chase [pp.192–193], *Escherichia coli* [p.192], ^{35}S and ^{32}P [p.193]

Boldfaced, Page-Referenced Terms

[p.190] deoxyribonucleic acid, DNA _____

_____

[p.192] bacteriophages _____

_____

Complete the Table

1. Complete the table below, which traces the discovery of DNA function.

Investigators	Year	Contribution
[p.190] a. Miescher	1868	
[p.192] b. Griffith	1928	
[p.192] c. Avery (also MacLeod and McCarty)	1944	
[pp.192–193] d. Hershey and Chase	1952	

Fill-in-the-Blanks

A bacteriophage is a kind of (2) _____ [p.192] that can infect (3) _____ [p.192] cells. Enzymes from the (2) take over enough of the host cell's metabolic processes to make substances that are necessary to construct new (4) _____ [p.192]. Some of these substances are (5) _____ [p.193] that can be labelled with a radioisotope of sulfur, ^{35}S. The genetic material of the bacteriophage used in the experiments by Hershey and Chase was labelled with the radioisotope of phosphorus known as (6) _____ [p.193]. When (7) _____ [p.193] particles were allowed to infect *Escherichia coli* cells, the (8) _____ [p.193] radiosotope remained outside the bacterial cells; the (9) _____ [p.193] radioisotope became part of the (10) _____ _____ [p.193] injected *into* the bacterial cells. Through many experiments, researchers accumulated strong evidence that (11) _____ [p.193], not (12) _____ [p.193], serves as the molecule of inheritance in all living cells.

12.2. DNA STRUCTURE [pp.194–195]
12.3. *Focus on Bioethics:* ROSALIND'S STORY [p.196]

Selected Words: E. Chargaff [p.194], R. Franklin and M. Wilkins [p.194], "helix" [p.195]

Boldfaced, Page-Referenced Terms

[p.194] nucleotide _____

_____

[p.194] adenine, A _____

_____

[p.194] guanine, G _____

_____

[p.194] thymine, T _____

_____

[p.194] cytosine, C _____

_____

[p.194] x-ray diffraction images _____

_____

Short Answer

1. List the three parts of a nucleotide. [p.194]

_____

Labeling

Four nucleotides are illustrated below. [All are from p.194] In the blank, label each nitrogen-containing base correctly as guanine, thymine, cytosine, or adenine. In the parentheses following each blank, indicate whether that nucleotide base is a purine (pu) or a pyrimidine (py). (See next section for definitions.)

2. _____ ( )   3. _____ ( )   4. _____ ( )   5. _____ ( )

Label–Match

Identify each indicated part of the DNA illustration below. Choose from these answers: phosphate group, purine, pyrimidine, nucleotide, and deoxyribose. Complete the exercise by matching and entering the letter of the proper structure description in the parentheses following each label.

The following DNA memory devices may be helpful: Use pyrCUT to remember that the single-ring **pyr**imidines are **c**ytosine, **u**racil, and **t**hymine; use purAG to remember that the double-ring **pur**ines are **a**denine and **g**uanine; pyrimidine is a long name for a narrow molecule; purine is a short name for a wide molecule; to recall the number of hydrogen bonds between DNA bases, remember that AT = 2 and CG = 3.

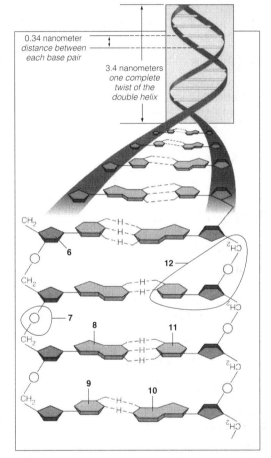

0.34 nanometer distance between each base pair

3.4 nanometers one complete twist of the double helix

6. _____ ( ) [p.195]

7. _____ _____ ( ) [p.195]

8. _____ ( ) [p.195]

9. _____ ( ) [p.195]

10. _____ ( ) [p.195]

11. _____ ( ) [p.195]

12. _____ ( ) [p.195]

A. The pyrimidine is thymine because it has two hydrogen bonds.
B. A five-carbon sugar joined to two phosphate groups in the upright portion of the DNA ladder.
C. The purine is guanine because it has three hydrogen bonds.
D. The pyrimidine is cytosine because it has three hydrogen bonds.
E. The purine is adenine because it has two hydrogen bonds.
F. Composed of three smaller molecules: a phosphate group, five-carbon deoxyribose sugar, and a nitrogenous base (in this case, a pyrimidine).
G. A chemical group that joins two sugars in the upright portion of the DNA ladder.

True–False

If the statement is true, write a T in the blank. If the statement is false, correct it by changing the underlined word(s) and writing the correct word(s) in the answer blank.

_____13. DNA is composed of <u>four</u> different types of nucleotides. [p.194]

_____14. In the DNA of every species, the amount of adenine present always equals the amount of <u>thymine</u>, and the amount of cytosine always equals the amount of <u>guanine</u> (A = T and C = G). [p.194]

_____15. In a nucleotide, the phosphate group is attached to the <u>nitrogen-containing base</u>, which is attached to the five-carbon sugar. [p.194]

_____16. Watson and Crick built their model of DNA in the early <u>1950s</u>. [p.190, recall]

_____17. Guanine pairs with <u>cytosine</u> and adenine pairs with <u>thymine</u> by forming hydrogen bonds between them. [p.195]

Fill-in-the-Blanks

Base (18) _____ [p.195] between the two nucleotide strands in DNA is (19) _____ [p.195] for all

species (A-T; G-C). The base (20) _____ [p.195] (determining which base follows the next in a

nucleotide strand) is (21) _____ [p.195] from species to species.

Short Answer

22. Explain why understanding the structure of DNA helps scientists understand how living organisms can have so much in common at the molecular level and yet be so diverse at the whole organism level. [p.195]

_____

_____

23. The term *antiparallel* means that parallel strands of a material run in opposite directions. Study Figure 12.7, read the descriptive messages, and identify the features of the DNA molecule that make biochemists describe its structure as antiparallel. [p.195] _____

_____

_____

_____

_____

12.4. DNA REPLICATION AND REPAIR [pp.196–197]

12.5. *Focus on Science:* DOLLY, DAISIES, AND DNA [p.198]

Selected Words: *semiconservative* replication [p.197], R. Okazaki [p.197], *continuous* assembly [p.197], *discontinuous* assembly [p.197], complementary [p.197], I. Wilmut [p.198], clone [p.198], "differentiated" [p.198], "uncommitted" [p.198], enucleated [p.198], R. Yanagimachi [p.198]

Boldfaced, Page-Referenced Terms

[p.196] DNA replication _____

_____

[p.197] DNA polymerases _____

[p.197] DNA ligases _____

[p.197] DNA repair _____

Labeling

1. The term *semiconservative replication* refers to the fact that each new DNA molecule resulting from the replication process is "half-old, half-new." In the illustration, complete the replication required in the middle of the molecule by adding the required letters representing the missing nucleotide bases. Recall that ATP, energy, and the appropriate enzymes are actually required in order to complete this process. [p.196]

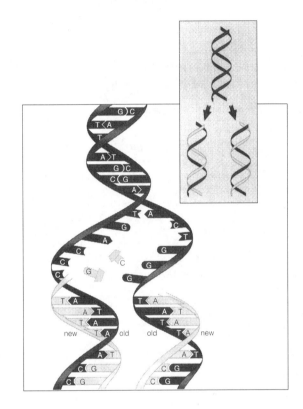

T-___		___-A	
G-___		___-C	
A-___		___-T	
C-___		___-G	
C-___		___-G	
C-___		___-G	
old	new	new	old

True–False

If the statement is true, write a T in the blank. If the statement is false, correct it by changing the underlined word(s) and writing the correct word(s) in the answer blank.

_____2. The hydrogen bonding of adenine to *guanine* is an example of complementary base pairing. [p.196]

_____3. The replication of DNA is considered a *semiconservative* process because *the same four nucleotides are used again and again during replication.* [pp.196–197]

_____4. Each parent strand *remains intact* during replication, and a new companion strand is assembled on each of those parent strands. [p.196]

_____5. Some of the enzymes associated with DNA assembly repair *errors* during the replication process. [p.197]

Self-Quiz

____ 1. Each DNA strand has a backbone that consists of alternating _____. [p.195]
   a. purines and pyrimidines
   b. nitrogen-containing bases
   c. hydrogen bonds
   d. sugar and phosphate molecules

____ 2. In DNA, complementary base pairing occurs between _____. [p.195]
   a. cytosine and uracil
   b. adenine and guanine
   c. adenine and uracil
   d. adenine and thymine

____ 3. Adenine and guanine are _____. [p.194; p.133 of this study guide]
   a. double-ringed purines
   b. single-ringed purines
   c. double-ringed pyrimidines
   d. single-ringed pyrimidines

____ 4. Franklin used the technique known as _____ to determine many of the physical characteristics of DNA. [pp.194,196]
   a. transformation
   b. cloning
   c. density-gradient centrifugation
   d. x-ray diffraction

____ 5. The significance of Griffith's experiment that used two strains of pneumonia-causing bacteria is that _____. [p.192]
   a. the conserving nature of DNA replication was finally demonstrated
   b. it demonstrated that harmless cells had become permanently transformed into pathogens through a change in the bacterial hereditary material
   c. it established that pure DNA extracted from disease-causing bacteria and injected into harmless strains transformed them into "pathogenic strains"
   d. it demonstrated that radioactively labeled bacteriophages transfer their DNA but not their protein coats to their host bacteria

____ 6. The significance of the experiments in which ^{32}P and ^{35}S were used is that _____. [p.193]
   a. the semiconservative nature of DNA replication was finally demonstrated
   b. it demonstrated that harmless cells had become permanently transformed

through a change in the bacterial hereditary system
   c. it established that pure DNA extracted from disease-causing bacteria transformed harmless strains into "killer strains"
   d. it demonstrated that radioactively labeled bacteriophages transfer their DNA but not their protein coats to their host bacteria

____ 7. Franklin's research contribution was essential in _____. [pp.195–196]
   a. establishing the double-stranded nature of DNA
   b. establishing the principle of base pairing
   c. establishing most of the principal structural features of DNA
   d. all of the above

____ 8. Chargaff's requirement that A = T and G = C suggested that _____. [p.195]
   a. cytosine molecules pair up with guanine molecules, and thymine molecules pair up with adenine molecules
   b. the two strands in DNA run in opposite directions (are antiparallel)
   c. the number of adenine molecules in DNA relative to the number of guanine molecules differs from one species to the next
   d. the replication process must necessarily be semiconservative

____ 9. A single strand of DNA with the base-pairing sequence C-G-A-T-T-G is compatible only with the sequence _____. [p.195]
   a. C-G-A-T-T-G
   b. G-C-T-A-A-G
   c. T-A-G-C-C-T
   d. G-C-T-A-A-C

____10. Rosalind Franklin's data indicated that the DNA molecule had to be long and thin with a width (diameter) that is 2 nanometers along its length. Watson and Crick declared that _____ ensured that the width of the DNA molecule must be uniform. [p.195]
   a. the antiparallel nature of DNA
   b. semiconservative replication processes
   c. hydrogen bonding of the sugar-phosphate backbones
   d. complementary base-pairing processes that match purine with pyrimidine

Chapter Objectives/Review Questions

1. Before 1952, _____ molecules and _____ molecules were suspected of housing the genetic code. [p.190]
2. The two scientists who assembled the clues to DNA structure and produced the first model were _____ and _____. [p.190]
3. Summarize the research carried out by Miescher, Griffith, Avery, and colleagues, and Hershey and Chase; state the specific advances made by each in the understanding of genetics. [pp.190,192–193]
4. Viruses called _____ were used in early research efforts to discover the genetic material. [p.192]
5. Summarize the specific research that demonstrated that DNA, not protein, governed inheritance. [p.193]
6. Draw the basic shape of a deoxyribose molecule and show how a phosphate group is joined to it when forming a nucleotide. [p.194]
7. Show how each nucleotide base would be joined to the sugar-phosphate combination drawn in objective 6. [p.194]
8. DNA is composed of double-ring nucleotides known as _____ and single-ring nucleotides known as _____ ; the two purines are _____ and _____ , whereas the two pyrimidines are _____ and _____. [p.194, p.133 of this study guide]
9. Assume that the two parent strands of DNA have been separated and that the base sequence on one parent strand is A-T-T-C-G-C; the base sequence that will complement that parent strand is _____. [p.195]
10. List the pieces of information about DNA structure that Rosalind Franklin discovered through her x-ray diffraction research. [pp.195–196]
11. Explain what is meant by the pairing of nitrogen-containing bases (base pairing), and explain the mechanism that causes bases of one DNA strand to join with bases of the other strand. [pp.195–197]
12. Describe how double-stranded DNA replicates from stockpiles of nucleotides. [pp.196–197]
13. Explain what is meant by "each parent strand is conserved in each new DNA molecule." [pp.196–197]
14. During DNA replication, enzymes called DNA _____ assemble new DNA strands. [pp.196–197]
15. Distinguish between continuous strand assembly and discontinuous strand assembly. [p.197]
16. Describe the process of making a genetically identical copy of yourself. [p.198]

Integrating and Applying Key Concepts

Review the stages of mitosis and meiosis, as well as the process of fertilization. Include what has now been learned about DNA replication and the relationship of DNA to a chromosome. As you cover the stages, be sure each cell receives the proper number of DNA threads.

13

FROM DNA TO PROTEINS

Interactive Exercises

Beyond Byssus [pp.200–201]

13.1. HOW IS DNA TRANSCRIBED INTO RNA? [pp.202–203]

Selected Words: byssus [p.200], genetic "code words" [p.201], 5' cap of pre-mRNA [p.202], template [p.202], 3' "poly-A tail" [p.203], to "pace" enzyme access [p.203], mRNA transcript [p.203]

Boldfaced, Page-Referenced Terms

[p.201] base sequence _____

_____

[p.201] transcription _____

_____

[p.201] translation _____

_____

[p.201] ribonucleic acid (RNA) _____

_____

[p.202] messenger RNA, mRNA _____

_____

[p.202] ribosomal RNA, rRNA _____

_____

[p.202] transfer RNA, tRNA _____

_____

[p.202] uracil _____

_____

[p.202] RNA polymerase _____

_____

[p.202] promoter _____

_____

[p.203] introns _____

_____

[p.203] exons _____

_____

Fill-in-the-Blanks

Which base follows the next in a strand of DNA is referred to as the base (1) _____ [p.201]. A region of DNA that calls for the assembly of specific amino acids into a polypeptide chain is a(n) (2) _____ [p.201]. The two steps from genes to proteins are called (3) _____ [p.201] and (4) _____ [p.201]. In (5) _____ [p.201], single-stranded molecules of RNA are assembled on DNA templates in the nucleus. In (6) _____ [p.201], the RNA molecules are shipped from the nucleus into the cytoplasm, where they are used as templates for assembling (7) _____ [p.201] chains. Following translation, one or more chains become (8) _____ [p.201] into the three-dimensional shape of protein molecules. Proteins have (9) _____ [p.201] and (10) _____ [p.201] roles in cells, including control of DNA.

Complete the Table

11. Three types of RNA are transcribed from DNA in the nucleus (from genes that code only for RNA). Complete the following table, which summarizes information about these molecules.

RNA Molecule	Abbreviation	Description/Function
[p.202] a. Ribosomal RNA		
[p.202] b. Messenger RNA		
[p.202] c. Transfer RNA		

Short Answer

12. List three ways in which a molecule of RNA is structurally different from a molecule of DNA. [p.202]

_____

_____

13. Cite two similarities in DNA replication and transcription. [p.202] _____

_____

14. What are the three key ways in which transcription differs from DNA replication? [p.202] _____

_____

Sequence

Arrange the steps of transcription in correct chronological sequence. Write the letter of the first step next to 15, the letter of the second step next to 16, and so on.

15. ___   A. The RNA strand grows along exposed bases until RNA polymerase meets a DNA base sequence that signals "stop." [p.203]

16. ___   B. RNA polymerase binds with the DNA promoter region to open up a local region of the DNA double helix. [p.203]

17. ___   C. An RNA polymerase enzyme locates the DNA bases of the promoter region of one DNA strand by recognizing DNA-associated proteins near a promoter. [p.203]

18. ___   D. RNA is released from the DNA template as a free, single-stranded transcript. [p.203]

19. ___   E. RNA polymerase moves stepwise along exposed nucleotides of one DNA strand; as it moves, the DNA double helix keeps unwinding. [p.203]

Completion

20. Suppose the line below represents the DNA strand that will act as a template for the production of mRNA through the process of transcription. Fill in the blanks below the DNA strand with the sequence of complementary bases that will represent the message carried from DNA to the ribosome in the cytoplasm. [See text Figure 13.4 on p.203]

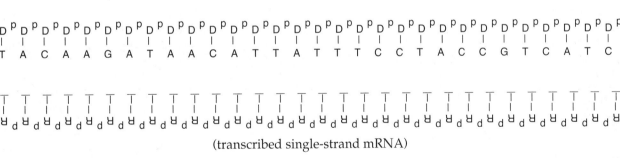

(transcribed single-strand mRNA)

Label–Match

Newly transcribed mRNA contains more genetic information than is necessary to code for a chain of amino acids. Before the mRNA leaves the nucleus for its ribosome destination, an editing process occurs as certain portions of nonessential information are snipped out. Identify each indicated part of the illustration below; use abbreviations for the nucleic acids. Complete the exercise by matching and entering the letter of the description in the parentheses following each label.

21. _____ ( ) [p.203]

22. _____ ( ) [p.203]

23. _____ ( ) [p.203]

24. _____ ( ) [p.203]

25. _____ ( ) [p.203]

26. _____ _____

_____ ( ) [p.203]

A. The actual coding portions of mRNA
B. Noncoding portions of the newly transcribed mRNA
C. Presence of cap and tail, introns snipped out, and exons spliced together
D. Acquiring of a poly-A tail by the modified mRNA transcript
E. The region of the DNA template strand to be copied
F. Reception of a nucleotide cap by the 5' end of mRNA (the first synthesized)

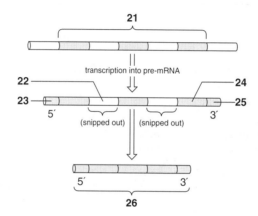

13.2. DECIPHERING THE mRNA TRANSCRIPTS [pp.204–205]

13.3. HOW IS mRNA TRANSLATED? [pp.206–207]

Selected Words: G. Khorana and M. Nirenberg [p.204], "three-bases-at-a-time" [p.204], START signal [p.204], tRNA "hook" [p.204], "wobble effect" [p.205], *initiation* [p.206], *elongation* [p.206], *termination* [p.206], polysome [p.207], polypeptide chains [p.207]

Boldfaced, Page-Referenced Terms

[p.204] codons _____

_____

[p.204] genetic code _____

_____

[p.204] anticodon _____

_____

Matching

Choose the most appropriate answer for each term.

1. ___codon [p.204]
2. ___three bases at a time [p.204]
3. ___sixty-one [p.204]
4. ___the genetic code [p.204]
5. ___molecular "hook" [p.204]
6. ___ribosome [p.205]
7. ___anticodon [p.204]
8. ___the "stop" codons [p.204]

A. Composed of two subunits, the small subunit with P and A amino acid binding sites as well as a binding site for mRNA
B. Reading frame of the nucleotide bases in mRNA
C. On tRNA, an attachment site for an amino acid
D. UAA, UAG, UGA
E. A sequence of three nucleotide bases that can pair with a specific mRNA codon
F. Name for each base triplet in mRNA
G. The number of codons that actually specify amino acids
H. Term for how the nucleotide sequences of DNA and then mRNA correspond to the amino acid sequence of a polypeptide chain

Complete the Table

9. Complete the following table, which distinguishes the stages of translation.

Translation Stage	Description
[p.206] a.	Special initiator tRNA loads onto small ribosomal subunit and recognizes AUG; small subunit binds with mRNA, and large ribosomal subunit joins small one.
[p.206] b.	Amino acids are strung together in sequence dictated by mRNA codons as the mRNA strand passes through the two ribosomal subunits; two tRNAs interact at P and A sites.
[p.206] c.	mRNA "stop" codon signals the end of the polypeptide chain; release factors detach the ribosome and polypeptide chain from the mRNA.

Completion

10. Given the following DNA sequence, deduce the composition of the mRNA transcript: [p.204]

TAC   AAG   ATA   ACA   TTA   TTT   CCT   ACC   GTC   ATC

___   ___   ___   ___   ___   ___   ___   ___   ___   ___
(mRNA transcript)

11. Deduce the composition of the tRNA anticodons that would pair with the above specific mRNA codons as these tRNAs deliver the amino acids (identified below) to the P and A binding sites of the small ribosomal subunit. [p.205]

___   ___   ___   ___   ___   ___   ___   ___   ___   ___
(tRNA anticodons)

12. From the mRNA transcript in question 10, use Figure 13.7 of the text to deduce the composition of the amino acids of the polypeptide sequence. [p.204]

___   ___   ___   ___   ___   ___   ___   ___   ___   ___
(amino acids)

Fill-in-the-Blanks

The order of (13) _____ _____ [p.205] in a protein is specified by a sequence of nucleotide bases. The genetic code is read in units of (14) _____ [p.205] nucleotides; each unit of three codes for (15) _____ [p.204] amino acid(s). In the table that showed which triplet specified a particular amino acid, the triplet code was incorporated in (16) _____ [p.204] molecules. Each of these triplets is referred to as a(n) (17) _____ [p.204]. (18) _____ [p.205] alone carries the instructions for assembling a particular sequence of amino acids from the DNA to the ribosomes in the cytoplasm, where (19) _____ [p.205] of the polypeptide occurs. (20) _____ [p.205] RNA acts as a shuttle molecule as each type brings its particular (21) _____ _____ [p.205] to the ribosome where it is to be incorporated into the growing (22) _____ [p.207]. A(n) (23) _____ [p.205] is a triplet on mRNA that forms hydrogen bonds with a(n) (24) _____ , [p.205], which is a triplet on tRNA.

13.4. DO MUTATIONS AFFECT PROTEIN SYNTHESIS? [pp.208–209]

Selected Words: sickle-cell anemia [p.208], Barbara McClintock [p.209], "frameshift mutation" [p.209]

Boldfaced, Page-Referenced Terms

[p.208] gene mutations _____

_____

[p.208] base-pair substitution _____

_____

[p.209] insertions _____

_____

[p.209] deletions _____

_____

[p.209] transposable elements _____

_____

[p.209] ionizing radiation _____

_____

[p.209] alkylating agents _____

_____

[p.209] carcinogens _____

_____

Short Answer

1. Cite several examples of mutagens. [p.209] _____

_____

Fill-in-the-Blanks

In addition to changes in chromosomes (crossing over, recombination, deletion, addition, translocation, and inversion), changes can also occur in the structure of DNA; these modifications are referred to as gene mutations. Complete the following exercise on types of spontaneous gene mutations.

Viruses, ultraviolet radiation, and certain chemicals are examples of environmental agents called

(2) _____ [p.209] that may enter cells and damage strands of DNA. If A becomes paired with C instead

of T during DNA replication, this spontaneous mutation is a base-pair (3) _____ [p.208]. Sickle-cell

anemia is a genetic disease whose cause has been traced to a single DNA base pair; the result is that one

(4) _____ _____ [p.208] is substituted for another in the beta chain of (5) _____ [p.208]. Some

DNA regions "jump" to new DNA locations and often inactivate the genes in their new environment; such

(6) _____ [p.209] elements may give rise to observable changes in the phenotype of an organism.

Label–Match

A summary of the flow of genetic information in protein synthesis is useful as an overview. Identify the indicated parts of the illustration on the next page by filling in the blanks with the names of the appropriate structures or functions. Choose from the following: DNA, mRNA, tRNA, polypeptide, rRNA subunits, intron, exon, mature mRNA transcript, new mRNA transcript, anticodon, amino acids, ribosome-mRNA complex. Complete the exercise by matching and entering the letter of the description in the parentheses following each label.

7. _____ (   ) [p.210]

8. _____ (   ) [p.210]
   (process)

9. _____ (   ) [p.203]

10. _____ (   ) [p.203]

11. _____ _____

_____ (   ) [p.210]

12. _____ (   ) [p.210]

13. _____ _____ (   ) [p.210]

14. _____ (   ) [p.210]

15. _____ (   ) [p.205]

16. _____ _____ (   ) [p.206]

17. _____ (   ) [p.205]

18. _____ - _____

_____ (   ) [p.206]

19. _____ (   ) [p.207]

A. Coding portion of mRNA that will translate into proteins
B. Carries a modified form of the genetic code from DNA in the nucleus to the cytoplasm
C. Transports amino acids to the ribosome and mRNA
D. The building blocks of polypeptides
E. Noncoding portions of newly transcribed mRNA
F. tRNA after delivering its amino acid to the ribosome-mRNA complex
G. Join when translation is initiated
H. Holds the genetic code for protein production
I. Place where translation occurs
J. DNA template creates new RNA transcript
K. A sequence of three bases that can pair with a specific mRNA codon
L. Snipping out of introns, only exons remaining
M. May serve as a functional protein (enzyme) or a structural protein

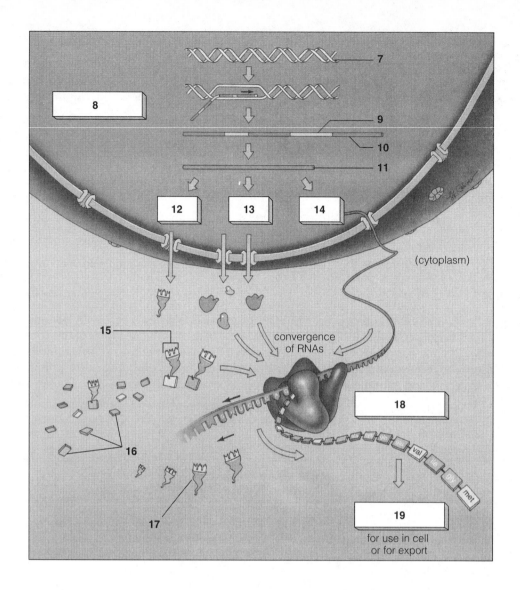

Self-Quiz

____ 1. Transcription _____. [p.202]
   a. occurs on the surface of the ribosome
   b. is the final process in the assembly of a protein
   c. occurs during the synthesis of any type of RNA by use of a DNA template
   d. is catalyzed by DNA polymerase

____ 2. _____ carry(ies) amino acids to ribosomes, where amino acids are linked into the primary structure of a polypeptide. [p.205]
   a. mRNA
   b. tRNA
   c. Introns
   d. rRNA

____ 3. Transfer RNA differs from other types of RNA because it _____. [p.205]
   a. transfers genetic instructions from cell nucleus to cytoplasm
   b. specifies the amino acid sequence of a particular protein
   c. carries an amino acid at one end
   d. contains codons

____ 4. _____ dominates the process of transcription. [p.202]
   a. RNA polymerase
   b. DNA polymerase
   c. Phenylketonuria
   d. Transfer RNA

___ 5. _____ and _____ are found in RNA but not in DNA. [p.202]
   a. Deoxyribose; thymine
   b. Deoxyribose; uracil
   c. Uracil; ribose
   d. Thymine; ribose

___ 6. Each "word" in the mRNA language consists of _____ letters. [p.204]
   a. three
   b. four
   c. five
   d. more than five

___ 7. If each kind of nucleotide is coded for only one kind of amino acid, how many different types of amino acids could be selected? [p.204, common sense]
   a. four
   b. sixteen
   c. twenty
   d. sixty-four

___ 8. The genetic code is composed of _____ codons. [p.204]
   a. three
   b. twenty
   c. sixteen
   d. sixty-four

___ 9. The cause of sickle-cell anemia has been traced to _____. [p.208]
   a. a mosquito-transmitted virus
   b. two DNA mutations that result in two incorrect amino acids in a hemoglobin chain
   c. three DNA mutations that result in three incorrect amino acids in a hemoglobin chain
   d. one DNA mutation that results in one incorrect amino acid in a hemoglobin chain

Chapter Objectives/Review Questions

1. State how RNA differs from DNA in structure and function, and indicate what features RNA has in common with DNA. [pp.201–202]
2. _____ RNA combines with certain proteins to form the ribosome; _____ RNA carries genetic information for protein construction from the nucleus to the cytoplasm; _____ RNA picks up specific amino acids and moves them to the area of mRNA and the ribosome. [p.202]
3. Describe the process of transcription, and indicate three ways in which it differs from replication. [pp.202–203]
4. What RNA code would be formed from the following DNA code: TAC-CTC-GTT-CCC-GAA? [p.202]
5. Transcription starts at a(n) _____, a specific sequence of bases on one of the two DNA strands that signals the start of a gene. [pp.202–203]
6. The first end of the mRNA to be synthesized is the _____ end; at the opposite end, the most mature transcripts acquire a(n) _____ tail. [pp.204–205]
7. Each base triplet in mRNA is called a(n) _____. [p.204]
8. State the relationship between the DNA genetic code and the order of amino acids in a protein chain. [pp.204–205]
9. Scrutinize Figure 13.7 in the text and decide whether the genetic code in this instance applies to DNA, mRNA, or tRNA. [p.204]
10. Explain how the DNA message TAC-CTC-GTT-CCC-GAA would be used to code for a segment of protein, and state what its amino acid sequence would be. [p.204]
11. Describe how the three types of RNA participate in the process of translation. [pp.206–207]
12. Cite an example of a change in one DNA base pair that has profound effects on the human phenotype. [p.208]
13. Briefly describe the spontaneous DNA mutations known as base-pair substitution, frameshift mutation, and transposable element. [pp.208–209]
14. List some of the environmental agents, or _____, that can cause mutations. [p.209]
15. Using a diagram, summarize the steps involved in the transformation of genetic messages into proteins [see text Figure 13.14 on p.210].

Integrating and Applying Key Concepts

Genes code for specific polypeptide sequences. Not every substance in living cells is a polypeptide. Explain how genes might be involved in the production of a storage starch (such as glycogen) that is constructed from simple sugars.

14

CONTROLS OVER GENES

Interactive Exercises

When DNA Can't Be Fixed [pp.212–213]

14.1. OVERVIEW OF GENE CONTROLS [p.214]

14.2. CONTROLS IN BACTERIAL CELLS [pp.214–215]

Selected Words: malignant melanoma [p.212], *xeroderma pigmentosum* [p.212], *benign* [p.213], *malignant* [p.213], CAP [p.215], cAMP [p.215]

Boldfaced, Page-Referenced Terms

[p.213] cancers _____

_____

[p.213] metastasis _____

_____

[p.214] regulatory proteins _____

_____

[p.214] negative control systems _____

_____

[p.214] positive control systems _____

_____

[p.214] promoter _____

_____

[p.214] operator _____

_____

[p.214] repressor _____

_____

[p.214] operon _____

_____

[p.215] activator proteins _____

_____

Complete the Table

1. All of the diploid cells in an organism possess the same genes, and every cell utilizes most of the same genes; yet specialized cells must activate only certain genes. Some agents of gene control have been discovered. Transcriptional controls are the most common. Complete the following table to summarize the agents of gene control.

Agents of Gene Control	*Method of Gene Control*
[pp.214–215] a. Repressor protein	
[p.214] b. Hormones	Major agents of vertebrate gene control; signaling molecules that move through the bloodstream to affect gene expression in target cells
[p.214] c. Promoters	
[p.214] d.	Short DNA base sequences between promoter and the start of a gene; a binding site for control agents
[p.215] e.	CAP is an example of these; CAP complexes with cAMP and now can adhere to the promoter and turn on transcription.

Label–Match

Escherichia coli, a bacterial cell living in mammalian digestive tracts, is able to exert a negative type of gene control over lactose metabolism. Use the numbered blanks to identify each part of the illustration below. Use abbreviations for nucleic acids. Complete the exercise by matching and entering the letter of the proper function description in the parentheses following each label. Choose from the following:

lactose   regulator gene   promoter   mRNA transcript   gene that codes for an enzyme
lactose operon   lactose-degrading enzymes   repressor protein   bound to lactose molecule
repressor protein   RNA polymerase   operator

2. _____ _____ ( )
   [p.215]

3. _____ _____ _____ ( )
   [p.215]
   _____ _____ _____ ( )

4. _____ _____ ( )
   [p.214]

5. _____ ( )
   [p.214]

6. _____ ( )
   [p.214]

7. _____ _____ ( )
   [p.215]

8. _____ _____ ( )
   [p.215]

9. _____ _____ _____ ( )
   [p.215]
   _____ _____ _____ ( )

10. _____ ( )
    [p.215]

11. _____ _____ ( )
    [p.215]

12. _____ - _____
    _____ ( ) [p.214]

A. Includes promoter, operator, and the lactose-degrading enzymes
B. Short DNA base sequence between promoter and the beginning of a gene
C. The nutrient molecule in the lactose operon
D. Major enzyme that catalyzes transcription
E. Binds to operator and overlaps promoter; this prevents RNA polymerase from binding to DNA and initiating transcription
F. Prevents repressor from binding to the operator
G. Genes that produce lactose-degrading enzymes
H. Catalyze the digestion of lactose
I. Carries genetic instructions to ribosomes for production of lactose enzymes
J. Specific base sequence that signals the beginning of a gene
K. Gene that contains coding for production of repressor protein

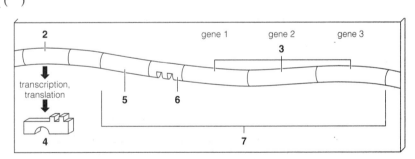

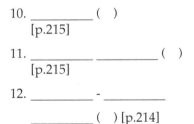

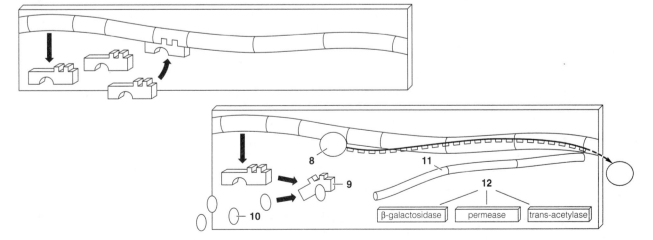

Fill-in-the-Blanks

(13) _____ _____ [p.214] is a species of bacterium that lives in mammalian digestive tracts and provided some of the first clues about gene control. A(n) (14) _____ [p.214] is any group of genes together with its promoter and operator sequence. Promoter and operator provide (15) _____ [p.214] that determine whether or not specific genes will be transcribed. The (16) _____ [p.215] codes for the formation of mRNA, which assembles a repressor protein. The affinity of the (17) _____ [p.215] for RNA polymerase dictates the rate at which a particular operon will be transcribed. Repressor protein allows (18) _____ _____ [p.215] over the lactose operon. Repressor binds with operator and overlaps promoter when lactose concentrations are (19) _____ [p.215]. This blocks (20) _____ _____ [p.215] from the genes that will process lactose. This (21) [choose one] ❐ blocks ❐ promotes [p.215] production of lactose-processing enzymes. When lactose is present, lactose molecules bind with the (22) _____ _____ [p.215]. Thus, repressor cannot bind to (23) _____ [p.215], and RNA polymerase has access to the lactose-processing genes. This gene control works well because lactose-degrading enzymes are not produced unless they are (24) _____ [p.215].

14.3. CONTROLS IN EUKARYOTIC CELLS [pp.216–217]
14.4. MANY LEVELS OF CONTROL [p.218]
14.5. *Focus on Science:* LOST CONTROLS AND CANCER [pp.218–219]

Selected Words: "spotting gene" [p.216], "mosaic" [p.216], *anhidrotic ectodermal dysplasia* [p.216], E. Tobin [p.217], troponin-1 [p.218], carcinogens [p.219], keratinocytes [p.219]

Boldfaced, Page-Referenced Terms

[p.216] cell differentiation _____

_____

[p.216] Barr body _____

_____

[p.217] hormones _____

_____

[p.217] enhancers _____

_____

[p.217] phytochrome _____

_____

[p.218] protein kinases _____

_____

[p.218] growth factors _____

_____

[p.218] ICE-like proteases _____

_____

[p.218] cancer _____

_____

[p.218] oncogene _____

_____

[p.218] proto-oncogenes _____

_____

[p.219] apoptosis _____

_____

Short Answer

1. Although a complex organism such as a human being arises from a single cell (the zygote), differentiation occurs in development. Define *differentiation*, and relate it to a definition of selective gene expression. [p.216]

_____

Fill-in-the-Blanks

All diploid cells in our bodies contain copies of the same (2) _____ [p.216]. These genetically identical

cells become structurally and functionally distinct from one another through a process called (3) _____

_____ [p.216], which arises through (4) _____ [p.216] gene expression in different cells. Cells

depend on (5) _____ [p.214, recall], which govern transcription, translation, and enzyme activity.

Controls that operate during transcription and transcript processing utilize (6) _____ [p.214, recall]

proteins, especially (7) _____ [p.215, recall] that are turned on and off by the addition and removal of

phosphate. A(n) (8) _____ [p.216] body is a condensed X chromosome. X chromosome inactivation

produces adult human females who are (9) _____ [p.216] for X-linked traits. This effect is shown in

human females with patches of skin that lack normal sweat glands, a disorder known as (10) _____

_____ _____ [p.216] and provides evidence for (11) _____ _____ [p.216] inactivation.

True–False

If the statement is true, write a T in the blank. If the statement is false, correct it by changing the underlined word(s) and writing the correct word(s) in the answer blank.

_____12. When cells become cancerous, cell populations <u>decrease</u> to very <u>low</u> densities and <u>stop</u> dividing. [p.218]

_____13. <u>All</u> abnormal growths and massings of new tissue in any region of the body are called tumors. [p.213, recall]

_____14. Malignant tumors have cells that <u>migrate and divide</u> in other organs. [p.213, recall]

_____15. Oncogenes are genes that <u>combat</u> cancerous transformations. [p.218]

_____16. Proto-oncogenes <u>rarely</u> trigger cancer. [p.218]

_____17. The normal expression of proto-oncogenes is vital, even though their <u>normal</u> expression may be lethal. [p.218]

Self-Quiz

___ 1. Any gene or group of genes together with its promoter and operator sequence is a(n) _____. [p.214]
   a. repressor
   b. operator
   c. promoter
   d. operon

___ 2. The operon model explains the regulation of _____ in prokaryotes. [p.215]
   a. replication
   b. transcription
   c. induction
   d. Lyonization

___ 3. _____ binds to operator whenever lactose concentrations are low. [p.215]
   a. Operon
   b. Repressor
   c. Promoter
   d. Operator

___ 4. In multicelled eukaryotes, cell differentiation occurs as a result of _____. [p.216]
   a. growth
   b. selective gene expression
   c. repressor molecules
   d. the death of certain cells

___ 5. _____ controls govern the rates at which mRNA transcripts that reach the cytoplasm will be translated into polypeptide chains at the ribosomes. [p.218]
   a. Transport
   b. Transcript processing
   c. Translational
   d. Transcriptional

For the next five questions, choose from these possibilities:
   a. CAP [p.215]
   b. cAMP [p.215]
   c. ICE-like protease [p.218]
   d. phytochrome [p.217]
   e. protein kinase [p.218]

___ 6. _____ is an enzyme that chops apart structural proteins in a cell and causes enough damage to kill the cell (apoptosis).

___ 7. _____ is a blue-green signaling molecule that helps plants to adjust their physiological activities to changing light conditions.

___ 8. _____ is a substance that links with and activates CAP and turns on transcription.

___ 9. A(n) _____ is an enzyme that can activate a protein by attaching a phosphate group to it.

___10. _____ is an activator protein that is part of a positive control system that helps to turn on an operon.

Chapter Objectives/Review Questions

1. The negative control of _____ proteins prevents the enzymes of transcription from binding to DNA; the positive control of _____ proteins enhances the binding of RNA polymerases to DNA. [pp.214–215]
2. The cells of *E. coli* manage to produce enzymes to degrade lactose when those molecules are _____ and to stop production of lactose-degrading enzymes when lactose is _____. [pp.214–215]
3. Explain how selective gene expression relates to cell differentiation in multicelled eukaryotes. [p.216]
4. Explain how X chromosome inactivation provides evidence for selective gene expression; use the example of anhidrotic ectodermal dysplasia. [p.216]
5. Describe the relationship of proto-oncogenes, environmental irritants, and oncogenes. [pp.218–219]

Integrating and Applying Key Concepts

Strange developmental disorders in frogs have been seen during the past several years in Minnesota and elsewhere. Can you create an hypothesis that connects damaged controls over gene function with frogs developing extra legs or misplaced eyes? How might those anomalies occur?

15

RECOMBINANT DNA
AND GENETIC ENGINEERING

Interactive Exercises

Mom, Dad, and Clogged Arteries [pp.222–223]

15.1. A TOOLKIT FOR MAKING RECOMBINANT DNA [pp.224–225]

15.2. PCR—A FASTER WAY TO AMPLIFY DNA [p.226]

Selected Words: high-density lipoproteins *(HDLs)* [p.222], low-density lipoproteins *(LDLs)* [p.222], *familial cholesterolemia* [p.222], *staggered* cuts [p.224], *Taq*I [p.224], *Eco*RI [p.224], *Not*I [p.224], *Thermus aquaticus* [p.224], "cloning factory" [p.225], amplify [p.225]

Boldfaced, Page-Referenced Terms

[p.222] gene therapy _____

_____

[p.223] recombinant DNA technology _____

_____

[p.223] genetic engineering _____

_____

[p.224] restriction enzymes _____

_____

[p.224] genome _____

_____

[p.224] DNA ligase _____

_____

[p.224] plasmid _____

_____

[p.225] DNA clone _____

_____

[p.225] cloning vector _____

_____

[p.225] cDNA _____

_____

[p.225] reverse transcriptase _____

_____

[p.226] PCR (polymerase chain reaction) _____

_____

[p.226] primers _____

_____

Short Answer

1. Outline a gene therapy procedure that has been used somewhat successfully to reverse the traumatic effects of familial cholesterolemia. [pp.222–223] _____

_____

_____

_____

2. Describe the bacterial chromosome and plasmids present in a bacterial cell, and distinguish between them. [pp.224–225] _____

_____

_____

True–False

If the statement is true, write T in the blank. If the statement is false, make it correct by changing the under-lined words and writing the correct word(s) in the answer blank.

_____3. Plasmids are <u>organelles on the surfaces of which amino acids are assembled into polypeptides</u>. [p.224]

_____4. Mix together DNA fragments cut by the same restriction enzyme and the sticky ends of any two fragments with complementary base sequences will base-pair and form a <u>recombinant DNA molecule</u>. [p.224]

Fill-in-the-Blanks

Genetic experiments have been occurring in nature for billions of years as a result of gene (5) _____ [p.223], crossing over and recombination, and other events. Humans now are causing genetic change by using (6) _____ _____ [p.223] technology in which researchers cut out and splice together gene regions from different (7) _____ [p.223], then greatly (8) _____ [p.223] the number of copies of the genes that interest them. The genes, and in some cases their (9) _____ [p.223] products, are produced in quantities that are large enough for (10) _____ [p.223] and for practical applications. (11) _____ _____ [p.222] involves isolating, modifying, and inserting particular genes back into the same organism or into a different one.

Matching

Choose the most appropriate answer for each term.

12. ___polymerase chain reaction (PCR) [p.226]
13. ___DNA ligase [p.224]
14. ___cloned DNA [p.225]
15. ___recombinant DNA technology [p.223]
16. ___plasmids [p.224]
17. ___restriction enzymes [p.224]
18. ___genome [p.224]

A. All the DNA in a haploid set of chromosomes
B. Process by which a gene is split into two strands and then copied over and over (most common type of gene amplification) by enzymes
C. Method of genetic engineering
D. Connects DNA fragments; is a modification enzyme
E. Small circular DNA molecules that carry only a few genes
F. Cuts DNA molecules; produces "sticky ends"
G. Multiple, identical copies of DNA fragments from an original chromosome

Fill-in-the-Blanks

Gene (19) _____ [p.225] does not automatically follow the successful taking-in of a modified gene by a host cell; genes must be suitably (20) _____ [p.225] first, which involves getting rid of the (21) _____ [p.225] from the mRNA transcripts, then using the enzyme reverse (22) _____ [p.225] to produce cDNA. The process goes like this:

a. A mature (23) _____ [p.225] transcript of a desired gene is used as a template for assembling a DNA strand. An enzyme, reverse (22), does the assembling. b. An mRNA- (24) _____ [p.225] hybrid molecule results. c. (25) _____ [p.225] action removes the mRNA and assembles a second strand of (26) _____ [p.225] on the first strand. d. The result is double-stranded (27) _____ [p.225], "copied" from an mRNA (28) _____ [p.225].

Bacterial host cells lack the proper (29) _____ _____ [p.225] to translate cloned genes unless the introns have been cut out.

Matching

Match the steps in the formation of a DNA library with the parts of the illustration below.

30. ___

31. ___

32. ___

33. ___

34. ___

35. ___

    A. Joining of chromosomal and plasmid DNA using DNA ligase [p.225]

    B. Restriction enzyme cuts chromosomal DNA at specific recognition sites [p.225]

    C. Cut plasmid DNA [p.225]

    D. Recombinant plasmids containing cloned library [p.225]

    E. Fragments of chromosomal DNA [p.225]

    F. Same restriction enzyme is used to cut plasmids [p.225]

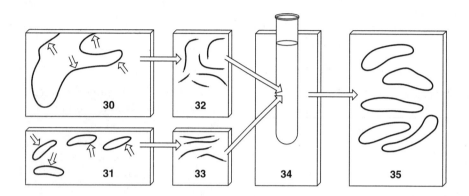

Complete the Table

36. Complete the table below, which summarizes some of the basic tools and procedures used in recombinant DNA technology.

Tool/Procedure	Definition and Role in Recombinant DNA Technology
[p.225] a. Cloned DNA	
[p.225] b. cDNA	
[p.224] c. DNA ligase	
[p.226] d. PCR	
[p.224] e. Restriction enzymes	
[p.225] f. Reverse transcription	

15.3. *Focus on Science:* DNA FINGERPRINTS [p.227]

15.4. HOW IS DNA SEQUENCED? [p.228]

15.5. FROM HAYSTACKS TO NEEDLES—ISOLATING GENES OF INTEREST [p.229]

Selected Words: tandem-repeat DNA fragments [p.227], restriction fragment length polymorphisms (RFLPs) [p.227], *Haemophilus influenzae* [p.228], T*, C*, A*, G* [p.228], *E. coli* genome [p.229], *genomic* library [p.229], *c*DNA library [p.229]

Boldfaced, Page-Referenced Terms

[p.227] DNA fingerprint _____

_____

[p.227] gel electrophoresis _____

_____

[p.228] automated DNA sequencing _____

_____

[p.229] gene library _____

_____

[p.229] probe _____

_____

[p.229] nucleic acid hybridization _____

_____

Complete the Table

1. Complete the table below, which summarizes some of the basic tools and procedures used in recombinant DNA technology.

Tool/Procedure	Definition and Role in Recombinant DNA Technology
[p.229] a. DNA library	
[p.229] b. DNA fingerprint	
[p.229] c. DNA probe	
[p.229] d. DNA sequencing	
[p.229] e. nucleic acid hybridization	
[p.227] f. RFLP	

Fill-in-the-Blanks

Each person has a genetic (2) _____ [p.227]: a unique pattern of DNA fragments inherited from each parent that can be used to map the human genome, apprehend criminals, and resolve cases of disputed paternity and maternity. More than 99 percent of the DNA is exactly the same in all human (3) _____ [p.227], but DNA fingerprinting focuses only on (4) _____ _____ [p.227], short regions of repeated DNA that differ substantially from person to person. A unique (2) can be prepared by separating the (4); an electric current applied to a buffered solution causes the (4) to migrate through a viscous medium in a process known as (5) _____ _____ [p.227].

Identification of the order and identity of nucleotides in DNA is called (6) _____ [p.228]. The process has been automated so that it can be applied to either a cloned or (7) _____ [p.228] DNA fragment. The procedure has enabled the entire DNA sequence of *Haemophilus influenzae,* a(n) (8) _____ [p.228] that causes upper respiratory infections in humans, to be determined. Host cells that take up modified genes may be identified by procedures involving DNA (9) _____ [p.229].

Short Answer

Probes can be used to locate a particular gene among many in a gene library. State the role of each item listed below in the probe procedure.

10. Bacterial clones derived from a single cell located on a culture plate [p.229] _____

_____

_____

11. Nylon or nitrocellulose filter [p.229] _____

_____

_____

12. Lysing solution [p.229] _____

_____

_____

13. Radioactively labeled probes [p.229] _____

_____

_____

14. X-ray film [p.229] _____

_____

_____

15. From the migration results shown, determine the sequence of nucleotides in the code molecule. The real nucleotide letters have been replaced by symbols. [p.228]

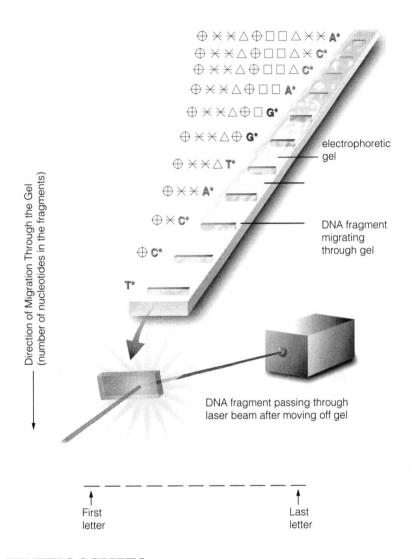

15.6. USING THE GENETIC SCRIPTS [p.230]

15.7. DESIGNER PLANTS [pp.230–231]

15.8. GENE TRANSFERS IN ANIMALS [pp.232–233]

15.9. *Focus on Bioethics:* WHO GETS ENHANCED? [p.233]

15.10. SAFETY ISSUES [p.234]

Selected Words: diabetics [p.230], insulin [p.230], F. Steward [p.230], *Southern corn leaf blight* [p.230], Ti plasmid [p.231], *Agrobacterium tumefaciens* [p.231], "biotech barnyards" [p.232], CFTR protein [p.232], TPA [p.232], *Dolly* [p.232], Human Genome Initiative [p.232], cDNA [p.232], C. Ventner and M. Adams [p.233], ESTs [p.233], gene therapy [p.233], *eugenic engineering* [p.233], "fail-safe" genes [p.234], "ice-minus bacteria" [p.234]

Fill-in-the-Blanks

Genetically engineered bacteria produce (1) _____ [p.230], a pancreatic hormone that sustains diabetics in a low-cost process. Other genetically engineered bacteria sponge up excess phosphates or (2) _____ _____ [p.230]. Deciphering bacterial genes helps us reconstruct the early millennia of life's (3) _____ [p.230] story.

Short Answer

4. Explain the goal of the Human Genome Initiative. [pp.232–233] _____

_____

In questions 5–10, summarize the results of the given experimentation dealing with genetic modifications of plants and animals.

5. Different cattle may soon be producing human collagen and human serum albumin: [p.232] _____

_____

6. The bacterium, *Agrobacterium tumefaciens:* [p.231] _____

_____

7. Cotton plants: [p.231] _____

_____

8. Introduction of the rat and human somatotropin gene into fertilized mouse eggs and embryos: [p.232] __

_____

9. "Ice-minus" bacteria and strawberry plants: [p.234] _____

_____

10. Bacteria that can degrade oil spills: [p.230] _____

_____

Fill-in-the-Blanks

The harmful gene is called the "ice-forming" gene, and the bacteria that lack it are known as (11) _____ - _____ [p.234] bacteria; genetic engineers were able to remove the harmful gene and test the modified bacterium on strawberry plants with no adverse effects. Inserting one or more genes into the (12) _____ _____ [p.222] of an organism for the purpose of correcting genetic defects is known as (13) _____ _____ [pp.222,233]. Attempting to modify a human trait by inserting genes into sperm or eggs is called (14) _____ _____ [p.233].

Self-Quiz

___ 1. Small circular molecules of DNA in bacteria are called _____. [p.224]
   a. plasmids
   b. desmids
   c. pili
   d. F particles
   e. transferins

___ 2. Base-pairing between nucleotide sequences from different sources is called _____. [p.229]
   a. nucleic acid hybridization
   b. reverse replication
   c. heterocloning
   d. plasmid formation

___ 3. Enzymes used to cut genes in recombinant DNA research are _____. [p.224]
   a. ligases
   b. restriction enzymes
   c. transcriptases
   d. DNA polymerases
   e. replicases

___ 4. The total DNA in a haploid set of chromosomes of a species is its _____. [p.224]
   a. plasmid
   b. enzyme potential
   c. genome
   d. DNA library
   e. none of the above

___ 5. An enzyme that heals random base-pairing of chromosomal fragments and plasmids is _____. [p.224]
   a. reverse transcriptase
   b. DNA polymerase
   c. cDNA
   d. DNA ligase

___ 6. A DNA library is _____. [p.229]
   a. a collection of DNA fragments produced by restriction enzymes and incorporated into plasmids
   b. cDNA plus the required restriction enzymes
   c. mRNA-cDNA
   d. composed of mature mRNA transcripts

___ 7. Amplification results in _____. [p.225]
   a. plasmid integration
   b. bacterial conjugation
   c. cloned DNA
   d. production of DNA ligase

___ 8. Any DNA molecule that is copied from mRNA is known as _____. [p.225]
   a. cloned DNA
   b. cDNA
   c. DNA ligase
   d. hybrid DNA

___ 9. The most commonly used method of DNA amplification is _____. [p.226]
   a. polymerase chain reaction
   b. gene expression
   c. genome mapping
   d. RFLPs (tandem-repeat DNA fragments)

___10. Restriction fragment length polymorphisms are valuable because _____. [p.227]
   a. they reduce the risks of genetic engineering
   b. they provide an easy way to sequence the human genome
   c. they allow fragmenting DNA without enzymes
   d. they provide DNA fragment sizes unique to each person

Chapter Objectives/Review Questions

1. Define *gene therapy* and *eugenic engineering*. [pp.222,233]
2. List the means by which natural genetic recombination occurs. [p.223]
3. Define *recombinant DNA technology*. [p.223]
4. _____ are small, circular, self-replicating molecules of DNA or RNA within a bacterial cell. [p.224]
5. Some bacteria produce _____ enzymes that cut apart DNA molecules injected into the cell by viruses; such DNA fragments or "_____ ends" often have staggered cuts capable of base-pairing with other DNA molecules cut by the same _____ enzymes. [p.224]
6. Base-pairing between chromosomal fragments and cut plasmids is made permanent by DNA _____. [p.224]
7. Multiple, identical copies of DNA fragments produced by restriction enzymes are known as _____ DNA. [pp.224–225]
8. A special viral enzyme, _____ _____, presides over the process by which mRNA is transcribed into DNA. [p.225]
9. Define *cDNA*. [p.225]
10. Why do researchers prefer to work with cDNA when working with human genes? [p.225]
11. List and define the two major methods of DNA amplification. [pp.225–226]
12. Be able to explain what a cDNA library is; review the steps used in creating such a library. [pp.225,229]
13. Polymerase chain reaction is the most commonly used method of DNA _____. [p.226]
14. List some practical genetic uses of RFLPs. [p.227]
15. Explain how gel electrophoresis is used to sequence DNA. [pp.227–228]
16. How is a cDNA probe used to identify a desired gene carried by a modified host cell? [p.229]
17. Explain why you believe that human "tinkering" with genes in different organisms is primarily a benefit or a disaster about to happen. Use examples from your text of recent publications. [pp.222–223,230–234]
18. Tell about the Human Genome Initiative and its implications. [pp.232–233]

Integrating and Applying Key Concepts

How could scientists guarantee that *Escherichia coli*, the human intestinal bacterium, will not be transformed into a severely pathogenic form and released into the environment if researchers use the bacterium in recombinant DNA experiments?

16

MICROEVOLUTION

Interactive Exercises

Designer Dogs [pp.238–239]

16.1. EARLY BELIEFS, CONFOUNDING DISCOVERIES [pp.240–241]

16.2. A FLURRY OF NEW THEORIES [pp.242–243]

16.3. DARWIN'S THEORY TAKES FORM [pp.244–245]

Selected Words: microevolution [p.239], alleles [p.239], "species" [p.240], theory of inheritance of acquired characteristics [p.242], "fluida" [p.242], H.M.S. *Beagle* [p.242], *Principles of Geology* [p.242], descent with modification [p.244], *artificial* selection [p.244], *natural* selection [p.244], "missing links" [p.245], *transitional* forms [p.245], *Archaeopteryx* [p.245]

Boldfaced, Page-Referenced Terms

[p.240] biogeography _____

[p.240] comparative anatomy _____

_____

[p.241] fossils _____

_____

[p.241] evolution _____

_____

[p.242] theory of catastrophism _____

_____

[p.243] theory of uniformity _____

_____

Matching

Choose the most appropriate answer for each term.

1. ___comparative anatomy [p.240]
2. ___biogeography [p.240]
3. ___fossils [p.241]
4. ___school of Hippocrates [p.240]
5. ___evolution [p.241]
6. ___Great Chain of Being [p.240]
7. ___Buffon [p.241]
8. ___species [p.240]
9. ___Aristotle [p.240]
10. ___sedimentary beds [p.241]

A. Rock layers deposited at different times; may contain fossils
B. Came to view nature as continuum of organization, from lifeless matter through complex forms of plant and animal life
C. Each kind of being; represent links in the Great Chain of Being
D. Modification of species over time
E. Extended from the lowest forms of life to humans and on to spiritual beings
F. Suggested that perhaps species had originated in more than one place and perhaps had been modified over time
G. Studies of body structure comparisons and patterning
H. Studies of the world distribution of plants and animals
I. Suggested that the gods were not the cause of the sacred disease
J. One type of direct evidence that organisms lived in the past

Choice

For questions 11–12, choose from the following:

a. Georges Cuvier   b. Jean-Baptiste Lamarck

11. ___ Stretching directed "fluida" to the necks of giraffes, which lengthened permanently [p.242]

12. ___ Acknowledged there were abrupt changes in the fossil record that corresponded to discontinuities between certain layers of sedimentary beds; thought this was evidence of change inpopulations of ancient organisms [p.242]

13. ___ The force for change in organisms is the drive for perfection. [p.242]

14. ___ There was only one time of creation that populated the world with all species. [p.242]

15. ___ Catastrophism [p.242]

16. ___ Permanently stretched giraffe necks were bestowed on offspring. [p.242]

17. ___ When a global catastrophe destroyed many organisms, a few survivors repopulated the world. [p.242]

18. ___ Theory of inheritance of acquired characteristics [p.242]

19. ___ The survivors of global catastrophe were not new species (may have differed from related fossils), but naturalists had not discovered them yet. [p.242]

20. ___ During a lifetime, environmental pressures and internal "needs" bring about permanent changes. [p.242]

21. ___ The force for change is a drive for perfection, up the Chain of Being. [p.242]

22. ___ The drive to change is centered in nerves that direct an unknown "fluida" to body parts in need of change. [p.242]

Complete the Table

23. Several key players and events in the life of Charles Darwin led him to his conclusions about natural selection and evolution. Summarize these influences by completing the table below.

Event/Person	Importance to Synthesis of Evolutionary Theory
[p.242] a.	Botanist at Cambridge University who perceived Darwin's real interests and arranged for Darwin to become a ship's naturalist
[p.242] b.	Where Darwin earned a degree in theology but also developed his love for natural history
[p.242] c.	British ship that carried Darwin (as a naturalist) on a five-year voyage around the world
[p.242] d.	Wrote *Principles of Geology*; advanced the theory of uniformity; suggested the Earth was much older than 6,000 years
[p.244] e.	Wrote an influential essay (read by Darwin) on human populations asserting that people tend to produce children faster than food supplies, living space, and other resources can be sustained
[pp.244–245] f.	Volcanic islands 900 kilometers from the South American coast where Darwin correlated differences in various species of finches with their environmental challenges
[p.244] g.	The key point in Darwin's theory of evolution; involves reproductive capacity, heritable variations, and adaptive traits
[p.245] h.	English naturalist contemporary with Darwin; independently developed Darwin's theory of evolution before Darwin published
[p.245] i.	Unearthed in 1861; the first transitional fossil (between reptiles and birds); provided evidence for Darwin's theory

16.4. INDIVIDUALS DON'T EVOLVE—POPULATIONS DO [pp.246–247]

16.5. *Focus on Science:* WHEN IS A POPULATION *NOT* EVOLVING? [pp.248–249]

Selected Words: *morphological* traits [p.246], *physiological* traits [p.246], *behavioral* traits [p.246], morphs [p.246], phenotype [p.246], *natural selection* [p.247], *gene flow* [p.247], *genetic drift* [p.247], *p, q, p², 2pq, q²* [p.248], Punnett square [p.248], genotypic frequencies [p.248]

Boldfaced, Page-Referenced Terms

[p.246] population _____

_____

[p.246] polymorphism _____

_____

[p.246] gene pool _____

_____

[p.246] alleles _____

_____

[p.246] allele frequencies _____

_____

[p.246] genetic equilibrium _____

_____

[p.247] microevolution _____

_____

[p.247] mutation rate _____

_____

[p.247] lethal mutations _____

_____

[p.247] neutral mutations _____

_____

[p.248] Hardy-Weinberg rule _____

_____

Labeling

The traits of individuals in a population are often classified as being morphological, physiological, or behavioral traits. In the list of traits below, enter "M" if the trait seems to be morphological, "P" if the trait is physiological, and "B" if the trait is behavioral.

___1.   Frogs have a three-chambered heart. [p.246]

___2.   The active transport of Na^+ and K^+ ions is unequal. [p.246]

___3.   Humans and orangutans possess an opposable thumb. [p.246]

___4.   An organism periodically seeking food. [p.246]

___5.   Some animals have a body temperature that fluctuates with the environmental temperature. [p.246]

___6.   The platypus is a strange mammal. It has a bill like a duck and lays eggs. [p.246]

___7.   Some vertebrates exhibit greater parental protection for offspring than others. [p.246]

___8.   During short photoperiods, the pituitary gland releases small quantities of gonadotropins. [p.246]

___9.   Red grouse defend large multipurpose territories where they forage, mate, nest, and rear young. [p.246]

___10.   Lampreys have an elongated cylindrical body without scales. [p.246]

Matching

Select the one most appropriate answer to match the sources of genetic variation.

11. ___ fertilization [p.246]

12. ___ changes in chromosome structure or number [p.246]

13. ___ crossing over at meiosis [p.246]

14. ___ gene mutation [p.246]

15. ___ independent assortment at meiosis [p.246]

A. Leads to mixes of paternal and maternal chromosomes in gametes
B. Produces new alleles
C. Leads to the loss, duplication, or alteration of alleles
D. Brings together combinations of alleles from two parents
E. Leads to new combinations of alleles in chromosomes

Short Answer

16. Briefly discuss the role of the environment and genetics in the production of an organism's phenotype. [p.246] _____

_____

_____

17. List the conditions (in any order) that must be met before genetic equilibrium (or nonevolution) will occur. [pp.246–247] _____

_____

Problems

18. For the following situation, assume that the conditions listed in question 17 do exist; therefore, there should be no change in gene frequency, generation after generation. Consider a population of hamsters in which dominant gene *B* produced black coat color and recessive gene *b* produces gray coat color (two alleles are responsible for color). The dominant gene has a frequency of 80 percent (or .80). It would follow that the frequency of the recessive gene is 20 percent (or .20). From this, the assumption is made that 80 percent of all sperm and eggs have gene B. Also 20 percent of all sperm and eggs carry gene *b*. [See pp.248–249 to solve the following problems.]

a. Calculate the probabilities of all possible matings in the Punnett square.
b. Summarize the genotype and phenotype frequencies of the F_1 generation.

Genotypes	Phenotypes
___ BB	
___ Bb	___ % black
___ bb	___ % gray

	Sperm	
	0.80	0.20
	B	b
Eggs 0.80 B	BB	Bb
0.20 b	Bb	bb

c. Further assume that the individuals of the F_1 generation produce another generation and the assumptions of the Hardy-Weinberg rule still hold. What are the frequencies of the sperm produced?

Parents (F_1)	B sperm	b sperm
___ BB	___	___
___ Bb	___	___
___ bb	___	___
Totals =	___	___

The egg frequencies may be similarly calculated. Note that the gamete frequencies of the F_2 are the same as the gamete frequencies of the last generation. Phenotype percentage also remains the same. Thus, the gene frequencies did not change between the F_1 and the F_2 generation. Again, given the assumptions of the Hardy-Weinberg equilibrium, gene frequencies do not change generation after generation.

19. In a population, 81 percent of the organisms are homozygous dominant, and 1 percent are homozygous recessive. Find the following [pp.248–249]

   a. the percentage of heterozygotes _____

   b. the frequency of the dominant allele _____

   c. the frequency of the recessive allele _____

20. In a population of 200 individuals, determine the following for a particular locus if $p = 0.80$. [pp.248–249]

   a. the number of homozygous dominant individuals _____

   b. the number of homozygous recessive individuals _____

   c. the number of heterozygous individuals if $p = 0.80$ _____

21. If the percentage of gene D is 70 percent in a gene pool, find the percentage of gene d. [pp.248–249]

_____

_____

22. If the frequency of gene R in a population is 0.60, what percentage of the individuals are heterozygous Rr? [pp.248–249]

_____

_____

Matching

Choose the most appropriate answer for each term.

23. ____ gene flow [p.247; glossary]

24. ____ allele frequencies [p.246]

25. ____ neutral mutations [p.247]

26. ____ microevolution [p.247]

27. ____ lethal mutation [p.247]

28. ____ alleles [p.246]

29. ____ mutation [p.247]

30. ____ genetic drift [p.247, glossary]

31. ____ genetic equilibrium [p.246]

32. ____ natural selection [p.249]

A. A heritable change in DNA
B. Zero evolution
C. Different molecular forms of a gene
D. Change in allele frequencies as individuals leave or enter a population
E. Change or stabilization of allele frequencies due to differences in survival and reproduction among variant members of a population
F. Random fluctuation in allele frequencies over time due to chance
G. Change in which expression of mutated gene always leads to the death of the individual
H. The abundance of each kind of allele in the entire population
I. Neither harmful nor helpful to the individual
J. Changes in allele frequencies brought about by mutation, genetic drift, gene flow, and natural selection

Fill-in-the-Blanks

A(n) (33) _____ [p.246] is a group of individuals of the same species that occupy a given area at a specific time. The (34) _____-_____ [p.246] principle allows researchers to establish a theoretical reference point (baseline) against which changes in allele frequency can be measured. Variation can be expressed in terms of (35) _____ _____ [p.246], which means the relative abundance of different alleles carried by the individuals in that population. The stability of allele ratios that would occur if all individuals had equal probability of surviving and reproducing is called (36) _____ _____ [p.246]. Over time, allele frequencies tend to change through infrequent but inevitable (37) _____ [p.247], which are the original source of genetic variation.

16.6. NATURAL SELECTION REVISITED [p.249]
16.7. DIRECTIONAL CHANGE IN THE RANGE OF VARIATION [pp.250–251]
16.8. SELECTION AGAINST OR IN FAVOR OF EXTREME PHENOTYPES [pp.252–253]
16.9. SPECIAL TYPES OF SELECTION [pp.254–255]

Selected Words: *Biston betularia* [p.250], H. B. Kettlewell [p.250], *mark-release-recapture* [p.250], *pest resurgence* [p.251], *biological controls* [p.251], *Eurosta solidaginis* [p.252], *Pyrenestes ostrinus* [p.253], sexual dimorphism [p.254], balancing selection [p.254], *sickle-cell anemia* [p.254], *malaria* [p.254], Hb^A/Hb^S [p.254], Hb^S/Hb^S [p.254], Hb^A/Hb^A [p.255]

Boldfaced, Page-Referenced Terms

[p.249] fitness _____

_____

[p.249] natural selection _____

_____

[p.250] directional selection _____

_____

[p.252] stabilizing selection _____

_____

[p.253] disruptive selection _____

_____

[p.254] sexual selection _____

_____

[p.254] balanced polymorphism _____

_____

Complete the Table

1. Complete the following table, which defines three types of natural selection effects.

Effect	Characteristics	Example
[p.252] a. Stabilizing selection		
[p.250] b. Directional selection		
[p.253] c. Disruptive selection		

Labeling

2. Identify the three curves below as stabilizing selection, directional selection, or disruptive selection. [pp.250,252–253]

 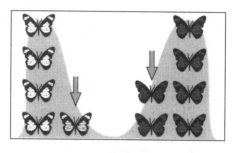 

a. _____   b. _____   c. _____

Fill-in-the-Blanks

When individuals of different phenotypes in a population differ in their ability to survive and reproduce, their alleles are subject to (3) _____ [p.249] selection. (4) _____ [p.252] selection favors the most common phenotypes in the population. (5) _____ [p.250] selection occurs when a specific change in the environment causes a heritable trait to occur with increasing frequency and the whole population tends to shift in a parallel direction. (6) _____ [p.253] selection favors the development of two or more distinct polymorphic varieties such that they become increasingly represented in a population and the population splits into different phenotypic variations.

(7) _____ _____ [p.252] provide an excellent example of stabilizing selection, because they have existed essentially unchanged for hundreds of millions of years. (8) _____ - _____ _____ [p.254], a genetic disorder that produces an abnormal form of hemoglobin, is an example of (9) _____ [p.252] selection that maintains a high frequency of the harmful (10) _____ [p.255] along with the normal gene. Such a genetic juggling act is called a (11) _____ _____ [p.255]. Differences in appearance between males and females of a species are known as (12) _____ _____ [p.254]. Among birds and mammals, females act as agents of (13) _____ [p.254] when they choose their mates. (14) _____ [p.254] selection is based on any trait that gives the individual a competitive edge in mating and producing offspring. The larger and more colorful and showy appearance of male birds of paradise when compared with females is the result of (15) _____ [p.254] selection.

Choice

For questions 16–26, choose from the following categories of natural selection. In some cases, two letters may be correct.

a. directional selection   b. stabilizing selection   c. disruptive selection

d. balanced polymorphism   e. sexual selection

16. ___ Sexual dimorphism [p.254]

17. ___ Phenotypic forms at both ends of the variation range are favored and intermediate forms are selected against. [p.253]

18. ___ May account for the persistence of intermediate phenotypes over time within a single species [p.252]

19. ___ Selection maintains two or more alleles for the same trait in a population in steady fashion, generation after generation. [p.254]

20. ___ The most common forms of a trait in a population are favored; over time, alleles for uncommon forms are eliminated. [p.252]

21. ___ Allele frequencies shift in a steady, consistent direction; this may be due to a change in the environment. [p.250]

22. ___ Counters the effects of mutation, genetic drift, and gene flow [p.252]

23. ___ Human newborns weighing an average of 7 pounds are favored. [common sense]

24. ___ In West Africa, finches known as black-bellied seedcrackers have either large or small bills. [p.253]

25. ___ The most frequent wing color of peppered moths shifted from a light form to a dark form as tree trunks became soot-darkened due to coal used for fuel during the industrial revolution. [p.251]

26. ___ Females of a species choosing mates to directly affect reproductive success [p.254]

16.10. GENE FLOW [p.255]

16.11. GENETIC DRIFT [pp.256–257]

Selected Words: *emigration* [p.255], *immigration* [p.255], *Ellis–van Creveld syndrome* [p.257], *endangered species* [p.257], *feline infectious peritonitis* [p.257]

Boldfaced, Page-Referenced Terms

[p.255] gene flow _____

_____

[p.256] genetic drift _____

_____

[p.256] sampling error _____

_____

[p.256] fixation _____

_____

[p.256] bottleneck _____

_____

[p.257] founder effect _____

_____

[p.257] inbreeding _____

_____

Choice

For questions 1–10, choose from the following microevolutionary forces that can change gene frequency:

a. genetic drift   b. gene flow

1. ___ A random change in allele frequencies over the generations, brought about by chance alone [p.256]
2. ___ Emigration [p.255]
3. ___ Prior to the turn of the century, hunters killed all but twenty of a large population of northern elephant seals [p.256]
4. ___ When allele frequencies change due to individuals leaving a population or new individuals enter it [p.255]
5. ___ Long ago, seabirds, winds, or ocean currents carried a few seeds from the Pacific Northwest to the Hawaiian Islands to establish plant populations [p.256]
6. ___ Sometimes result in endangered species [p.256]
7. ___ Founder effects and bottlenecks are two extreme cases [p.256]
8. ___ Immigration [p.255]
9. ___ Only 20,000 cheetahs have survived to the present [p.256]
10. ___ Scrubjays make hundreds of round trips carrying acorns from oak trees as much as a mile away to soil in their home territories for winter storage; this introduces new alleles to oak stands [p.255]

Short Answer

11. Distinguish between the two extreme cases of genetic drift: the founder effect and bottlenecks. [pp.256–257] _____

_____

Self-Quiz

___ 1. An acceptable definition of evolution is _____. [p.241]
   a. changes in organisms that are extinct
   b. changes in organisms since the flood
   c. changes in organisms over time
   d. changes in organisms in only one place

___ 2. The two scientists most closely associated with the concept of evolution are _____. [p.245]
   a. Lyell and Malthus
   b. Henslow and Cuvier
   c. Henslow and Malthus
   d. Darwin and Wallace

___ 3. Studies of the distribution of organisms on Earth is known as _____. [p.240]
   a. the Great Chain of Being
   b. stratification
   c. geological evolution
   d. biogeography

___ 4. Buffon suggested that _____. [p.241]
   a. tail bones in a human have no place in a perfectly designed body
   b. perhaps species originated in more than one place and have been modified over time
   c. the force for change in organisms was a built-in drive for perfection, up the Chain of Being
   d. gradual processes now molding the Earth's surface had also been at work in the past

___ 5. In Lyell's book *Principles of Geology*, he suggested that _____. [p.243]
   a. tail bones in a human have no place in a perfectly designed body
   b. perhaps species originated in more than one place and have been modified over time
   c. the force for change in organisms was a built-in drive for perfection, up the Chain of Being
   d. gradual processes now molding the Earth's surface had also been at work in the past

___ 6. One of the central ideas of Lamarck's theory of inheritance of acquired characteristics was that _____. [p.242]
   a. tail bones in a human have no place in a perfectly designed body
   b. perhaps species originated in more than one place and have been modified over time
   c. the force for change in organisms was a built-in drive for perfection, up the Chain of Being
   d. gradual processes now molding the Earth's surface had also been at work in the past

___ 7. The theory of catastrophism is associated with _____. [p.242]
   a. Darwin
   b. Cuvier
   c. Buffon
   d. Lamarck

___ 8. The idea that any population tends to outgrow its resources and that its members must compete for what is available belonged to _____. [p.244]
   a. Malthus
   b. Darwin
   c. Lyell
   d. Henslow

___ 9. The ideas that all natural populations have the reproductive capacity to exceed the resources required to sustain them and that members of a natural population show great variation in their traits are associated with _____. [p.244]
   a. catastrophism
   b. desired evolution
   c. natural selection
   d. Malthus's idea of survival

___10. *Archaeopteryx* is evidence for _____. [p.245]
   a. the theory that birds descended from reptiles
   b. evolution
   c. an evolutionary "link" between two major groups of organisms
   d. the existence of fossils
   e. all of the above

Chapter Objectives/Review Questions

1. It was the study of _____ that raised the following question: If all species were created at the same time in the same place, why were certain species found in only some parts of the world and not others? [p.240]
2. Give examples of the type of evidence found by comparative anatomists that suggested living things may have changed with time. [pp.240–241]
3. Be able to state the significance of the following: Henslow, H.M.S. *Beagle*, Lyell, Darwin, Malthus, Wallace, and *Archaeopteryx*. [pp.242–245]
4. What conclusion was Darwin led to when he considered the various species of finches living on the separate islands of the Galápagos? [pp.244–245]
5. A _____ is a group of individuals occupying a given area and belonging to the same species. [p.246]
6. Distinguish between morphological, physiological, and behavioral traits. [p.246]
7. All of the genes of an entire population belong to a _____ _____. [p.246]
8. Each kind of gene usually exists in one or more molecular forms, called _____. [p.246]
9. Review the five categories through which genetic variation occurs among individuals. [p.246]
10. The abundance of each kind of allele in the whole population is referred to as allele _____. [p.246]
11. Be able to list the five conditions that must exist before conditions for the Hardy-Weinberg "rule" are met. [p.246]
12. A point at which allele frequencies for a trait remain stable through the generations is called genetic _____. [p.246]
13. Changes in allele frequencies brought about by mutation, genetic drift, gene flow, and natural selection are called _____. [p.247]
14. A _____ is a random, heritable change in DNA. [p.247]
15. Distinguish a lethal mutation from a neutral mutation. [p.247]
16. Be able to calculate all allele and genotype frequencies when provided with the *genotype* frequency of the recessive allele (q^2). [p.248]
17. Be able to define and provide an example of directional selection, stabilizing selection, and disruptive selection. [pp.250–253]
18. Define *balanced polymorphism*. [p.254]
19. The occurrence of phenotypic differences between males and females of a species is called _____ _____. [p.254]
20. _____ selection is based on any trait that gives an individual a competitive edge in mating and producing offspring. [p.254]
21. Allele frequencies change as individuals leave or enter a population; this is gene _____. [p.255]
22. Random fluctuations in allele frequencies over time, due to chance, is called _____ _____. [p.256]
23. Distinguish the founder effect from a bottleneck. [pp.256–257]
24. Relate bottlenecks to endangered species. [pp.256–257]

Integrating and Applying Key Concepts

Can you imagine any way in which directional selection may have occurred or may be occurring in humans? Which factors do you suppose are the driving forces that sustain the trend? Do you think the trend could be reversed? If so, by what factor(s)?

17

SPECIATION

Interactive Exercises

The Case of the Road-Killed Snails [pp.260–261]

17.1. ON THE ROAD TO SPECIATION [pp.262–263]

17.2. SPECIATION IN GEOGRAPHICALLY ISOLATED POPULATIONS [pp.264–265]

Selected Words: *Helix aspersa* [p.260], *behavioral isolation* [p.263], *temporal isolation* [p.263], *mechanical isolation* [p.263], *ecological isolation* [p.263], *gametic mortality* [p.263]

Boldfaced, Page-Referenced Terms

[p.261] speciation _____

_____

[p.262] species _____

_____

[p.262] biological species concept _____

_____

[p.262] gene flow _____

_____

[p.262] genetic divergence _____

_____

[p.262] reproductive isolating mechanisms _____

_____

[p.264] allopatric speciation _____

_____

[p.265] archipelago _____

_____

Choice

For questions 1–10, choose from the following:

a. Mayr's biological species   b. genetic divergence   c. isolating mechanisms

1. ___ Heritable aspect of body form, physiology, or behavior that prevents gene flow between genetically divergent populations [p.262]

2. ___ Groups of interbreeding natural populations [p.262]

3. ___ May occur as a result of stopping gene flow [p.262]

4. ___ Phenotypically very different but can interbreed and produce fertile offspring [p.262]

5. ___ Defined as the buildup of differences in the separated pools of alleles of populations [p.262]

6. ___ A "kind" of living thing [p.262]

7. ___ Factors that prevent horses and zebras from mating in the wild [p.262]

8. ___ Prezygotic [p.262]

9. ___ Does not work well with organisms that reproduce solely by asexual means or those known only from the fossil record [p.262]

10. ___ Postzygotic [p.262]

Matching

Select the most appropriate answer to match the isolating mechanisms; complete the exercise by entering "pre" in the parentheses if the mechanism is prezygotic and "post" if the mechanism is postzygotic.

11. ___ecological ( ) [p.263]

12. ___temporal ( ) [p.263]

13. ___hybrid inviability ( ) [p.263]

14. ___mechanical ( ) [p.263]

15. ___zygote mortality ( ) [p.263]

16. ___gametic mortality ( ) [p.263]

17. ___behavioral ( ) [p.263]

18. ___hybrid offspring ( ) [p.263]

A. Potential mates occupy overlapping ranges but reproduce at different times.
B. The first-generation hybrid forms but shows very low fitness.
C. Potential mates occupy different local habitats within the same area.
D. Potential mates meet but cannot figure out what to do about it.
E. Sperm is transferred but the egg is not fertilized (gametes die or gametes are incompatible).
F. The hybrid is sterile or partially so.
G. Potential mates attempt engagement, but sperm cannot be successfully transferred.
H. The egg is fertilized, but the zygote or embryo dies.

Choice

For questions 19–25, choose from the following isolating mechanisms:

   a. temporal   b. behavioral   c. mechanical   d. gametic   e. ecological   f. postzygotic

19. ___ Two sage species; each has its flower petals arranged as a "landing platform" for a different pollinator [p.263]

20. ___ Two species of cicada; one matures, emerges, and reproduces every thirteen years; the other, every seventeen years [p.263]

21. ___ Two different populations of manzanita have different built-in physiological mechanisms that enhance water conservation during the dry season [p.263]

22. ___ Molecularly mismatched gametes of different sea urchin species [p.263]

23. ___ Prior to copulation, male and female birds engage in complex courtship rituals recognized only by birds of their own species [p.263]

24. ___ Sterile mules [p.263]

Identification

25. The illustration below depicts the divergence of one species as time passes. Each horizontal line represents a population of that species and each vertical line, a point in time. Answer the questions below the illustration.

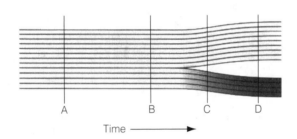

Time ⟶

   a. How many species are represented at time A? _____[p.262]

   b. How many species exist at time D? _____[p.262]

   c. Between what times (letters) does divergence begin? _____[p.262]

   d. What letter represents the time that complete divergence is reached? _____[p.262]

   e. Between what two times (letters) is divergence clearly underway but not complete? _____[p.262]

Fill-in-the-Blanks

(26) _____ [p.261] is the process whereby species are formed. A(n) (27) _____ [p.262] is composed

of one or more populations of individuals who can interbreed under natural conditions and produce fertile,

reproductively isolated offspring. (28) _____ [p.262] can be described as a process in which differences

in alleles accumulate between populations. Any aspect of structure, function, or behavior that prevents

interbreeding is a(n) (29) _____ _____ [p.262] mechanism. Physical barriers that prevent gene

flow as in the case of large rivers changing course or forests giving way to grasslands are (30) _____

[p.264] barriers.

17.3. MODELS FOR OTHER SPECIATION ROUTES [pp.266–267]

17.4. PATTERNS OF SPECIATION [pp.268–269]

Selected Words: sym– [p.266], "dosage compensation" [p.266], para– [p.267], *subspecies* [p.267], ana– [p.268], *branch* [p.268], *branch point* [p.268], *physical* access [p.269], *evolutionary* access [p.269], *ecological* access [p.269]

Boldfaced, Page-Referenced Terms

[p.266] sympatric speciation _____

_____

[p.266] polyploidy _____

_____

[p.267] parapatric speciation _____

_____

[p.267] hybrid zone _____

_____

[p.268] cladogenesis _____

_____

[p.268] anagenesis _____

_____

[p.268] evolutionary tree _____

_____

[p.268] gradual model of speciation _____

_____

[p.268] punctuation model of speciation _____

_____

[p.268] adaptive radiation _____

_____

[p.268] adaptive zones _____

_____

[p.269] key innovations _____

_____

[p.269] extinction _____

_____

[p.269] mass extinction _____

_____

Complete the Table

1. Complete the following table, which defines three speciation models.

Speciation Model	Description
[p.266] a.	Suggests that species can form within the range of an existing species, in the absence of physical or ecological barriers; a controversial model; evidence comes from crater lake fishes
[p.264] b.	Emphasizes disruption of gene flow; perhaps the main speciation route; through divergence and evolution of isolating mechanisms, separated populations become reproductively incompatible and cannot interbreed
[p.267] c.	Occurs where populations share a common border; the borders may be permeable to gene flow; in some, gene exchange is confined to a common border, the hybrid zone

Dichotomous Choice

Circle one of two possible answers given between parentheses in each statement.

2. Instant speciation by polyploidy in many flowering plant species represents (parapatric/sympatric) speciation. [p.266]
3. In the past, hybrid orioles existed in a hybrid zone between eastern Baltimore orioles and western Bullock orioles; today, hybrid orioles are becoming less frequent in this example of (allopatric/parapatric) speciation. [p.267]
4. In the absence of gene flow between geographically separate populations, genetic divergence may become sufficient to bring about (allopatric/parapatric) speciation. [p.264]
5. The uplifted ridge of land we now call the Isthmus of Panama divides the Atlantic Ocean from the Pacific Ocean and formed a natural laboratory for biologists to study (allopatric/sympatric) speciation. [p.264]
6. In the 1800s, a major earthquake changed the course of the Mississippi River; this isolated some populations of insects that could not swim or fly and set the stage for (parapatric/allopatric) speciation. [p.264]

Short Answer

7. Study the illustration below, which shows the possible course of wheat evolution; similar genomes (a genome is a complete set of chromosomes) are designated by the same letter. Answer the questions below the diagram.

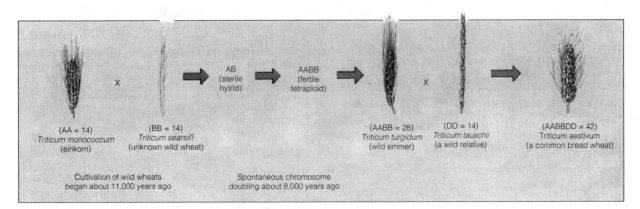

a. How many pairs of chromosomes are found in *Triticum monococcum*? _____ [p.267]

b. How are the *T. monococcum* chromosomes designated? _____ [p.267]

c. How many pairs of chromosomes are found in the unknown wild wheat? _____ [p.267]

d. How are the unknown's chromosomes designated? _____ [p.267]

e. What cellular event must have occurred to create the *T. turgidum* genome? _____ [p.267]

f. What cellular event must have occurred to create the *T. aestivum* genome? _____ [p.267]

g. What is the source of the A genome in *T. aestivum*? The source of the B genome in *T. aestivum*? The source of the D genome in *T. aestivum*? _____ [p.267]

Wheat is an example of a species formed by (h) _____ and (i) _____. [p.267]

Complete the Table

8. Complete the table below to summarize information to interpret evolutionary tree diagrams. [Refer to text Figure 17.11 on p.268.] Ask yourself, "How would the interpretation be illustrated on an evolutionary tree diagram?"

Branch Form	Interpretation
a.	Traits changed rapidly around the time of speciation
b.	An adaptive radiation occurred
c.	Extinction
d.	Speciation occurred through gradual changes in traits over geologic time
e.	Traits of the new species did not change much thereafter
f.	Evidence of this presumed evolutionary relationship is only sketchy

Matching

Choose the most appropriate answer for each term.

9. ___cladogenesis [p.268]

10. ___evolutionary tree diagrams [p.268]

11. ___gradual model of speciation [p.268]

12. ___punctuation model of speciation [p.268]

13. ___adaptive radiation [p.268]

14. ___adaptive zones [pp.268-269]

15. ___key innovation [p.269]

16. ___mass extinction [p.269]

17. ___physical, evolutionary, and ecological access [p.269]

18. ___anagenesis [p.268]

A. Modification of some structure or function that permits a lineage to exploit the environment in more efficient ways.

B. Conditions allowing a lineage to radiate into an adaptive zone

C. Speciation in which most morphological changes are compressed into a brief period of hundreds or thousands of years when populations first start to diverge; evolutionary tree branches make abrupt ninety-degree turns

D. Ways of life such as "burrowing in the seafloor"

E. Speciation with slight angles on the evolutionary tree, which conveys many small changes in form over long time spans

F. An abrupt rise in the rates of species disappearance above background level

G. Speciation occurring within a single, unbranched line of descent; peppered moth populations, for example

H. A method of summarizing information about the continuities of relationships among species

I. A branching speciation pattern; isolated populations diverge

J. A burst of microevolutionary activity within a lineage; results in the formation of new species in a wide range of habitats

Self-Quiz

____ 1. Of the following, _____ is (are) not a major consideration of the biological species concept. [p.262]
   a. groups of populations
   b. reproductive isolation
   c. asexual reproduction
   d. interbreeding natural populations
   e. production of fertile offspring

____ 2. When something prevents gene flow between two populations or subpopulations, _____ may occur. [p.262]
   a. genetic drift, natural selection, and mutation
   b. genetic divergence
   c. a buildup of differences in the separated gene pools of alleles
   d. all of the above

____ 3. When potential mates occupy different local habitats within the same area, it is termed _____ isolation. [p.263]
   a. temporal
   b. behavioral
   c. mechanical
   d. ecological

____ 4. "Potential mates occupy overlapping ranges but reproduce at different times" is a description of _____ isolation. [p.263]
   a. temporal
   b. behavioral
   c. mechanical
   d. ecological

____ 5. Of the following, _____ is a postzygotic isolating mechanism. [p.263]
   a. behavioral isolation
   b. gametic mortality
   c. hybrid inviability
   d. mechanical isolation

For questions 6–8, choose from the following answers:
   a. parapatric speciation
   b. sympatric speciation
   c. allopatric speciation

____ 6. When the Isthmus of Panama was formed, it provided contemporary scientists with an ideal natural laboratory to study _____. [p.264]

____ 7. Polyploidy and cichlid fishes of crater lakes provide evidence for _____. [p.266]

____ 8. The identification of a hybrid zone between the ranges of eastern Baltimore orioles and western Bullock orioles is an example of _____. [p.267]

____ 9. The branching speciation pattern revealed by the fossil record is called _____. [p.268]
   a. anagenesis
   b. the gradual model
   c. cladogenesis
   d. the punctuation model

____ 10. Species formation by many small changes over long time spans fit the _____ model of speciation; rapid speciation with morphological changes compressed into a brief period of population divergence describes the _____ model of speciation. [p.268]
   a. anagenesis; punctuation
   b. punctuation; gradual
   c. gradual; cladogenesis
   d. gradual; punctuation

____ 11. Adaptive radiation is _____. [p.268]
   a. a burst of microevolutionary activity within a lineage
   b. the formation of new species in a wide range of habitats
   c. the spreading of lineages into unfilled adaptive zones
   d. a lineage radiating into an adaptive zone when it has physical, ecological, or evolutionary access to it
   e. all of the above

____ 12. Hawaiian honeycreepers and Galápagos finches are excellent examples of _____. [p.265]
   a. allopatric speciation
   b. genetic divergence in isolated subpopulations
   c. adaptive radiations
   d. colonization of isolated archipelagos
   e. all of the above

For questions 13–14, choose from the following answers:

    a. stabilizing selection
    b. divergence
    c. a reproductive isolating mechanism
    d. polyploidy

___13. Two species of sage plants with differently shaped floral parts that prevent pollination by the same pollinator serve as an example of _____. [p.263]

___14. _____ occurs when isolated populations accumulate allele frequency differences between them over time. [p.262]

Chapter Objectives/Review Questions

1. The process by which species are formed is known as _____. [p.261]
2. Be able to outline the biological species concept as stated by Ernst Mayr. [p.262]
3. _____ _____ is a buildup of differences in separated pools of alleles. [p.262]
4. The evolution of reproductive _____ _____ pave the way for genetic divergence and speciation. [p.262]
5. Be able to briefly define the major categories of prezygotic and postzygotic isolating mechanisms. [p.263, and text Table 17.1 on p.270]
6. _____ speciation occurs when daughter species form gradually by divergence in the absence of gene flow between geographically separate populations. [p.264]
7. In _____ speciation, daughter species arise, sometimes rapidly, from a small proportion of individuals within an existing population. [p.266]
8. When daughter species form from a small proportion of individuals along a common border between two populations, it is called _____ speciation. [p.267]
9. Explain why sympatric speciation by polyploidy is a rapid method of speciation. [p.266]
10. In some cases of parapatric speciation, gene exchange between two species is confined to a(n) _____ zone. [p.267]
11. Distinguish cladogenesis from anagenesis. [p.268]
12. Be able to explain the use of evolutionary tree diagrams and the symbolism used. [p.268]
13. Explain why branches on an evolutionary tree that have slight angles indicate the _____ model of speciation. [p.268]
14. Branches on an evolutionary tree that turn abruptly with 90-degree turns are consistent with the _____ model of speciation. [p.268]
15. An adaptive radiation is a burst of _____ activity within a lineage that results in the formation of new species in a variety of habitats. [p.268]
16. Cite examples of adaptive zones. [p.269]

Integrating and Applying Key Concepts

Systematics is defined as the practice of describing, naming, and classifying living things; this includes the comparative study of organisms and all relationships among them. Plant systematists who work with flowering plants readily accept Ernst Mayr's definition of a "biological species" as described in this chapter. However, in real practice, systematists often must also work with a concept known as the "morphological species." For example, the statement is sometimes made that "two plant specimens belong to the same morphological species but not to the same biological species." Explain the meaning of that statement. What type of experimental evidence would be necessary as a basis for this statement? From your study of this chapter, can you suggest a reason that application of the term "morphological species" is sometimes necessary? What difficulties and inaccuracies, in terms of identifying a species, might the use of both species concepts present?

18

THE MACROEVOLUTIONARY PUZZLE

Interactive Exercises

Of Floods and Fossils [pp.272–273]

18.1. FOSSILS—EVIDENCE OF ANCIENT LIFE [pp.274–275]

Selected Words: *species* [p.273], ammonites [p.273], homology [p.273], *trace* fossils [p.274], *Cooksonia* [p.274], Proterozoic [p.275], Archean [p.275], Paleozoic era [p.275], Mesozoic era [p.275], Cenozoic era [p.275]

Boldfaced, Page-Referenced Terms

[p.273] lineage _____

_____

[p.273] macroevolution _____

_____

[p.274] fossils _____

_____

[p.274] fossilization _____

_____

[p.275] stratification _____

_____

[p.275] geologic time scale _____

_____

Matching

Choose the most appropriate answer for each term.

1. ___fossilization [p.274]
2. ___stratification [p.275]
3. ___trace fossils [p.274]
4. ___lineage [p.273]
5. ___macroevolution [p.273]
6. ___fossils [p.274]
7. ___biased fossil record [p.275]
8. ___formation of sedimentary deposits [p.275]
9. ___oldest fossils [p.275]
10. ___rupture or tilt of sedimentary layers [p.275]

A. Evolutionary connectedness between organisms, past to present
B. Found in the deepest rock layers
C. Affected by movements of the Earth's crust, hardness of organisms, population sizes, and certain environments
D. Produced by gradual deposition of silt, volcanic ash, and other materials
E. Refers to large-scale patterns, trends, and rates of change among groups of species
F. A process favored by rapid burial in the absence of oxygen
G. Evidence of later geologic disturbance
H. Recognizable, physical evidence of ancient life; the most common are bones, teeth, shells, spore capsules, seeds, and other hard parts
I. Includes the indirect evidence of coprolites and imprints of leaves, stems, tracks, trails, and burrows
J. A layering of sedimentary deposits

Choice

For questions 11–15, choose from the following:

a. Archean and earlier   b. Cenozoic era   c. Mesozoic era   d. Paleozoic era   e. Proterozoic era

11. ___ The "modern" era that began 65 million years ago; dominated by birds, mammals, and flowering plants [p.275]

12. ___ Stretched from 240,000,000 to 65,000,000 years ago; dominated by dinosaurs and gymnosperms [p.275]

13. ___ Life originated during this time [p.275]

14. ___ Stretched from 550,000,000 to 240,000,000 years ago, includes first amphibians and reptiles, first land plants [p.275]

15. ___ From 2,500,000,000 to 550,000,000 years ago; first eukaryotic cells, first multicellular animals [p.275]

18.2. EVIDENCE FROM COMPARATIVE MORPHOLOGY [pp.276–277]

18.3. EVIDENCE FROM PATTERNS OF DEVELOPMENT [pp.278–279]

18.4. EVIDENCE FROM COMPARATIVE BIOCHEMISTRY [pp.280–281]

Selected Words: *morpho–* [p.276], "stem" reptile [p.276], key innovation [p.276], *homo–* [p.276], homologous [p.277], "converged" [p.277], analogous [p.277], *Delphinium decorum* [p.278], *Delphinium nudicaule* [p.278], allometric changes [p.279], cytochrome *c* [p.280], "hybrid" molecules [p.280]

Boldfaced, Page-Referenced Terms

[p.276] comparative morphology _____

_____

[p.276] morphological divergence _____

_____

[p.276] homology _____

_____

[p.277] morphological convergence _____

_____

[p.277] analogy _____

_____

[p.280] nucleic acid hybridization _____

_____

[p.280] neutral mutations _____

_____

[p.281] molecular clock _____

_____

True–False

If the statement is true, write a T in the blank. If the statement is false, correct it by changing the underlined word(s) and writing the correct word(s) in the answer blank.

_____1. The extent to which a strand of DNA or RNA isolated from individuals of one species will base-pair with a comparable strand from individuals of a different species is a rough measure of the evolutionary <u>distance</u> between them. [p.280]

_____2. The amino acid sequence of cytochrome *c* is very similar in such distantly related organisms as (1) yeast, (2) wheat, and (3) humans; the probability that this striking resemblance resulted from chance alone is very <u>high</u>. [p.280]

_____3. Analogous structures in two different lineages indicate strong evidence of morphological <u>divergence</u>. [p.277]

_____4. Body parts that have similar form and function in different lineages yet develop from different tissues in their embryos are analogous structures that provide very strong evidence of morphological <u>divergence</u>. [p.277]

_____5. Biochemical <u>similarities</u> are fewest among the most closely related species and most numerous among the most distantly related. [p.280]

Choice

For questions 6–10, choose from the following:

a. analogy   b. homology   c. an example of a lineage
d. a specific example of morphological convergence   e. a specific example of morphological divergence

6. ___ Sharks, penguins, and porpoises [p.277]

7. ___ Jawless fishes, land-dwelling stem reptiles, mammalian ancestor, porpoise [p.277]

8. ___ Body parts that were once quite different among distantly related lineages but are now quite similar in function and in apparent structure [p.277]

9. ___ The human arm and hand, the porpoise's front flipper, and the bird's wing [p.276]

10. ___ Body parts of different organisms that develop from similar embryonic parts yet might result in quite different-appearing adult structures that would probably not have similar functions [p.276]

Labeling

Evidence for macroevolution comes from comparative morphology and comparative biochemistry. For each of the following listed items, place an "M" in the blank if the evidence comes from comparative morphology, a "B" if the evidence comes from comparative biochemistry, and a "D" if the evidence comes from developmental evidence.

11. ___ Early in the vertebrate developmental program, embryos of different lineages proceed through strikingly similar stages. [p.279]

12. ___ Comparison of the developmental rates of changes in a chimpanzee skull and a human skull. [p.279]

13. ___ The evolution of different rates of flower development in two larkspur species also led to the coevolution of two different pollinators, honeybees and hummingbirds. [p.278]

14. ___ Using accumulated neutral mutations to date the divergence of two species from a common ancestor. [p.280]

15. ___ The amino acid sequence in the cytochrome c of humans precisely matches the sequence in chimpanzees. [p.280]

16. ___ Establishing a rough measure of evolutionary distance between two organisms by use of DNA hybridization studies that demonstrate the extent to which the DNA from one species base-pairs with another [p.280]

17. ___ Regulatory genes control the rate of growth of different body parts and can produce large differences in the development of two very similar embryos. [p.279]

18. ___ Results from DNA-DNA hybridization studies supported an earlier view that giant pandas, bears, and red pandas are related by common descent. [pp.280–281]

19. ___ Distantly related vertebrates, such as sharks, penguins, and porpoises, show a similarity to one another in their proportion, position, and function of body parts. [p.277]

20. ___ Finding similarities in vertebrate forelimbs when comparing the wings of pterosaurs, birds, bats, and porpoise flippers. [p.276]

Short Answer

21. List three methods used to determine the patterns of diversity in an evolutionary context. [pp.280,287]

_____

_____

_____

_____

_____

18.5. IDENTIFYING SPECIES, PAST AND PRESENT [pp.282–283]

Selected Words: relative degrees of similarity [p.282], R. Whittaker [p.282]

Boldfaced, Page-Referenced Terms

[p.282] taxonomy _____

_____

[p.282] binomial system _____

_____

[p.282] genus (plural, genera; adjectival, generic) _____

_____

[p.282] specific name _____

_____

[p.282] classification scheme _____

_____

[p.282] phylogeny _____

_____

[p.282] Monera _____

_____

[p.283] Protista _____

_____

[p.283] Fungi _____

_____

[p.283] Plantae _____

_____

[p.283] Animalia _____

_____

[p.283] Eubacteria _____

_____

[p.283] Archaebacteria _____

_____

[p.283] six-kingdom classification scheme _____

_____

Matching

Choose the most appropriate answer for each term.

1. ___ Monera [p.282]
2. ___ Protista [p.282]
3. ___ Fungi [p.283]
4. ___ Plantae [p.283]
5. ___ Animalia [p.283]

A. Multicelled heterotrophs that feed by extracellular digestion and absorption; many are decomposers; some parasites and pathogens
B. Single-celled prokaryotes, some autotrophs, others heterotrophs
C. Diverse multicelled heterotrophs, including predators and parasites
D. Multicelled photosynthetic autotrophs
E. Diverse single-celled eukaryotes, some photosynthetic autotrophs, many heterotrophs

Short Answer

6. Arrange the following jumbled taxa in proper order, with the most inclusive first: family, class, species, kingdom, genus, phylum (or division), and order. [p.282] _____

_____

_____

7. Arrange the following jumbled taxa and their categories in proper order, the most inclusive first: genus: *Archibaccharis*; kingdom: Plantae; order: Asterales; species: *lineariloba*; class: Dicotyledonae; division: Anthophyta; family: Asteraceae. [p.282] _____

_____

_____

Matching

Choose the most appropriate answer for each term.

8. ___black bear [p.282]
9. ___phylogeny [p.282]
10. ___binomial system [p.282]
11. ___*Pinus strobus* and *Pinus banksiana* [p.282]
12. ___species [p.283]
13. ___genus [p.282]
14. ___family [p.283]
15. ___taxa [p.282]
16. ___*Ursus americanus* (black bear) and *Homarus americanus* (Atlantic lobster) [p.282]
17. ___Linnaeus [p.282]

A. A group of similar species
B. The taxon most often studied; the real entity in a classification scheme
C. Evolutionary relationships among species, reflected in classification schemes
D. The organizing units of classification schemes
E. Classification level that includes similar genera
F. Originated the modern practice of providing two scientific names for organisms
G. An example of a common name
H. Two species belonging to one genus
I. An example of two very different organisms having the same specific name
J. System of assigning a two-part Latin name to a species

Short Answer

18. Present the reasons why the Kingdom Monera has been divided into two kingdoms: Archaebacteria and Eubacteria. [p.283]_____

_____

_____

_____

18.6. EVIDENCE OF A CHANGING EARTH [pp.284–285]

18.7. *Focus on Science:* DATING PIECES OF THE MACROEVOLUTIONARY PUZZLE
[p.286]

Selected Words: continental drift [p.284], Pangea [p.284], continents [p.284], mantle [p.284], magnetism [p.284], *superplumes* [p.285], "hot spots" [p.285], Gondwana [p.285], *Glossopteris* [p.285], *unstable* [p.286], ^{14}C [p.286]

Boldfaced, Page-Referenced Terms

[p.284] theory of uniformity _____

_____

[p.284] theory of plate tectonics _____

_____

[p.286] radiometric dating _____

_____

[p.286] half-life _____

_____

Fill-in-the-Blanks

According to the theory of (1) _____ [p.284], mountain building and erosion had operated on Earth's surface in the same ways repeatedly across vast stretches of time, yet geologists noticed that the southern (2) _____ [p.284] seemed to fit together like pieces of a jigsaw puzzle. A. Wegener pondered the idea that (2) could drift about and collide with each other to form a(n) (3) _____ [p.286] named Pangea. (2) are thinner and less dense than the underlying (4) _____ [p.284]. Rocks associated with (2) are often rich in (5) _____ [p.284], the particles of which adopt a north-south (6) _____ [p.284] aligned with Earth's magnetic poles as the rocks form. The discovery that the polarities of North American and European rocks only became aligned when sketches of the two (2) were joined together caused geologists to (7) [choose one] ❏ accept ❏ reject [p.284] Wegener's idea.

Deep-sea probes also revealed evidence of (8) _____ _____ [p.284]. Driving these movements are massive upwelling (9) _____ [p.284] of molten rock convected from the interior. (9) that occurred in the past ruptured the (10) _____ [p.284] at ridges on the ocean floor; (9) also cause deep splitting within the (2), such as that occurring in (11) _____ [p.284], at Lake Baikal in Russia, and along the east African rift zone. Displacement caused by (8) moves (12) _____ [p.284] away from the ridges and forces other parts of a (12) under an adjoining (12), which lifts up as a result. The Appalachians, Sierras, Cascades, and (13) _____ [p.284] are examples of mountain ranges paralleling coasts that were formed this way.

The discovery of (14) _____ _____ [p.286] enabled geologists to convert relative ages of rocks into absolute ones. Relative ages had been established by noting the particular (15) _____ [p.286] of rock where specific fossils had been found. Using the hypothesis that deeper (15) of sedimentary rock were always older than rock layers closer to the surface, a(n) (16) _____ _____ _____ [p.286] was constructed that defined the boundaries of time in terms of the fossils and chemical and physical composition of rocks associated with specified layers.

By a method called (17) _____ _____ [p.286], researchers now measure the amount of an original unstable (18) _____ [p.286] trapped in the newly formed rock and compare it with the amount of a more stable (18) formed from it in the same rock. The (19) [choose one] ❐ older ❐ more recent [p.286] the rock, the (20) [choose one] ❐ more ❐ less [p.286] of the original (18) exists relative to its daughter (18).

Self-Quiz

___ 1. Phylogeny is _____. [p.282]
   a. the identification of organisms and assigning names to them
   b. producing a "retrieval system" of several levels
   c. an evolutionary history of organism(s), both living and extinct
   d. a study of the adaptive responses of various organisms

___ 2. The significant contribution by Linnaeus was to develop _____. [p.282]
   a. the idea that organisms should be placed in two kingdoms
   b. the binomial system
   c. the theory of evolution
   d. the idea that a species could have several common names

___ 3. The process of grouping organisms according to similarities derived from a common ancestor is known as _____. [p.282]
   a. stratification
   b. classification
   c. a binomial system
   d. macroevolution

___ 4. Early embryos of vertebrates strongly resemble one another *because* _____. [pp.278–279]

   a. the genes that guide early embryonic development are the same (or similar) in all vertebrates
   b. the analogous structures they each contain are evidence of morphological convergence
   c. the homologous structures they each contain provide very strong evidence of morphological divergence
   d. DNA-DNA hybridization techniques reveal many structural alterations that have resulted from gene mutations in early embryos

___ 5. All living organisms have eukaryotic cell structure except the _____. [pp.282–283]
   a. Animalia
   b. Plantae
   c. Fungi
   d. Monera (= Archaebacteria + Eubacteria)
   e. Protista

For questions 6–7, choose from the following answers:
   a. convergence
   b. divergence

___ 6. The wings of pterosaurs, birds, and bats serve as examples of _____. [p.276]

___ 7. Penguins and porpoises serve as examples of _____ . [p.277]

For questions 8–12, choose from the following answers:

    a. DNA-DNA hybridization
    b. homologous
    c. macroevolution
    d. analogous
    e. genes that regulate timing of development

___ 8. _____ structures resemble one another due to common descent. [p.276]

___ 9. A rough measure of the evolutionary distance between two species is provided by _____ . [p.280]

___ 10. _____ control the rate of growth in different body parts. [p.279]

___ 11. Examples of _____ structures are the shark fin and the penguin wing. [p.277]

___ 12. _____ refers to changes in groups of species. [p.273]

Chapter Objectives/Review Questions

1. _____ refers to large-scale patterns, trends, and rates of change among groups of species. [p.273]
2. Give a generalized definition of a "fossil"; cite examples. [p.274]
3. _____ begins with rapid burial in sediments or volcanic ash. [p.274]
4. What is the name given to the layering of sedimentary deposits? [p.275]
5. Are the oldest fossils found in the deepest layers of sedimentary rocks or nearer the surface? [p.275]
6. The fossil record is heavily _____ toward certain environments and certain types of organisms. [p.275]
7. The macroevolutionary pattern of change from a common ancestor is known as morphological _____ . [p.276]
8. Distinguish homologies from analogies as applied to organisms. [pp.276–277]
9. A pattern of long-term change in similar directions among remotely related lineages is called morphological _____ . [p.277]
10. Explain why two species of larkspurs have flowers with different shapes that attract different pollinators. [p.278]
11. Early in the vertebrate developmental program, _____ of different lineages proceed through strikingly similar stages. [p.278]
12. Mutation of _____ genes explains the differences in the timing of growth sequences of chimpanzees and humans (even though they have nearly identical genes). [p.279]
13. Some nucleic acid comparisons are based on _____ hybridization. [p.280]
14. Define *neutral mutations*, and discuss their use as molecular clocks. [pp.280–281]
15. Describe the binomial system as developed by Linnaeus; why are Latin and Greek used as the languages of binomials? [p.282]
16. Be able to arrange the classification groupings correctly, from most to least inclusive. [pp.282–283]
17. Modern classification schemes reflect _____ , the evolutionary relationships among species, starting with the most ancestral and including all the branches leading to their descendants. [p.282]
18. List the six kingdoms of life in the new system and cite examples of the organisms in each. [pp.282–283]
19. Describe some of the geologic forces that have changed Earth's crust and the distribution of oceans during the past 3.8 billion years. [p.284]
20. Explain how some of the changes that you mentioned in #19 could have affected the evolution of living organisms. [p.285]
21. Describe how the time when a particular rock was formed can be estimated by radiometric dating. [p. 286]
22. Explain why ^{14}C (carbon 14) is of interest to biologists who want to know when an organism lived on Earth. If ^{14}C only has a useful range of 60,000 years, how do scientists know that the ammonites shown in text Figure 18.1 [p.273] lived and died 65,000,000 years ago? Did they use ^{14}C? [p.286]

Integrating and Applying Key Concepts

Water crowfoot plants (family Ranunculaceae) often have two or more distinct leaf types within the same species and on the same plant. One type is capillary (hair-like), found submerged in the water or growing well above the water. The other type is laminate (flat and expanded), and is found floating on the water or submerged. In some *Ranunculus* species three different leaf shapes form on a plant in a sequence. Suppose two taxonomists describe the same crowfoot plant as being two different and distinct species due to these two different leaf shapes. Can you think of difficulties that might arise when other researchers are required to refer to these "two species" in the course of their work?

19

THE ORIGIN AND EVOLUTION OF LIFE

In the Beginning . . .

CONDITIONS ON THE EARLY EARTH
  Origin of the Earth
  The First Atmosphere
  Synthesis of Organic Compounds

EMERGENCE OF THE FIRST LIVING CELLS
  Origin of Agents of Metabolism
  Origin of Self-Replicating Systems
  Origin of the First Plasma Membranes

ORIGIN OF PROKARYOTIC AND EUKARYOTIC CELLS

Focus on Science: **WHERE DID ORGANELLES COME FROM?**

LIFE IN THE PALEOZOIC ERA

LIFE IN THE MESOZOIC ERA
  Speciation on a Grand Scale
  Rise of the Ruling Reptiles

Focus on Science: **HORRENDOUS END TO DOMINANCE**

LIFE IN THE CENOZOIC ERA

SUMMARY OF EARTH AND LIFE HISTORY

Interactive Exercises

In the Beginning . . . [pp.290–291]

19.1. CONDITIONS ON THE EARLY EARTH [pp.292–293]

19.2. EMERGENCE OF THE FIRST LIVING CELLS [pp.294–295]

Selected Words: galaxy [p.290], precursor [p.292], Stanley Miller [p.293], template [p.293], condensation [p.293], Sidney Fox [p.293], "protenoids" [p.293], *metabolism* [p.294], photosynthesis [p.294], porphyrin [p.294], coenzyme [p.294], lipid bilayer [p.295]

Boldfaced, Page-Referenced Terms

[p.290] big bang _____

_____

[p.292] crust (of Earth) _____

_____

[p.292] mantle (of Earth) _____

_____

[p.295] RNA world _____

_____

[p.295] proto-cells _____

_____

Choice

Answer questions 1–20 by choosing from the following probable stages of the physical and chemical evolution of life.

a. Formation of Earth
b. Early Earth and the first atmosphere
c. The synthesis of organic compounds
d. Origin of agents of metabolism
e. Origin of self-replicating systems
f. Origin of the first plasma membranes

1. ___ Most likely, the proto-cells were little more than membrane sacs protecting information-storing templates and various metabolic agents from the environment. [p.295]

2. ___ Simple systems (of this type) of RNA, enzymes, and coenzymes have been created in the laboratory. [p.294]

3. ___ Four billion years ago, Earth was a thin-crusted inferno. [p.292]

4. ___ Sunlight, lightning, or heat escaping from Earth's crust could have supplied the energy to drive their condensation into complex organic molecules. [p.293]

5. ___ During the first 600 million years of Earth's history, enzymes, ATP, and other molecules could have assembled spontaneously at the same locations. [p.294]

6. ___ Were gaseous oxygen (O_2) and water also present? Probably not. [p.292]

7. ___ Sidney Fox heated amino acids under dry conditions to form protein chains, which he placed in hot water. The cooled chains self-assembled into small, stable spheres. The spheres were selectively permeable. [p.295]

8. ___ Stanley Miller mixed hydrogen, methane, ammonia, and water in a reaction chamber that recirculated the mixture and bombarded it with a spark discharge. Within a week, amino acids and other small organic compounds had been formed. [p.293]

9. ___ Without an oxygen-free atmosphere, the organic compounds that started the story of life never would have formed on their own. Free oxygen would have attacked them. [p.292]

10. ___ Much of Earth's inner rocky material melted. [p.292]

11. ___ Imagine an ancient sunlit estuary, rich in clay deposits. Countless aggregations of organic molecules stick to the clay. [p.294]

12. ___ In other experiments, fatty acids and glycerol combined to form long-tail lipids under conditions that simulated evaporating tidepools. The lipids self-assembled into small, water-filled sacs, which in many were like cell membranes. [p.295]

13. ___ We still don't know how DNA entered the picture. [p.295]

14. ___ Rocks collected from Mars, meteorites, and the moon contain chemical precursors that must have been present on early Earth. [pp.292–293]

15. ___ This differentiation resulted in the formation of a crust of basalt, granite, and other types of low-density rock, a rocky region of intermediate density (the mantle), and a high-density, partially molten core of nickel and iron. [p.292]

16. ___ When the crust finally cooled and solidified, water condensed into clouds and the rains began. For millions of years, runoff from rains stripped mineral salts and other compounds from Earth's parched rocks. [p.292]

17. ___ The close association of enzymes, ATP, and other molecules would have promoted chemical interactions. Sunlight energy alone could have driven the spontaneous formation of RNA molecules. [p.294]

18. ___ Even if amino acids did form in the early seas, they wouldn't have lasted long. In water, the favored direction of most spontaneous reactions is toward hydrolysis, not condensation. [p.293]

19. ___ Between 4.6 and 4.5 billion years ago, the outer regions of the cloud cooled. [p.292]

20. ___ At first, it probably consisted of gaseous hydrogen (H_2), nitrogen (N_2), carbon monoxide (CO), and carbon dioxide (CO_2). [p.292]

19.3. ORIGIN OF PROKARYOTIC AND EUKARYOTIC CELLS [pp.296–297]

19.4. *Focus on Science:* WHERE DID ORGANELLES COME FROM? [pp.298–299]

19.5. LIFE IN THE PALEOZOIC ERA [pp.300–301]

Selected Words: "Precambrian" [p.297], fermentation [p.297], *Euglena* [p.298], *Nitrobacter* [p.299], Lynn Margulis [p.299], endocytic vesicles [p.299], "mitochondrial code" [p.299], *Cyanophora paradoxa* [p.299], trilobites [p.300], Gondwana [p.300], nautiloids [p.300], mass extinction [p.300]

Boldfaced, Page-Referenced Terms

[p.296] Archean _____

_____

[p.296] prokaryotic cells _____

_____

[p.296] eubacteria _____

_____

[p.296] archaebacteria _____

_____

[p.296] eukaryotic cells _____

_____

[p.296] stromatolites _____

_____

[p.297] Proterozoic _____

_____

[p.299] endosymbiosis _____

_____

[p.299] protistans _____

_____

[p.300] Paleozoic _____

_____

[p.300] Ediacarans _____

_____

Choice

For questions 1–15, choose from the following.

a. Archean   b. Proterozoic

1. ___ By 2.5 billion years ago, the noncyclic pathway of photosynthesis had evolved in some eubacterial species. [p.297]

2. ___ By 1.2 billion years ago, eukaryotes had originated. [p.297]

3. ___ The first prokaryotic cells emerged. [p.296]

4. ___ Oxygen, one of the by-products of photosynthesis, began to accumulate. [p.297]

5. ___ There was an absence of free oxygen. [p.296]

6. ___ There was a divergence of the original prokaryotic lineage into three evolutionary directions. [p.296]

7. ___ About 800 million years ago, stromatolites began a dramatic decline. [p.297]

8. ___ Between 3.5 and 3.2 billion years ago, the cyclic pathway of photosynthesis evolved in some species of eubacteria. [p.296]

9. ___ Fermentation pathways were the most likely sources of energy. [p.296]

10. ___ An oxygen-rich atmosphere stopped the further chemical origin of living cells. [p.297]

11. ___ Food was available, predators were absent, and biological molecules were free from oxygen attacks. [p.296]

12. ___ Aerobic respiration became the dominant energy-releasing pathway. [p.297]

13. ___ Archaebacteria and eubacteria arose. [p.296]

14. ___ The first cells may have originated in tidal flats or on the seafloor, in muddy sediments warmed by heat from volcanic vents. [p.296]

15. ___ Near the shores of the supercontinent Laurentia, small, soft-bodied animals were leaving tracks and burrows on the seafloor. [p.297]

Fill-in-the-Blanks

The most important feature of eukaryotic cells is the abundance of membrane-bound (16) _____ [p.298] in the cytoplasm. The origin of these structures remains a mystery. Some probably evolved gradually, through (17) _____ [p.298] mutations and natural selection. Some extant prokaryotic cells do possess infoldings of the (18) _____ [p.298] membrane, a site where enzymes and other metabolic agents are embedded. In prokaryotic cells that were ancestral to eukaryotic cells, similar membranous infoldings may have served as a(n) (19) _____ [p.298] from the surface to deep within the cell. These membranes may have evolved into (20) _____ [p.298] channels and into an envelope around the DNA. The advantage of such membranous enclosures may have been to protect (21) _____ [p.298] and their products from foreign invader cells. Evidence for this is that yeasts, simple nucleated eukaryotic cells, have an abundance of (22) _____ [p.298], which they transfer from cell to cell much as prokaryotic cells do.

As the evolution of eukaryotes proceeded, accidental partnerships between different (23) _____ [p.298] species must have formed countless times. Some partnerships resulted in the origin of mitochondria, chloroplasts, and other organelles. This is the theory of (24) _____ [p.298], developed in greatest detail by Lynn Margulis. According to this theory, (25) _____ [p.299] arose after the noncyclic pathway of photosynthesis emerged and oxygen had accumulated to significant levels.

(26) _____ [p.299] transport systems had been expanded in some bacteria to include extra cytochromes to donate electrons to (27) _____ [p.299] in a system of aerobic respiration. By 1.2 billion years ago or earlier, amoebalike ancestors of eukaryotes were engulfing aerobic (28) _____ [p.299] and perhaps forming endocytic vesicles around the food for delivery to the cytoplasm for digestion. Some aerobic bacteria resisted digestion and thrived in a new, protected, nutrient-rich environment. In time they released extra (29) _____ [p.299], which the hosts came to depend on for growth, greater activity, and assembly of other structures. The guest aerobic bacteria came to depend on (30) _____ [p.299] metabolic functions. The aerobic and anaerobic cells were now (31) _____ [p.299] of independent existence. The guests had become (32) _____ [p.299], supreme suppliers of (33) _____ [p.299].

Mitochondria are bacteria-size and replicate their own (34) _____ [p.299], dividing independently of the host cell's division process. In addition, (35) _____ [p.299] may be descended from aerobic eubacteria that engaged in oxygen-producing photosynthesis. Such cells may have been engulfed, resisted digestion, and provided their respiring (36) _____ [p.299] cells with needed oxygen. Chloroplasts resemble some (37) _____ [p.299] in metabolism and overall nucleic acid sequence. Chloroplasts have self-replicating DNA, and they divide independently of the host cell's division.

New cells appeared on the evolutionary stage that were equipped with a nucleus, ER membranes, and mitochondria or chloroplasts (or both); they were the first eukaryotic cells, the first (38) _____ [p.299].

Sequence

Refer to text Figure 19.20 on p.306. Study of the geologic record reveals that as the major events in the evolution of the Earth and its organisms occurred, there were periodic major *extinctions* of organisms followed by major *radiations* of organisms. Using the list below, arrange the letters of the extinctions and radiations in the approximate order in which they occurred, from *youngest* to *oldest*.

39. ___   A. Pangea, worldwide ocean forms; shallow seas squeezed out. Major radiations of reptiles, gymnosperms. [pp.306–307]

40. ___   B. Origin of animals with hard parts. Simple marine communities. [pp.300–301]

41. ___   C. Mass extinction of many marine invertebrates, most fishes. [pp.300–301]

42. ___   D. Gondwana moves south. Major radiations of marine invertebrates, early fishes. [pp.300–301]

43. ___   E. Laurasia forms. Gondwana moves north. Vast swamplands, early vascular plants. Radiation of fishes continues. Origin of amphibians. [pp.300–301]

44. ___   F. Mass extinction. Nearly all species in seas and on land perish. [pp.300–301]

45. ___   G. Tethys sea forms. Recurring glaciations. Major radiations of insects, amphibians. Spore-bearing plants dominant; gymnosperms present; origin of reptiles. [pp.300–301]

Chronology of Events—Geologic Time

Refer to text Figure 19.20 on p.306. From the list of evolutionary events above (A–G), select the letters occurring in a particular era of time by circling the appropriate letters. [pp.300–301]

46. Border of Mesozoic-Paleozoic: A - B - C - D - E - F - G
47. Paleozoic: A - B - C - D - E - F - G

19.6. LIFE IN THE MESOZOIC ERA [pp.302–303]

19.7. *Focus on Science:* **HORRENDOUS END TO DOMINANCE** [p.304]

19.8. LIFE IN THE CENOZOIC ERA [p.305]

19.9. SUMMARY OF EARTH AND LIFE HISTORY [pp.306–307]

Selected Words: *Lystrosaurus* [p.302], plumes [p.302], asteroids [p.302], *Triceratops* [p.302], *Velociraptor* [p.302], superplume [p.302], *Smilodon* [p.305], Laurentia [p.306], Gondwana [p.306], Laurasia [p.306]

Boldfaced, Page-Referenced Terms

[p.302] Mesozoic _____

_____

[p.302] gymnosperms _____

_____

[p.302] angiosperms _____

_____

[p.302] dinosaurs _____

_____

[p.302] asteroids _____

_____

[p.304] K-T asteroid impact theory _____

_____

[p.304] global broiling hypothesis _____

_____

[p.305] Cenozoic _____

_____

Sequence

Earth's history has been divided into four great eras that are based on four abrupt transitions in the fossil record. The oldest era has been subdivided. Arrange the eras in correct chronological sequence from the *oldest* to the *youngest*.

1. ___   A. Mesozoic [p.306]

2. ___   B. Cenozoic [p.306]

3. ___   C. Proterozoic [p.306]

4. ___   D. Archean [p.306]

5. ___   E. Paleozoic [p.306]

Sequence

Refer to text Figure 19.20 on p.306. Study of the geologic record reveals that as the major events in the evolution of Earth and its organisms occurred, there were periodic major *extinctions* of organisms followed by major *radiations* of organisms. Using the list below, arrange the letters of the extinctions and radiations in the approximate order in which they occurred, from *youngest* to *oldest*.

6. ___   A. Pangea breakup begins. Rich marine communities. Major radiations of dinosaurs. [pp.306–307]

7. ___   B. Asteroid impact? Mass extinction of all dinosaurs and many marine organisms. [pp.306–307]

8. ___   C. Recovery, radiations of marine invertebrates, fishes, dinosaurs. Gymnosperms the dominant land plants. Origin of mammals. [pp.306–307]

9. ___   D. Major glaciations. Modern humans emerge and begin what may be the greatest mass extinction of all time on land, starting with Ice Age hunters. [pp.306–307]

10. ___   E. Unprecedented mountain building as continents rupture, drift, collide. Major climatic shifts; vast grasslands emerge. Major radiations of flowering plants, insects, birds, mammals. Origin of earliest human forms. [pp.306–307]

11. ___   F. Pangea breakup continues, broad inland seas form. Major radiations of marine invertebrates, fishes, insects, dinosaurs, origin of angiosperms (flowering plants). [pp.306–307]

12. ___   G. Asteroid impact? Mass extinction of many organisms in seas, some on land; dinosaurs, mammals survive. [pp.306–307]

Chronology of Events—Geologic Time

Refer to text Figure 19.20 on p.306. From the list of evolutionary events above (A–G), select the letters occurring in a particular era of time by circling the appropriate letters. [pp.306–307]

13. Cenozoic: A - B - C - D - E - F - G
14. Border of Cenozoic-Mesozoic: A - B - C - D - E - F - G
15. Mesozoic: A - B - C - D - E - F - G

Complete the Table

16. Refer to text Figure 19.20 on p.306. To review some of the events that occurred in the geologic past, complete the table below by entering the geologic era (Archean, Proterozoic, Paleozoic, Mesozoic, and Cenozoic) and the *approximate* time in millions of years since the time of the events. [pp.306–307]

Era	Time	Events
a.		A few reptile lineages give rise to mammals and the dinosaurs.
b.		Formation of Earth's crust, early atmosphere, oceans; chemical evolution leading to the origin of life.
c.		Origin of amphibians.
d.		Origin of animals with hard parts.
e.		Rocks 3.5 billion years old contain fossils of well-developed prokaryotic cells that probably lived in tidal mud flats.
f.		Flowering plants emerge, gymnosperms begin their decline.
g.		In the Carboniferous, there were major radiations of insects and amphibians; gymnosperms present, origin of reptiles.
h.		Oxygen accumulated in the atmosphere.
i.		Before the close of this era, the first photosynthetic bacteria had evolved.
j.		Insects, amphibians, and early reptiles flourished in the swamp forests of the Permian.
k.		Most of the major animal phyla evolved in rather short order.
l.		Dinosaurs ruled.
m.		Origin of aerobic metabolism; origin of protistans algae, fungi, animals.
n.		Grasslands emerge and serve as new adaptive zones for plant-eating mammals and their predators.
o.		Humans destroy habitats and many species.
p.		The first ice age initiated the first global mass extinction; reef life everywhere collapsed.
q.		The invasion of land begins; small stalked plants establish themselves along muddy margins, and the lobe-finned fishes ancestral to amphibians move onto land.

Self-Quiz

___ 1. Between 4.6 and 4.5 billion years ago, _____. [p.292]
    a. Earth was a thin-crusted inferno
    b. the outer regions of the cloud from which our solar system formed was cooling
    c. life originated
    d. the atmosphere was laden with oxygen

___ 2. More than 3.8 billion years ago, _____. [p.296]
    a. Earth was a thin-crusted inferno
    b. the outer regions of the cloud from which our solar system formed was cooling
    c. life originated
    d. the atmosphere was laden with oxygen

For questions 3–10, choose from these answers:
a. Archean   d. Paleozoic
b. Cenozoic   e. Proterozoic
c. Mesozoic

___ 3. Dinosaurs and gymnosperms were the dominant forms of life during the _____ era. [p.302]

___ 4. The Alps, Andes, Himalayas, and Cascade Range were born during major reorganization of land masses early in the _____ era. [p.305]

___ 5. The composition of Earth's atmosphere changed during the _____ era from one that was anaerobic to one that was aerobic. [p.297]

___ 6. Invertebrates, primitive plants, and primitive vertebrates were the principal groups of organisms on Earth during the _____ era. [pp.300–301]

___ 7. Before the close of the _____ era, the first photosynthetic bacteria had evolved. [p.296]

___ 8. The _____ era ended with the greatest of all extinctions, the Permian extinction. [p.301]

___ 9. Late in the _____ era, flowering plants arose and underwent a major radiation. [p.302]

___10. The _____ era included adaptive zones into which plant-eating mammals and their predators radiated. [p.305]

Chapter Objectives/Review Questions

1. Cloudlike remnants of stars are mostly _____, but they also contain water, iron, silicates, hydrogen cyanide, methane, ammonia, formaldehyde, and many other simple inorganic and organic substances. [p.292]
2. Describe the formation of early Earth prior to the formation of the first atmosphere. [p.292]
3. Be able to list the probable chemical constituents of Earth's first atmosphere. [p.292]
4. _____ and liquid water were probably not present in early Earth's atmosphere. [p.292]
5. Describe experimental evidence provided by Stanley Miller (and others) that the formation of biological molecules from simple precursor molecules might have occurred on early Earth. [p.293]
6. _____ might have been the medium on which condensation reactions yielding complex organic compounds occurred. [p.293]
7. During the first _____ years of Earth's history, enzymes, ATP, and other molecules could have assembled at the same location; their close association would have promoted chemical interactions—and the beginning of _____ pathways. [p.294]
8. Although simple, self-replicating systems of RNA, enzymes, and coenzymes have been created in the laboratory, the chemical ancestors of _____ and DNA remain unknown. [p.295]
9. Describe the experiments performed by Sidney Fox that aided understanding of the origin of plasma membranes. [p.295]
10. The first era of the geologic time scale is now called the _____, "the beginning." [p.296]

11. List the three evolutionary directions of the original prokaryotic lineage during the early Archean era. [p.296]
12. Populations of some early anaerobic eubacteria utilized the cyclic photosynthetic pathway; their populations formed huge mats called _____. [p.296]
13. Describe how the endosymbiosis theory may help to explain the origin of eukaryotic cells. [pp.298–299]
14. Be able to list the five geologic eras in proper order, from oldest to youngest. [pp.306–307]
15. Be able to generally discuss the important geological and biological events occurring throughout the Archean, Proterozoic, Paleozoic, Mesozoic, and Cenozoic eras. [pp.306–307]

Integrating and Applying Key Concepts

As Earth becomes increasingly loaded with carbon dioxide and various industrial waste products, how do you think living forms on Earth will evolve to cope with these changes?

20

BACTERIA, VIRUSES, AND PROTISTANS

Interactive Exercises

The Unseen Multitudes [pp.310–311]

20.1. CHARACTERISTICS OF BACTERIA [pp.312–313]

20.2. BACTERIAL DIVERSITY [pp.314–315]

20.3. BACTERIAL GROWTH AND REPRODUCTION [p.316]

Selected Words: Bacillus [p.310], *Didinium* [p.310], *Paramecium* [p.310], *Escherichia coli* [p.312], *photoautotrophic* [p.312], *chemoautotrophic* [p.312], *photoheterotrophic* [p.312], *chemoheterotrophic* [p.312], *Helicobacter pylori* [p.313], coccus, bacillus, and spirillum [p.313], *Staphylococcus aureus* [p.313], *Gram-positive* [p.313], *Gram-negative* [p.313], *Methanococcus jannaschii* [p.314], *Anabaena* [p.314], *Lactobacillus* [p.314], *Rhizobium* [p.314], *Clostridium botulinum* [p.315], *botulism* [p.315], *C. tetani* [p.315], *tetanus* [p.315], *Borrelia burgdorferi* [p.315], *Lyme disease* [p.315], myxobacteria [p.315]

Boldfaced, Page-Referenced Terms

[p.310] microorganisms _____

[p.311] pathogens _____

_____

[p.312] numerical taxonomy _____

_____

[p.312] eubacteria _____

_____

[p.312] archaebacteria _____

_____

[p.313] prokaryotic cells _____

_____

[p.314] methanogens _____

_____

[p.314] halophiles _____

_____

[p.314] extreme thermophiles _____

_____

[p.314] heterocysts _____

_____

[p.315] endospore _____

_____

[p.316] prokaryotic fission _____

_____

[p.316] plasmid _____

_____

Choice

For questions 1–5, choose from the following:

a. archaebacteria   b. chemoautotrophic eubacteria   c. chemoheterotrophic eubacteria
d. photoautotrophic eubacteria   e. photoheterotrophic eubacteria

1. ___ Use CO_2 from the environment as a source of carbon atoms *and* use electrons, hydrogen, and energy released from reactions between various inorganic substances to assemble chains of carbon (food storage) [pp.312,314]

2. ___ *Use CO_2 and H_2O from the environment* as sources of carbon, hydrogen, and oxygen atoms *and use sunlight* to power the assembly of food storage molecules [pp.312,314]

3. ___ *Cannot* use CO_2 from the environment to construct own carbon chains; instead, obtain nutrients from the products, wastes, or remains of other organisms; can break down glucose to pyruvate and follow it with fermentation of some sort or another [pp.312,314]

4. ___ *Cannot* use CO_2 from the environment to construct own cellular molecules but *can* absorb sunlight and transfer some of that energy to the bonds of ATP; must obtain food molecules (carbon chains) produced by other organisms to construct their own molecules [p.312]

5. ___ Cell walls lack peptidoglycan [p.314]

Fill-in-the-Blanks

Bacteria are microscopic, (6) _____ [p.313] cells having one bacterial chromosome and, often, a number of smaller (7) _____ [p.312]. The cells of nearly all bacterial species have a(n) (8) _____ [p.312] around the plasma membrane, and a(n) (9) _____ [p.313] or slime layer surrounding the cell wall. Typically, the width or length of these cells falls between 1 and 10 (10) [choose one] ❐ millimeters ❐ nanometers ❐ centimeters ❐ micrometers [p.312]. Most bacteria reproduce by (11) _____ _____ [p.312]. Spherical bacteria are (12) _____ [p.313], rod-shaped bacteria are (13) _____ [p.313], and helical bacteria are (14) _____ [p.313]. (15) Gram-_____ [p.313] bacteria retain the purple stain when washed with alcohol.

In many respects, the cell structure, metabolism, and nucleic acid sequences are unique to the (16) _____ [p.314]; some extreme (17) _____ [p.314] live in environments as hot as 110°C. On the other hand, (18) _____ [p.314] are far more common than the three rather unusual types of (1). The most common photoautotrophic bacteria are the (19) _____ [p.314] (also called the blue-green algae). *Anabaena* and others of the (19) group produce oxygen during photosynthesis. Heterocysts are cells in *Anabaena* that carry out (20) _____ - _____ [p.314]. Many species of chemoautotrophic eubacteria affect the global cycling of nitrogen, (21) _____ [p.314] and other nutrients. Sugarcane and corn plants benefit from a(n) (22) _____-fixing [p.314] spirochete, *Azospirillum*. Beans and other legumes benefit from the nitrogen-fixing activities of (23) _____ [p.314], which dwells in their roots.

When environmental conditions become adverse, many bacteria form (24) _____ [p.315], which resist moisture loss, irradiation, disinfectants, and even acids.

Anaerobic bacteria can live in canned food, reproduce, and produce deadly toxins. Two examples of pathogenic (disease-causing) bacteria that form endospores harmful to humans are (25) _____ _____ [p.315] and (26) _____ _____ [p.315]. (27) _____ _____ [p.315] may be the most common tick-borne disease in the United States by now; tick bites deliver the (28) _____ [p.315] from one host to another. Bacterial behavior depends on (29) _____ _____ [p.315], which change shape when they absorb or connect with chemical compounds. Cyanobacteria require (30) _____ [p.315] as their source of energy that drives their metabolic activities. Most of the world's bacteria are (31) [choose one] ❐ producers ❐ consumers ❐ decomposers, so we think of them as "good" heterotrophs. *Escherichia coli*, which dwell in our gut, synthesize vitamin (32) _____ [p.314] and substances useful in digesting fats.

Matching

Match each of the items below with a lowercase letter designating its principal bacterial group and an uppercase letter denoting its best descriptor from the right-hand column.

a. Archaebacteria   b. Chemoautotrophic eubacteria   c. Chemoheterotrophic eubacteria
d. Photoautotrophic eubacteria

33. ___, ___ *Anabaena*, "blue green algae" [p.314]

34. ___, ___ *Bacillus, Clostridium* [p.315]

35. ___, ___ *Borrelia burgdorferi* [p.315]

36. ___, ___ extreme thermophiles [p.314]

37. ___, ___ *Escherichia coli* [p.314]

38. ___, ___ halophiles [p.314]

39. ___, ___ *Lactobacillus* [p.314]

40. ___, ___ *Methanococcus jannaschii* [p.314]

41. ___, ___ *Staphylococcus, Streptococcus* [p.313]

A. Lives in sewage, stockyards, oxygen-free habitats, and in the animal gut; chemosynthetic; affects the global cycling of carbon
B. Spirochete that causes Lyme disease
C. Endospore-forming rods and cocci that live in the soil and in the animal gut; some major pathogens included
D. Gram-positive cocci that live on the skin and mucous membranes of animals; some major pathogens
E. Gram-positive nonsporulating rods that ferment plant and animal material; some are important in dairy industry; others contaminate milk, cheese
F. In acidic soil, hot springs, hydrothermal vents on the seafloor; may use sulfur as a source of electrons for the ATP formation
G. Live in extremely salty water; have a unique form of photosynthesis
H. Mostly in lakes and ponds; cyanobacteria; produce O_2 from water as an electron donor
I. Gram-negative anaerobic rod that inhabits the human colon, where it produces vitamin K

20.4. CHARACTERISTICS OF VIRUSES [p.317]

20.5. VIRAL MULTIPLICATION CYCLES [pp.318–319]

20.6. *Focus on Health:* INFECTIOUS PARTICLES TINIER THAN VIRUSES [p.319]

Selected Words: *helical, polyhedral, enveloped,* and *complex* viruses [p.317], influenza [p.317], *lytic* [p.318], *lysogenic* [p.318], Type I *Herpes simplex* [p.318], *cold sores* [p.318], HIV [p.319], retrovirus [p.319], *kuru* [p.319], *Creutzfeldt-Jakob* disease [p.319], BSE (bovine spongiform encephalopathy) or *mad cow disease* [p.319]

Boldfaced, Page-Referenced Terms

[p.317] virus _____

_____

[p.317] bacteriophages _____

_____

[p.318] lysis _____

_____

[p.319] viroids _____

_____

[p.319] prions _____

_____

Short Answer

1. a. State the principal characteristics of viruses. [p.317] _____

_____

    b. Describe the structure of viruses. [p.317] _____

_____

    c. Distinguish between the ways viruses replicate themselves. [pp.317–318] _____

_____

2. a. List two specific viruses that cause human illness. [pp.317–318] _____

_____

    b. Describe how each virus in (a) does its dirty work. [pp.317–319] _____

_____

Fill-in-the-Blanks

A(n) (3) _____ [p.317] is a noncellular, nonliving infectious agent, each of which consists of a central

(4) _____ _____ [p.317] core surrounded by a protective (5) _____ _____ [p.317].

(6) _____ [p.317] contain the blueprints for making more of themselves but cannot carry on metabolic

activities. (7) _____ [p.318] refers to cytoplasm leakage and cell death that occurs when either the cell

wall or plasma membrane, or both, are damaged. Naked strands or circles of RNA that lack a protein coat

are called (8) _____ [p.319]. (9) _____ [p.319] are RNA viruses that infect animal cells, cause AIDS,

and follow (10) _____ [pp.318–319] pathways of replication. (11) _____ [p.317] are the usual units

of measurement with which to measure viruses, while microbiologists measure bacteria and protistans in

terms of (12) _____ [recall p.312]. A bacterium 86 micrometers in length is (13) _____ [recall metric

system knowledge, p.310] nanometers long. During a period of (14) _____ [p.318], viral genes remain

inactive inside the host cell and any of its descendants. Pathogenic protein particles are called

(15) _____ [p.319].

Identification

Identify the virus that causes the following illnesses by writing the name of the agent in the blank preceding
the disease. Then, tell whether it is a DNA virus, an RNA virus, a prion, or a viroid.

16. _____ _____ BSE, mad cow disease [p.319]
17. _____ _____ AIDS [p.319]
18. _____ _____ Cold sores [p.318]

Choice

For questions 19–23, choose from the following:

a. bacteriophage   b. lysogenic pathway   c. lytic pathway   d. pathogen   e. virus

19. ___ A virus that infects a bacterium [p.317]

20. ___ Damage and destruction to host cells occurs quickly [p.318]

21. ___ Noncellular infectious agent that must take over a living cell in order to reproduce itself [p.317]

22. ___ Viral nucleic acid is integrated into the nucleic acid system of the host cell and replicated during this time [p.318]

23. ___ Any disease-causing organism or agent [recall p.311]

20.7. KINGDOM AT THE CROSSROADS [p.320]

20.8. PARASITIC OR PREDATORY MOLDS [pp.320–321]

20.9. THE ANIMAL-LIKE PROTISTANS [pp.322–323]

20.10. THE NOTORIOUS SPOROZOANS [p.324]

20.11. *Focus on Science:* THE NATURE OF INFECTIOUS DISEASES [p.325]

Selected Words: "saprobes" [p.320], *Saprolegnia* [p.320], *downy mildew* [p.320], *late blight* [p.320], cellular slime molds [p.320], plasmodial slime molds [p.321], *Dictyostelium discoideum* [p.321], *Entamoeba histolytica* [p.322], *amoebic dysentery* [p.322], pseudopods [p.322], *Paramecium* [p.322], *micro*nucleus [p.323], *macro*nucleus [p.323], *Giardia lamblia* [p.323], *Trichomonas vaginalis* [p.323], *African sleeping sickness* [p.323], *Chagas disease* [p.323], *Cryptosporidium* [p.324], *Plasmodium* [p.324], *malaria* [p.324], *contagious, sporadic,* and *endemic* diseases [p.325], *Ebola* [p.325]

Boldfaced, Page-Referenced Terms

[p.320] protistans _____

_____

[p.320] chytrids _____

_____

[p.320] water molds _____

_____

[p.320] slime molds _____

_____

[p.322] protozoans _____

_____

[p.322] binary fission _____

_____

[p.322] cyst _____

_____

[p.322] plankton _____

_____

[p.323] contractile vacuoles _____

_____

[p.324] sporozoan _____

_____

[p.325] infection _____

_____

[p.325] disease _____

_____

[p.325] epidemic _____

_____

Choice

1. For each structure, indicate with a P if it is a prokaryotic (bacterial) characteristic, and indicate with an E if it is a eukaryotic characteristic.

[p.320] a. double-membraned nucleus	
[p.320] b. mitochondria present	
[p.322] c. reproduce by binary fission	
[p.320] d. engage in mitosis	
[p.312] e. circular chromosome present	
[p.320] f. endoplasmic reticulum present	
[p.320] g. cilia or flagella with 9+2 core	

Fill-in-the-Blanks

(2) _____ [p.320] and (3) _____ _____ [p.320] and (4) _____ _____ [p.320] are heterotrophic and produce spores; many are also saprobic or parasitic, but not photosynthetic; in some respects they resemble fungi. Water molds are only distantly related to other fungi and are major (5) _____ [p.320] of aquatic habitats. The cells of some (6) _____ _____ [p.321] differentiate and form (7) _____ _____ [p.321], stalked structures that produce gametes or (8) _____ [p.321]. The cytoplasm of a (9) _____ [p.321] slime mold flows throughout the entire mass, distributing nutrients and oxygen within the organism which may grow to become several square (10) _____ (units) [p.321] in size .

Several parasitic water molds cause difficulties for their hosts or for humans. *Phytophthora infestans*, known as (11) _____ _____ [p.320], rots potato and tomato plants. *Saprolegnia* destroys tissues of aquarium (12) _____ [p.320]. *Plasmopara viticola* causes (13) _____ _____ [p.320] in grapes, threatening large vineyards in France and North America. When marsh grasses die, break off, and travel out to sea on the tides, bacteria and (14) _____ [p.320] decompose the dead grasses by secreting (15) _____ [p.320] that digest the organic matter to a form that other detritis feeders can use.

Amoebas move by sending out (16) _____ [p.322], which surround food and engulf it. (17) _____ [p.322] secrete a hard exterior covering of calcareous material that is peppered with tiny holes through which sticky, food-trapping pseudopods extend. Accumulated shells of (18) _____ [p.322], which generally have a skeleton of silica (glass), and foraminiferans are key components of many oceanic sediments. Almost all (18) drift with the ocean currents, as part of (19) _____ [p.322].

Paramecium is a ciliate that lives in (20) _____ [p.322] environments and depends on (21) _____ _____ [p.323] for eliminating the excess water constantly flowing into the cell. *Paramecium* has a (22) _____ [p.322], a cavity that opens to the external watery world. Once inside the cavity, food particles become enclosed in (23) _____-_____ _____ [p.322], where digestion takes place.

Examples of flagellated protozoans that are parasitic include the (24) _____ [p.323], two species of which cause African sleeping sickness and Chagas disease.

(25) _____ [p.324] is a famous sporozoan that causes malaria. When a particular (26) _____ [p.324] draws blood from an infected individual, (27) _____ [p.324] of the parasite fuse to form zygotes, which eventually develop within the mosquito.

Matching

Choose as many appropriate answers as are applicable for each term.

28. ___ *Entamoeba histolytica* [p.322]
29. ___ foraminiferans [p.322]
30. ___ *Giardia lamblia* [p.323]
31. ___ *Paramecium* [pp.322–323]
32. ___ *Plasmodium* [p.324]
33. ___ *Trichomonas vaginalis* [p.323]
34. ___ *Trypanosoma brucei* [p.323]

A. Ciliophora
B. Mastigophora
C. Sarcodina
D. Sporozoa
E. Amoeboid protozoans
F. Animal-like flagellates
G. African sleeping sickness
H. Malaria
I. Amoebic dysentery
J. Red tide
K. Primary component of many ocean sediments

Matching

Match the pictures with the names below.

35. ____
36. ____
37. ____
38. ____
39. ____

A. *Didinium*, a ciliate [recall p.311]
B. *cellular slime mold* [p.321]
C. Flagellated protozoans [p.323]
D. Radiolarian [p.322]
E. *Paramecium* [p.323]

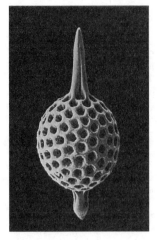

35. _____

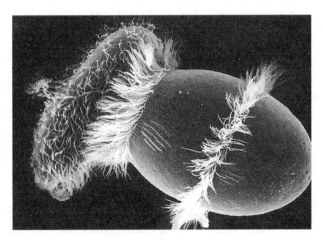

36. _____

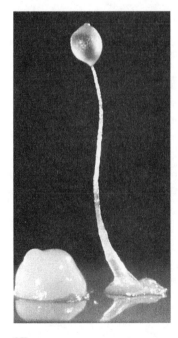

37. _____

38. _____

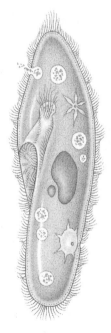

39. _____

Matching

Choose the most appropriate answer for each term.

40. ___antibiotic [p.325]
41. ___endemic [p.325]
42. ___epidemic [p.325]
43. ___microorganism [pp.310,325]
44. ___sporadic [p.325]

A. Any organism too small to be seen without a microscope
B. Disease that breaks out irregularly, affects few organisms
C. Chemical substance that interferes with gene expression or other normal functions of bacteria
D. Disease abruptly spreads through large portions of a population
E. Disease that occurs continuously, but is localized to a relatively small portion of the population

20.12. A SAMPLING OF THE (MOSTLY) SINGLE-CELLED ALGAE [pp.326–327]

20.13. RED ALGAE [p.327]

20.14. BROWN ALGAE [p.328]

20.15. GREEN ALGAE [pp.328–329]

*Selected Words: phyto*plankton [p.326], *Euglena* [p.326], "eyespot" [p.326], chlorophylls a, b, c_1, c_2 [p.326], carotenoids [p.326], fucoxanthin [p.326], diatoms [p.326], golden algae [p.326], phycobilins [p.327], *Pfiesteria* [p.327], *Gymnodinium* [p.327], *Eucheuma* [p.327], *Porphyra* [p.327]

Boldfaced, Page-Referenced Terms

[p.326] euglenoids _____

_____

[p.326] chrysophytes _____

_____

[p.327] red tides _____

_____

[p.327] red algae _____

_____

[p.328] brown algae _____

_____

[p.328] green algae _____

_____

Fill-in-the-Blanks

Euglenoids contain (1) _____ [p.326], which enable them to carry out photosyntheses. A(n) (2) _____ [p.326] of carotenoid pigment granules partly shields a light-sensitive receptor and enables *Euglena* to remain where light is optimal for its activities. Some strains of *Euglena* can be converted from photosynthetic, chloroplast-containing forms to strains that are (3) _____ [p.326, Table 23.2].

The term (4) "_____" [p.326] no longer has formal classification significance, because organisms once lumped under that term are now assigned to different kingdoms. (5) _____ [p.326] include 600 species of "yellow-green algae," about 500 species of "golden algae," and more than 5,600 existing species of golden-brown (6) _____ [p.326]. Photosynthetic chrysophytes contain xanthophylls and (7) _____ [p.326]; those pigments mask the green color of chlorophyll in golden algae and diatoms. Diatom cells have external thin, overlapping "shells" of (8) _____ [p.326] that fit together like a pill box. 270,000 metric tons of (9) _____ _____ [p.326] are extracted annually from a quarry near Lompoc, California, and are used to make abrasives, (10) _____ [p.326] and insulating materials. Dinoflagellates are photosynthetic members of marine (11) _____ [p.327] ecosystems; some forms, for example, stages from the *Pfiesteria* life cycle, are also heterotrophic. Some (12) _____ [p.327] undergo explosive population growth and color the seas red or brown, causing a red tide that may kill hundreds or thousands of fish and, occasionally, people.

Several species of red algae secrete (13) _____ [p.327] (used in culture media) as part of their cell walls. Most red algae live in (14) _____ [p.327] habitats where they are major producers. Some red algae are eaten as (15) _____ [p.327], participate in coral reef building, and are stabilizers in paints, dairy products, and other emulsions. The (16) _____ [p.328] algae live offshore or in intertidal zones and have many representatives with large sporophytes known as kelps; some species produce (17) _____ [p.328], valuable thickening agents. Green algae are thought to be ancestral to more complex plants, because they have the same types and proportions of (18) _____ [p.328] pigments, have (19) _____ [p.328] in their cell walls, and store their carbohydrates as (20) _____ [p.326].

Complete the Table

21. Complete the table below.

Type of Alga	Typical Pigments	Division (Phylum)	Uses by Humans	Representatives
[p.327] Red algae	a.	b.	c.	d.
[p.328] Brown algae	e.	f.	g.	h.
[pp.328–329] Green algae	i.	j.	k.	l.

Label–Match

Identify each indicated part of the illustration below by entering its name in the appropriate numbered blank. Choose from the following terms: cytoplasmic fusion, asexual reproduction, resistant zygote, fertilization, zygote, meiosis and germination, spore mitosis, gametes meet. Complete the exercise by matching from the list below, entering the correct letter in the parentheses following each label.

22. _____ ( ) [p.329]

23. _____ _____ ( ) [p.329]

24. _____ and _____ ( ) [p.329]

25. _____ _____ ( ) [p.329]

26. _____ _____ ( ) [p.329]

27. _____ _____ ( ) [p.329]

28. _____ _____ ( ) [p.329]

29. _____ ( ) [p.329]

A. Fusion of two gametes of different mating types
B. A device to survive unfavorable environmental conditions
C. More spore copies are produced
D. Fusion of two haploid nuclei
E. Haploid cells form smaller haploid gametes when nitrogen levels are low
F. Formed after fertilization
G. Two haploid gametes coming together
H. Reduction of the chromosome number

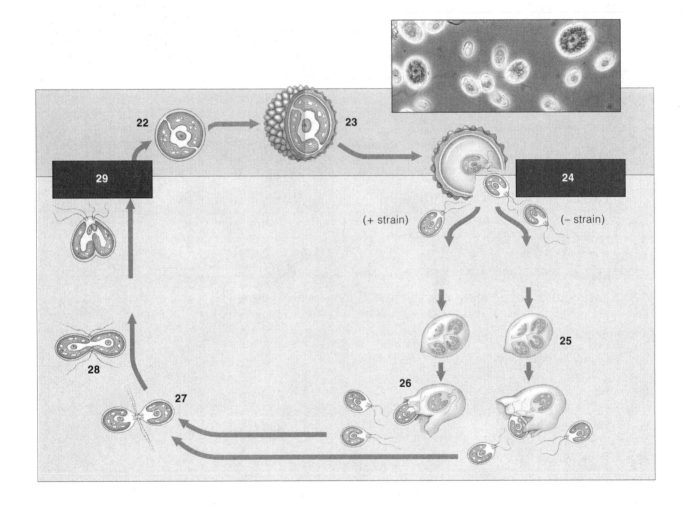

30. For each group below, indicate with a "+" if it has chloroplasts, and a "−" if members of that group lack chloroplasts and the ability to do photosynthesis.

a. Brown algae [p.328]	f. Protozoans [p.322]
b. Chytrids [pp.320–321]	g. Red algae [p.327]
c. Chrysophytes [p.326]	h. Slime molds [pp.320–321]
d. Dinoflagellates [p.327]	i. Sporozoans [p.324]
e. Green algae [p.328]	j. Water molds [pp.320–321]

Self-Quiz

___ 1. Bacteriophages are _____ . [p.317]
   a. viruses that parasitize bacteria
   b. bacteria that parasitize viruses
   c. bacteria that phagocytize viruses
   d. composed of a protein core surrounded by a nucleic acid coat

___ 2. Many biologists believe that chloroplasts are descendants of _____ that were able to live symbiotically within a predatory host cell. [p.320, recall section 19.4]
   a. aerobic bacteria with "extra" cytochromes
   b. endosymbiont prokaryotic autotrophs
   c. dinoflagellates
   d. bacterial heterotrophs

___ 3. Which of the following specialized structures is not correctly paired with a function? [p.323]
   a. gullet—ingestion
   b. cilia—food gathering
   c. contractile vacuole—digestion
   d. anal pore—waste elimination

___ 4. Population "blooms" of _____ cause "red tides" and extensive fish kills. [p.327]
   a. *Euglena*
   b. specific dinoflagellates
   c. diatoms
   d. *Plasmodium*

___ 5. Which of the following protists does *not* cause great misery to humans? [pp.321, 322–324]
   a. *Dictyostelium discoideum*
   b. *Entamoeba histolytica*
   c. *Plasmodium*
   d. *Trypanosoma brucei*

___ 6. Incurable AIDS, caused by the emerging pathogen HIV, rapidly became classified as a(n) _____ disease. [p.325]
   a. endemic
   b. epidemic
   c. pandemic
   d. sporadic

___ 7. Clouds of bacteria that form the basis of the hydrothermal vent food webs are included in the _____ category. [p.314]
   a. archaebacteria
   b. chemoautotrophic eubacteria
   c. chemoheterotrophic eubacteria
   d. photoautotrophic eubacteria
   e. photoheterotrophic eubacteria

___ 8. Red algae _____ . [p.327]
   a. are primarily marine organisms
   b. are thought to have developed from green algae
   c. contain xanthophyll as their main accessory pigments
   d. all of the above

___ 9. Stemlike structure, leaflike blades, and gas-filled floats are found in the species of _____. [p.328]
   a. red algae
   b. brown algae
   c. bryophytes
   d. green algae

___ 10. Because of pigmentation, cellulose walls, and starch storage similarities, the _____ algae are thought to be ancestral to more complex plants. [p.328]
   a. red
   b. brown
   c. blue-green
   d. green

Matching

Match all applicable letters with the appropriate terms. A letter may be used more than once, and a blank may contain more than one letter.

11. ___*Amoeba proteus* [p.322]

12. ___diatoms [p.326]

13. ___*Dictyostelium* [pp.320–321]

14. ___foraminifera [p.322]

15. ___*Gymnodinium breve* (red tide) [p.327]

16. ___*Paramecium* [p.322]

17. ___*Plasmodium* [p.324]

18. ___*Volvox* [p.328]

A. Protista
B. Slime mold
C. Green algae (Chlorophyta)
D. Dinoflagellates
E. Obtain food by using pseudopodia
F. Causes malaria
G. A sporozoan
H. A ciliate
I. Live in "glass" houses
J. Live in hardened shells that have thousands of tiny holes, through which pseudopodia protrude

Matching

Match the pictures with the names below.

19. ___

20. ___

21. ___

22. ___

23. ___

24. ___

25. ___

26. ___

A. *Amoeba proteus* [p.322]
B. *Bacillus* [p.310]
C. bacteriophage [p.317]
D. Cyanobacterium (Anabaena) [p.315]
E. diatom [p.326]
F. HIV [p.317]
G. *Methanococcus jannaschii* [p.314]
H. *Paramecium* [p.323]

19.

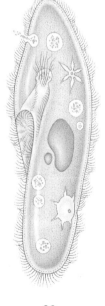

20.

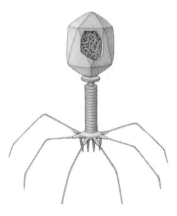

21.

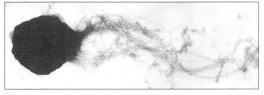

22.

23.

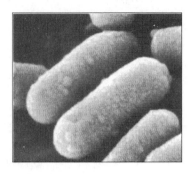

24.

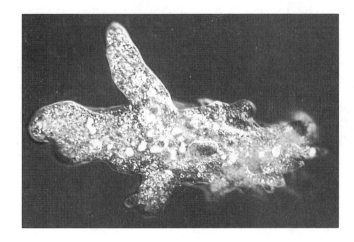

25.

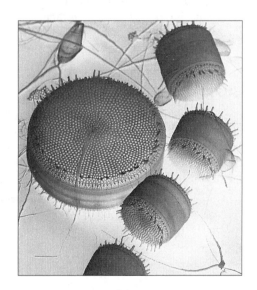

26.

Matching

Match all applicable letters with the appropriate terms. A letter may be used more than once, and a blank may contain more than one letter.

27. ___*Anabaena* [pp.324–325]

28. ___*Clostridium botulinum* [p.315]

29. ___*Escherichia coli* [p.313]

30. ___*Herpes simplex* [p.318]

31. ___HIV [p.319]

32. ___*Lactobacillus* [p.314]

33. ___*Methanococcus* [p.314]

34. ___*Staphylococcus* [p.313]

A. Eubacteria
B. Virus
C. Cyanobacteria
D. Archaebacteria
E. Cause cold sores and a type of venereal disease
F. Associated with AIDS, ARC

Chapter Objectives/Review Questions

1. Describe the principal body forms of monerans (inside and outside). [pp.312–313]
2. Distinguish chemoautotrophs from photoautotrophs. [pp.312,314]
3. Explain how, with no nucleus, or few if any, membrane-bound organelles, bacteria reproduce themselves and obtain energy to carry on metabolism. [pp.312,314,316]
4. State the ways in which archaebacteria differ from eubacteria. [p.313]
5. Name two different diseases caused by viruses. [p.317]
6. Describe the general structure of viruses and tell how they are classified into two main groups. [pp.317–318]
7. Describe and distinguish between the two pathways used by viruses to replicate themselves. [pp.318–319]
8. _____ protistans generally make their own food by photosynthesis, but _____ protistans act as decomposers, predators, or parasites in order to obtain the energy that fuels life. [p.320]
9. State the principal characteristics of the amoebas, radiolarians, and foraminiferans. Indicate how they generally move from one place to another and how they obtain food. [p.322]
10. List the features common to most ciliated protozoans. [pp.322–323]
11. Two flagellated protozoans that cause human misery are _____ and _____. [p.323]
12. Characterize the sporozoan group, identify the group's most prominent representative, and describe the life cycle of that organism. [p.324]
13. How do golden algae resemble diatoms? [p.326]
14. Explain what causes red tides. [p.327]
15. State the outstanding characteristics of organisms of the red, brown, and green algae divisions. [pp.327–328]

Integrating and Applying Key Concepts

The textbook identifies natural gas as a nonrenewable fuel resource, yet there is a group of archaebacteria that produce methane, the burning of which can serve as a fuel for heating, cooking, or both. Recall or imagine how these bacteria could be incorporated into a system that could serve human societies by generating methane in a cycle that is renewable. Why did your text categorize natural gas as a nonrenewable resource? Is methane a constituent of natural gas? Why or why not?

21

FUNGI

Interactive Exercises

Ode to the Fungus Among Us [pp.332–333]

21.1. CHARACTERISTICS OF FUNGI [p.334]

Selected Words: Lobaria [p.332], "fungus-root" [p.332], "imperfect fungi" [p.334]

Boldfaced, Page-Referenced Terms

[p.332] symbiosis _____

_____

[p.332] mutualism _____

_____

[p.332] lichen _____

_____

[p.332] mycorrhiza _____

_____

[p.332] decomposers _____

_____

[p.332] extracellular digestion and absorption _____

_____

[p.334] fungi _____

_____

[p.334] saprobes _____

_____

[p.334] parasites _____

_____

[p.334] zygomycetes _____

_____

[p.334] sac fungi _____

_____

[p.334] club fungi _____

_____

[p.334] spores _____

_____

[p.334] mycelium (plural, mycelia) _____

_____

[p.334] hypha (plural, hyphae) _____

_____

Fill-in-the-Blanks

(1) _____ [p.332] refers to species that live closely together. In cases of (1) called (2) _____ [p.332], their interaction benefits both partners or does one of them no harm. A(n) (3) _____ [p.332] is a vegetative body in which a fungus has become intertwined with one or more photosynthetic organisms. Fungi that enter into mutualistic interactions with young tree roots form a(n) (4) _____ [p.332], which means "fungus root." Plants benefit from fungi as they are premier (5) _____ [p.332] that break down organic compounds in their surroundings.

Matching

Choose the most appropriate answer for each term.

6. ___saprobes [p.334]

7. ___parasites [p.334]

8. ___mycelium [p.334]

9. ___fungi [p.334]

10. ___hypha [p.334]

11. ___zygomycetes, sac fungi, club fungi [p.334]

12. ___spores [p.334]

13. ___extracellular digestion and absorption [p.332]

A. Nonmotile reproductive cells or multicelled structures; often walled and germinate following dispersal from the parent body
B. Represent major lineages of fungal evolution
C. A mesh of branching fungal filaments that grows over and into organic matter, secretes digestive enzymes, and functions in food absorption
D. Fungi that obtain nutrients from nonliving organic matter and so cause its decay
E. Fungi that extract nutrients from tissues of a living host
F. External secretion of enzymes prior to nutrient intake
G. Each filament in a mycelium; consists of cells of interconnecting cytoplasm and chitin-reinforced walls
H. A richly diverse group of heterotrophs that are premier decomposers

21.2. CONSIDER THE CLUB FUNGI [pp.334-335]

Selected Words: *Agaricus brunnescens* [p.334], *Armillaria bulbosa* [p.334], *Ramaria* [p.334], *Polyporus* [p.334], *Amanita muscaria* [p.335], *dikaryotic* mycelium [p.335]

Boldfaced, Page-Referenced Terms

[p.335] mushrooms _____

_____

[p.335] basidiospores _____

_____

Matching

Choose the most appropriate answer for each term.

1. ___rust and smut fungi [p.334]

2. ___*Agaricus brunnescens* [p.334]

3. ___*Armillaria bulbosa* [p.334]

4. ___*Amanita muscaria* [p.335]

5. ___*Amanita phalloides* [p.335]

A. Fly agaric mushroom, causes hallucinations when eaten; ritualistic use
B. Among the oldest and largest of the club fungi
C. Destroys entire fields of wheat, corn, and other major crops
D. Common cultivated mushroom; multimillion dollar business
E. Death cap mushroom; kills humans by liver and kidney degeneration

Fill-in-the-Blanks

The numbered items in the illustration of a club fungus life cycle below represent missing information; complete the blanks in the following narrative to supply that missing information.

The mature mushroom is actually a short-lived (6) _____ [p.335] body; its living (7) _____ [p.335]
is buried in soil or decaying wood. Each mushroom consists of a cap and a stalk. Spore-producing
(8) _____ [p.335] -shaped structures develop on the gills. Each bears two haploid (*n* + *n*) nuclei.
(9) _____ [p.335] fusion occurs within the club-shaped structures, which yields a(n) (10) _____
[p.335] stage. (11) _____ [p.335] occurs within the club-shaped structures and four haploid
(12) _____ [p.335] emerge at the tip of each. The spores are released and each may germinate into a
haploid (*n*) mycelium. When hyphae of two compatible mating strains meet, (13) _____ [p.335] fusion
occurs. Following this, a(n) (14) "_____" [p.335] (*n* + *n*) mycelium forms and gives rise to the spore-
bearing mushrooms.

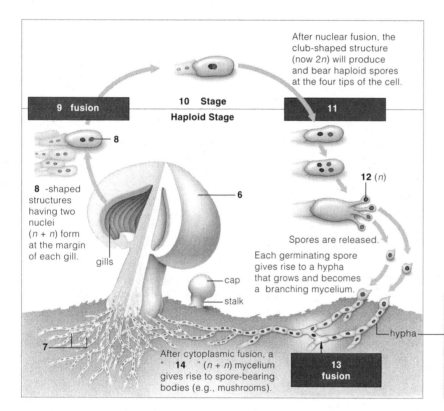

After nuclear fusion, the club-shaped structure (now 2*n*) will produce and bear haploid spores at the four tips of the cell.

9 fusion

10 Stage
Haploid Stage

11

8

8 -shaped structures having two nuclei (*n* + *n*) form at the margin of each gill.

gills

6

cap

stalk

12 (*n*)

Spores are released.
Each germinating spore gives rise to a hypha that grows and becomes a branching mycelium.

hypha

7

After cytoplasmic fusion, a " 14 " (*n* + *n*) mycelium gives rise to spore-bearing bodies (e.g., mushrooms).

13 fusion

hypha in mycelium

21.3. SPORES AND MORE SPORES [pp.336–337]

21.4. *Focus on Science:* A LOOK AT THE UNLOVED FEW [p.337]

Selected Words: *Rhizopus stolonifer* [p.336], zygosporangium [p.336], asci [p.336], *Candida albicans* [p.337], *Penicillium* [p.337], *Aspergillus* [p.337], *Neurospora crassa* [p.337], *Cryphonectria parasitica* [p.337], *Ajellomyces capsulatus* [p.337], histoplasmosis [p.337], *Claviceps purpurea* [p.337], ergotism [p.337], *Epidermophyton floccosum* [p.337], *Venturia inequalis* [p.337]

Boldfaced, Page-Referenced Terms

[p.336] ascospores _____

Fill-in-the-Blanks

The numbered items in the illustration below (*Rhizopus* life cycle) represent missing information; complete the blanks in the following narrative to supply missing information about zygomycetes.

The sexual phase begins when haploid (1) _____ [p.336] of two different mating strains (+ and –) grow into each other and fuse due to a chemical attraction. Two (2) _____ [p.336] form between two hyphae, and several haploid nuclei are produced inside each. Later, their nuclei fuse, forming a zygote with a thick protective wall called a(n) (3) _____ [p.336]. This structure may remain dormant for several months. Meiosis proceeds and haploid sexual (4) _____ [p.336] are produced when this structure germinates. Each gives rise to stalked structures, each with a spore sac on its tip, that can produce many spores, each of which can be the start of an extensive (5) _____ [p.336].

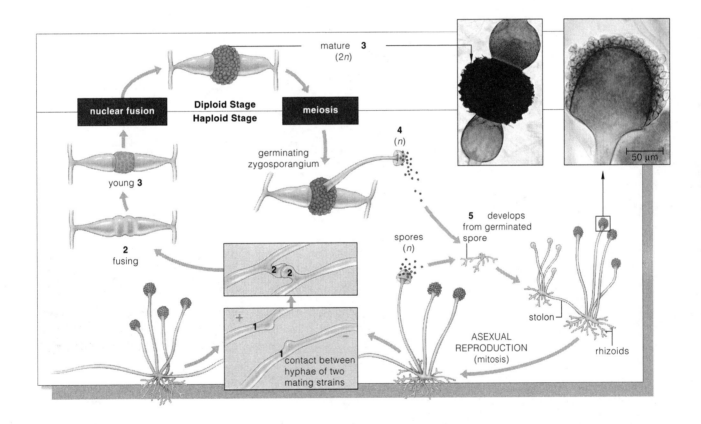

Matching

Choose the most appropriate answer for each term.

6. ___truffles and morels [p.336]
7. ___yeasts [p.337]
8. ___*Candida albicans* [p.337]
9. ___*Penicillium* spp. [p.337]
10. ___*Aspergillis* spp. [p.337]
11. ___*Neurospora crassa* [p.337]

A. "Flavor" Camembert and Roquefort cheeses; produce antibiotics
B. Has uses in genetic research
C. Fermenting by-products put to use by bakers and vintners
D. Causes vexing infections in humans
E. Highly prized edibles
F. Make citric acid for candies and soft drinks; ferment soybeans for soy sauce

Complete the Table

12. After reading *Focus on Science*, "A Look at the Unloved Few" [text p.337], complete the following table, which deals with five pathogenic and toxic fungi.

Fungus Latin Name	Description
[p.337] a.	Turned most of eastern North America's chestnut trees into stubby versions of their former selves
[p.337] b.	Histoplasmosis
[p.337] c.	Athlete's foot
[p.337] d.	Apple scab
[p.337] e.	Ergotism

21.5. THE SYMBIONTS REVISITED [p.338]

Selected Words: lichen [p.338], *myco*biont [p.338], *photo*biont [p.338], mycorrhizae [p.338], *ecto*mycorrhiza [p.338], *endo*mycorrhizae [p.338]

Fill-in-the-Blanks

(1) _____ [p.338] refers to species that live in close ecological association. In cases of symbiosis called

(2) _____ [p.338], interaction benefits both partners or does one of them no harm. A(n) (3) _____

[p.338] is commonly called a mutualistic interaction between a fungus and one or more photosynthetic

species. The fungal part of a lichen is known as the (4) _____ [p.338] and the photosynthetic

component is the (5) _____ [p.338]. Of about 13,500 known types of lichens, nearly half incorporate

(6) _____ [p.338] fungi. Lichens can colonize places that are too (7) _____ [p.338] for most

organisms. Almost always, the (8) _____ [p.338] is the largest component of the lichen. The fungus

benefits by having a long-term source of nutrients that it absorbs from cells of the (9) _____ [p.338].

Nutrient withdrawals affect the (10) _____ [p.338] growth a bit, but the lichen may help (11) _____

[p.338] it. If more than one fungus is present in the lichen, it might be a mycobiont, a (12) _____ [p.338],

or even an opportunist that is using the lichen as a substrate. A lichen forms after the tip of a fungal

(13) _____ [p.338] binds with a suitable host cell. Both lose their wall and their (14) _____ [p.338]

fuses or the hypha induces the host cell to cup around it. The (15) _____ [p.338] and the (16) _____

[p.338] grow and multiply together. The lichen commonly has distinct (17) _____ [p.338]. The overall

pattern of growth may be leaflike, flattened, pendulous, or erect.

Fungi also are mutualists with young tree roots, as (18) _____ [p.338]. Many plants do not grow as

(19) _____ [p.338] without this association. In a(n) (20) _____ [p.338], hyphae form a dense net

around living cells in the roots but do not penetrate them. Ectomycorrhizae are common in (21) _____

[p.338] forests and help trees withstand seasonal changes in temperature and rainfall. About 5,000 fungal

species, mostly (22) _____ [p.338] fungi, enter into such associations. The more common (23)

_____ [p.338] form in about 80 percent of all vascular plants. These fungal hyphae (24) _____

[p.338] plant cells, as they do in lichens. Fewer than 200 species of (25) _____ [p.338] serve as the fungal

partner. Their hyphae branch extensively, forming tree-shaped (26) _____ [p.338] structures in plant

cells. Hyphae also extend several centimeters into the (27) _____ [p.338].

Self-Quiz

Label–Match

In the blank beneath each illustration below (1–8), identify the organism by common name (or scientific name if a common name is unavailable). Then match each organism with the appropriate item (may be used more than once) from the list below the illustrations by entering the letter in the parentheses.

1. _____ ( )
   [p.334]

2. _____ ( )
   [p.339]

3. _____ ( )
   [p.335]

4. _____ ( )
   [p.337]

A. Sac fungi
B. Zygomycetes
C. Club fungi
D. Imperfect fungi
E. Algae and fungi

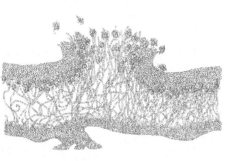

5. _____ (   )
   [p.338]

6. _____ (   )
   [p.338]

7. _____ (   )
   [p.333]

8. _____ (   )
   [p.337]

___ 9. Most true fungi send out cellular filaments
   called _____. [p.334]
   a. mycelia
   b. hyphae
   c. mycorrhizae
   d. asci

___10. Heterotrophic species of fungi can be
   _____. [pp.334,338]
   a. saprobic
   b. parasitic
   c. mutualistic
   d. all of the above

For questions 11–20, choose from the following:
a. club fungi   c. sac fungi
b. imperfect fungi   d. zygomycetes

___11. The group that includes *Rhizopus stolonifer*,
   the notorious black bread mold is the
   _____. [p.336]

___12. The group that includes delectable morels
   and truffles but also includes bakers' and
   brewers' yeasts is the _____. [p.336]

___13. The group that includes shelf fungi, which
   decompose dead and dying trees, and mycorrhizal symbionts that help trees extract
   mineral ions from the soil is the _____.
   [pp.334,338]

___14. The group that includes the commercial mushroom *Agaricus brunnescens*, as well as the death cap mushroom, *Amanita phalloides*, is the _____. [pp.334–335,338]

___15. The group that includes *Penicillium*, which has a variety of species that produce penicillin and substances that flavor Camembert and Roquefort cheeses, is the _____. [p.337]

___16. The group whose spore-producing structures (asci) are shaped like flasks, globes, and shallow cups is the _____. [p.336]

___17. The group that forms a thin, clear covering around the zygote, the zygosporangium, is the _____. [p.336]

___18. The group whose spore-producing structures are club-shaped is the _____. [p.335]

___19. The groups that are symbiotic with young roots of shrubs and trees in mycorrhizal associations are _____ and _____. [p.338]

___20. A group of puzzling kinds of fungi whose members are lumped together but do not form a formal taxonomic group. [p.334]

Chapter Objectives/Review Questions

1. Fungi are heterotrophs; most are _____ and obtain nutrients from nonliving organic matter and so cause its decay. [p.334]
2. Other fungi are _____; they extract nutrients from tissues of a living host. [p.334]
3. The common names of the three major lineages of fungi are the _____, the _____ fungi, and the _____ fungi. [p.334]
4. Distinguish between the meanings of the following terms: hypha, hyphae, mycelium, and mycelia. [p.334]
5. Describe the diverse appearances of the fungi classified as club fungi. [p.334]
6. List the reproductive modes of fungi. [p.334]
7. The short-lived reproductive bodies of most club fungi are known as _____. [p.335]
8. The spores produced by members of the club fungi are known as _____. [p.335]
9. A(n) _____ mycelium is one in which the hyphae have undergone cytoplasmic fusion but not nuclear fusion. [p.335]
10. Be able to review the generalized life cycle of a club fungus. [p.335]
11. What is the importance of the club fungi in the genus *Amanita*? [p.335]
12. Zygomycetes form sexual spores by way of _____. [p.336]
13. Be able to review the life cycle of *Rhizopus*. [p.336]
14. Most sac fungi produce sexual spores called _____. [p.336]
15. What are some typical shapes of the reproductive structures enclosing the asci? [p.336]
16. Flavoring cheeses, producing citric acid for candies and soft drinks, food spoilage, genetic research, uses by bakers and vintners are all activities associated with fungi known as the _____ fungi. [p.337]
17. Give the name of the fungus that causes the disease known as ergotism; list the symptoms of ergotism. [p.337]
18. Name the cause and describe the symptoms of histoplasmosis. [p.337]
19. Define *mutualism* and explain why a lichen fits that definition. [p.338]
20. Distinguish the mycobiont from the photobiont. [p.338]
21. Of what value are lichens in conditions of deteriorating environmental conditions? [p.332]
22. Describe the fungus-plant root association known as mycorrhizae. [p.338]
23. Distinguish ectomycorrhizae from endomycorrhizae. [p.338]
24. What is the effect of pollution on mycorrhizae? [p.332]

Integrating and Applying Key Concepts

Suppose humans acquired a few well-placed fungal genes that caused them to reproduce in the manner of a "typical" fungus [text Figure 21.4, p.335]. Try to imagine the behavioral changes that humans would likely undergo. Would their food supplies necessarily be different? Table manners? Stages of their life cycle? Courtship patterns? Habitat? Would the natural limits to population increase be the same? Would their body structure change? Would there necessarily have to be separate sexes? Compose a descriptive science-fiction tale about two mutants who find each other and set up "housekeeping" together.

22

PLANTS

Interactive Exercises

Pioneers in a New World [pp.340–341]

22.1. EVOLUTIONARY TRENDS AMONG PLANTS [pp.342–343]

Selected Words: *non*vascular plants [p.342], *seedless* vascular plants [p.342], *seed-bearing* vascular plants [p.342], lignin [p.342], *haploid (n)* phase [p.342], *diploid (2n)* phase [p.342], *hetero*spory [p.343], *homo*spory [p.343]

Boldfaced, Page-Referenced Terms

[p.342] vascular plants _____

_____

[p.342] bryophytes _____

_____

[p.342] gymnosperms _____

_____

[p.342] angiosperms _____

_____

[p.342] root systems _____

_____

[p.342] shoot systems _____

_____

[p.342] lignin _____

_____

[p.342] xylem _____

_____

[p.342] phloem _____

_____

[p.342] cuticle _____

_____

[p.342] stomata _____

_____

[p.342] gametophytes _____

_____

[p.342] sporophyte _____

_____

[p.342] spores _____

_____

[p.343] pollen grains _____

_____

[p.343] seed _____

_____

Matching

Choose the most appropriate answer for each term.

1. ___seedless vascular plants [p.342]
2. ___angiosperms [p.342]
3. ___vascular plants [p.342]
4. ___bryophytes [p.342]
5. ___gymnosperms [p.342]

A. Liverworts, hornworts, and mosses
B. Seed-bearing plants that include cycads, ginkgo, gnetophytes, and conifers
C. Whisk ferns, lycophytes, horsetails, and ferns
D. A group of plants that bear flowers and seeds
E. In general, a large number of plants having internal tissues that conduct water and solutes through the plant body

Complete the Table

6. As plants evolved, several key evolutionary developments occurred that solved the problems of living in new land environments. Complete the following table to summarize these events. Choose from heterospory, xylem and phloem, seed, shoot systems, sporophytes, spores, lignin, gametophytes, root systems, stomata, pollen grains, and cuticle.

Evolutionary Trends	Survival Problem Solved
[p.342] a.	Provides a large surface area for rapidly taking up soil water and dissolved mineral ions; often anchors the plant
[p.342] b.	Consist of stems and leaves that function efficiently in the absorption of sunlight energy and CO_2 from the air
[p.342] c.	Allows extensive growth of stems and branches; a very hard organic substance that strengthens cell walls, thus allowing erect plant parts to display leaves to sunlight
[p.342] d.	Provides cellular pipelines to distribute water, dissolved ions, and solutes such as dissolved sugars to all living plant cells
[p.342] e.	A waxy coat on many plant organs that helps conserve water on hot, dry days
[p.342] f.	Tiny openings across the surfaces of stems and some leaves that help control the absorption of CO_2 and restrict evaporative water loss
[p.342] g.	The haploid (n) phase that dominates the life cycles of algae
[p.342] h.	The diploid ($2n$) phase that dominates the life cycles of most plants
[p.342] i.	Haploid cells produced by meiosis in sporophyte plants; later divide by mitosis to give rise to the gametophytes
[p.343] j.	Condition in some seedless plant species and seed-bearing plants where two kinds of spores are produced
[p.343] k.	Developed from one type of spore in gymnosperms and angiosperms; in turn develops into mature, sperm-bearing male gametophytes
[p.343] l.	Consists of an embryo sporophyte, nutritive tissues, and a protective coat

22.2. BRYOPHYTES [pp.344–345]

Selected Words: Polytrichum [p.344], *Sphagnum* [p.345]

Boldfaced, Page-Referenced Terms

[p.344] mosses _____

_____

[p.344] liverworts _____

_____

[p.344] hornworts _____

_____

[p.345] peat bogs _____

_____

True–False

If the statement is true, write a T in the blank. If the statement is false, make it correct by changing the underlined word(s) and writing the correct word(s) in the answer blank.

_____ 1. Mosses are especially sensitive to <u>water</u> pollution. [p.344]

_____ 2. Bryophytes <u>have</u> leaflike, stemlike, and rootlike parts although they do not contain xylem or phloem. [p.344]

_____ 3. Most bryophytes have <u>rhizomes</u>, elongated cells or threads that attach gametophytes to soil and serve as absorptive structures. [p.344]

_____ 4. Bryophytes are the simplest plants to exhibit a cuticle, cellular jackets around parts that produce sperms and eggs, and large gametophytes that do not depend on <u>sporophytes</u> for nutrition. [p.344]

_____ 5. The true <u>liverworts</u> are the most common bryophytes. [p.344]

_____ 6. Following fertilization, zygotes give rise to <u>gametophytes</u>. [p.345]

_____ 7. Each <u>sporophyte</u> consists of a stalk and a jacketed structure in which spores will develop. [p.345]

_____ 8. <u>Club</u> moss is a bog moss whose large, dead cells in their leaflike parts soak up five times as much water as cotton does. [p.345]

_____ 9. Bryophyte sperm reach eggs by movement through <u>air</u>. [p.345]

_____10. Bryophytes are <u>vascular</u> plants with flagellated sperm. [p.345]

Fill-in-the-Blanks

The numbered items in the illustration below represent missing information about a typical moss life cycle; complete the blanks in the following narrative to supply that missing information.

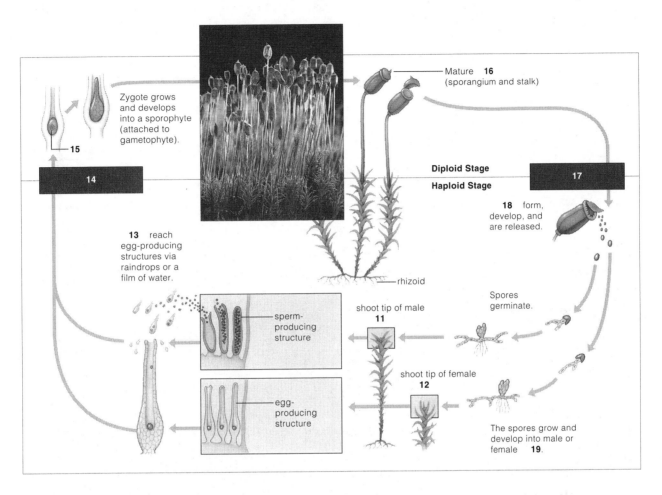

The gametophytes are the green leafy "moss plants." Sperms develop in jacketed structures at the shoot tip of the male (11) _____ [pp.344–345] and eggs develop in jacketed structures at the shoot tip of the female (12) _____ [pp.344–345]. Raindrops or a film of water on plant surfaces transport (13) _____ [pp.344–345] to the egg-producing structure. (14) _____ [pp.344–345] occurs within the egg-producing structure. The (15) _____ [pp.344–345] grows and develops into a mature (16) _____ [pp.344–345] (with sporangium and stalk) while attached to the gametophyte. (17) _____ [pp.344–345] occurs within the sporangium of the sporophyte where haploid (18) _____ [pp.344–345] form, develop, and are released. The released spores germinate with some growing and developing into male or female (19) _____ [pp.344–345].

22.3. *Focus on the Environment:* ANCIENT CARBON TREASURES [p.345]
22.4. EXISTING SEEDLESS VASCULAR PLANTS [pp.346–347]

Selected Words: lycophyte trees [p.345], *Calamites* [p.345], "fossil fuel" [p.345], *Cooksonia* [p.346], *Psilotum* [p.346], *Lycopodium* [p.346], *Selaginella* [p.346], *Equisetum* [p.346], epiphyte [p.347]

Boldfaced, Page-Referenced Terms

[p.345] coal _____

_____

[p.346] whisk ferns _____

_____

[p.346] lycophytes _____

_____

[pp.346–347] horsetails _____

_____

[pp.346–347] ferns _____

_____

[p.346] rhizomes _____

_____

[p.346] strobilus (plural, strobili) _____

_____

Choice

For questions 1–20, choose from the following:

      a. whisk ferns   b. lycophytes   c. horsetails   d. ferns   e. applies to a, b, c, and d

1. __ A group in which only the genus *Equisetum* survives [p.346]

2. __ Familiar club mosses growing as mats on forest floors [p.346]

3. __ Seedless vascular plants [p.346]

4. __ *Psilotum* [p.346]

5. __ Rust-colored patches, the sori, are on the lower surface of their fronds. [p.347]

6. __ Some tropical species are the size of trees. [p.347]

7. __ Ancestral plants living in the Carboniferous; through time, heat, and pressure these plants became peat and coal, the "ancient carbon treasures" [p.345]

8. __ Stems were used by pioneers of the American West to scrub cooking pots. [p.347]

9. __ The sporophytes have no roots or leaves. [p.346]

10. __ Mature leaves are usually divided into leaflets. [p.347]

11. __ The sporophyte has vascular tissues. [pp.346–347]

12. __ When the spore chamber snaps open, spores catapult through the air. [p.347]

13. ___ Grow in mud soil of streambanks and in disturbed habitats, such as roadsides and vacant lots [p.346]

14. ___ Sporophytes have rhizomes and hollow, photosynthetic, aboveground stems with scalelike leaves. [p.347]

15. ___ The sporophyte is the larger, longer lived phase of the life cycle. [p.346]

16. ___ *Lycopodium* [pp.346–347]

17. ___ The young leaves are coiled into the shape of a fiddlehead. [p.347]

18. ___ A germinating spore develops into a small, green, heart-shaped gametophyte. [p.347]

19. ___ *Selaginella* is heterosporous and produces two spore types. [p.346]

20. ___ "Amphibians" of the plant kingdom; life cycles require water [pp.346–347]

Fill-in-the-Blanks

The numbered items in the illustration of a generalized fern life cycle below represent missing information; complete the blanks in the following narrative to supply that missing information.

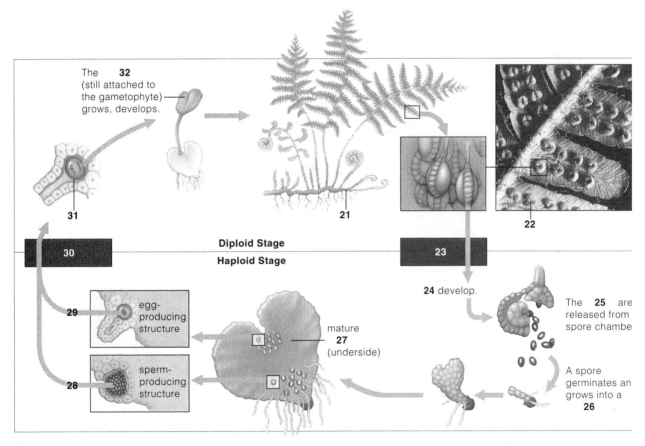

Fern leaves (fronds) of the sporophyte are usually divided into leaflets. The underground stem of the sporophyte is termed a(n) (21) _____ [p.347]. On the undersides of many fern fronds, rust-colored patches, each of which is called a(n) (22) _____ [p.347], is composed of spore chambers. (23) _____ [p.347] of diploid cells within each sporangium produces haploid (24) _____ [p.347]. The (25) _____ [p.347] are catapulted into the air when each spore chamber snaps open. A spore may germinate and grow into a (26) _____ [p.347] that is small, green, and heart-shaped. Jacketed structures develop on the underside of the mature (27) _____ [p.347]. Each male jacketed structure produces many (28) _____ [p.347], whereas each female jacketed structure produces a single (29) _____ [p.347]. These gametes meet in (30) _____ [p.347]. The diploid (31) _____ [p.347] is first formed inside the female jacketed structure; it divides to form the developing (32) _____ [p.347], still attached to the gametophyte.

22.5. THE RISE OF THE SEED-BEARING PLANTS [pp.348–349]

22.6. *Focus on the Environment:* GOOD-BYE, FORESTS [p.349]

Selected Words: Pinus ponderosa [p.348]

Boldfaced, Page-Referenced Terms

[p.348] microspores _____

_____

[p.349] pollination _____

_____

[p.349] megaspores _____

_____

[p.349] ovules _____

_____

[p.349] deforestation _____

_____

Matching

Choose the most appropriate answer for each term.

1. ___microspores [p.348]
2. ___pollination [p.349]
3. ___megaspore [p.349]
4. ___ovule [p.349]
5. ___seed ferns [p.348]

A. Spore type found in seed-bearing plants that develop within ovules
B. Female reproductive parts, which are seeds at maturity
C. Spore type found in seed-bearing plants that gives rise to pollen grains
D. Formerly dominant seed-bearing plant group along with gymnosperms and the later angiosperms
E. The name for the arrival of pollen grains on the female reproductive structures

Fill-in-the-Blanks

The numbered items in the illustration below represent missing information; complete the blanks in the following narrative to supply that missing information.

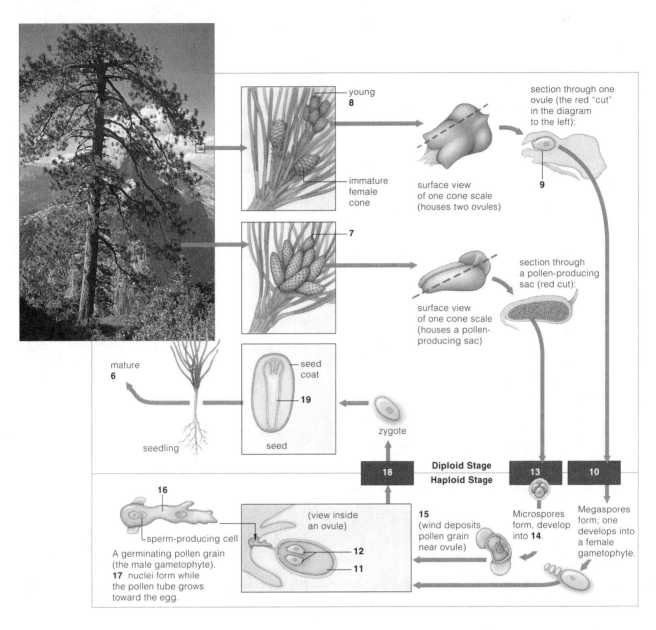

young
8

section through one
ovule (the red "cut"
in the diagram
to the left):

immature
female
cone

surface view
of one cone scale
(houses two ovules)

9

7

section through
a pollen-producing
sac (red cut):

surface view
of one cone scale
(houses a pollen-
producing sac)

mature
6

seed
coat

19

zygote

seedling

seed

Diploid Stage

18

13

10

Haploid Stage

16

sperm-producing cell

A germinating pollen grain
(the male gametophyte).
17 nuclei form while
the pollen tube grows
toward the egg.

(view inside
an ovule)

12

11

15
(wind deposits
pollen grain
near ovule)

Microspores
form, develop
into **14**.

Megaspores
form; one
develops into
a female
gametophyte.

The familiar pine tree, a conifer, represents the mature (6) _____ [p.348]. Pine trees produce two kinds of spores in two kinds of cones. Pollen grains are produced in male (7) _____ [p.348]. Ovules are produced in young female (8) _____ [p.348]. Inside each (9) _____ [p.348], (10) _____ [p.348] occurs to produce haploid megaspores; one develops into a many-celled female (11) _____ [p.348] that contains haploid (12) _____ [p.348]. Diploid cells within pollen sacs of male cones undergo (13) _____ [p.348] to produce haploid microspores. Microspores develop into (14) _____ [p.348] grains. (15) _____ [p.348] occurs when spring air currents deposit pollen grains near ovules of female cones. A pollen (16) _____ [p.348] representing a male gametophyte grows toward the female gametophyte. (17) _____ [p.348] nuclei form within the pollen tube as it grows toward the egg.

(18) _____ [p.348] follows and the ovule becomes a seed that is composed of an outer seed coat, the

(19) _____ [p.348] diploid sporophyte plant, and nutritive tissue.

22.7. GYMNOSPERM DIVERSITY [p.350]

Selected Words: gymnos [p.350], *sperma* [p.350], *evergreen* [p.350], *deciduous* [p.350], *Zamia* [p.350], *Ginkgo biloba* [p.350], *Gnetum* [p.350], *Ephedra* [p.350], *Welwitschia mirabilis* [p.350]

Boldfaced, Page-Referenced Terms

[p.350] conifers _____

_____

[p.350] cycad _____

_____

[p.350] ginkgos _____

_____

[p.350] gnetophytes _____

_____

Choice

For questions 1–12, choose from the following:

    a. cycads   b. ginkgos   c. gnetophytes   d. conifers   e. gymnosperms (includes a, b, c, d)

1. ___ Fleshy-coated seeds of female trees produce an awful stench. [p.350]

2. ___ Includes pines and redwoods. [p.350]

3. ___ Seeds and a flour made from the trunk are edible following removal of poisonous alkaloids. [p.350]

4. ___ Only a single species survives, the maidenhair tree. [p.350]

5. ___ Includes *Welwitschia* of hot deserts of south and west Africa, *Gnetum* of humid tropical regions, and *Ephedra* of California deserts and other arid regions. [p.350]

6. ___ Form pollen-bearing cones and seed-bearing cones on separate plants; leaves superficially resemble those of palm trees. [p.350]

7. ___ Seeds are mature ovules. [p.350]

8. ___ The favored male trees are now widely planted; they have attractive, fan-shaped leaves and are resistant to insects, disease, and air pollutants. [p.350]

9. ___ Their ovules and seeds are not covered; they are borne on surfaces of spore-producing reproductive structures. [p.350]

10. ___ Some sporophyte plants in this group mainly have a deep-reaching taproot; the exposed part is a woody disk-shaped stem bearing cones and one or two strap-shaped leaves that split lengthwise repeatedly as the plant ages. [p.350]

11. ___ Most species are "evergreen" trees and shrubs with needlelike or scalelike leaves. [p.350]

12. ___ Includes conifers, cycads, ginkgos, and gnetophytes. [p.350]

22.8. ANGIOSPERMS—THE FLOWERING, SEED-BEARING PLANTS [p.351]
22.9. VISUAL OVERVIEW OF FLOWERING PLANT LIFE CYCLES [p.352]
22.10. *Commentary:* SEED PLANTS AND PEOPLE [p.353]

Selected Words: angeion [p.351], *Nymphaea* [p.351], *Arceuthobium* [p.351], *Monotropa uniflora* [p.351], *Eucalyptus* trees [p.351], large sporophyte [p.351], endosperm [p.351], seeds [p.351], fruits [p.351], *Lilium* [p.352], "double" fertilization [p.352], embryo [p.352], *Homo erectus* [p.353], *Agave* [p.353], Mexican cockroach plants [p.353], neem tree leaves [p.353], *Digitalis purpurea* [p.353], *Aloe vera* [p.353], *Triticum* [p.353], *Secale* [p.353], *Saccharum officinarum* [p.353], *Theobroma cacao* [p.353]

Boldfaced, Page-Referenced Terms

[p.351] flowers _____

_____

[p.351] pollinators _____

_____

[p.351] dicots _____

_____

[p.351] monocots _____

_____

Matching

Choose the most appropriate answer for each term.

1. ___examples of monocot plants [p.351]
2. ___flowers [p.351]
3. ___pollinators [p.351]
4. ___double fertilization [p.352]
5. ___fruits [p.351]
6. ___endosperm [p.352]
7. ___examples of dicot plants [p.351]
8. ___seed [p.351]

A. Nutritive tissue found within a seed.
B. Orchids, palms, lilies, and grasses, including rye, sugarcane, corn, rice, and wheat.
C. Unique angiosperm reproductive structures.
D. Mature ovaries; protect and help disperse plant embryos.
E. Insects, bats, birds, and other animals that withdraw nectar or pollen from a flower and, in so doing, transfer pollen to its female reproductive parts.
F. Among all plants, unique to flowering plant life cycles; one sperm fertilizes the egg, the other sperm fertilizes a cell that gives rise to endosperm.
G. Packaged in fruits; each covered by a protective tissue and containing an embryo and nutritive tissue
H. Most herbaceous plants, such as cabbages and daisies; most flowering shrubs and trees, such as oaks and apple trees; water lilies and cacti.

Self-Quiz

Complete the Table

1. Complete the table below to compare the plant groups studied in this chapter.

Plant Group	Dominant Generation	Vascular Tissue Present?	Seeds Present?
[pp.344–345] a. Bryophytes			
[p.346] b. Lycophytes			
[pp.346–347] c. Horsetails			
[p.347] d. Ferns			
[pp.348–349] e. Gymnosperms			
[pp.351–352] f. Angiosperms			

___ 2. Members of the plant kingdom probably evolved from _____ more than 400 million years ago. [p.340]
   a. unicellular brown algae
   b. multicellular green algae
   c. unicellular green algae
   d. multicellular red algae

___ 3. The _____ is *not* a trend in the evolution of plants. [pp.342–343]
   a. evolution of roots, stems, and leaves
   b. shift from homospory to heterospory
   c. shift from diploid to haploid dominance
   d. development of xylem and phloem
   e. development of cuticles and stomata

___ 4. Existing nonvascular plants do *not* include _____. [p.344]
   a. horsetails
   b. mosses
   c. liverworts
   d. hornworts

___ 5. Plants possessing xylem and phloem are called _____ plants. [p.342]
   a. gametophyte
   b. nonvascular
   c. vascular
   d. seedless

___ 6. Bryophytes _____. [p.344]
   a. have vascular systems that enable them to live on land
   b. include lycopods, horsetails, and ferns
   c. have true roots but not stems
   d. include mosses, liverworts, and hornworts

___ 7. _____ are not seedless vascular plants. [p.346]
   a. Lycophytes
   b. Gymnosperms
   c. Horsetails
   d. Whisk ferns
   e. Ferns

___ 8. In horsetails, lycopods, and ferns, _____. [pp.346–347]
   a. spores give rise to gametophytes
   b. the dominant plant body is a gametophyte
   c. the sporophyte bears sperm- and egg-producing structures
   d. all of the above

___ 9. _____ are seed plants. [pp.350–351]
   a. Cycads and ginkgos
   b. Conifers
   c. Angiosperms
   d. all of the above

___10. In complex land plants, the diploid stage is resistant to adverse environmental conditions such as dwindling water supplies and cold weather. The diploid stage progresses through this sequence: _____. [pp.342–343]
   a. gametophyte ⟶ male and female gametes
   b. spores ⟶ sporophyte
   c. zygote ⟶ sporophyte
   d. zygote ⟶ gametophyte

___11. Monocots and dicots are groups of _____. [p.351]
   a. gymnosperms
   b. club mosses
   c. angiosperms
   d. horsetails

Chapter Objectives/Review Questions

1. Every plant you see today is a descendant of ancient species of _____ algae that lived near the water's edge or made it onto land. [p.340]
2. Most of the members of the plant kingdom are _____ plants, with internal tissues that conduct water and solutes through roots, stems, and leaves. [p.342]
3. Distinguish between the seed-bearing vascular plants known as gymnosperms and those known as angiosperms. [p.342]
4. State the general functions of the root systems and shoot systems of vascular plants. [p.342]
5. Explain the significance of "cells with lignified walls." [p.342]
6. _____ tissue distributes water and dissolved ions to all of the plant's living cells; _____ tissue distributes dissolved sugars and other photosynthetic products. [p.342]
7. Be able to name the plant structures that protect leaves and young stems from water loss and also name the plant structures serving as routes for absorbing carbon dioxide and controlling evaporative water loss. [p.342]
8. Give the reasons that diploid dominance allowed plants to successfully exploit the land environment. [pp.342–343]
9. As plants evolved two spore types (heterospory), one spore type develops into pollen grains that become mature sperm-bearing male _____; the other spore type develops into female _____, where eggs form and later become fertilized. [p.343]
10. The combination of a plant embryo, nutritive tissues, and protective tissues constitutes a _____. [p.343]
11. Mosses, liverworts, and hornworts belong to a plant group called the _____. [p.344]
12. Describe and state the functions of rhizoids. [p.344]
13. What group of plants first displayed cuticles, cellular jackets around the parts that produce sperms and eggs, and large gametophytes that retain sporophytes? [p.344]
14. The remains of peat mosses accumulate into compressed, excessively moist mats called _____ _____. [p.345]
15. Describe the sources of and the formation of coal. [p.345]
16. Why is coal said to be a nonrenewable source of energy? [p.345]
17. Be able to list the four groups of seedless vascular plants. [p.346]
18. The _____ is the larger, longer lived phase of the seedless vascular plants. [p.346]
19. Be able to describe structural characteristics of *Psilotum*, *Lycopodium*, *Equisetum*, and a fern; be generally familiar with their life cycles. [pp.346–347]
20. _____ are underground, branching, short, mostly horizontal absorptive stems. [p.346]
21. Some sporophytes of club mosses have nonphotosynthetic, cone-shaped clusters of leaves known as a _____ that bears spore-producing structures. [p.346]
22. Explain the meaning of the general term *epiphyte*. [p.347]
23. Microspores give rise to _____ _____. [p.348]
24. Be generally familiar with the life cycle of *Pinus*, a somewhat typical gymnosperm. [p.348]
25. Define the term *pollination*. [p.349]
26. An _____ contains the female gametophyte, surrounded by nutritive tissue, and a jacket of cell layers. [p.349]

27. _____ develop inside ovules. [p.349]
28. _____ is the removal of all trees from large tracts, as by clear-cutting. [p.349]
29. Conifers, cycads, ginkgos, and gnetophytes are all members of the _____ lineage. [p.350]
30. Describe the structure of a typical conifer cone. [p.350]
31. Briefly characterize plants known as cycads, ginkgos, and gnetophytes. [p.350]
32. List reasons why *Ginkgo biloba* is a unique plant. [p.350]
33. *Gnetum*, *Ephedra*, and *Welwitschia* represent genera of _____. [p.350]
34. Only angiosperms produce reproductive structures known as _____. [p.351]
35. Most flowering plants coevolved with _____ such as insects, bats, birds, and other animals that withdraw nectar or pollen from a flower. [p.351]
36. Name and cite examples of the two classes of flowering plants. [p.351]
37. _____, a nutritive tissue, surrounds embryo sporophytes inside the seeds of flowering plants. [p.351]
38. As seeds develop, tissues of most ovaries and other structures mature into _____. [p.351]

Integrating and Applying Key Concepts

Considering the pine life cycle, how many haploid (*n*) and diploid (*2n*) phases are represented within a pine seed?

23

ANIMALS: THE INVERTEBRATES

Interactive Exercises

Madeleine's Limbs [pp.356–357]

23.1. OVERVIEW OF THE ANIMAL KINGDOM [pp.358–359]
23.2. PUZZLES ABOUT ORIGINS [p.360]
23.3. SPONGES—SUCCESS IN SIMPLICITY [pp.360–361]

Selected Words: anterior end [p.359], *posterior* end [p.359], *dorsal* surface [p.359], *ventral* surface [p.359], peritoneum [p.359], pseudocoel [p.359], *Paramecium* [p.360], *Volvox* colonies [p.360], *Trichoplax adhaerens* [p.360]

Boldfaced, Page-Referenced Terms

[p.358] animals _____

_____

[p.358] ectoderm _____

_____

[p.358] endoderm _____

_____

[p.358] mesoderm _____

_____

[p.358] vertebrates _____

_____

[p.358] invertebrates _____

_____

[p.358] radial symmetry _____

_____

[p.358] bilateral symmetry _____

_____

[p.359] cephalization _____

_____

[p.359] gut _____

_____

[p.359] coelom _____

_____

[p.360] placozoan _____

_____

[p.360] sponges _____

_____

[p.360] collar cells _____

_____

[p.361] larva, (plural, larvae) _____

_____

Matching

Choose the most appropriate answer for each term.

1. ___animals [p.358]
2. ___ventral surface [p.359]
3. ___ectoderm, endoderm, mesoderm [p.358]
4. ___anterior end [p.359]
5. ___vertebrates [p.358]
6. ___invertebrates [p.358]
7. ___radial symmetry [p.358]
8. ___bilateral symmetry [p.358]
9. ___dorsal surface [p.359]
10. ___gut [p.359]
11. ___coelom [p.359]
12. ___thoracic cavity [p.359]
13. ___abdominal cavity [p.359]
14. ___posterior end [p.359]
15. ___segmentation [p.359]
16. ___cephalization [p.359]

A. All animals whose ancestors evolved before backbones did
B. An evolutionary process whereby sensory structures and nerve cells became concentrated in a head
C. The back surface
D. Animal body cavity lined with a peritoneum—found in most bilateral animals; some worms lack this cavity, other worms have a false cavity
E. Head end
F. Animals having body parts arranged regularly around a central axis, like spokes of a bike wheel
G. Upper coelom cavity holding a heart and lungs
H. Primary tissue layers that give rise to all adult animal tissues and organs
I. Surface opposite the dorsal surface
J. Region inside animal body in which food is digested
K. Animals having right and left halves that are mirror images of each other
L. Series of animal body units that may or may not be similar to one another
M. End opposite the anterior end
N. Lower coelom cavity holding a stomach, intestines, and other organs
O. Multicellular organisms with tissues forming organs and organ systems, diploid body cells, heterotrophic, aerobic respiration, sexual reproduction, sometimes asexual, most are motile in some part of the life cycle, and the life cycle shows embryonic development
P. Animals with a "backbone"

Complete the Table

17. Complete the table below by filling in the appropriate phylum or representative group name.

Phylum	Some Representative	Number of Known Species
[p.358] a.	*Trichoplax*; simplest animal	1
[p.358] b. Porifera		8,000
[p.358] c.	Hydrozoans, jellyfishes, corals, sea anemones	11,000
[p.358] d. Platyhelminthes		15,000
[p.358] e.	Pinworms, hookworms	20,000
[p.358] f. Mollusca		110,000
[p.358] g.	Leeches, earthworms, polychaetes	15,000
[p.358] h. Arthropoda		1,000,000
[p.358] i.	Sea stars, sea urchins	6,000

Choice

For questions 18–25 about animal origins, choose from the following:

a. *Paramecium*   b. *Volvox*   c. *Trichoplax adhaerens*
d. different animal lineages arose from more than one group of protistanlike ancestors

18. ___ The only known placazoan [p.360]

19. ___ Similar ciliate forerunners may have had multiple nuclei within a single-celled body [p.360]

20. ___ Similar to colonies that became flattened and crept on the seafloor [p.360]

21. ___ The answer to animal origins might require more than one answer [p.360]

22. ___ As simple as animals get [p.360]

23. ___ By another hypothesis, multicelled animals arose from spherical colonies of a number of flagellated cells, perhaps like *Volvox* colonies [p.360]

24. ___ A soft-bodied marine animal, shaped a bit like a tiny pita bread [p.360]

25. ___ Has no symmetry, no tissues, and no mouth [p.360]

Matching

Choose the most appropriate answer for each term.

26. ___fragmentation [p.361]

27. ___sponge phylum [p.360]

28. ___adult [p.361]

29. ___larva [p.361]

30. ___amoebalike cells [pp.360–361]

31. ___*Trichoplax* [p.360]

32. ___sponge skeletal elements [p.360]

33. ___food-trapping microvilli [p.360]

34. ___collar cells [p.360]

35. ___water entering a sponge body [p.360]

36. ___gemmules [p.361]

A. Flagellated cells that absorb and move water through a sponge as well as engulf food
B. Reside in a gelatin-like substance between inner and outer cell linings
C. An organism whose cell layers resemble those of a sponge
D. Sexually mature form of a species
E. Form the "collars" of collar cells
F. Clusters of sponge cells capable of germinating and establishing new colonies
G. Microscopic pores and chambers
H. Random chunks of sponge tissue break off and grow into more sponges
I. Spongin fibers, glasslike spicules of silica or calcium carbonate, or both
J. Sexually immature form that grows and develops into an adult
K. Porifera

23.4. CNIDARIANS—TISSUES EMERGE [pp.362–363]

Selected Words: *Hydra* [p.362], *Chrysaora* [p.362], mesoglea [p.362], *Obelia* [p.363], dinoflagellates [p.363], *Physalia* [p.363]

Boldfaced, Page-Referenced Terms

[p.362] Cnidaria _____

_____

[p.362] nematocysts _____

_____

[p.362] medusa (plural, medusae) _____

_____

[p.362] polyp _____

_____

[p.362] epithelium _____

_____

[p.362] nerve cells _____

_____

[p.362] contractile cells _____

_____

[p.363] hydrostatic skeleton _____

_____

[p.363] gonads _____

_____

[p.363] planulas _____

_____

Fill-in-the-Blanks

All members of the phylum (1) _____ [p.362] are tentacled, radial animals; they include jellyfishes, sea anemones, corals, and animals such as *Hydra*. Most of these animals live in the sea, and they alone produce (2) _____ [p.362], which are capsules capable of discharging threads that entangle or pierce prey. Cnidarians have two body plans that are most common—the (3) _____ [p.362], which look like bells or upside-down saucers, and the (4) _____ [p.362], which has a tubelike body with a tentacle-fringed mouth at one end. The saclike cnidarian gut processes food with its (5) _____ [p.362], a sheetlike lining with many glandular cells that secrete digestive enzymes. The (6) _____ [p.362] lines the rest of the body's surfaces [p.362]. Each of these linings is referred to as a(n) (7) _____ [p.362]. (8) _____ [p.362] cells form a "nerve net" that coordinates responses to stimulation. Working together with (9) _____ [p.362] cells that detect and signal, and (10) _____ [p.362] cells to carry out responses, the nerve net controls movement and shape changes.

The (11) _____ [p.362] is a layer of secreted material that lies between the epidermis and gastrodermis [p.362]. Most polyps have little mesoglea and use water in their gut as a(n) (12) _____ [p.363] skeleton, a fluid-filled cavity or cell mass. Reef-forming (13) _____ [p.363] consist of interconnected polyps that secrete calcium-reinforced external skeletons that, over time, build reefs. Many cnidarians have only a polyp or a medusa stage in the life cycle with the medusa being the sexual form. They have simple (14) _____ [p.363] that rupture and release gametes. Zygotes formed at fertilization develop into (15) _____ [p.363], a type of swimming or creeping larva that usually has ciliated epidermal cells. Eventually a mouth opens at one end of a (15), transforming the larva into a polyp or a medusa, and the cycle begins anew. (16) _____ [p.363] live as mutualists in coral tissues.

Labeling

Identify each indicated part of the illustration below.

17. _____ _____ [p.363]

18. _____ _____ [p.363]

19. _____ _____ [p.363]

20. _____ _____ [p.363]

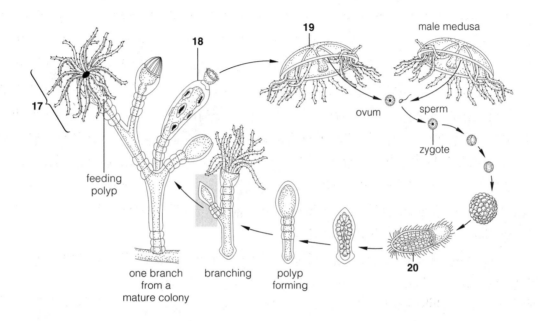

23.5. ACOELOMATE ANIMALS—AND THE SIMPLEST ORGAN SYSTEMS [pp.364–365]

Selected Words: *organ-system* level of construction [p.364], *transverse* fission [p.364], *definitive* host [p.365], *intermediate* host [p.365]

Boldfaced, Page-Referenced Terms

[p.364] organ _____

_____

[p.364] organ system _____

_____

[p.364] flatworms _____

_____

[p.364] hermaphrodites _____

_____

[p.365] proglottids _____

_____

Choice

For questions 1–25, choose from the following. You may use more than one letter for some questions.

a. turbellarians   b. flukes   c. tapeworms

1. ___ Parasitic worms [pp.364–365]

2. ___ Flame cells [p.364]

3. ___ Parasitize the intestines of vertebrates [p.365]

4. ___ Planarians that reproduce asexually by transverse fission [p.364]

5. ___ Possess a scolex [p.365]

6. ___ Ancestral forms probably had a gut but later lost it during their evolution in animal intestines. [p.365]

7. ___ Water-regulating systems have one or more tiny, branched tubes called protonephridia. [p.364]

8. ___ Only planarians and a few others live in freshwater. [p.364]

9. ___ Their life cycles have sexual and asexual phases and at least two kinds of hosts. [pp.364–365]

10. ___ Flame cells, each with a tuft of cilia, that drive out excess water to the surroundings [p.364]

11. ___ Thrive in predigested food in vertebrate intestines [p.365]

12. ___ After division, each half regenerates the missing parts. [p.364]

13. ___ Proglottids are new units of the body that bud just behind the head. [p.365]

14. ___ These parasites attach to the intestinal wall by a scolex. [p.365]

15. ___ Possess a structure equipped with suckers, hooks, or both [p.365]

16. ___ They reach sexual maturity in an animal that serves as their primary host. [p.365]

17. ___ Older proglottids store fertilized eggs; they break off and leave the body in feces. [p.365]

18. ___ Larval stages use an intermediate host, in which they either develop or become encysted. [p.365]

Labeling

Identify the parts of the animals shown dissected in the accompanying drawings.

19. _____ _____ [p.364]

20. _____ [p.364]

21. _____ [p.364]

22. _____ _____ [p.364]

23. _____ [p.364]

24. _____ [p.364]

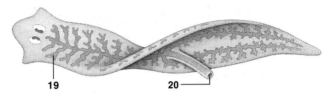

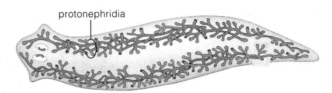

Answer questions 25–28 for the drawing of the accompanying dissected animal.

25. What is the common name (or genus) of the animal dissected? _____ [p.364]

26. Is the animal parasitic? _____ [p.364]

27. Is the animal hermaphroditic? _____ [p.364]

28. Name the coelom type exhibited by this animal. _____ [p.364]

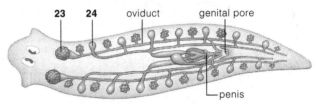

23.6. ROUNDWORMS [p.365]

23.7. *Focus on Health:* A ROGUE'S GALLERY OF WORMS [pp.366–367]

Selected Words: *Paratylenchus* [p.365], *Caenorhabditis elegans* [p.365], *schistosomiasis* [p.366], *Schistosoma japonicum* [p.366], *Enterobius vermicularis* [p.366], *Taenia saginata* [p.366], hookworms [p.367], *Trichinella spiralis* [p.367], *Wuchereria bancrofti* [p.367], *elephantiasis* [p.367]

Boldfaced, Page-Referenced Terms

[p.365] roundworms _____

_____

Answer questions 1–3 for the drawing of the dissected animal below.

1. What is the common name of the animal dissected? _____ [p.365]

2. Is the animal hermaphroditic? _____ [p.365]

3. Name the coelom type exhibited by this animal. _____ [p.365]

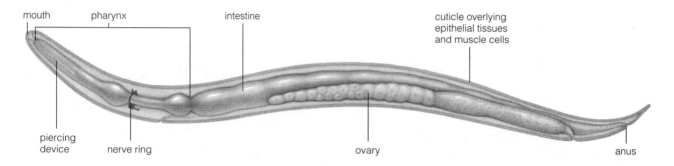

Fill-in-the-Blanks

Roundworms are (4) _____ [p.365] worms. They thrive nearly everywhere and may be the most

(5) _____ [p.365] animals alive. A round worm has (6) _____ [p.365] symmetry. But its body is

(7) _____ [p.365], typically tapered at both ends, and covered by a protective cuticle. A roundworm is

the simplest animal equipped with a complete (8) _____ [p.365] system. Between the gut and body

wall is a false (9) _____ [p.365], typically filled with reproductive organs. The cells in every tissue

absorb (10) _____ [p.365] from the coelomic fluid and give up wastes to it. (11) _____ [p.365]

species can severely damage their hosts, which include, humans, cats, dogs, cows, sheep, soybeans,

potatoes, and other crop plants.

Most roundworms are harmless, free (12) -_____ [p.365] types that do beneficial work, as when

they help cycle nutrients in a variety of communities. One species, *Caenorhabditis elegans*, is a(n)

(13) _____ [p.365] organism that has been used in research on inheritance, development, and aging.

Schistosoma japonicum is the Southeast Asian blood fluke responsible for (14) _____ [p.366]. *Enterobius*

vermicularis is a(n) (15) _____ [p.366] of temperate regions that parasitizes humans. Adult hookworms

live in the small intestine; (16) _____ [p.367] forms enter the bare skin of a host. Human infection by

Trichinella spiralis occurs when eating insufficiently cooked (17) _____ [p.367] or some game animals.

Wucheria bancrofti is an infection associated with a condition called (18) _____ [p.367].

Fill-in-the-Blanks

The numbered items on the illustration below (blood-fluke life cycle) represent missing information; complete the corresponding numbered blanks in the narrative to supply missing information about flukes.

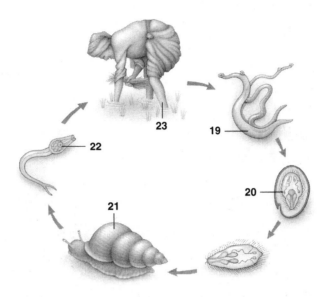

The life cycle of the Southeast Asian blood fluke *(Schistosoma japonicum)* requires a human primary host standing in water in which the fluke larvae can swim. This life cycle also requires an aquatic snail as an intermediate host. Flukes reproduce (19) _____ [p.366], and eggs mature in a human body. Eggs leave the body in feces, then hatch into ciliated, swimming (20) _____ [p.366] that burrow into a(n) (21) _____ [p.366] and multiply asexually. In time, many fork-tailed (22) _____ [p.366] develop. These leave the snail and swim until they contact (23) _____ [p.366] skin. They bore inward and migrate to thin-walled intestinal veins, and the cycle begins anew. In infected humans, white blood cells that defend the body attack the masses of fluke eggs, and grainy masses form in tissues. In time, the liver, spleen, bladder, and kidneys deteriorate.

Fill-in-the-Blanks

The numbered items on the illustration below (beef tapeworm life cycle) represent missing information; complete the corresponding numbered blanks in the narrative to supply missing information about the beef tapeworm.

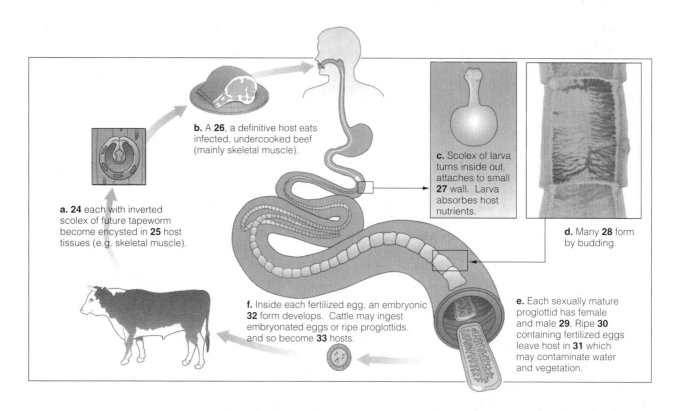

a. 24 each with inverted scolex of future tapeworm become encysted in **25** host tissues (e.g. skeletal muscle).

b. A **26**, a definitive host eats infected, undercooked beef (mainly skeletal muscle).

c. Scolex of larva turns inside out, attaches to small **27** wall. Larva absorbes host nutrients.

d. Many **28** form by budding.

e. Each sexually mature proglottid has female and male **29**. Ripe **30** containing fertilized eggs leave host in **31** which may contaminate water and vegetation.

f. Inside each fertilized egg, an embryonic **32** form develops. Cattle may ingest embryonated eggs or ripe proglottids. and so become **33** hosts.

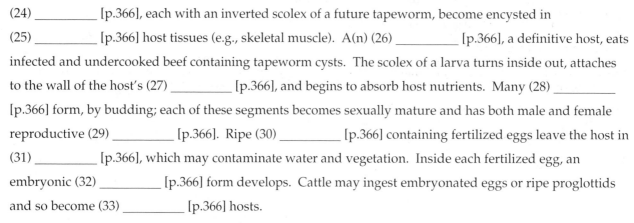

(24) _____ [p.366], each with an inverted scolex of a future tapeworm, become encysted in

(25) _____ [p.366] host tissues (e.g., skeletal muscle). A(n) (26) _____ [p.366], a definitive host, eats

infected and undercooked beef containing tapeworm cysts. The scolex of a larva turns inside out, attaches

to the wall of the host's (27) _____ [p.366], and begins to absorb host nutrients. Many (28) _____

[p.366] form, by budding; each of these segments becomes sexually mature and has both male and female

reproductive (29) _____ [p.366]. Ripe (30) _____ [p.366] containing fertilized eggs leave the host in

(31) _____ [p.366], which may contaminate water and vegetation. Inside each fertilized egg, an

embryonic (32) _____ [p.366] form develops. Cattle may ingest embryonated eggs or ripe proglottids

and so become (33) _____ [p.366] hosts.

23.8. TWO MAJOR DIVERGENCES [p.367]

23.9. A SAMPLING OF MOLLUSKS [pp.368–369]

Selected Words: spiral *cleavage* [p.367], radial *cleavage* [p.367], *molluscus* [p.368], *Dosidiscus* [p.368], *jet propulsion* [p.369]

Boldfaced, Page-Referenced Terms

[p.367] protostomes _____

_____

[p.367] deuterostomes _____

_____

[p.368] mollusk _____

_____

[p.368] mantle _____

_____

[p.369] torsion _____

_____

Choice

For questions 1–10, choose from the following:

a. protostomes   b. deuterostomes

1. ___ The first external opening in these embryos becomes the anus; the second becomes the mouth. [p.367]
2. ___ Animals having a developmental pattern in which the early cell divisions (cuts) are made parallel and perpendicular to the axis [p.367]
3. ___ A coelom arises from spaces in the mesoderm [p.367]
4. ___ Radial cleavage [p.367]
5. ___ The first external opening in these embryos becomes the mouth [p.367]
6. ___ A coelom forms from outpouchings of the gut wall [p.367]
7. ___ Spiral cleavage [p.367]
8. ___ Animal having a developmental pattern in which early cell divisions (cuts) are at oblique angles relative to the genetically prescribed body axis [p.367]
9. ___ Echinoderms and chordates [p.367]
10. ___ Mollusks, annelids, and arthropods [p.367]

Matching

Identify the animals pictured below and on p.263 by matching each with the appropriate description.

11. ___Animal A [p.368]
12. ___Animal B [p.369]
13. ___Animal C [p.369]

I.   Bivalve
II. Cephalopod
III. Gastropod

Labeling

Identify each numbered part in the drawings by writing its name in the appropriate blank.

14. _____ [p.368]
15. _____ [p.368]
16. _____ [p.368]
17. _____ [p.368]
18. _____ [p.368]
19. _____ [p.368]
20. _____ [p.368]
21. _____ [p.368]

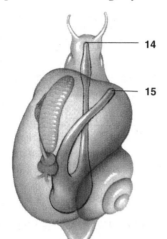

Animal A

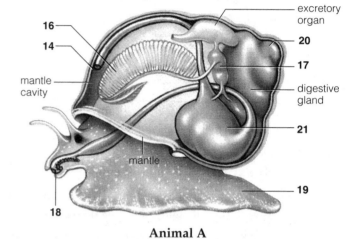

Animal A

22. _____ [p.369]
23. _____ [p.369]
24. _____ [p.369]
25. _____ [p.369]
26. _____ [p.369]

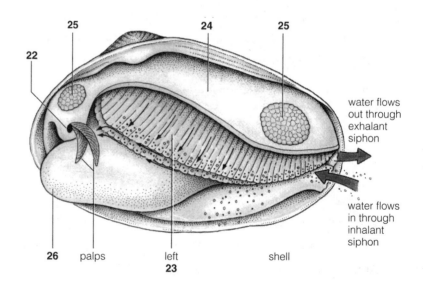

Animal B

27. _____ [p.369]

28. _____ _____ [p.369]

29. _____ [p.369]

30. _____ _____ [p.369]

31. _____ [p.369]

32. _____ _____ [p.369]

33. _____ [p.369]

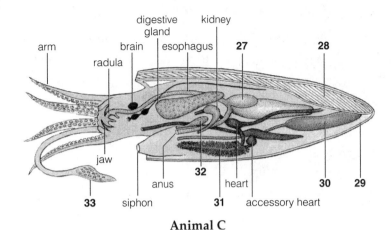

Animal C

Fill-in-the-Blanks

A(n) (34) _____ [p.368] is a bilateral animal having a small coelom and a fleshy soft body. Most have a (35) _____ [p.368] of calcium carbonate and protein, which were secreted from cells of a tissue that drapes like a skirt over the body mass. This tissue, the (36) _____ [p.368], is unique to mollusks. Special (37) _____ [p.368] with thin-walled leaflets serve in gas exchange. Most mollusks have a fleshy (38) _____ [p.368]. Many have a(n) (39) _____ [p.368], a tonguelike, toothed organ that by rhythmic protractions and retractions, rasps small algae and other food from substrates and draws it into the mouth. Mollusks with a well-developed head have (40) _____ [p.368] and (41) _____ [p.368], but not all have a head. There is great diversity in the phylum and here we sample four classes: chitons, gastropods, bivalves, and cephalopods.

Choice

For questions 42–65, choose from the following classes of the phylum Mollusca:

a. chitons   b. gastropods   c. bivalves   d. cephalopods

42. ___ Class with the swiftest invertebrates, the squids [p.368]

43. ___ The "belly foots" [p.368]

44. ___ The largest class with 90,000 species [p.368]

45. ___ Possess a dorsal shell divided into eight plates [p.368]

46. ___ Class with the smartest invertebrates, the octopuses [p.368]

47. ___ So named because their soft foot spreads out as they crawl [p.368]

48. ___ Animals having a "two-valved shell" [p.368]

49. ___ Class with the largest invertebrates, the giant squids [p.368]

50. ___ Coiling compacts the organs into a mass that can be balanced above the body, rather like a backpack [p.368]

51. ___ Class with the most complex invertebrates in the world, the octopuses and squids, with a memory and a capacity for learning [p.368]

52. ___ Highly active predators of the seas [p.368]

53. ___ Many have spirally coiled or conical shells. [p.368]

54. ___ The only mollusks with a closed circulatory system [p.368]

55. ___ Includes clams, scallops, oysters, and mussels [p.368]

56. ___ Includes aquatic snails, land snails, and sea slugs [p.368]

57. ___ Move rapidly with a system of jet propulsion [p.369]

58. ___ Water is drawn into the mantle cavity through one siphon and leaves through the other, carrying wastes. [p.369]

59. ___ Most can discharge dark fluid from an ink sac, perhaps to confuse predators. [p.369]

60. ___ The anus dumps wastes near the mouth. [p.369]

61. ___ Being highly active, they have great demands for oxygen [p.368]

62. ___ The only class of mollusks where torsion operates [p.369]

63. ___ Class containing the chambered nautilus [p.369]

64. ___ The shells of many are lined with iridescent mother-of-pearl. [p.368]

65. ___ As the embryo develops, a cavity between the mantle and shell twists 180° counterclockwise as does nearly all of the visceral mass. [p.369]

23.10. ANNELIDS—SEGMENTS GALORE [pp.370–371]

Selected Words: setae or *chaetae* [p.370], *oligo–* [p.370], *poly–* [p.370], *Hirudo medicinalis* [p.370]

Boldfaced, Page-Referenced Terms

[p.370] annelids _____

_____

[p.371] nephridia (singular, nephridium) _____

_____

[p.371] brain _____

_____

[p.371] nerve cords _____

_____

[p.371] ganglion _____

_____

Matching

Choose the appropriate answer for each term.

1. ___cuticle [p.371]
2. ___annelids [p.370]
3. ___earthworms [pp.370–371]
4. ___nephridia with cells similar to flame cells [p.371]
5. ___marine polychaetes [p.370]
6. ___brain [p.371]
7. ___nephridia [p.371]
8. ___nerve cords [p.371]
9. ___setae or chaetae [p.370]
10. ___advantage of segmentation [p.370]
11. ___hydrostatic skeleton [p.371]
12. ___leeches [p.370]
13. ___earthworm crawling and burrowing [pp.370–371]
14. ___ganglion [p.371]

A. Fluid-cushioned coelomic chambers against which muscles act
B. Oligochaete scavengers with a closed circulatory system and few bristles per segment
C. Paired, each a bundle of long extensions of nerve cell bodies leading away from the brain; pathways for rapid communication
D. Possess many bristles per segment
E. Muscle contraction with protraction and retraction of segment bristles
F. Secreted wrapping around the body surface of most annelids that permits respiratory exchange
G. A rudimentary aggregation of nerve cell bodies that integrate sensory input and commands for muscle responses for the whole body
H. The only group of annelids lacking bristles on all segments
I. Implies an evolutionary link between flatworms and annelids
J. Different body parts can evolve separately and specialize in different tasks
K. Except for leeches, chitin-reinforced bristles on each side of the body on nearly all segments
L. The term means "ringed forms"
M. An enlargement of the paired nerve cords in each segment of an earthworm; controls local activity
N. Regulates volume and composition of body fluids; often begins with a funnel-shaped structure in each segment

Labeling

Identify each indicated part of the illustrations below.

15. _____ [p.371]

16. _____ [p.371]

17. _____ _____ [p.371]

18. _____ [p.371]

19. _____ [p.371]

20. _____ _____ [p.371]

21. _____ [p.371]

22. _____ [p.371]

23. _____ [p.371]

24. _____ [p.371]

25. _____ _____ [p.371]

26. Name this animal. _____ [p.371]

27. Name this animal's phylum. _____ [p.370]

28. Name two distinguishing characteristics of this group. _____ _____ [pp.370–371]

29. Is this animal segmented? ( ) yes ( ) no [p.370]

30. Symmetry of adult: ( ) radial ( ) bilateral [p.370]

31. Does this animal have a true coelom? ( ) yes ( ) no [p.371]

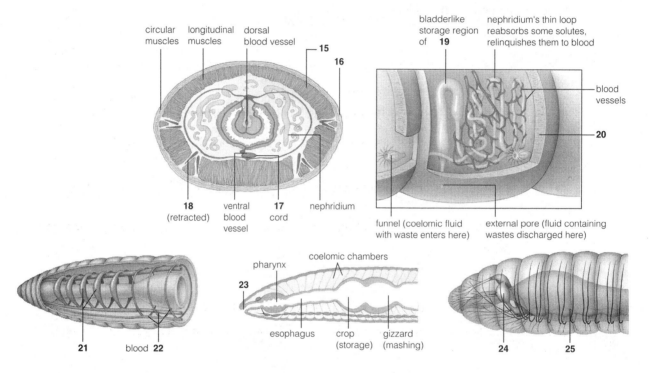

23.11. ARTHROPODS—THE MOST SUCCESSFUL ORGANISMS ON EARTH [p.372]

Selected Words: *joints* [p.372], *jointed foot* [p.372], *juvenile* [p.372], *division of labor* [p.372]

Boldfaced, Page-Referenced Terms

[p.372] arthropods _____

_____

[p.372] exoskeleton _____

_____

[p.372] molting _____

_____

[p.372] metamorphosis _____

_____

Choice

For questions 1–13, choose from the following six adaptations that contributed to the success of arthropods:

    a. hardened exoskeletons   b. fused and modified segments   c. jointed appendages
       d. respiratory structures   e. specialized sensory structures   f. division of labor

1. ___ Intricate eyes and other sensory organs that contributed to arthropod success [p.372]

2. ___ The new individual is a *juvenile*, a miniaturized form of the adult that simply changes in size and proportion until reaching sexual maturity. [p.372]

3. ___ A cuticle of chitin, proteins, and surface waxes that may be impregnated with calcium carbonate [p.372]

4. ___ In most of their existing descendants, however, the serial repeats of them body wall and organs are masked, for many fused-together, modified segments perform more specialized functions. [p.372]

5. ___ Metamorphosis from embryo to adult forms [p.372]

6. ___ Might have evolved as defenses against predation [p.372]

7. ___ Immature stages, such as caterpillars, specialize in feeding and growing in size [p.372]

8. ___ Gills of aquatic arthropods [p.372]

9. ___ In the ancestors of insects, different segments became combined into a head, a thorax, and an abdomen. [p.372]

10. ___ A jointed exoskeleton was a key innovation that led to appendages as diverse as wings, antennae, and legs. [p.372]

11. ___ Numerous species have a wide angle of vision and can process visual information from many directions. [p.372]

12. ___ Air-conducting tubes evolved among insects and other land-dwellers. [p.372]

13. ___ Their waxy surfaces restrict evaporative water loss and can support a body deprived of water's buoyancy. [p.372]

Short Answer

14. Why are the arthropods said to be the most biologically successful organisms on Earth? [p.372]

_____

_____

_____

_____

_____

23.12. A LOOK AT SPIDERS AND THEIR KIN [p.373]
23.13. A LOOK AT THE CRUSTACEANS [pp.374–375]
23.14. *HOW MANY LEGS?* [p.375]

Selected Words: *open* circulatory system [p.373], *Scutigera* [p.375]

Boldfaced, Page-Referenced Terms

[p.375] millipedes _____

_____

[p.375] centipedes _____

_____

Matching

Select the most appropriate answer for each term.

1. ___ticks [p.373]
2. ___arachnid forebody appendages [p.373]
3. ___arachnids [p.373]
4. ___arachnid hindbody appendages [p.373]
5. ___chelicerates [p.373]
6. ___spider, internal organs [p.373]
7. ___efficient predatory arachnids [p.373]

A. Scorpions and spiders that sting, bite, and may subdue prey with venom
B. An open circulatory system and book lungs
C. A group including mites, horseshoe crabs, sea spiders, spiders, ticks, and chigger mites
D. Spin out silk thread for webs and egg cases
E. Some transmit bacterial agents of Rocky Mountain spotted fever or Lyme disease to humans
F. A familiar group including scorpions, spiders, ticks, and chigger mites
G. Four pairs of legs, a pair of pedipalps that have mainly sensory functions, and a pair of chelicerae that can inflict wounds and discharge venom

Dichotomous Choice

Circle one of two possible answers given between parentheses in each statement.

8. Nearly all arthropods possess (strong claws/an exoskeleton). [p.374]

9. The giant crustaceans are (lobsters and crabs/barnacles and pillbugs). [p.374]

10. The simplest crustaceans have many pairs of (different/similar) appendages along their length. [p.374]

11. (Barnacles/Lobsters and crabs) have strong claws that collect food, intimidate other animals, and sometimes dig burrows. [p.374]

12. (Barnacles/Lobsters and crabs) have feathery appendages that comb microscopic bits of food from the water. [p.374]

13. (Barnacles/Copepods) are the most numerous animals in aquatic habitats, maybe even in the world. [p.375]

14. Of all arthropods, only (lobsters and crabs/barnacles) have a calcified "shell." [p.375]

15. Adult (barnacles/copepods) cement themselves to wharf pilings, rocks, and similar surfaces. [p.375]

16. As is true of other arthropods, crustaceans undergo a series of (rapid feedings/molts) and so shed the exoskeleton during their life cycle. [p.375]

17. As (millipedes/centipedes) develop, pairs of segments fuse, so each segment in the cylindrical body ends up with two pairs of legs. [p.375]

18. Adult (millipedes/centipedes) have a flattened body, are fast-moving, and have a pair of walking legs on every segment except two. [p.375]

19. (Millipedes/Centipedes) mainly scavenge for decaying vegetation in soil and forest litter. [p.375]

20. (Millipedes/Centipedes) are fast-moving, aggressive predators, outfitted with fangs and venom glands. [p.375]

Labeling

Identify each numbered part of the animal pictured at right.

21. _____ [p.374]

22. _____ [p.374]

23. _____ [p.374]

24. _____ [p.374]

25. _____ [p.374]

26. _____ [p.374]

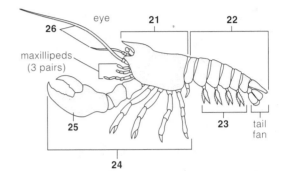

Answer questions 27–32 for the animal pictured at the right.

27. Name the animal pictured. _____ [p.374]

28. Name the subgroup of arthropods to which this animal belongs. _____ [p.374]

29. Name two distinguishing characteristics of this group. _____ _____ _____ [p.374]

30. Is this animal segmented? (   ) yes (   ) no [p.374]

31. Symmetry of adult: (   ) radial (   ) bilateral [p.374]

32. Does this animal have a true coelom? (   ) yes (   ) no [p.374]

Identify each numbered body part in this illustration, then answer question 38.

33. _____ _____ [p.373]

34. _____ [p.373]

35. _____ [p.373]

36. _____ [p.373]

37. _____ _____ [p.373]

38. Name the subgroup of arthropods to which the animal belongs. _____ [p.373]

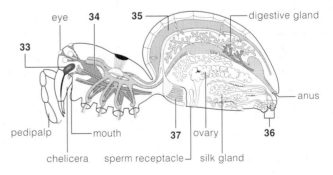

23.15. A LOOK AT INSECT DIVERSITY [pp.376–377]

Selected Words: incomplete metamorphosis [p.376], *complete* metamorphosis [p.376]

Boldfaced, Page-Referenced Terms

[p.376] Malpighian tubules _____

_____

[p.376] nymphs _____

_____

[p.376] pupae _____

_____

Matching

Choose the most appropriate answer for each term.

1. ___complete metamorphosis [p.376]

2. ___the most successful species of insects [p.376]

3. ___incomplete metamorphosis [p.376]

4. ___insect success based on aggressive competition with humans [pp.376–377]

5. ___Malpighian tubules [p.376]

6. ___metamorphosis [p.376]

7. ___a variation of insect headparts [p.376]

8. ___insect activities considered beneficial to humans [p.377]

9. ___insect life-cycle stages [p.376]

10. ___shared insect adaptations [p.376]

A. Post-embryonic resumption of growth and transformation into an adult form
B. Destruction of crops, stored food, wool, paper, and timber, drawing blood from humans and their pets, transmitting pathogenic microorganisms
C. Head, thorax, and abdomen, paired sensory antennae and mouthparts, three pairs of legs and two pairs of wings
D. Chewing, sponge up, siphon, piercing, and sucking
E. Involves gradual, partial change from the first immature form until the last molt
F. Winged insects, also the only winged invertebrates
G. Pollination of flowering plants and crop plants; parasitizing or attacking plants man would rather do without
H. Larva - nymph - pupa - adult
I. Structures that collect nitrogen-containing wastes from blood and convert them to harmless crystals of uric acid that are eliminated with feces
J. Tissues of immature forms are destroyed and replaced before emergence of the adult

23.16. THE PUZZLING ECHINODERMS [pp.378–379]

Boldfaced, Page-Referenced Terms

[p.378] echinoderms _____

_____

[p.379] water-vascular system _____

_____

Fill-in-the-Blanks

The second lineage of coelomate animals is referred to as the (1) _____ [p.378]. The major invertebrate members of this lineage include (2) _____ [p.378]. The body wall of all echinoderms has protective spines, spicules, or plates made rigid with (3) _____ _____ [p.378]. Most echinoderms also have a well-developed internal (4) _____ [p.378], which is composed of calcium carbonate and other substances secreted from specialized cells. Oddly, adult echinoderms have (5) _____ [p.378] symmetry with some bilateral features, but some produce larvae with (6) _____ symmetry [p.378]. Adult echinoderms have no (7) _____ [p.378] but a decentralized (8) _____ [p.378] system allows them to respond to information about food, predators, and so forth that is coming from different directions. For example, any (9) _____ [pp.378–379] of a sea star that senses the shell of a tasty scallop can become the leader, directing the remainder of the body to move in a direction suitable for prey capture.

The (10) _____ [p.379] feet of sea stars are used for walking, burrowing, clinging to a rock, or gripping a meal of clam or snail. These "feet" are part of a(n) (11) _____ [p.379] vascular system unique to echinoderms. Each foot contains a(n) (12) _____ [p.379] that acts like a rubber bulb on a medicine dropper as it contracts and forces fluid into a foot that then lengthens. Tube feet change shape constantly as (13) _____ [p.379] action redistributes fluid through the water vascular system. Some sea stars are simply able to swallow their prey (14) _____ [p.379]; others can push part of their stomach outside the mouth and around their prey, then start (15) _____ [p.379] their meal even before swallowing it. Coarse, indigestible remnants are regurgitated back through the mouth. Their small (16) _____ [p.379] is of no help in getting rid of empty clam or snail shells.

Labeling

Identify each indicated part of the two illustrations below.

17. _____ _____ [p.379]

18. _____ _____ [p.379]

19. _____ [p.379]

20. _____ [p.379]

21. _____ [p.379]

22. _____ _____ [p.379]

23. _____ [p.379]

24. _____ _____ [p.379]

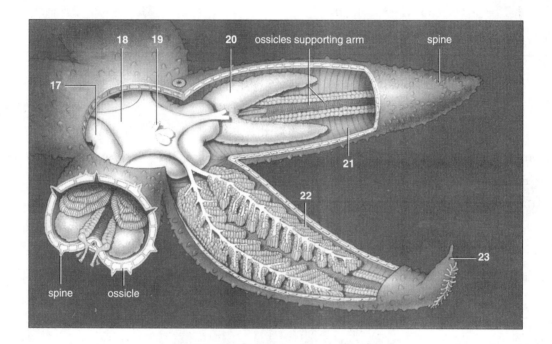

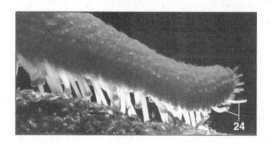

Answer questions 25–26 for the animal pictured above.

25. Name the animal shown. _____ [p.379]

26. Protostome (   ) or deuterostome (   ) [p.378]

Identifying

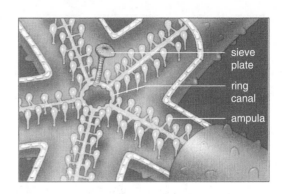

sieve
plate

ring
canal

ampula

27. Name the system shown at the right. _____
_____ system [p.379]

28. Identify each creature below by its common name (a–d).

29. Name the phylum of the animals pictured below.
_____ [p.378]

30. Name two distinguishing characteristics of this group.
_____ [pp.378–379]

31. Symmetry of adults represented below: (   ) radial (   )
bilateral [p.378]

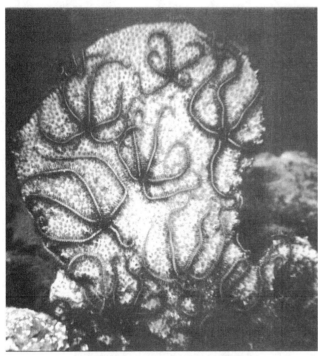

a. _____ [p.378]

b. _____ [p.378]

c. _____ [p.378]

d. _____ [p.378]

Self-Quiz

___ 1. Which of the following is *not* true of sponges? They have no _____. [p.360]
   a. distinct cell types
   b. nerve cells
   c. muscles
   d. gut

___ 2. Which of the following is *not* a protostome? [p.367]
   a. earthworm
   b. crayfish or lobster
   c. sea star
   d. squid

___ 3. Deuterostomes undergo _____ cleavage; protostomes undergo _____ cleavage. [p.367]
   a. radial; spiral
   b. radial; radial
   c. spiral; radial
   d. spiral; spiral

___ 4. Bilateral symmetry is characteristic of _____. [p.364]
   a. cnidarians
   b. sponges
   c. jellyfish
   d. flatworms

___ 5. Flukes and tapeworms are parasitic _____. [pp.364–365]
   a. leeches
   b. flatworms
   c. jellyfish
   d. roundworms

___ 6. Insects include _____. [p.377]
   a. spiders, mites, ticks
   b. centipedes and millipedes
   c. termites, aphids, and beetles
   d. all of the above

___ 7. The simplest animals to exhibit the organ-system level of construction are the _____. [p.364]
   a. sponges
   b. arthropods
   c. cnidarians
   d. flatworms

___ 8. Torsion is a process characteristic of _____. [p.369]
   a. chitons
   b. bivalves
   c. gastropods
   d. cephalopods
   e. echinoderms

___ 9. The _____ body plan is characterized by bilateral symmetry, a flattened body, cephalization, a digestive system with a pharynx for feeding, and hermaphroditism. [p.364]
   a. annelid
   b. roundworm
   c. echinoderm
   d. flatworm

___ 10. The _____ have bilateral symmetry, cylindrical bodies tapered on both ends, a tough protective cuticle, a false coelom, and present the simplest example of a complete digestive system. [p.365]
   a. roundworms
   b. cnidarians
   c. flatworms
   d. echinoderms

___ 11. _____ with a peritoneum separates the gut and body wall of most bilateral animals. [p.367]
   a. A coelom
   b. Mesoderm
   c. A mantle
   d. A water-vascular system

___ 12. In some annelid groups, a _____ is composed of cells similar to flatworm flame cells. [p.371]
   a. trachea
   b. nephridium
   c. mantle
   d. parapodium

Matching

Match each phylum below with the corresponding characteristics (a–i) and representatives (A–L). A phylum may match with more than one letter from the group of representatives.

___ , ___ 13. Annelida [pp.370–371]

___ , ___ 14. Arthropoda [p.372]

___ , ___ 15. Cnidaria [pp.362–363]

___ , ___ 16. Echinodermata [pp.378–379]

___ , ___ 17. Mollusca [pp.368–369]

___ , ___ 18. Nematoda [p.365]

___ , ___ 19. Platyhelminthes [pp.358–359]

___ , ___ 20. Porifera [pp.360–361]

a. choanocytes (= collar cells) + spicules
b. jointed legs + an exoskeleton
c. gill slits in pharynx + dorsal, tubular nerve cord + notochord
d. soft body + mantle; may or may not have radula or shell
e. bilateral symmetry, simplest organ-system animals
f. radial symmetry, medusae, and polyps; stinging cells
g. body compartmentalized into repetitive segments; coelom containing   nephridia (= primitive kidneys)
h. tube feet + calcium carbonate structures in skin
i. complete gut + bilateral symmetry + cuticle; includes many parasitic species, some of which are harmful to humans

A. Corals, sea anemones, and *Hydra*
B. Tapeworms and planaria
C. Insects
D. Jellyfish and the Portuguese man-of-war
E. Sand dollars and starfishes
F. Earthworms and leeches
G. Lobsters, shrimp, and crayfish
H. Organisms with spicules and choanocytes
I. Scorpions and millipedes
J. Octopuses and oysters
K. Flukes
L. Hookworm, trichina worm

Chapter Objectives/Review Questions

1. Be able to list the six general characteristics that define an "animal." [p.358]
2. _____ cells are the forerunners of the primary tissue layers, the ectoderm, endoderm, and, in most species, mesoderm. [p.358]
3. List the primary characteristic that separates vertebrates from invertebrates. [p.358]
4. Distinguish radial symmetry from bilateral symmetry and generally describe various animal gut types. [pp.358–359]
5. Describe meanings of the following terms relating to aspects of an animal body: anterior and posterior ends, dorsal and ventral surfaces, and cephalization. [p.359]
6. The _____ is a tubular or saclike region in the body in which food is digested, then absorbed into the internal environment. [p.359]
7. List two benefits that the development of a coelom brings to an animal. [p.359]
8. *Trichoplax* is as simple as an animal can get, having only _____ distinct layers of cells. [p.360]
9. List the characteristics that distinguish sponges from other animal groups. [pp.360–361]
10. Describe the processes involved in sponge reproduction, both asexual and sexual. [p.361]
11. State what nematocysts are used for and explain how they function. [p.362]
12. Two cnidarian body types are the _____ and the _____. [p.362]
13. Describe the structure typical of a cnidarian using terms such as epithelium, nerve cells, contractile cells, nerve net, mesoglea, a hydrostatic skeleton. [pp.362–363]
14. Describe the life cycle of *Obelia*. [p.363]
15. Define *organ-system* level of construction, and relate this to flatworms. [p.364]
16. List the three main types of flatworms and briefly describe each; name the groups that are parasitic. [pp.364–365]

17. Describe the body plan of roundworms, comparing its various systems with those of the flatworm body plan. [p.365]
18. Southeast Asian blood flukes, tapeworms, and pinworms are examples of _____ parasites. [pp.366–367]
19. Define, by their characteristics, *protostome* and *deuterostome* lineages and cite examples of animal groups belonging to each lineage. [p.367]
20. List and generally describe the major groups of mollusks and their members. [p.368]
21. Define *mantle,* and tell what role it plays in the molluscan body. [p.368]
22. Describe the process of torsion that occurs only in gastropods. [p.368]
23. Explain why cephalopods came to have such well-developed sensory and motor systems and are able to learn. [p.369]
24. Describe the advantages of segmentation and tell how this relates to the development of specialized internal organs. [p.370]
25. Be generally familiar with the external and internal structure of a typical annelid, the earthworm. [pp.370–371]
26. List four different lineages of arthropods; be able to briefly describe each. [p.372]
27. Be able to list the six arthropod adaptations that led to their success. [p.372]
28. List four different lineages of arthropods; be able to briefly describe each. [p.370]
29. List the groups of organisms known as the familiar chelicerates. [p.373]
30. Be able to describe the characteristics of and the significance of the arachnid lifestyle. [p.373]
31. Name the most obvious characteristic shared by most crustaceans. [p.374]
32. Name some common types of crustaceans. [pp.374–375]
33. Name the characteristics that all insects share, even though they may appear very dissimilar. [p.376]
34. The developmental process of many insects proceeds through very different post-embryonic stages, and then to an adult form by a process known as _____. [p.376]
35. The major invertebrate members of the _____ lineage are them echinoderms; be able to list their major characteristics. [p.378]
36. List five examples of animals known as echinoderms. [p.378]
37. Describe how locomotion and eating occurs in sea stars. [p.379]

Integrating and Applying Key Concepts

Scan text Table 23.32 on p.380 to verify that most highly evolved animals have a complete gut, a closed blood vascular system, both central and peripheral nervous systems, and are dioecious. Why do you suppose having two sexes in separate individuals is considered to be more highly evolved than the monoecious condition utilized by many flatworms? Wouldn't it be more efficient if all individuals in a population could produce both kinds of gametes? Cross-fertilization would then result in both individuals being able to produce offspring.

24

ANIMALS: THE VERTEBRATES

Interactive Exercises

Making Do (Rather Well) with What You've Got [pp.382–383]

24.1. THE CHORDATE HERITAGE [p.384]

24.2. INVERTEBRATE CHORDATES [pp.384–385]

24.3. EVOLUTIONARY TRENDS AMONG THE VERTEBRATES [pp.386–387]

24.4. EXISTING JAWLESS FISHES [p.387]

24.5. EXISTING JAWED FISHES [pp.388–389]

Selected Words: "invertebrate chordates" [p.384], Urochordata [p.384], Cephalochordata [p.384], metamorphosis [p.384], hemichordates [p.386], larvae [p.386], Agnatha [p.387], dermis [p.388], lungfish [p.389], coelacanth [p.389]

Boldfaced, Page-Referenced Terms

[p.384] chordates _____

_____

[p.384] vertebrates _____

_____

[p.384] notochord _____

_____

[p.384] nerve cord _____

_____

[p.384] pharynx _____

_____

[p.384] tunicates _____

_____

[p.384] filter feeders _____

_____

[p.384] gill slits _____

_____

[p.384] lancelets _____

_____

[p.386] vertebrae (singular, vertebra) _____

_____

[p.386] jaws _____

_____

[p.386] fins _____

_____

[p.386] gills _____

_____

[p.387] lungs _____

_____

[p.387] ostracoderms _____

_____

[p.387] placoderms _____

_____

[p.387] hagfishes _____

_____

[p.387] lampreys _____

_____

[p.388] swim bladder _____

_____

[p.388] cartilaginous fishes _____

_____

[p.388] scales _____

_____

[p.388] bony fishes _____

_____

[p.389] lobe-finned fish _____

_____

Fill-in-the-Blanks

Four major features distinguish the embryos of chordates from those of all other animals: a hollow dorsal (1) _____ _____ [p.384], a(n) (2) _____ [p.384] with slits in its wall, a(n) (3) _____ [p.384], and a tail that extends past the anus at least during part of their lives. In some chordates, the (4) _____ [p.384] chordates, the notochord is *not* divided into a skeletal column of separate, hard segments; in others, the (5) _____ [p.384], it is.

Invertebrate chordates living today are represented by tunicates and (6) _____ [p.384], which obtain their food by (7) _____ - _____ [p.384]; they draw in plankton-laden water through the mouth and pass it over sheets of mucus, which trap the particulate food before the water exits through the (8) _____ _____ [p.384] in the pharynx. (9) _____ [p.384] are among the most primitive of all living chordates; when they are larvae, they look and swim like (10) _____ [p.384]. A rod of stiffened tissue, the (11) _____ [p.384], runs much of the length of the larval body; it functions like a(n) (12) _____ _____ [p.384]. Most adults remain attached to rocks and hard substrates. Even though the adult forms of hemichordates, echinoderms, and tunicates look very different, they have similar embryonic developmental patterns. Chordates may have developed from a mutated ancestral deuterostome (13) _____ [p.386] of an attached, filter-feeding adult. A larva is an immature, motile form of an organism, but if a(n) (14) _____ [p.386] occurred that caused (15) _____ _____ [p.386] to become functional in the larval body, then a motile larva that could reproduce would have been more successful in finding food and avoiding (16) _____ [p.386] than a attached adult; in time, the attached stages in the species would be eliminated.

The ancestors of the vertebrate line may have been mutated forms of their closest relatives, the (17) _____ [pp.384–385], in which the notochord became segmented and the segments became hardened (18) _____ [p.386]. The vertebral column was the foundation for fast-moving (19) _____ [p.386], some of which were ancestral to all other vertebrates. The evolution of (20) _____ [p.386] intensified the competition for prey and the competition to avoid being preyed upon; animals in which mutations expanded the nerve cord into a(n) (21) _____ [p.386] that enabled the animal to compete effectively survived more frequently than their duller-witted fellows and passed along their genes into the next generations. Fins became (22) _____ [p.386]; in some fishes, those became (23) _____ [p.386] and equipped with skeletal supports; these forms set the stage for the development of legs, arms, and wings in later groups. As the ancestors of land vertebrates began spending less time immersed and more of their lives exposed to air, use of gills declined and (24) _____ [p.387] evolved; more elaborate and efficient (25) _____ [p.387] systems evolved along with more complex and efficient lungs.

Questions 26–27 refer to the illustrations below.

26.   Name the creature in illustration B. [p.385] _____

27.   Name the creature in illustration C. [p.385] _____

Labeling

Name the structures numbered below.

28. _____ [p.385]

29. _____ with _____ _____ [p.385]

30. _____ [p.385]

31. _____ [p.385]

32. _____ [p.385]

33. _____ [by deduction]

34. _____ _____ [p.385]

35. _____ _____ [p.385]

36. _____ [p.385]

37. _____ _____ _____ [p.385]

38. _____ [p.385]

39. _____ [p.385]

40. _____ _____ _____ _____
     [p.385]

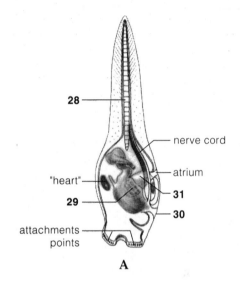

28

nerve cord

"heart"

atrium

29

31

30

attachments
points

A

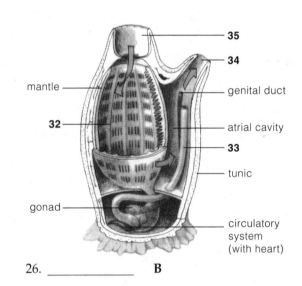

35

34

mantle

genital duct

32

atrial cavity

33

tunic

gonad

circulatory
system
(with heart)

26. _____   **B**

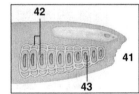

42

41

43

41. _____   **D**

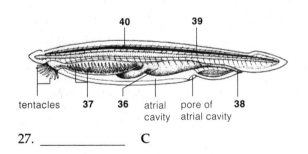

40   39

tentacles   37   36   atrial   pore of   38
                      cavity   atrial cavity

27. _____   **C**

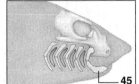

45

44. _____   **E**

280   Chapter Twenty-Four

Labeling (continued)

41. Name the creature whose head is shown in illustration D. [p.386] _____
42. Name the structures. [p.386] _____ _____
43. Name the structures. [p.386] _____ _____
44. Name the creature shown in illustration E. [p.386] _____
45. What structures occupy the front of its head space? [p.386] _____

Complete the Table

Provide common names wherever (*) appears, and provide the class name wherever ❐ appears.

Phylum: Hemichordata   *Phylum: Chordata*

46. * [p.386]	*Subphylum: Urochordata*	*Subphylum: Cephalochordata*	*Subphylum: Vertebrata*	
	47.* [p.384]	48.* [p.384]	Class: *Agnatha*	49.* [p.384]
			Class: *Placodermi*	50.* [p.384]
			Class: *Chondrichthyes*	51.* [p.384]
			Class: *Osteichthyes*	52.* [p.384]
			Class: 53. ❐	amphibians [p.384]
			Class: 54. ❐	reptiles [p.384]
			Class: 55. ❐	birds [p.384]
			Class: 56. ❐	mammals [p.384]

Fill-in-the-Blanks

Cartilaginous fishes include about 850 species of rays, skates, (57) _____ [p.388], and chimaeras. They

have conspicuous fins and five to seven (58) _____ _____ [p.388] on both sides of the pharynx.

Ninety-six percent of the existing species of fishes are (59) _____ [p.388]. Their ancestors arose during

the Silurian period perhaps as early as 450 million years ago and soon gave rise to three lineages: the

(60) _____-_____ [p.388] fishes, the lobe-finned fishes, and lungfishes.

Matching

Match the numbered item with its letter. One letter is used twice.

61. ___
62. ___
63. ___
64. ___
65. ___
66. ___
67. ___
68. ___
69. ___
70. ___
71. ___
72. ___
73. ___
74. ___
75. ___
76. ___

A. anal fin [p.389]
B. anus [p.389]
C. brain [p.389]
D. caudal fin [p.389]
E. dorsal fin [p.389]
F. gallbladder [p.389]
G. heart [p.389]
H. intestine [p.389]
I. kidney [p.389]
J. liver [p.389]
K. pectoral fin (paired) [p.389]
L. pelvic fin (paired) [p.389]
M. stomach [p.389]
N. swim bladder [p.389]
O. urinary bladder [p.389]

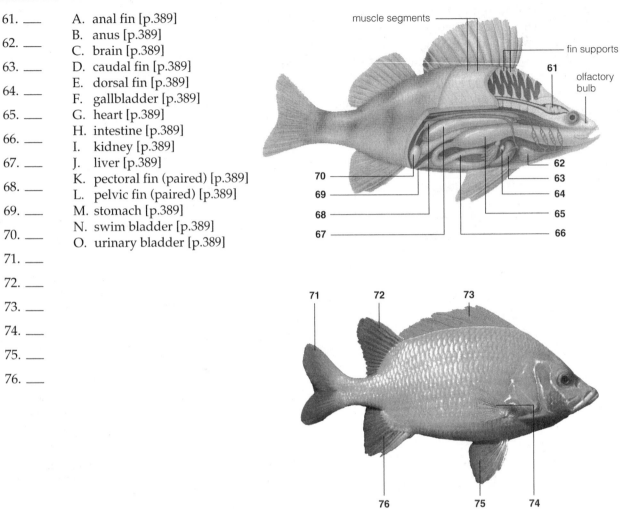

Analysis and Short Answer

Text Figure 24.11 on p.390 mentions features that distinguish the amphibian-lungfish lineage from the lineage that leads to sturgeons and other bony fishes. At least one of those features may have developed near ⑤ in this evolutionary tree.

Questions 77–87 refer to the evolutionary diagram illustrated on the next page.

77. What single feature do the lampreys, hagfishes, and extinct ostracoderms have in common that is different from the placoderms? [p.387] _____

78. How did ostracoderms feed? [p.387] _____

79. A mutation in ostracoderm stock led to the development of what feature in all organisms that descended from ①? [p.387] _____

80. Mutation at ② led to the development of an endoskeleton made of what? [p.388] _____

81. Mutations at ③ led to an endoskeleton of what? [p.388] _____

82. Mutations at ④ led to which spectacularly diverse fishes that have delicate fins originating from the dermis? [pp.388–389] _____

83. Mutations at ⑤ led to which fishes whose fins incorporate fleshy extensions from the body? [pp.389–390]
_____

84. Which branch, ④ or ⑤, gave rise to the amphibians? [p.390] _____

85. Which branch gave rise to the modern bony fishes? [p.388] _____

86. In which period did three distinctly different lineages of *bony* fishes appear in the fossil record? [p.388]
_____

87. Approximately how many million years ago did the fork in the evolutionary path that led to the amphibians occur? (Estimate from evolutionary diagram.) _____

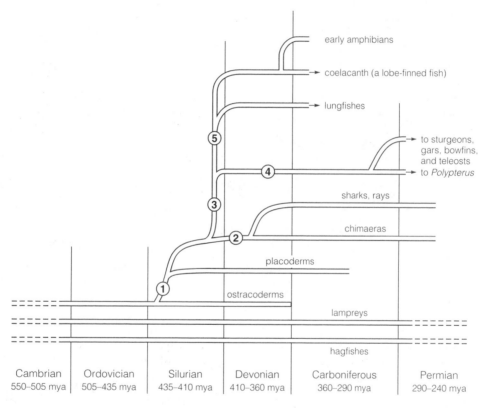

mya = million years ago

24.6. AMPHIBIANS [pp.390–391]

24.7. THE RISE OF REPTILES [pp.392–393]

24.8. BIRDS [pp.394–395]

24.9. THE RISE OF MAMMALS [pp.396–397]

Selected Words: cerebral cortex [p.392], dinosaurs [p.392], crocodilians [p.392], turtles [p.392], tuataras [p.392], snakes [p.392], lizards [p.392], third "eye" [p.393], *Archaeopteryx* [p.394], synapsids [p.396], therapsids [p.396], therians [p.396], marsupials [p.397]

Boldfaced, Page-Referenced Terms

[p.390] amphibian _____

_____

[p.391] salamanders _____

_____

[p.391] frogs _____

_____

[p.391] caecilians _____

_____

[p.392] reptiles _____

_____

[p.392] amniote eggs _____

_____

[p.394] birds _____

_____

[p.394] feathers _____

_____

[p.396] mammals _____

_____

[p.396] dentition _____

_____

[p.396] placenta _____

_____

[p.397] convergent evolution _____

_____

Fill-in-the-Blanks

Natural selection acting on lobe-finned fishes during the Devonian period favored the evolution of ever more efficient (1) _____ [p.390] used in gas exchange and stronger (2) _____ [p.390] used in locomotion. Without the buoyancy of water, an animal traveling over land must support its own weight against the pull of gravity. (3) _____ _____ [p.390] inside the lobed fins of certain fishes evolved over long periods of time into the limb bones of early amphibians. The (4) _____ [p.390] of early amphibians underwent dramatic modifications that involved evaluating signals related to vision, hearing, and (5) _____ [p.390]. Changes in the (6) _____ [p.390] system converted the two-chambered heart of fishes into a three-chambered one, and more efficient changes in (7) _____ [p.390] distribution patterns delivered oxygen to cells throughout the body. The Devonian period brought humid, forested swamps with an abundance of aquatic invertebrates and (8) _____ [p.390]—ideal prey for amphibians.

There are three groups of existing amphibians: (9) _____ [p.391], frogs and toads, and caecilians. Amphibians require free-standing (10) _____ [p.390] or at least a moist habitat in order to (11) _____ [p.390]. Amphibian skin generally lacks scales but contains many glands, some of which produce (12) _____ [p.391].

In the late Carboniferous, (13) _____ [p.392] began a major adaptive radiation into the lush habitats on land, and only amphibians that mutated and developed certain (14) _____ [p.392] features were able to follow them and exploit an abundant food supply. Several features helped: modification of (15) _____ [p.392] bones favored swiftness, modification of teeth and jaws enabled them to feed efficiently on a variety of prey items, and the development of a(n) (16) _____ [p.392] egg protected the embryo inside from drying out, even in dry habitats. Today's reptiles include (17) _____ [p.392], crocodilians, snakes, and (18) _____ [p.392]. All rely on (19) _____ [p.392] fertilization, and most lay leathery-shelled eggs.

(For questions 20–28, consult the time line and the figure for the Analysis and Short Answer exercise on the next page.)

Although amphibians originated during Devonian times, ancestral "stem" reptiles appeared during the (20) _____ [p.392] period, about 340 million years ago. Reptilian groups living today that have existed on Earth longest are the (21) _____ [see next page]; their ancestral path diverged from that of the "stem" reptiles during the (22) _____ [see next page] period. Crocodilian ancestors appeared in the early (23) _____ [see next page] period, about 220 million years ago. Snake ancestry diverged from lizard stocks during the late (24) _____ [see next page] period, about 140 million years ago. (25) _____ [p.393] are more closely related to extinct dinosaurs and crocodiles than to any other existing vertebrates; they, too, have a (26) _____-chambered [p.395] heart. Mammals have descended from therapsids, which in turn are descended from the (27) _____ [p.396] group of reptiles, which diverged earlier from the stem reptile group during the (28) _____ [see next page] period, approximately 320 million years ago.

Interpretation

In blanks 29–35, arrange the following groups in sequence, from earliest to latest, according to their first appearance in the fossil record. (Consult the evolutionary diagram on the next page]

A. Birds   B. Crocodilians   C. Dinosaurs   D. Early ancestors of mammals (= synapsid reptiles)
    E. Early ancestors of turtles (= anapsid reptiles)   F. Lizards and snakes   G. "Stem" reptiles

29. ____ 30. ____ 31. ____ 32. ____ 33. ____ 34. ____ 35. ____

Interpretation

In blanks 36–42, select from the choices immediately below the geologic period in which each group (29–35 in the preceding exercise) first appeared, and write it in the correct space. For example, the group in blank 29 first appeared when? Write the answer in blank 36. (Consult the evolutionary diagram below and interpret.)

Paleozoic Era		Mesozoic Era		
A. Carboniferous	B. Permian	C. Triassic	D. Jurassic	E. Cretaceous
360–290 mya	290–240 mya	240–205 mya	205–138 mya	138–65 mya

36. ____   37. ____   38. ____   39. ____   40. ____   41. ____   42. ____

Analysis and Short Answer

43. Text p.396 mentions the features that distinguish, say, the mammals from all of the remaining groups on the right-hand side of the diagram; therefore, at some time during the evolution of mammals, mutations that produced those features appeared. Consult the evolutionary diagram below and imagine what sort of mutation(s) occurred at the numbered places.

   a. What may have occurred at ①? [p.392] _____

   b. What may have occurred at ②? [p.392] _____

   c. What may have occurred at ③? [p.394] _____

   d. What may have occurred at ④? [p.396] _____

   e. What may have occurred at ⑤? [p.393] _____

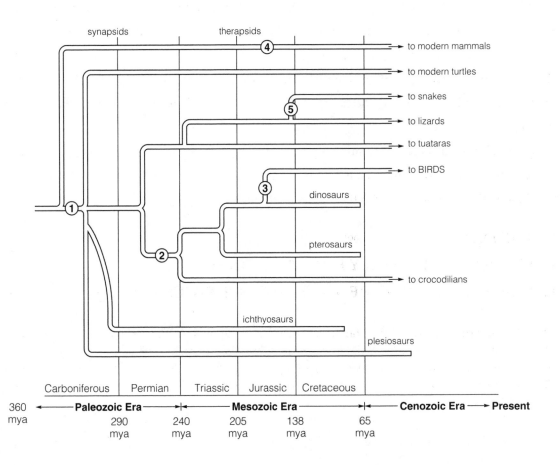

Labeling

Label the structures pictured at the right.

44. _____ [p.394]

45. _____ [p.394]

46. _____ _____ [p.394]

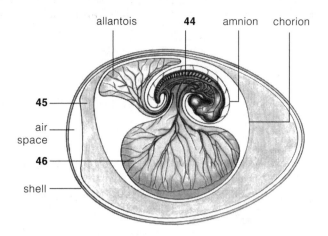

allantois   **44**   amnion   chorion

45

air space

46

shell

Fill-in-the-Blanks

Birds descended from (47) _____ [p.394] that ran around on two legs some 160 million years ago. All birds have (48) _____ [p.394] that insulate and help get the bird aloft. Generally, birds have a greatly enlarged (49) _____ [p.395] to which flight muscles are attached. Bird bones contain (50) _____ _____ [p.395], and air flows through sacs, not into and out of them, for gas exchange in the lungs. Birds have a large, durable (51) _____-_____ _____ [p.395] that pumps oxygen-poor blood to the lungs and oxygen-enriched blood back to the heart and to the rest of its body, just as you do. Many birds display (52) _____ [p.394] behavior: a recurring pattern of relocation between two places in response to environmental rhythms. Birds also lay hard-shelled (53) _____ [p.394] eggs, have complex courtship behaviors, and generally nurture their offspring.

There are three groups of existing mammals: those that lay eggs (examples are the (54) _____ [p.396] and the spiny anteater), those that are (55) _____ [p.397] (examples are the opossum and the kangaroo), and those that are (56) _____ [p.396] mammals (there are more than 4,500 species of these). Mammals regulate their body temperature, have a(n) (57) _____-chambered [p.395] heart, and show a high degree of parental nurture. Most mammals have (58) _____ [p.396] as a means of insulation, and mammalian mothers generally suckle their young with milk.

24.10. THE PRIMATES [pp.398–399]
24.11. EMERGENCE OF EARLY HUMANS [pp.400–401]
24.12. *Focus on Science:* OUT OF AFRICA—ONCE, TWICE, OR . . . [p.402]

Selected Words: prosimians [p.398], tarsioids [p.398], anthropoids [p.398], tarsiers [p.398], monkeys [p.398], apes [p.398], Paleocene [p.398], "Lucy" [p.399], hominoids [p.399], hominids [p.399], *Australopithecus afarensis* [p.399], bipedal [p.399], *Homo habilis* [p.400], *Homo erectus* [p.400], *Homo sapiens* [p.401], Neandertal [p.401], multiregional model [p.402], African emergence model [p.402]

Boldfaced, Page-Referenced Terms

[p.398] primates _____

_____

[p.398] hominids _____

_____

[p.398] culture _____

_____

[p.399] australopiths _____

_____

[p.400] humans _____

_____

Fill-in-the-Blanks

During the Cenozoic Era (65 million years ago to the present), birds, (1) _____ [recall pp.305–307], and flowering plants evolved to dominate Earth's assemblage of organisms. (1) are warm-blooded vertebrates with (2) _____ [recall p.396] that began their evolution more than 200 million years ago. There are many groups, or (3) _____ [p.398], within the class Mammalia; each order has its distinctive array of characteristics. Humans, apes, monkeys, and prosimians are all (4) _____ [p.398]; members of this order have excellent (5) _____ [p.398] perception as a result of their forward-directed eyes. They also have hands that are (6) _____ [p.399]; that is, they are adapted for (7) _____ [p.399] instead of running. Primates rely less on their sense of (8) _____ [p.398] and more on daytime vision. Their (9) _____ [p.398] became larger and more complex; this trend was accompanied by refined technologies and the development of (10) _____ [p.398]: the collection of behavior patterns of a social group, passed from generation to generation by learning and by symbolic behavior (language). Modification in the (11) _____ [p.398] and muscles led to increased dexterity and manipulative skills. Changes in primate (12) _____ [p.398] indicate that there was a shift from eating insects to fruit and leaves and on to a mixed diet. Primates began to evolve from ancestral mammals more than (13) _____ [p.398] million years ago, during the Paleocene. The first primates resembled small (14) _____ [p.398] or tree shrews; they foraged at night for (15) _____ [p.398], seeds, buds, and eggs on the forest floor, and they could clumsily climb trees searching for safety and sleep.

During the (16) _____ [p.399] (25–5 million years ago), the continents began to assume their current positions, and climates became cooler and (17) _____ [p.399]. Under these conditions, forests began to give way to mixed woodlands and (18) _____ [p.399]; an adaptive radiation of apelike forms—the first hominoids—took place as subpopulations of apes became reproductively isolated within the shrinking stands of trees. Between (19) _____ [p.399] million and (20) _____ [p.399] million years ago, three divergences occurred that gave rise to the ancestors of modern gorillas, chimpanzees, and humans.

The family Hominidae (hominids) includes all species on the genetic path leading to humans since the time when that path diverged from the path leading to the (21) _____ [p.399]; hominids emerged between (22) _____ [p.399] million and (23) _____ [p.399] million years ago, during the late Miocene. (24) _____-million-year-old [p.399] fossils of humanlike forms have been discovered in Africa, and they all were bipedal and omnivorous and had an expanded brain. Lucy was one of the earliest (25) _____ [p.399], a collection of forms that combined ape and human features; they were fully two-legged, or (26) _____ [p.399], with essentially human bodies and ape-shaped heads.

The oldest fossils of the genus *Homo*, makers of stone tools, date from approximately (27) _____ [p.400] million years ago. Between 1.8 million and 400,000 years ago, during the Pleistocene periods of glaciation, there were also intermittent periods of warming; during these interglacial times, a larger- brained human species, (28) _____ _____ [p.400], migrated out of Africa and into China, Southeast Asia, and Europe. Over time, (28) became better at toolmaking and learned how to control (29) _____ [p.401] as members of the species became adapted to a wide range of habitats and were associated with abundant (30) _____ [p.400] artifacts. The (31) _____ [p.401] were a distinct hominid population that appeared 200,000 years ago in Southern France, Central Europe, and the Near East; their cranial capacity was indistinguishable from our own, and they had a complex culture.

Anatomically modern humans evolved from (32) _____ _____ [p.401]. Analysis of (33) _____ [p.402] and (34) _____ [p.402] comparisons and specimens from the fossil record support the model of human origins that hypothesizes that *Homo erectus* migrated out of Africa, then formed distinctive subpopulations of modern humans as an outcome of genetic divergence in different geographic regions. From (35) _____ [p.401] years ago to the present, human evolution has been almost entirely cultural rather than biological. By 30,000 years ago, there was only one remaining hominid species: (36) _____ _____ [p.401].

Choice

For questions 37–46, choose from the choices below. Choose the single *smallest* group to which each belongs. If no choice fits, let the blank remain empty.

    a. Anthropoids   b. Hominids   c. Hominoids   d. Prosimians   e. Tarsioids

37. ___ australopiths [p.399]
38. ___ chimpanzees [pp.398–399]
39. ___ gibbons [p.398]
40. ___ gorillas [pp.398–399]
41. ___ humans [p.399]
42. ___ kangaroo [p.397]
43. ___ New World monkeys [p.398]
44. ___ Old World monkeys [p.398]
45. ___ orangutans [p.398]
46. ___ tarsiers [p.398]

Choice

For questions 47–54, choose from the following. (All answers are from text Figure 24.26, p.401.)

a. anthropoids   b. apes   c. hominids   d. hominoids   e. monkeys

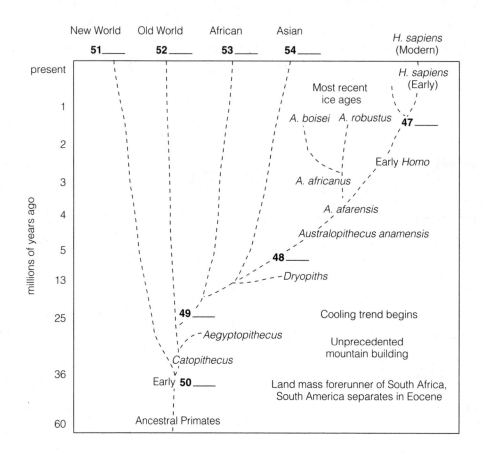

Matching

Suppose you are a student who wants to be chosen to accompany a paleontologist who has spent forty years teaching and roaming the world in search of human ancestors. Above the desk in her office, the paleontologist keeps reconstructions of many skull types. Five are shown on this page, and you have heard that she chooses the graduate students who accompany her on her summer safaris on the basis of their ability on a short matching quiz. The quiz is presented below. *Without consulting your text (or any other)* while you are taking the quiz, match up all five skulls with *all* applicable letters. Seventeen correctly placed letters win you a place on the expedition. Fifteen correct answers put you on a waiting list. Fourteen or fewer, and the paleontologist suggests that you may wish to investigate dinosaur fossils instead of human ancestry. Which will it be for you?

55.

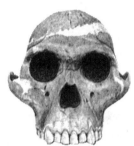

56.

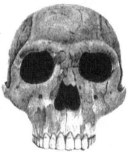

57.

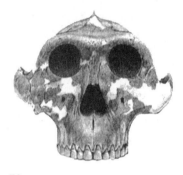

58.

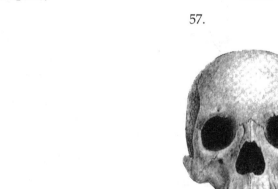

59.

55. _____ [p.399]*

56. _____ [p.399]*

57. _____ [pp.400–401]

58. _____ [p.399]*

59. _____ [p.401]

A. Gracile form(s) of australopiths
B. Robust form(s) of australopiths
C. The most recent hominid to appear
D. Has a chin
E. First controlled use of fire
F. Lived 4 million years ago
G. From a lineage that lived some time during the time span from approximately 3 million years ago until about 1.25 million years ago
H. Lived 2 million years ago
I. Lived 2 million years ago until at least 27,000 years ago
J. A species of *Australopithecus*
K. *Homo erectus*
L. *Homo sapiens*

*See also preceding page of this text. [p.290]

Self-Quiz

___ 1. Filter-feeding chordates rely on _____, which have cilia that create water currents and mucous sheets that capture nutrients suspended in the water. [p.384]
   a. notochords
   b. differentially permeable membranes
   c. filiform tongues
   d. gill slits

___ 2. In true fishes, the gills serve primarily _____ function. [p.386]
   a. a gas-exchange
   b. a feeding
   c. a water-elimination
   d. both a feeding and a gas-exchange

___ 3. The feeding behavior of true fishes selected for highly developed _____. [p.386]
   a. parapodia
   b. notochords
   c. sense organs
   d. gill slits

___ 4. Which of the following is *not* considered to have been a key character in early primate evolution? [p.398]
   a. Eyes adapted for discerning color and shape in a three-dimensional field
   b. Body and limbs adapted for tree climbing
   c. Bipedalism and increased cranial capacity
   d. Eyes adapted for discerning movement in a three-dimensional field

___ 5. Primitive primates generally live _____. [p.398]
   a. in tropical and subtropical forest canopies
   b. in temperate savanna and grassland habitats
   c. near rivers, lakes, and streams in the East African Rift Valley
   d. in caves where there are abundant supplies of insects

___ 6. All the ancestral placental mammals apparently arose from ancestral forms of a group _____. [p.396]
   a. that includes omnivorous shrews and mouselike forms
   b. that includes dogs, cats, and seals
   c. that includes mice and beavers
   d. that includes the koala ("teddy bear") and flying phalanger (resembles the flying squirrel)

___ 7. The hominid evolutionary line stems from a divergence from the ape line that apparently occurred _____. [p.398]
   a. somewhere between 10 million and 5 million years ago
   b. about 3 million years ago
   c. during the Pliocene epoch
   d. less than 2 million years ago

___ 8. Donald Johanson, from the University of California, discovered Lucy (named for the Beatles tune "Lucy in the sky with Diamonds"), who was a(n) _____. [p.398]
   a. dryopith
   b. australopith
   c. member of *Homo*
   d. prosimian

___ 9. A hominid of Europe and Asia that became extinct nearly 30,000 years ago was _____. [p.401]
   a. a dryopith
   b. *Australopithecus*
   c. *Homo habilis*
   d. Neandertal

Matching

Match the following groups and classes with the corresponding characteristics (a–i) and representatives (A–I).

10. __ , __ Amphibians
11. __ , __ Birds
12. __ , __ Bony fishes
13. __ , __ Cartilaginous fishes
14. __ , __ Cephalochordates
15. __ , __ Jawless fishes
16. __ , __ Mammals
17. __ , __ Reptiles
18. __ , __ Urochordates

a. hair + vertebrae [p.396]
b. feathers + hollow bones [p.395]
c. jawless + cartilaginous skeleton (in existing species); vertebrae [p.387]
d. two pairs of limbs (usually) + glandular skin + "jelly"-covered eggs [pp.390–391]
e. amniote eggs + scaly skin + bony skeleton [p.392]
f. invertebrate + an attached adult (usually) that cannot swim [p.384]
g. jaws + cartilaginous skeleton + vertebrae [p.388]
h. in adult, notochord stretches from head to tail; mostly burrowed-in, adult can swim [p.385]
i. bony skeleton + skin covered with scales, fins generally high manuverable [p.389]

A. lancelet
B. loons, penguins, and eagles
C. tunicates, sea squirts
D. sharks and manta rays
E. lampreys and hagfishes (and ostracoderms)

F. true eels and sea horses
G. lizards and turtles
H. caecilians and salamanders
I. platypuses and opossums

Identifying

Provide the common name and the major chordate group to which each creature pictured below and on p.294 belongs.

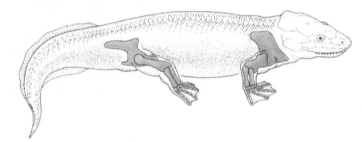

19. [p.390] _____ , _____

20. [p.397] _____ , _____

21. [p.389] _____ , _____

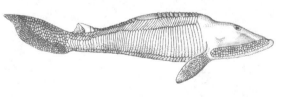

22. [p.386] _____ , _____

(Continued on p.278)

23. [p.394] _____, _____

24. [p.393] _____, _____

25. [p.388] _____, _____

26. [p.389] _____, _____

27. [p.388] _____, _____

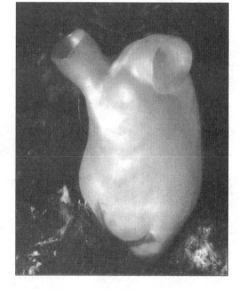

28. [p.385] _____, _____

29. [p.385] _____, _____

Matching

Choose the most appropriate answer for each term.

30. ___anthropoids [p.398]

31. ___australopiths [p.399]

32. ___Cenozoic [recall pp.305, 307]

33. ___hominids [p.398]

34. ___hominoids [p.399]

35. ___Miocene [p.399]

36. ___primates [p.398]

37. ___prosimians [p.398]

A. A group that includes apes and humans
B. Organisms in a suborder that includes New World and Old World monkeys, apes, and humans
C. An era that began 65 to 63 million years ago; characterized by the evolution of birds, mammals, and flowering plants
D. A group that includes humans and their most recent ancestors
E. An epoch of the Cenozoic era lasting from 25 million to 5 million years ago; characterized by the appearance of primitive apes, whales, and grazing animals of the grasslands
F. Organisms in a suborder that includes tree shrews, lemurs, and others
G. A group that includes prosimians, tarsioids, and anthropoids
H. Bipedal organisms living from about 4 million to 1 million years ago, with essentially human bodies and ape-shaped heads; brains no larger than those of chimpanzees

Chapter Objectives/Review Questions

1. List four characteristics found only in chordates. [p.384]
2. Describe the adaptations that sustain the sessile or sedentary life-style seen in primitive chordates such as tunicates and lancelets. [pp.384–385]
3. State what sort of changes occurred in the primitive chordate body plan that could have encouraged the emergence of vertebrates. [pp.386–387]
4. Describe the differences between primitive and advanced fishes in terms of skeleton, jaws, special senses, and brain. [pp.387–389]
5. Describe the changes that enabled aquatic fishes to give rise to swamp dwellers. [pp.390–391]
6. List the principal characteristics that distinguish each major group of four-limbed vertebrates. [pp.384,390,392,394, and 396]
7. Where do tarsier survivors dwell today on Earth? [p.398]
8. Beginning with the primates most closely related to humans, list the main groups of primates in order by decreasing closeness of relationship to humans. [pp.398,401]
9. Five key characters of primate evolution are _____ , _____ , _____ , _____ , and _____ . [p.398]
10. Describe the general physical features and behavioral patterns attributed to early primates. [p.398]
11. Trace primate evolutionary development through the Cenozoic era. Describe how Earth's climates were changing as primates changed and adapted. Be specific about times of major divergence. [pp.398–401]
12. State which anatomical features underwent the greatest changes along the evolutionary line from early anthropoids to humans. [pp.398–400]
13. Explain how you think *Homo sapiens* arose. Make sure your theory incorporates existing paleontological (fossil), biochemical, morphological, and immunological data. [pp.401–402]

Integrating and Applying Key Concepts

Suppose someone told you that some time between 12 million and 6 million years ago, early tree-dwelling apes were forced by larger predatory members of the cat family to flee the forests and take up residence in estuarine, riverine, and sea coastal habitats where they could take refuge in the nearby water to evade the tigers. Those that, through mutations, became naked, developed an upright stance, developed subcutaneous fat deposits as insulation, and developed a bridged nose that had advantages in watery habitats (features that other dryopiths that remained inland never developed) survived and expanded their populations.

As time went on, predation by the big cats and competition with other animals for available food caused most of the terrestrial tree apes to become extinct, but the water-habitat varieties survived as scattered remnant populations, adapting to easily available shellfish and fish, wild rice and oats, and various tubers, nuts, and fruits. It was in these aquatic habitats that the first food-getting tools (baskets, nets, and pebble tools) were developed, as well as the first words that signified different kinds of food. How does such a story fit with current speculations about and evidence of human origins? How could such a story be shown to be true or false?

25

PLANT TISSUES

Interactive Exercises

Plants Versus the Volcano [pp.406–407]

25.1. OVERVIEW OF THE PLANT BODY [pp.408–409]

Selected Words: "annuals" [p.408], *nonwoody* [p.408], "biennials" [p.408], "perennials" [p.408], *Lycopersicon* [p.408], *primary* meristems [p.409], *lengthening* of stems [p.409], *thickening* of stems [p.409]

Boldfaced, Page-Referenced Terms

[p.408] shoots _____

_____

[p.408] roots _____

_____

[p.408] ground tissue system _____

_____

[p.408] vascular tissue system _____

_____

[p.408] dermal tissue system _____

_____

[p.409] meristems _____

_____

[p.409] apical meristem _____

_____

[p.409] vascular cambium _____

_____

[p.409] cork cambium _____

_____

Label–Match

Identify each part of the accompanying illustration. Choose from dermal tissues (epidermis), root system, ground tissues, shoot system, and vascular tissues. Complete the exercise by matching and entering the letter of the proper description in the parentheses following each label.

1. _____ _____ ( ) [p.408]
2. _____ _____ ( ) [p.408]
3. _____ _____ ( ) [p.408]
4. _____ _____ ( ) [p.408]
5. _____ _____ ( ) [p.408]

A. Typically consists of stems, leaves, flowers (reproductive shoots)
B. Typically grows below ground, anchors aboveground parts, absorbs soil water and minerals, stores and releases food, and anchors the aboveground parts
C. Tissues that make up the bulk of the plant body
D. Covers and protects the plant's surfaces
E. Two conducting tissues that distribute water and solutes through the plant body

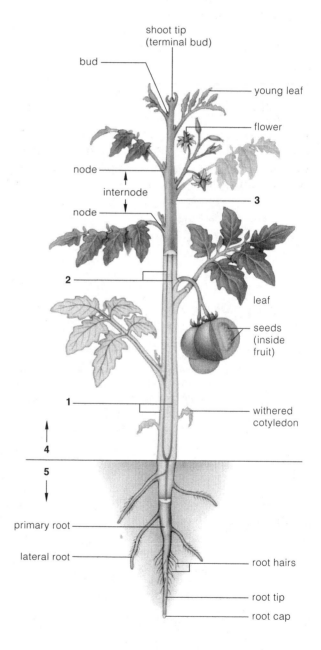

Matching

Choose the most appropriate answer for each term.

6. ___secondary growth [p.409]

7. ___vascular cambium and cork cambium [p.409]

8. ___meristems [p.409]

9. ___apical meristems [p.409]

10. ___primary growth [p.409]

11. ___transitional meristems [p.409]

A. Represented by a lengthening of stems and roots originating from cell divisions at apical meristems

B. Localized regions of dividing cells

C. Cell populations forming from apical meristems; include protoderm, ground meristem, and procambium

D. Lateral meristems giving rise to, respectively, secondary vascular tissues, and a sturdier plant covering that replaces epidermis

E. Represented by a thickening of stems and roots due to activity of the lateral meristems

F. Located in the dome-shaped tips of all shoots and roots; responsible for lengthening those organs

Labeling

It is important to understand the terms that identify the thin sections (slices) of plant organs and tissues that are prepared for study. Label each of the three diagrams below. Choose from radial section (cut along the radius of the organ), transverse or cross section (cuts perpendicular to the long axis of the organ), or tangential section (cuts made at right angles to the radius of the organ). Note: The dark area indicates the slice.

12. _____ section [p.409]

13. _____ section [p.409]

14. _____ section [p.409]

Label–Match

Identify each part of the accompanying illustration. Choose from shoot apical meristem, root apical meristem, root transitional meristems, shoot transitional meristems, and lateral meristems. Complete the exercise by matching and entering the letter of the proper description in the parentheses following each label.

15. _____ _____ _____ ( ) [p.409]

16. _____ _____ _____ ( ) [p.409]

17. _____ _____ _____ ( ) [p.409]

18. _____ _____ _____ ( ) [p.409]

19. _____ _____ ( ) [p.409]

A. Near all root tips, also gives rise to three transitional meristems from which the root's primary tissue systems develop (lengthening)

B. These embryonic descendants of apical meristem (protoderm, ground meristem, and procambium) divide grow, and differentiate into a shoot's primary tissue systems

C. Found only inside the older stems and roots of woody plants; sources of secondary growth (increases in diameter)

D. A region of embryonic cells near the dome-shaped tip of all shoots; the source of primary growth (lengthening)

E. Embryonic descendants of the root apical meristem (the protoderm, ground meristem, and procambium) that divide, grow, and differentiate into the root's primary tissue systems

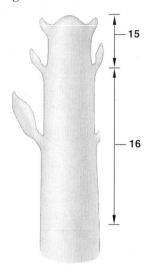

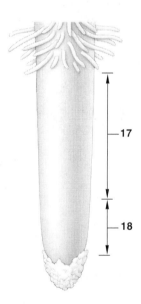

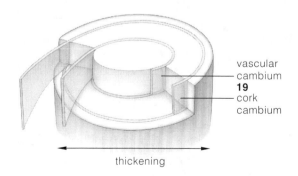

vascular cambium
19
cork cambium

thickening

Complete the Table

20. The cells of the transitional meristems and the lateral meristems give rise to tissues, either primary or secondary. Complete the table below, which summarizes the activity of these meristems.

Meristem	Tissue(s)	Primary or Secondary
[p.409] a. Protoderm		
[p.409] b. Ground meristem		
[p.409] c. Procambium		
[p.409] d. Vascular cambium		
[p.409] e. Cork cambium		

25.2. TYPES OF PLANT TISSUES [pp.410–411]

Selected Words: mesophyll [p.410], *fibers* [p.409], *sclereids* [p.410], *vessel members* [p.411], *tracheids* [p.411], *sieve-tube members* [p.411], *companion cells* [p.411]

Boldfaced, Page-Referenced Terms

[p.410] parenchyma _____

_____

[p.410] collenchyma _____

_____

[p.410] sclerenchyma _____

_____

[p.411] xylem _____

_____

[p.411] phloem _____

_____

[p.411] epidermis _____

_____

[p.411] cuticle _____

_____

[p.411] stoma (plural, stomata) _____

_____

[p.411] periderm _____

_____

[p.411] dicots _____

_____

[p.411] monocots _____

_____

Labeling

Label the following illustrations as either collenchyma, (uneven wall thickenings, often appearing like thickened corners); parenchyma (thin-walled with intercellular spaces); or sclerenchyma (very thick walls in cross-section, includes fibers and sclereids such as stone cells).

1. _____
   [p.410]

2. _____
   [p.410]

3. _____
   [p.410]

thick secondary wall

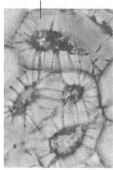

4. _____
   [p.410]

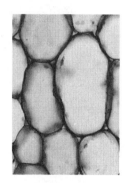

5. _____
   [p.410]

6. _____
   [p.410]

Choice

For questions 7–17, choose from the following:

a. parenchyma   b. collenchyma   c. sclerenchyma

7. ___ Patches or cylinders of this tissue are found near the surface of lengthening stems. [p.410]

8. ___ The cells have thick, lignin-impregnated walls. [p.410]

9. ___ Cells are alive at maturity and retain the capacity to divide. [p.410]

10. ___ Some types specialize in storage, secretion, and other tasks. [p.410]

11. ___ Cells are mostly elongated, with unevenly thickened walls. [p.410]

12. ___ Provides flexible support for primary tissues [p.410]

13. ___ Abundant air spaces are found around the cells. [p.410]

14. ___ Supports mature plant parts and often protects seeds [p.410]

15. ___ Cells are thin-walled, pliable, and many-sided. [p.410]

16. ___ Mesophyll is specialized for photosynthesis. [p.410]

17. ___ Fibers and sclereids [p.410]

Labeling

Identify the cell types and cellular structures in the accompanying illustrations by entering the correct name in the blanks. Complete the exercise by entering the name of the complex tissue (xylem or phloem) in which that cell or structure is found.

18. _____ (_____) [p.410]

19. _____ (_____) [p.410]

20. _____ (_____) [p.410]

21. _____ (_____) [p.410]

22. _____ (_____) [p.410]

23. _____ (_____) [p.410]

24. _____ (_____) [p.410]

25. _____ (_____) [p.410]

26. _____ (_____) [p.410]

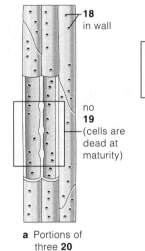

18 in wall

no **19** (cells are dead at maturity)

a Portions of three **20**

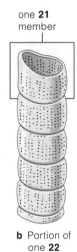

one **21** member

b Portion of one **22**

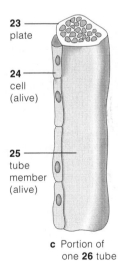

23 plate

24 cell (alive)

25 tube member (alive)

c Portion of one **26** tube

Complete the Table

27. The two types of vascular tissues, xylem and phloem, occur in strands called vascular bundles within the plant body (each of these complex tissues also has fibers and parenchyma). Complete the table below, which summarizes important information about these vascular tissues.

Vascular Tissue	Major Conducting Cells	Cells Alive?	Function(s)
[p.411] a. Xylem			
[p.411] b. Phloem			

28. Two types of dermal tissues cover the plant body. Complete the table below, which summarizes information about the dermal tissues.

Dermal Tissue	Primary or Secondary Plant Body	Function(s)
[p.411] a. Epidermis		
[p.411] b. Periderm		

29. There are two classes of flowering plants, monocots and dicots. Complete the table below, which summarizes information about the two groups of flowering plants.

Class	Number of Cotyledons	Number of Floral Parts	Leaf Venation	Pollen Grains	Vascular Bundles
[p.411] a. Monocots					
[p.411] b. Dicots					

25.3. PRIMARY STRUCTURE OF SHOOTS [pp.412–413]

Selected Words: Coleus [p.412], Medicago [p.413], Zea mays [p.413]

Boldfaced, Page-Referenced Terms

[p.412] bud _____

_____

[p.412] vascular bundles _____

_____

[p.412] cortex _____

_____

[p.412] pith _____

_____

Labeling

Name the structures numbered in the illustrations of leaf development and leaf forms below.

1. _____ [p.412]

2. _____ _____ [p.412]

3. _____ [p.412]

4. _____ [p.412]

5. _____ [p.412]

6. _____ [p.412]

7. _____ [p.412]

8. _____ [p.412]

9. _____ _____ [p.412]

10. _____ [p.412]

11. _____ [p.412]

12. _____ [p.412]

13. _____ [p.412]

14. _____ _____ [p.412]

15. _____ _____ [p.412]

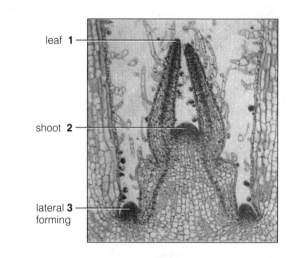

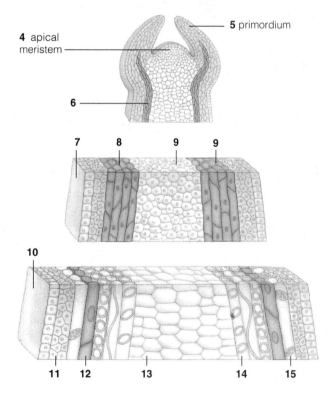

Labeling

Name the structures numbered in the illustrations of the monocot stem below. Choose from air space, epidermis, sclerenchyma cells, vessel, vascular bundle, sieve tube, ground tissue, and companion cell.

16. _____ [p.413]

17. _____ _____ [p.413]

18. _____ _____ [p.413]

19. _____ _____ [p.413]

20. _____ _____ [p.413]

21. _____ _____ [p.413]

22. _____ _____ [p.413]

23. _____ _____ [p.413]

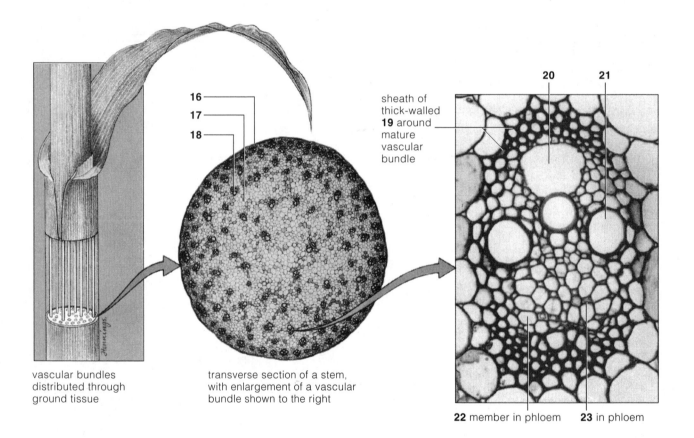

vascular bundles distributed through ground tissue

transverse section of a stem, with enlargement of a vascular bundle shown to the right

sheath of thick-walled **19** around mature vascular bundle

22 member in phloem   **23** in phloem

Labeling

Name the structures numbered in the illustrations of the herbaceous dicot stem below. Choose from sieve tube and companion cells in phloem, vascular bundle, xylem vessels, cortex, vascular cambium, pith, phloem fibers, and epidermis.

24. _____ [p.413]

25. _____ [p.413]

26. _____ _____ [p.413]

27. _____ [p.413]

28. _____ [p.413]

29. _____ _____ [p.413]

30. _____ _____ [p.413]

31. _____ [p.413]

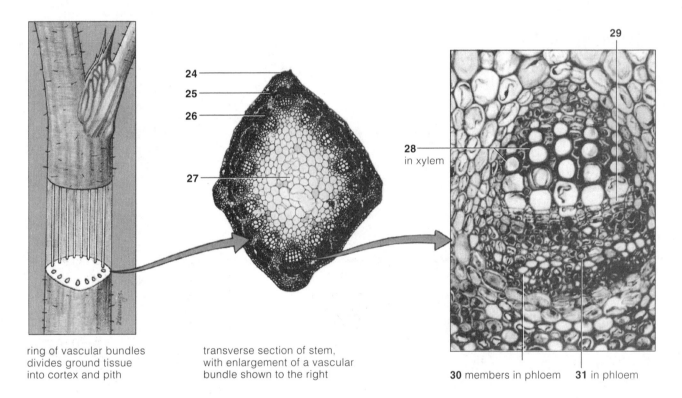

ring of vascular bundles divides ground tissue into cortex and pith

transverse section of stem, with enlargement of a vascular bundle shown to the right

30 members in phloem   **31** in phloem

25.4. A CLOSER LOOK AT LEAVES [pp.414–415]

Selected Words: *deciduous* [p.414], *simple* leaves [p.414], *compound* leaves [p.414], *Agapanthus* [p.414], *Phaseolus* [p.415], *palisade* mesophyll [p.415], *spongy* mesophyll [p.415], *Nicotiana* [p.415], *Cannabis sativa* [p.415], *Erythroxylum* [p.415], *Hyoscyamus niger* [p.415]

Boldfaced, Page-Referenced Terms

[p.414] leaf _____

_____

[p.414] mesophyll _____

_____

[p.415] veins _____

_____

Labeling

Name the structures numbered in the illustrations of leaf development and leaf forms.

1. _____ [p.414]
2. _____ [p.414]
3. _____ _____ [p.414]
4. _____ [p.414]
5. _____ [p.414]
6. _____ [p.414]

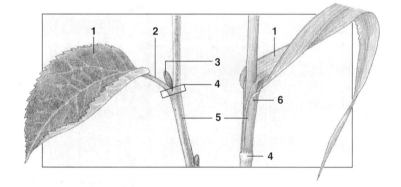

Dichotomous Choice

Circle one of two possible answers given between parentheses in each statement.

7. The leaf illustrated on the left above is a (monocot/dicot). [p.414]
8. The leaf illustrated on the right above is a (monocot/dicot) [p.414]

Label–Match

Identify each numbered part of the accompanying illustration [see text Figure 25.16 on p.415]. Complete the exercise by matching and entering the letter of the proper function description in the parentheses following each label.

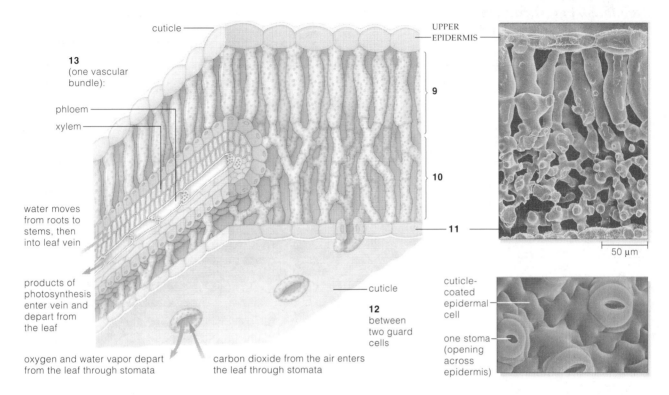

9. _____ _____ ( ) [p.415]

10. _____ _____ ( ) [p.415]

11. _____ _____ ( ) [p.415]

12. _____ ( ) [p.415]

13. _____ _____ ( ) [p.415]

A. Lowermost cuticle-covered cell layer
B. Loosely packed photosynthetic parenchyma cells just above the lower epidermal layer
C. Allows movement of oxygen and water vapor out of leaves and allows carbon dioxide to enter
D. Photosynthetic parenchyma cells just beneath the upper epidermis
E. Move water and solutes to photosynthetic cells and carry products away from them

Matching

Choose the most appropriate answer for each term.

14. ___nightshade leaves [p.415]

15. ___tobacco plants [p.415]

16. ___marijuana [p.415]

17. ___coca leaves [p.415]

18. ___henbane [p.415]

A. Smoked by Mayan priests to carry their priestly thoughts to the gods
B. Produces mind-altering products; linked with low sperm counts
C. Romeo dropped dead after sipping this potion
D. Poured in Hamlet's ear to cause his death
E. Source of cocaine; has mind-bending properties with devastating social and economic effects

25.5. PRIMARY STRUCTURE OF ROOTS [pp.416–417]

Selected Words: *adventitious* structures [p.416], *Eschscholzia* [p.416], *Zea mays* [p.416], *Ranunculus* [p.417], *Salix* [p.417]

Boldfaced, Page-Referenced Terms

[p.416] primary root _____

_____

[p.416] lateral roots _____

_____

[p.416] taproot system _____

_____

[p.416] fibrous root system _____

_____

[p.416] root hairs _____

_____

[p.417] vascular cylinder _____

_____

Label–Match

Identify each numbered part of the accompanying illustration. Complete the exercise by matching the letter of the proper description in the parentheses following each label. Some choices are used more than once.

1. _____ _____ ( ) [p.416]

2. _____ ( ) [p.416]

3. _____ ( ) [p.416]

4. _____ ( ) [p.416]

5. _____ ( ) [p.416]

6. _____ _____ _____ ( ) [p.416]

7. _____ _____ ( ) [p.416]

8. _____ ( ) [p.416]

9. _____ ( ) [p.416]

10. _____ _____ ( ) [p.416]

A. Dome-shaped cell mass produced by the apical meristem

B. Part of the vascular cylinder; gives rise to lateral roots

C. Part of the vascular cylinder; transports photosynthetic products

D. Ground tissue region surrounding the vascular cylinder

E. The absorptive interface with the root's environment

F. The region of apical and primary meristems

G. Innermost part of the root cortex; helps control water and mineral movement into the vascular column

H. Epidermal cell extensions; greatly increases the surface available for taking up water and solutes

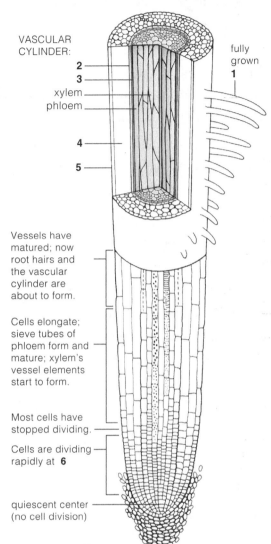

VASCULAR CYLINDER:

2
3
xylem
phloem

4

5

fully grown
1

Vessels have matured; now root hairs and the vascular cylinder are about to form.

Cells elongate; sieve tubes of phloem form and mature; xylem's vessel elements start to form.

Most cells have stopped dividing.

Cells are dividing rapidly at 6

quiescent center (no cell division)

7

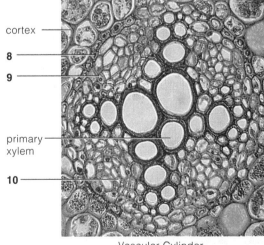

cortex

8

9

primary xylem

10

Vascular Cylinder

Fill-in-the-Blanks

The first part of a seedling to emerge from the seed coat is the (11) _____ [p.416] root. In most dicot seedlings, the primary root increases in diameter while it grows downward. Later, (12) _____ [p.416] roots begin forming in internal tissues at an angle perpendicular to the primary root's axis, then they erupt through epidermis. Oak trees, carrots, poppies, and dandelions are examples of plants whose primary root and lateral branchings represent a(n) (13) _____ [p.416] system. In monocots such as grasses, the primary root is short-lived; in its place, numerous (14) _____ [p.416] roots arise from the stem of the young plant. Such roots and their branches are somewhat alike in length and diameter and form a(n) (15) _____ [p.416] root system.

Some root epidermal cells send out absorptive extensions called root (16) _____ [p.416]. Vascular tissues form a(n) (17) _____ [p.417] cylinder, a central column inside the root consisting of primary xylem and phloem and one or more layers of parenchyma called the (18) _____ [p.417]. Ground tissues surrounding the cylinder are called the root (19) _____ [p.417]. A monocot's vascular cylinder divides the ground tissue system into cortex and (20) _____ [p.417] regions. There are many (21) _____ [p.417] spaces in between cells of the ground tissue system, and (22) _____ [p.417] can easily diffuse through them. Water entering the root moves from cell to cell until it reaches the (23) _____ [p.417], the innermost cell layer of the root cortex. Abutting walls of its cells are waterproof, so they force incoming water to pass through the cytoplasm. This arrangement helps (24) _____ [p.417] the movement of water and dissolved substances into the vascular cylinder. Just inside the endodermis is the (25) _____ [p.417]. This part of the vascular cylinder gives rise to (26) _____ [p.417] roots, which then erupt through the (27) _____ [p.417] and epidermis.

25.6. ACCUMULATED SECONDARY GROWTH—THE WOODY PLANTS [pp.418–419]

Selected Words: *nonwoody* plants [p.418], *woody* plants [p.418], *inner* face [p.418], *outer* face [p.418], secondary xylem [p.418], *girdling* [p.418], *primary* growth [p.418], *secondary* growth [p.418], *early* wood [p.419], *late* wood [p.419], "tree rings" [p.419]

Boldfaced, Page-Referenced Terms

[p.418] annuals _____

_____

[p.418] perennials _____

_____

[p.418] cork _____

_____

[p.418] bark _____

_____

[p.418] heartwood _____

_____

[p.419] sapwood _____

_____

[p.419] growth rings _____

_____

[p.419] hardwoods _____

_____

[p.419] softwood _____

_____

Matching

Choose the most appropriate answer for each term.

1. ___ tree rings [p.419]
2. ___ heartwood [pp.418–419]
3. ___ sapwood [p.419]
4. ___ early wood [p.419]
5. ___ nonwoody plants [p.418]
6. ___ woody plants [p.418]
7. ___ softwood [p.419]
8. ___ girdling [p.418]
9. ___ bark [p.418]
10. ___ cork [p.418]
11. ___ late wood [p.419]
12. ___ annuals [p.418]
13. ___ hardwoods [p.419]
14. ___ perennials [p.418]

A. Produced by cells of the cork cambium
B. Includes all tissues external to the vascular cambium
C. Alternating bands of early and late wood, which reflect light differently; show secondary growth during two or more growing seasons
D. Complete the life cycle in a single growing season and are generally herbaceous
E. Another term for herbaceous plants
F. The first xylem cells produced at the start of the growing season; tend to have large diameters and thin walls
G. Produced by conifers that lack fibers and vessels in their xylem
H. Term applied to the wood of dicot trees that evolved in temperate and tropical regions; possess vessels, tracheids, and fibers in their xylem
I. Some monocots and many dicots; they put on extensive secondary growth over two or more growing seasons
J. Formed in dry summer, has xylem cells with smaller diameters and thicker walls
K. A dumping ground at the center of older stems and roots for metabolic wastes such as resins, oils, gums, and tannins
L. Removal of a band of secondary phloem around a trunk's circumference; the tree will eventually die
M. Secondary growth located between heartwood and the vascular cambium; wet, usually pale
N. Plants that add secondary growth during two or more growing seasons

Label–Match

Identify each numbered part of the accompanying illustration. Complete the exercise by matching and entering the letter of the proper description in the parentheses following each label.

15. _____ ( ) [p.419]

16. _____ _____ ( ) [p.419]

17. _____ _____ ( ) [p.419]

18. _____ ( ) [p.419]

19. year _____ ( ) [p.419]

20. and 21. years _____ and _____ ( ) [p.419]

22. _____ _____ ( ) [p.419]

A. All secondary growth
B. Produced later in the growing season; vessels have smaller diameters with thick walls
C. Includes the primary growth and some secondary growth
D. Includes all tissues external to the vascular cambium
E. Large-diameter conducting cell in the xylem
F. Produced early in the growing season; vessels have large diameters and thin walls
G. Produces secondary vascular tissues, xylem, and phloem

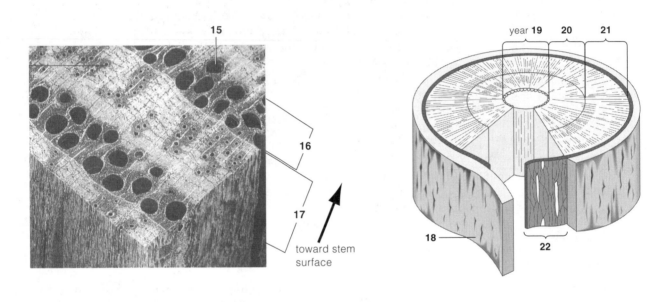

toward stem surface

year **19** **20** **21**

Self-Quiz

___ 1. _____ covers and protects the plant's surfaces. [p.408]
   a. Ground tissue
   b. Dermal tissue
   c. Vascular tissue
   d. Pericycle

___ 2. Which of the following is *not* considered a type of simple tissue? [p.410]
   a. Epidermis
   b. Parenchyma
   c. Collenchyma
   d. Sclerenchyma

___ 3. Of the following cell types, which one does not appear in vascular tissues? [p.411]
   a. vessel members
   b. cork cells
   c. tracheids
   d. sieve-tube members
   e. companion cells

___ 4. The _____ is a leaflike structure that is part of the embryo; monocot embryos have one, dicot embryos have two. [p.411]
   a. shoot tip
   b. root tip
   c. cotyledon
   d. apical meristem

___ 5. Each part of the stem where one or more leaves are attached is a(n) _____. [p.412]
a. node
b. internode
c. vascular bundle
d. cotyledon

___ 6. Which of the following structures is *not* considered to be meristematic? [p.409]
a. vascular cambium
b. lateral meristem
c. cork cambium
d. endodermis

___ 7. Which of the following statements about monocots is *false*? [p.411]
a. They are usually herbaceous.
b. They develop one cotyledon in their seeds.
c. Their vascular bundles are scattered throughout the ground tissue of their stems.
d. They have a single central vascular cylinder in their stems.

___ 8. New plants grow and older plant parts lengthen through cell divisions at _____ meristems present at root and shoot tips; older roots and stems of woody plants increase in diameter through cell divisions at _____ meristems. [p.409]
a. lateral; lateral
b. lateral; apical
c. apical; apical
d. apical; lateral

___ 9. Vascular bundles called _____ form a network through a leaf blade. [p.415]
a. xylem
b. phloem
c. veins
d. cuticles
e. vessels

___10. A primary root and its lateral branchings represent a(n) _____ system. [p.416]
a. lateral root
b. adventitious root
c. taproot
d. branch root

___11. Plants whose vegetative growth and seed formation continue year after year are _____ plants. [p.418]
a. annual
b. perennial
c. nonwoody
d. herbaceous

___12. The _____ layer of a root divides to produce lateral roots. [p.417]
a. endodermis
b. pericycle
c. xylem
d. cortex
e. phloem

Chapter Objectives/Review Questions

1. The aboveground parts of flowering plants are called _____; the plants' descending parts are called _____. [p.408]
2. Distinguish between the ground tissue system, the vascular tissue system, and the dermal tissue system. [p.408]
3. Plants grow at localized regions of self-perpetuating embryonic cells called _____. [p.409]
4. Lengthening of stems and roots originates at _____ meristems and all dividing tissues derived from them; this is called _____ growth. [p.409]
5. Cell populations of protoderm, ground meristem, and procambium are derived from the apical meristem and are known as _____ meristems. [p.409]
6. Increases in the thickness of a plant originate at _____ meristems. [p.409]
7. Describe the role of vascular cambium and cork cambium in producing secondary tissues of the plant body. [p.409]
8. Be able to visually identify and generally describe the simple tissues called parenchyma, collenchyma, and sclerenchyma. [p.410]
9. Fibers and sclereids are both types of _____ cells. [p.410]

10. Name the cell wall compound that was necessary for the evolution of rigid and erect land plants. [p.410]
11. _____ tissue conducts soil water and dissolved minerals, and it mechanically supports the plant. [p.411]
12. _____ tissue transports sugars and other solutes. [p.411]
13. Name and describe the functions of the conducting cells in xylem and phloem. [p.411]
14. All surfaces of primary plant parts are covered and protected by a dermal tissue system called _____ and a surface coating called a _____. [p.411]
15. What is the general function of guard cells and stomata found within the epidermis of young stems and leaves? [p.411]
16. The cork cells of _____ replace the epidermis of stems and roots showing secondary growth. [p.411]
17. Distinguish between monocots and dicots by listing their characteristics and citing examples of each group. [p.411]
18. A(n) _____ is an undeveloped shoot of mostly meristematic tissue, often protected by scales; it gives rise to new stems, leaves, and flowers. [p.412]
19. The primary xylem and phloem develop as vascular _____. [p.412]
20. Distinguish between the stem's cortex and its pith. [p.412]
21. Be able to visually distinguish between monocot stems and dicot stems, as seen in cross-section. [p.413]
22. Describe the principal difference between deciduous and evergreen plants. [p.414]
23. How does the "simple" leaf type differ from "compound" leaves? [p.414]
24. Describe the structure (cells and layers) and major functions of leaf epidermis, mesophyll, and vein tissue. [pp.414–415]
25. Be able to list the names of three plants and their negative effects as given in the section on *Curious and Misguided Uses of Leaves.* [p.415]
26. The _____ root is the first to poke through the coat of a germinating seed; later, _____ roots erupt through the epidermis. [p.416]
27. How does a taproot system differ from a fibrous root system? [p.416]
28. Define the term *adventitious*. [p.416]
29. Describe the origin and function of root hairs. [pp.416–417]
30. A _____ _____ consists of primary xylem and primary phloem and one or more layers of parenchyma cells called the pericycle. [p.417]
31. Describe the passage of soil water through root epidermis to the xylem of the vascular cylinder; include the role of the endodermis. [p.417]
32. Define these categories of flowering plants: *annuals* and *perennials*. [p.418]
33. Distinguish a woody plant from a nonwoody plant in terms of secondary growth. [p.418]
34. Each growing season, new tissues that increase the girth of woody plants originate at their _____ meristems. [p.418]
35. Describe the formation of cork and bark. [p.418]
36. Distinguish early wood from late wood; heartwood from sapwood. [pp.418–419]
37. Explain the origin of the annual growth layers (tree rings) seen in a cross-section of a tree trunk. [p.419]
38. Hardwood trees possess _____ and _____ but softwood trees lack these cells. [p.419]

Integrating and Applying Key Concepts

Try to imagine the specific behavioral restrictions that might be imposed if the human body resembled the plant body in having (1) open growth with apical meristematic regions, (2) stomata in the epidermis, (3) cells with chloroplasts, (4) excess carbohydrates stored primarily as starch rather than as fat, and (5) dependence on the soil as a source of water and inorganic compounds.

26

PLANT NUTRITION AND TRANSPORT

Flies for Dinner

SOIL AND ITS NUTRIENTS
- Properties of Soil
- Nutrients Essential for Plant Growth
- Leaching and Erosion

HOW DO ROOTS ABSORB WATER AND MINERAL IONS?
- Absorption Routes
- Specialized Absorptive Structures

HOW IS WATER TRANSPORTED THROUGH PLANTS?
- Transpiration Defined
- Cohesion-Tension Theory of Water Transport

HOW DO STEMS AND LEAVES CONSERVE WATER?
- The Water-Conserving Cuticle
- Controlled Water Loss at Stomata

HOW ARE ORGANIC COMPOUNDS DISTRIBUTED THROUGH PLANTS?
- Translocation
- Pressure Flow Theory

Interactive Exercises

Flies for Dinner [pp.422–423]

26.1. SOIL AND ITS NUTRIENTS [pp.424–425]

Selected Words: *Dionaea muscipula* [p.422], *Darlingtonia californica* [p.423], *profile* properties [p.424], *macro*nutrients [p.425], *micro*nutrients [p.425]

Boldfaced, Page-Referenced Terms

[p.422] carnivorous plants _____

_____

[p.422] plant physiology _____

_____

[p.424] soil _____

_____

[p.424] humus _____

_____

[p.424] loams _____

_____

[p.424] topsoil _____

_____

[p.424] nutrients _____

_____

[p.425] leaching _____

_____

[p.425] erosion _____

_____

Matching

Choose the most appropriate answer for each term.

1. ___soil [p.424]
2. ___humus [p.424]
3. ___loams [p.424]
4. ___profile properties [p.424]
5. ___topsoil [p.424]
6. ___nutrients [p.424]
7. ___macronutrients [p.425]
8. ___micronutrients [p.425]
9. ___leaching [p.425]
10. ___erosion [p.425]

A. Elements essential for a given organism because, directly or indirectly, they have roles in metabolism that are unable to be fulfilled by any other element
B. Refers to the layered characteristics of soils, which are in different stages of development in different places
C. Refers to the decomposing organic material in soil
D. The movement of land under the force of wind, running water, and ice
E. Elements other than the macronutrients that are essential for plant growth
F. Consists of particles of minerals mixed with variable amounts of decomposing organic material
G. Refers to the removal of some of the nutrients in soil as water percolates through it
H. Uppermost part of the soil that is highly variable in depth (A horizon); the most essential layer for plant growth
I. Soils having more or less equal proportions of sand, silt, and clay; best for plant growth
J. Nine of the essential elements required for plant growth

Fill-in-the-Blanks

The three essential elements that plants use as their main metabolic building blocks are oxygen, carbon, and

(11) _____ [p.424]. Besides these elements, plants depend on the uptake of at least (number)

(12) _____ [p.425] others. These are typically available to plants in (13) _____ [p.425] forms. Plants

give up (14) _____ [p.425] ions to the clay in exchange for weakly bound elements such as Ca^{++} and K^+.

Nine essential elements are (15) _____ [p.425] and are required in amounts above 0.5 percent of the

plant's dry weight. The other elements are (16) _____ [p.425]; they make up traces of the plant's dry

weight.

Complete the Table

17. Thirteen essential elements are available to plants as mineral ions. Complete the table below, which summarizes information about these important plant nutrients.

Mineral Element	Macronutrient or Micronutrient	Known Functions
[p.425] a.		Roles in chlorophyll synthesis, electron transport
[p.425] b.		Activation of enzymes, role in maintaining water-solute balance
[p.425] c.		Role in chlorophyll synthesis; coenzyme activity
[p.425] d.		Role in root, shoot growth; role in photolysis
[p.425] e.		Role in chlorophyll synthesis; coenzyme activity
[p.425] f.		Component of enzyme used in nitrogen metabolism
[p.425] g.		Component of proteins, nucleic acids, coenzymes, chlorophyll
[p.425] h.		Component of most proteins, two vitamins
[p.425] i.		Roles in flowering, germination, fruiting, cell division, nitrogen metabolism
[p.425] j.		Role in formation of auxin, chloroplasts, and starch; enzyme component
[p.425] k.		Roles in cementing cell walls, regulation of many cell functions
[p.425] l.		Component of several enzymes
[p.425] m.		Component of nucleic acids, phospholipids, ATP

26.2. HOW DO ROOTS ABSORB WATER AND MINERAL IONS? [pp.426–427]

Selected Words: *Rhizobium* [p.427], *Bradyrhizobium* [p.427]

Boldfaced, Page-Referenced Terms

[p.426] vascular cylinder _____

_____

[p.426] endodermis _____

_____

[p.426] Casparian strip _____

_____

[p.426] exodermis _____

_____

[p.426] root hairs _____

_____

[p.426] mutualism _____

_____

[p.427] nitrogen fixation _____

_____

[p.427] root nodules _____

_____

[p.427] mycorrhizae (singular, mycorrhiza) _____

_____

Label–Match

Identify each numbered part of the illustration below that deals with the control of nutrient uptake by plant roots. Choose from the following: epidermis, water movement, cytoplasm, root hair, vascular cylinder, endodermal cell wall, exodermis, endodermis, cortex, and Casparian strip. Complete the exercise by matching from the list below and entering the correct letter in the parentheses following each label.

1. _____ (   ) [p.426]

2. _____ _____ (   ) [p.426]

3. _____ (   ) [p.426]

4. _____ _____ (   ) [p.426]

5. _____ (   ) [p.426]

6. _____ (   ) [p.426]

7. _____ (   ) [p.426]

8. _____ _____ (   ) [p.426]

9. _____ _____ (   ) [p.426]

10. _____ _____ _____ (   ) [p.426]

A. Cellular area through which water and dissolved nutrients must move due to Casparian strips

B. A layer of cortex cells just inside the epidermis; also equipped with Casparian strips

C. Cellular area of a root between the exodermis (if present) and the endodermis

D. Waxy band acting as an impermeable barrier between the walls of abutting endodermal cells; forces water and dissolved nutrients through the cytoplasm of endodermal cells

E. Tiny extensions of the root epidermal cells that greatly increase absorptive capacity of a root

F. Specific location of the waxy strips known as Casparian strips

G. Substance whose diffusion occurs through the cytoplasm of endodermal cells due to Casparian strips

H. Outermost layer of root cells

I. A cylindrical layer of cells that wraps around the vascular column

J. Tissues include the xylem, phloem, and pericycle

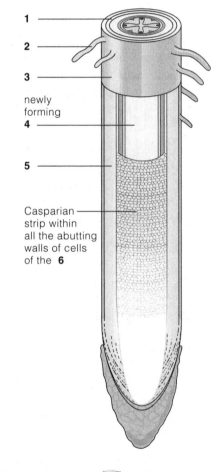

newly forming
4

5

Casparian strip within all the abutting walls of cells of the **6**

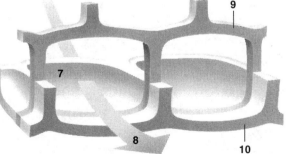

7

8

9

10

Sequence

Arrange in correct chronological sequence the path that water and nutrients take from the soil to cells in living plant tissues. Write the letter of the first step next to 11, the letter of the second step next to 12, and so on.

11. ___ [p.426] A. Cortex cells lacking Casparian strips

12. ___ [p.426] B. Endodermis cells with Casparian strips

13. ___ [p.426] C. Root epidermis and root hairs

14. ___ [p.426] D. Exodermis cells with Casparian strips (in some plants)

15. ___ [p.426] E. Vascular cylinder

Dichotomous Choice

Circle one of two possible answers given between parentheses in each statement.

16. A two-way flow of benefits between species, is a symbiotic interaction known as (mutualism/parasitism). [p.426]
17. (Gaseous nitrogen/Nitrogen "fixed" by bacteria) represents the chemical form of nitrogen plants can use in their metabolism. [p.427]
18. Nitrogen-fixing bacteria reside in localized swellings on legume plants known as (root hairs/root nodules). [p.427]
19. Mycorrhizae represent symbiotic relationships in which fungi and the roots they cover both benefit; the roots receive (sugars and nitrogen-containing compounds/scarce minerals that the fungus is better able to absorb). [p.427]
20. In a mycorrhizal interaction between a young root and a fungus, the fungus, benefits by (obtaining sugars and nitrogen-containing compounds/scarce minerals that the fungus is better able to absorb). [p.427]

26.3. HOW IS WATER TRANSPORTED THROUGH PLANTS? [pp.428–429]

Selected Words: cohesion [p.428], *tension* [p.428]

Boldfaced, Page-Referenced Terms

[p.428] transpiration _____

_____

[p.428] xylem _____

_____

[p.428] tracheids _____

_____

[p.428] vessel members _____

_____

[p.428] cohesion–tension theory _____

_____

Fill-in-the-Blanks

The cohesion–tension theory explains (1) _____ [p.428] transport to the tops of, often very high, plants. It travels pipelines in the (2) _____ [p.428], which are formed by water-conducting cells called (3) _____ [p.428] and (4) _____ [p.428]. The process begins with the drying power of air, which causes (5) _____ [p.428], the evaporation of water from plant parts exposed to air. The collective strength of (6) _____ [p.428] bonds between water molecules in the narrow, tubular xylem cells, imparts (7) _____ [p.428]. This provides unbroken, fluid columns of water. The xylem water columns, as they are pulled upward, are under (8) _____ [p.428]. This force extends from the veins inside leaves, down through the stems, and on into the young (9) _____ [p.428] where water is being absorbed. As long as water molecules continue to escape from the plant, the continuous tension inside the (10) _____ [p.428] permits more molecules to be pulled upward from the roots to replace them.

26.4. HOW DO STEMS AND LEAVES CONSERVE WATER? [pp.430–431]

Selected Words: inward diffusion [p.430], *outward* diffusion [p.430], *Opuntia* [p.430], *Commelina communis* [p.431]

Boldfaced, Page-Referenced Terms

[p.430] cuticle _____

_____

[p.430] cutin _____

_____

[p.430] stomata (singular, stoma) _____

_____

[p.430] guard cells _____

_____

[p.431] turgor pressure _____

_____

[p.431] CAM plants _____

_____

True–False

If the statement is true, write a T in the blank. If the statement is false, make it correct by changing the underlined word(s) and writing the correct word(s) in the answer blank.

_____ 1. Of the water moving into a leaf, 2 percent or more is lost by evaporation into the surrounding air. [p.430]

_____ 2. When evaporation exceeds water uptake by roots, plant tissues wilt and water-dependent activities are seriously disrupted. [p.430]

_____ 3. The surfaces of all plant epidermal cell walls have an outer layer of waxes embedded in cutin, the cuticle, which restricts water loss, restricts inward diffusion of carbon dioxide, and limits outward diffusion of oxygen by-products. [p.430]

_____ 4. Plant epidermal layers are peppered with tiny openings called <u>guard</u> <u>cells</u> through which water leaves the plant and carbon dioxide enters. [p.430]

_____ 5. When a pair of guard cells swells with turgor pressure, the opening between them <u>closes</u>. [p.431]

_____ 6. In most plants, stomata remain <u>open</u> during the daylight photosynthetic period; water is lost from plants but they accumulate carbon dioxide. [p.431]

_____ 7. Stomata stay <u>closed</u> at night in most plants. [p.431]

_____ 8. Photosynthesis starts when the sun comes up; as the morning progresses, carbon dioxide levels <u>increase</u> in cells, including guard cells. [p.431]

_____ 9. As the morning progresses, a drop in carbon dioxide level within guard cells triggers an inward active transport of potassium ions; water follows by osmosis and the fluid pressure <u>closes</u> the stoma. [p.431]

_____10. When the sun goes down and photosynthesis stops, carbon dioxide levels rise; stomata <u>close</u> when potassium, then water, moves out of the guard cells. [p.431]

_____11. CAM plants such as cacti and other succulents open stomata during the <u>day</u> when they fix carbon dioxide by way of a special C4 metabolic pathway. [p.431]

_____12. CAM plants use carbon dioxide in photosynthesis the following <u>night</u> when stomata are closed. [p.431]

26.5. HOW ARE ORGANIC COMPOUNDS DISTRIBUTED THROUGH PLANTS?
[pp.432–433]

Selected Words: source [p.432], sink [p.432], Sonchus [p.433]

Boldfaced, Page-Referenced Terms

[p.432] phloem _____

_____

[p.432] sieve tubes _____

_____

[p.432] companion cells _____

_____

[p.432] translocation _____

_____

[p.432] pressure flow theory _____

_____

Fill-in-the-Blanks

Sucrose and other organic products resulting from (1) _____ [p.432] are used throughout the plant organs. Most plant cells store their carbohydrates as (2) _____ [p.432] in plastids. The (3) _____ [p.432] and fats, which cells synthesize from carbohydrates and the amino acids distributed to them, are stored in many (4) _____ [p.432]. Avocados and some other fruits also accumulate (5) _____ [p.432]. (6) _____ [p.432] molecules are too large to cross cell membranes and too insoluble to be transported to other regions of the plant body. Overall, (7) _____ [p.432] are too large and fats are too insoluble to be transported from the storage sites. Plant cells convert storage forms of organic compounds to (8) _____ [p.432] of smaller size that are more easily transported through the phloem. For example, the cells degrade starch to glucose monomers. When one of these monomers combines with fructose, the result is (9) _____ [p.432], an easily transportable sugar. Experiments with phloem-embedded aphid mouthparts revealed that (10) _____ [p.432], an easily transportable sugar, was the most abundant carbohydrate being forced out of those tubes.

Matching

Choose the most appropriate answer for each term.

11. ___translocation [p.432]

12. ___sieve-tube members [p.432]

13. ___companion cells [p.432]

14. ___aphids [p.432]

15. ___source [p.432]

16. ___sink [p.432]

17. ___pressure flow theory [pp.432–433]

A. Any region where organic compounds are being loaded into the sieve tubes

B. Nonconducting cells adjacent to sieve-tube members that supply energy to load sucrose at the source

C. Any region of the plant where organic compounds are being unloaded from the sieve-tube system and used or stored

D. Process occurring in phloem that distributes sucrose and other organic compounds through the plant (apparently under pressure)

E. Internal pressure builds up at the source end of a sieve-tube system and pushes the solute-rich solution toward a sink, where they are removed

F. Passive conduits for translocation within vascular bundles; water and organic compounds flow rapidly through large pores on their end walls

G. Insects used to verify that in most plant species sucrose is the main carbohydrate translocated under pressure

Self-Quiz

___ 1. The three elements that are present in carbohydrates, lipids, proteins, and nucleic acids are _____. [p.424]
  a. oxygen, carbon, and nitrogen
  b. oxygen, hydrogen, and nitrogen
  c. oxygen, carbon, and hydrogen
  d. carbon, nitrogen, and hydrogen

___ 2. Macronutrients are the nine dissolved mineral ions that _____. [p.425]
  a. furnish the basic ingredients for photosynthesis
  b. occur in only small traces in plant tissues
  c. become detectable in plant tissues
  d. can function only without the presence of micronutrients
  e. both a and c

___ 3. Gaseous nitrogen is converted to a plant-usable form by _____. [p.427]
  a. root nodules
  b. mycorrhizae
  c. nitrogen-fixing bacteria
  d. Venus flytraps

___ 4. _____ prevent(s) inward-moving water from moving past the abutting walls of the root endodermal cells. [p.426]
  a. Cytoplasm
  b. Plasma membranes
  c. Osmosis
  d. Casparian strips

___ 5. Most of the water moving into a leaf is lost through _____. [p.428]
  a. osmotic gradients being established
  b. evaporation of water from plant parts exposed to air
  c. pressure-flow forces
  d. translocation

___ 6. Stomata remain _____ during daylight, when photosynthesis occurs, but remain _____ during the night when carbon dioxide accumulates through aerobic respiration. [pp.430–431]
  a. open; open
  b. closed; open
  c. closed; closed
  d. open; closed

___ 7. By control of _____ levels inside the guard cells of stomata, the activity of stomata is controlled when leaves are losing more water than roots can absorb. [p.431]
  a. oxygen
  b. potassium
  c. carbon dioxide
  d. ATP

___ 8. Without _____, plants would rapidly wilt and die during hot, dry spells. [p.430]
  a. a cuticle
  b. mycorrhizae
  c. phloem
  d. cotyledons

___ 9. The _____ theory of water transport states that hydrogen bonding allows water molecules to maintain a continuous fluid column as water is pulled from roots to leaves. [p.428]
  a. pressure flow
  b. cohesion-tension
  c. evaporation
  d. abscission

___ 10. Leaves represent _____ regions; growing leaves, stems, fruits, seeds, and roots represent _____ regions. [p.432]
  a. source; source
  b. sink; source
  c. source; sink
  d. sink; sink

Chapter Objectives/Review Questions

1. Explain how carnivorous plants such as Venus flytraps and bladderworts accomplish their nutritional needs. [pp.422–423]
2. _____ _____ is the study of adaptations by which plants function in their environment. [p.423]
3. Define *soil*. [p.424]
4. Plants do best in _____, which are the soils having more or less equal proportions of sand, silt, and clay. [p.424]
5. Name the soil layer that is the most essential for plant growth. [p.424]
6. Define *nutrients* in terms of plant nutrition. [p.424]
7. Name the three elements considered essential for plant nutrition. [pp.425–426]
8. Distinguish between macronutrients and micronutrients in relation to their role in plant nutrition. [p.425]
9. _____ refers to the removal of some of the nutrients in soil as water percolates through it. [p.425]
10. The movement of land under the force of wind, running water, and ice is called _____. [p.425]
11. Be able to trace the path of water and mineral ions into roots; name the structure and function of plant structures involved. [p.426]
12. Differentiate between the endodermis and the exodermis of the root cortex. [p.426]
13. Due to the presence of the _____ strip in the walls of endodermal cells, water can move into the vascular cylinder only by crossing the plasma membrane and diffusing through the cytoplasm. [p.426]
14. Explain why root hairs are so valuable in root absorption. [p.426]
15. Define *mutualism*. [p.426]
16. Describe the roles of root nodules and mycorrhizae in plant nutrition. [p.427]
17. The evaporation of water from leaves as well as from stems and other plant parts is known as _____. [p.428]
18. Henry Dixon's _____-_____ theory explains how water moves upward in an unbroken column through xylem to the tops of tall trees. [p.428]
19. In a plant's vascular tissues, water moves through pipelines called _____. [p.428]
20. Be able to give the key points of Dixon's explanation of upward water transport in plants. [pp.428–429]
21. Even mildly stressed plants would rapidly wilt and die if it were not for the _____ covering their parts; describe additional functions of this structure. [p.430]
22. Describe the chemical compounds found in cutin. [p.430]
23. Evaporation from plant parts occurs mostly at _____, tiny epidermal passageways of leaves and stems. [p.430]
24. Explain the mechanism by which stomata open during daylight and close during the night. [p.431]
25. Define *turgor pressure*, and explain its role in controlling water loss at stomata. [p.431]
26. Describe the mechanisms by which CAM plants conserve water. [p.431]
27. Describe the role of phloem sieve-tube members and companion cells in translocation. [p.432]
28. _____ is the main form in which sugars are transported through most plants. [p.432]
29. Define *translocation*. [p.432]
30. According to the _____ _____ theory, pressure builds up at the source end of a sieve tube system and pushes solutes toward a sink, where they are removed. [pp.432–433]
31. Companion cells supply the _____ that loads sucrose at the source. [p.433]
32. Name the gradients responsible for continuous flow of organic compounds through phloem. [pp.432–433]

Integrating and Applying Key Concepts

How do you think maple syrup is made from maple trees? Which specific systems of the plant are involved, and why are maple trees tapped only at certain times of the year?

27

PLANT REPRODUCTION AND DEVELOPMENT

Interactive Exercises

A Coevolutionary Tale [pp.436–437]

27.1. REPRODUCTIVE STRUCTURES OF FLOWERING PLANTS [pp.438–439]

27.2. *Focus on the Environment:* POLLEN SETS ME SNEEZING [p.439]

Selected Words: *Angraecum sesquipedale* [p.437], *Prunus* [p.438], *Rosa* [p.439], *perfect* flowers [p.439], *imperfect* flowers [p.439], *allergic rhinitis* [p.439]

Boldfaced, Page-Referenced Terms

[p.436] flower _____

_____

[p.436] coevolution _____

_____

[p.436] pollinator _____

_____

[p.438] sporophyte _____

_____

[p.438] gametophytes _____

_____

[p.439] stamens _____

_____

[p.439] pollen grains _____

_____

[p.439] carpels _____

_____

[p.439] ovary _____

_____

Label–Match

Identify each numbered part of the accompanying illustration. Complete the exercise by matching and entering the letter of the proper description in the parentheses following each label.

1. _____ (   ) [p.438]
2. _____ (   ) [p.438]
3. _____ (   ) [p.438]
4. _____ (   ) [p.438]
5. _____ (   ) [p.438]
6. _____ (   ) [p.438]

A. An event that produces a young sporophyte
B. A reproductive shoot produced by the sporophyte
C. The "plant"; a vegetative body that develops from a zygote
D. Cellular division event occurring within flowers to produce spores
E. Produces haploid eggs by mitosis
F. Produces haploid sperm by mitosis

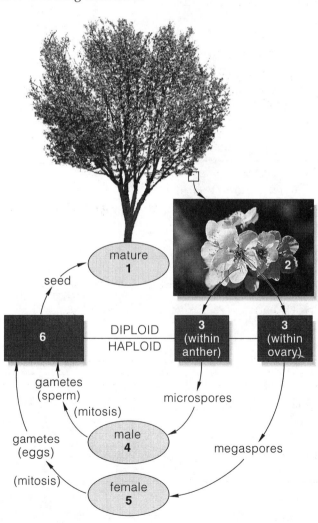

Labeling

Identify each numbered part of the accompanying illustration.

7. _____ [p.438]

8. _____ [p.438]

9. _____ [p.438]

10. _____ [p.438]

11. _____ [p.438]

12. _____ [p.438]

13. _____ [p.438]

14. _____ [p.438]

15. _____ [p.438]

16. _____ [p.438]

17. _____ [p.438]

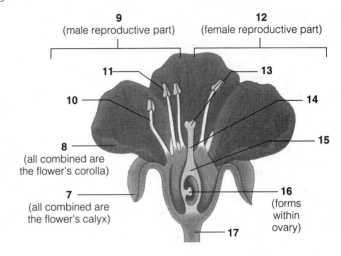

Matching

Choose the most appropriate answer for each term.

18. ___sepals [p.438]

19. ___petals [p.438]

20. ___stamens [p.439]

21. ___ovule [pp.438–439]

22. ___pollen grain [p.439]

23. ___carpel [p.439]

24. ___ovary [p.439]

25. ___perfect flowers [p.439]

26. ___imperfect flowers [p.439]

A. Have both male and female parts
B. Mature haploid spore whose contents develop into a male gametophyte
C. Collectively, the flower's "corolla"
D. Have male or female parts, but not both
E. Female reproductive part; includes stigma, style, and ovary
F. Structure inside the ovary; matures to become a seed
G. Found just inside the flower's corolla, the male reproductive parts
H. Lower portion of the carpel where egg formation, fertilization, and seed development occur
I. Outermost leaflike whorl of floral organs; collectively, the calyx

27.3. A NEW GENERATION BEGINS [pp.440–441]

Selected Words: Prunus [p.440]

Boldfaced, Page-Referenced Terms

[p.440] microspores _____

_____

[p.440] ovule _____

_____

[p.440] megaspores _____

_____

[p.440] endosperm _____

_____

[p.440] pollination _____

_____

[p.441] double fertilization _____

_____

Fill-in-the-Blanks

The numbered items in the illustration below represent missing information; complete the blanks in the following narrative to supply that information.

Within each (1) _____ [p.440], mitotic divisions produce four masses of spore-forming cells, each mass forming within a(n) (2) _____ _____ [p.440]. Each one of these diploid cells is known as a(n) (3) _____ _____ [p.440] cell and undergoes (4) _____ [p.440] to produce four haploid (5) _____ [p.440]. Mitosis within each haploid microspore results in a two-celled haploid body, the immature male gametophyte. One of these cells will give rise to a(n) (6) _____ _____ [p.440]; the other cell will develop into a(n) (7) _____-_____ [p.440] cell. Mature microspores are eventually released from the pollen sacs of the anther as (8) _____ [p.440]. Pollination occurs and after the pollen lands on a(n) (9) _____ [p.440] of a

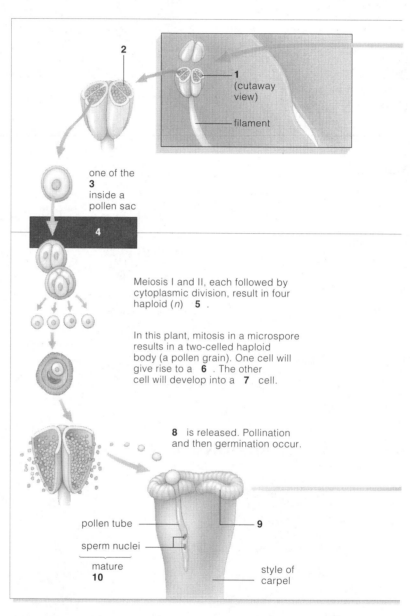

2

1
(cutaway view)

filament

one of the
3
inside a
pollen sac

4

Meiosis I and II, each followed by cytoplasmic division, result in four haploid (n) 5 .

In this plant, mitosis in a microspore results in a two-celled haploid body (a pollen grain). One cell will give rise to a 6 . The other cell will develop into a 7 cell.

8 is released. Pollination and then germination occur.

pollen tube

sperm nuclei

mature
10

9

style of carpel

carpel, the pollen tube develops from one of the cells in the pollen grain; the other cell within the pollen grain divides to form two sperm cells. As the pollen tube grows through the carpel tissues, it contains the two sperm cells and a tube nucleus. The pollen tube with its two sperm cells and the tube nucleus is known as the mature (10) _____ _____ [p.440].

Fill-in-the-Blanks

The numbered items on the illustration below represent missing information; complete the numbered blanks in the following narrative to supply that information.

In the carpel of the flower, one or more dome-shaped, diploid tissue masses develop on the inner wall of the ovary. Each mass is the beginning of a(n) (11) _____ [p.441]. A tissue forms inside a domed mass as it grows, and one or two protective layers called (12) _____ [p.441] form around it. Inside each mass, a diploid cell (the megaspore mother cell) divides by (13) _____ [p.441] to form four haploid spores known as (14) _____ [p.441]. Commonly, all but one (15) _____ [p.441] disintegrates. The remaining (16) _____ [p.441] undergoes (17) _____ [p.441] three times without cytoplasmic division. At first, this structure is a cell with (18) _____ [p.441] haploid nuclei. Cytoplasmic division results in a seven-cell (19) _____ _____ [p.441], which represents the mature (20) _____ _____ [p.441]. Six of those cells have a single nucleus, but one cell has (number) (21) _____ [p.441]

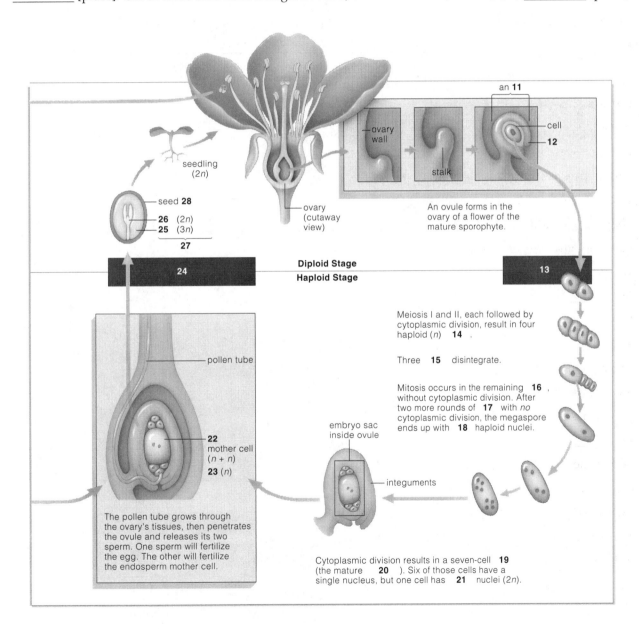

nuclei (2*n*) and represents the (22) _____ [p.441] mother cell (*n* + *n*). Another haploid cell within the embryo sac is the (23) _____ [p.441]. Following (24) _____ _____ [p.441] with one sperm, the *n* + *n* cell will help form the 3*n* (25) _____ [p.441], a nutritive tissue for the forthcoming embryo. The other sperm involved in this unique fertilization process fertilizes the haploid egg; this combination forms the diploid (26) _____ [p.441]. Thus, the ovule is transformed to a(n) (27) _____ [p.441] that is composed of three parts, a seed (28) _____ [p.441], an embryo, and nourishment for the embryo, the endosperm.

Short Answer

29. What guides the growth of the pollen tube down through the female floral tissues toward the chamber holding the egg? [p.441] _____

_____

30. Describe the site of double fertilization, an event known only in flowering plants. [p.441] _____

_____

_____

Complete the Table

31. Complete the table below to summarize the unique double fertilization occurring only in flowering plant life cycles.

Double Fertilization Products	Origin	Produces?	Function
[p.441] a. Zygote (2*n*) nucleus			
[p.441] b. Endosperm (3*n*) nucleus			

27.4. FROM ZYGOTE TO SEEDS AND FRUITS [pp.442–443]

Selected Words: *Capsella* [p.442], *simple* fruits [p.442], *aggregate* fruits [p442], *multiple* fruits [p.442], *Ananas* [p.443], *Fragaria* [p.443], *Malus* [p.443], *Acer* [p.443], *seed dispersal* [p.443]

Boldfaced, Page-Referenced Terms

[p.442] fruit _____

_____

[p.442] cotyledons _____

_____

[p.442] seed _____

_____

Complete the Table

1. Complete the following table, which summarizes concepts associated with seeds and fruits.

Structure	Origin
[p.442] a. Cotyledons	
[p.442] b. Seeds	
[p.442] c. Seed coat	
[p.442] d. Fruit	

Labeling

Identify each numbered part of the accompanying illustration.

2. _____ [p.442]   9. _____ [p.442]

3. _____ [p.442]   10. _____ [p.442]

4. _____ [p.442]   11. _____ [p.442]

5. _____ [p.442]   12. _____ _____ [p.442]

6. _____ [p.442]   13. _____ [p.442]

7. _____ _____ [p.442]   14. _____ [p.442]

8. _____ _____ [p.442]

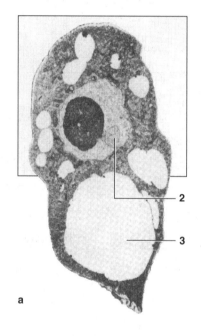

a

b Upper part of **4**
gives rise to embryo

c Globular
5 stage

d Heart-shaped
stage of **6**

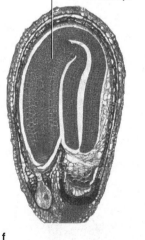

7

embryo's
8

embryo's
9 (two)

10

11

12

e

13 embryo

f

A **14** (a mature ovary)
cut open to show
seeds (the mature
ovules). The embryo
in each seed is at
stage (f).

g

Fill-in-the-Blanks

Following fertilization, the newly formed zygote initiates a course of (15) _____ [p.442] cell divisions that lead to a mature embryo (16) _____ [p.442]. The embryo develops as part of a(n) (17) _____ [p.442] and is accompanied by the formation of a(n) (18) _____ [p.442], a mature ovary.

By the time a *Capsella* embryo is near maturity, two (19) _____ [p.442], or seed leaves, have begun to develop from two lobes of meristematic tissue. Dicot embryos, such as *Capsella*, have (number) (20) _____ [p.442] cotyledon(s) and monocot embryos have (number) (21) _____ [p.442] cotyledon(s). Like most dicots, the *Capsella* embryo has rather thick cotyledons that absorb nutrients from the (22) _____ [p.442] of the seed and stores them in its cotyledons. In corn, wheat and most other monocots, endosperm is not tapped until the seed (23) _____ [p.442]. Digestive enzymes become stockpiled inside the (24) _____ [p.442] cotyledons of monocot embryos. When the (25) _____ [p.442] do become active, nutrients stored in the endosperm will be released and transferred to the growing (26) _____ [p.442].

From the time a zygote forms until a mature embryo develops, a parent sporophyte plant transfers nutrients to tissues of the (27) _____ [p.442]. Food reserves accumulate in (28) _____ [p.442] or cotyledons. Eventually, the ovule separates from the (29) _____ [p.442] wall, and its integuments thicken and harden into a seed (30) _____ [p.442]. The embryo, food reserves, and seed coat are a self-contained package—a(n) (31) _____ [p.442], which is defined as a mature (32) _____ [p.442]. While seeds are forming, changes in other parts of the flower are forming (33) _____ [p.442]. They may be fleshy or (34) _____ [p.442]; they may consist of one or more (35) _____ [p.442], and they may incorporate tissues besides those of the ovary, such as tissues of the (36) _____ [p.442].

Choice

For questions 37–46, choose from the following:

         a. wind-dispersed fruits   b. fruits dispersed by animals   c. water-dispersed fruits

37. ___ heavy wax coats [p.443]

38. ___ coconut palms [p.443]

39. ___ seed coats assaulted by digestive enzymes to assist in releasing embryos [p.443]

40. ___ maples [p.443]

41. ___ orchids [p.443]

42. ___ air sacs [p.443]

43. ___ hooks, spines, hairs, and sticky surfaces [p.443]

44. ___ cacao [p.443]

45. ___ dandelions [p.443]

46. ___ cockleburs, bur clover, and bedstraw [p.443]

Choice

For questions 47–54, choose from the following fruit types.

a. simple dry dehiscent   b. simple dry indehiscent   c. simple fleshy fruits   d. aggregate fruits
e. multiple fruits   f. accessory fruits

47. ___ pineapple [p.442]

48. ___ nuts, grains [p.442]

49. ___ raspberries [p.442]

50. ___ grapes, tomatoes [p.442]

51. ___ apples, pears, strawberries [p.442]

52. ___ sunflowers and carrots [p.442]

53. ___ oranges [p.442]

54. ___ peaches, cherries, and other drupes [p.442]

27.5. ASEXUAL REPRODUCTION OF FLOWERING PLANTS [pp.444–445]

Selected Words: *Populus tremuloides* [p.444], *Larrea divaricata* [p.444], parthenogenesis [p.444], *Daucus carota* [p.444]

Boldfaced, Page-Referenced Terms

[p.444] vegetative growth _____

_____

[p.444] tissue culture propagation _____

_____

Matching

Match the following asexual reproductive modes of flowering plants.

1. ___corm [p.444]

2. ___bulb [p.444]

3. ___parthenogenesis [p.444]

4. ___runner [p.444]

5. ___vegetative propagation on modified stems [p.444]

6. ___rhizome [p.444]

7. ___tuber [p.444]

8. ___tissue culture propagation (induced propagation) [p.444]

9. ___vegetative growth [p.444]

A. In a general sense, new plants develop from tissues or organs that drop or separate from parent plants

B. New shoots arise from axillary buds (enlarged tips of slender underground rhizomes)

C. New plants arise from cells in parent plant that were not irreversibly differentiated; a laboratory technique

D. New plants arise at nodes of underground horizontal stem

E. New plant arises from axillary bud on short, thick vertical underground stem

F. New plants arise at nodes on an aboveground horizontal stem

G. New bulb arises from an axillary bud on a short underground stem

H. Involves asexual reproduction utilizing runners, rhizomes, corms, tubers, and bulbs

I. Embryo develops without nucleus or cellular fusion

Matching

Choose the most appropriate example for each modified stem.

10. ___bulb [p.444]
11. ___rhizome [p.444]
12. ___tuber [p.444]
13. ___runner [p.444]
14. ___corm [p.444]

A. Potato
B. Strawberry
C. Gladiolus
D. Onion, lily
E. Bermuda grass

27.6. PATTERNS OF EARLY GROWTH AND DEVELOPMENT—AN OVERVIEW
[pp.446–447]

Selected Words: *Phaseolus vulgaris* [p.446], *Zea mays* [p.446], imbibition [p.446]

Boldfaced, Page-Referenced Terms

[p.446] germination _____

_____

Fill-in-the-Blanks

Before or after seed dispersal, the growth of the (1) _____ [p.446] idles. For seeds, (2) _____ [p.446] is the resumption of growth by an immature stage in the life cycle after a time of arrested development. Germination depends on (3) _____ [p.446] factors, such as soil temperature, moisture, and oxygen levels, and the number of seasonal daylight hours available. By a process known as (4) _____ [p.446], water molecules move into the seed. As more water moves in, the seed swells, and its coat (5) _____ [p.446]. Once the seed coat splits, more oxygen reaches the embryo, and (6) _____ [p.446] respiration moves into high gear. The embryo's (7) _____ [p.446] cells begin to divide rapidly. In general, the (8) _____ [p.446] meristem is the first to be activated. Its meristematic descendants divide, elongate, and give rise to the (9) _____ _____ [p.446]. When this structure breaks through the seed coat, (10) _____ [p.446] is over. For both monocots and dicots, the patterns of germination, growth, and development that unfold have a(n) (11) _____ [p.446] basis; they are dictated by the plant's (12) _____ [p.446]. All cells in the new plant arise from the same cell, the (13) _____ [p.446]. Thus, all cells inherit the same (14) _____ [p.446]. Unequal (15) _____ [p.446] divisions between daughter cells lead to differences in their (16) _____ [p.447] equipment and output. Activities in daughter cells start to vary as a result of (17) _____ [p.447] gene expression. As example, genes governing the synthesis of growth-stimulating (18) _____ [p.447] are activated in some cells but not others. (19) _____ [p.447] among genes, hormones, and the environment govern how an individual plant grows and develops.

Labeling

Identify each numbered part of the accompanying illustration.

20. _____ [p.447]   26. _____ [p.447]   32. _____ [p.447]

21. _____ [p.447]   27. _____ [p.447]   33. _____ [p.447]

22. _____ [p.447]   28. _____ [p.447]   34. _____ [p.447]

23. _____ [p.447]   29. _____ [p.447]   35. _____ [p.447]

24. _____ [p.447]   30. _____ [p.447]   36. _____ [p.447]

25. _____ [p.447]   31. _____ [p.447]   37. _____ [p.447]

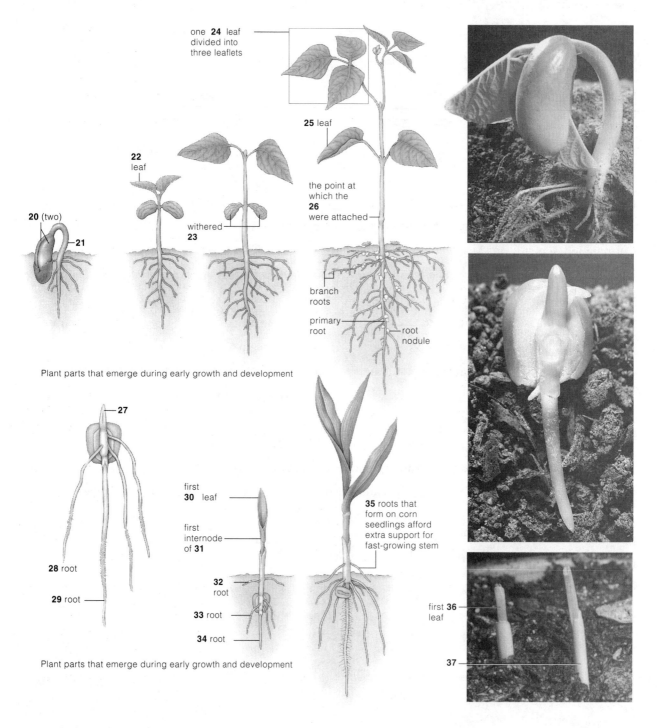

one **24** leaf divided into three leaflets

25 leaf

the point at which the **26** were attached

22 leaf

withered **23**

20 (two)

21

branch roots

primary root

root nodule

Plant parts that emerge during early growth and development

27

28 root

29 root

first **30** leaf

first internode of **31**

32 root

33 root

34 root

35 roots that form on corn seedlings afford extra support for fast-growing stem

first **36** leaf

37

Plant parts that emerge during early growth and development

27.7. EFFECTS OF PLANT HORMONES [pp.448–449]

27.8. *Focus on Science*: FOOLISH SEEDLINGS! [p.449]

Selected Words: *quantitative* terms [p.448], *qualitative* terms [p.448], *herbicides* [p.448], *Gibberella fujikuroi* [p.449], *Vitis* [p.449]

Boldfaced, Page-Referenced Terms

[p.448] hormones _____

_____

[p.448] growth _____

_____

[p.448] development _____

_____

[p.448] gibberellins _____

_____

[p.448] auxins _____

_____

[p.449] cytokinins _____

_____

[p.449] abscisic acid _____

_____

[p.449] ethylene _____

_____

Choice

For questions 1–14, choose from the following:

a. auxins   b. gibberellins   c. cytokinins   d. abscisic acid   e. ethylene

1. ___ Ancient Chinese burned incense to hurry fruit ripening. [p.449]

2. ___ Natural and synthetic versions are used to prolong the shelf life of cut flowers, lettuces, mushrooms, and other vegetables. [p.449]

3. ___ Orchardists spray trees with IAA to thin out overcrowded seedlings in the spring. [p.448]

4. ___ Inhibits cell growth, induces bud dormancy, and prevents seeds from germinating prematurely; causes stomata to close when a plant is water stressed [p.449]

5. ___ IAA, the most important naturally occurring compound of its type [p.448]

6. ___ Exposed to oranges and other citrus fruits to brighten their color of their rind before being displayed in the market [p.449]

7. ___ Influences stem lengthening; influence plant responses to gravity and light and promote coleoptile lengthening [p.448]

8. ___ Cause stems to lengthen; help buds and seeds break dormancy and resume growth in the spring [p.448]

9. ___ Used to prevent premature fruit drop—all fruit can be picked at the same time [p.448]

10. ___ Stimulate cell division [p.449]

11. ___ Used by food distributors to ripen green fruit after it is shipped to grocery stores [p.449]

12. ___ Most abundant in root and shoot meristems and in the tissues of maturing fruit [p.449]

13. ___ Under its influence, fruit ripens and flowers, fruits, and leaves drop away from plants at prescribed times of the year. [p.449]

14. ___ Some synthetic forms serve as herbicides. [p.448]

Matching

Choose the most appropriate answer for each term.

15. ___2,4-D [p.448]

16. ___hormone [p.448]

17. ___apical dominance [p.449]

18. ___herbicide [p.448]

19. ___growth [p.448]

20. ___ "foolish seedling" effect on rice plants [p.449]

21. ___IAA [p.448]

22. ___development [p.448]

23. ___2,4,5-T [p.448]

24. ___coleoptile [p.448]

25. ___target cell [p.448]

A. Mixed with 2,4-D to produce Agent Orange for defoliation in Vietnam

B. The emergence of specialized, morphologically different body parts; measured in qualitative terms

C. Gibberellin from fungal extracts

D. A cell with receptors to bind a given signaling molecule

E. Hormonal effect that blocks lateral bud growth

F. Synthetic auxin used as an herbicide to selectively kill broadleaf weeds that compete with vigorously

G. An increase in the number, size, and volume of cells; measured in quantitative terms

H. A signaling molecule released from one cell that travels to target cells and stimulates or inhibits gene activity

I. The most important naturally occurring auxin

J. Any synthetic auxin compound used to kill some plants but not others

K. A thin sheath around the primary shoot of grass seedlings, such as corn plants

27.9. ADJUSTMENTS IN THE RATE AND DIRECTION OF GROWTH [pp.450–451]

Selected Words: trope [p.450], auxein [p.451], thigma [p.451]

Boldfaced, Page-Referenced Terms

[p.450] plant tropism _____

_____

[p.450] gravitropism _____

_____

[p.450] statoliths _____

_____

[p.450] phototropism _____

_____

[p.451] thigmotropism _____

_____

Choice

For questions 1–10, choose from the following:

        a. phototropism   b. gravitropism   c. thigmotropism   d. mechanical stress

1. ___ More intense sunlight on one side of a plant—stems curve toward the light [pp.450–451]
2. ___ Vines climbing around a fencepost as they grow upward [p.451]
3. ___ Plants grown outdoors have shorter stems than plants grown in a greenhouse. [p.451]
4. ___ A root turned on its side will curve downward. [p.450]
5. ___ Briefly shaking a plant daily inhibits the growth of the entire plant. [p.451]
6. ___ A potted seedling turned on its side—the growing stem curves upward [p.450]
7. ___ Leaves turn until their flat surfaces face light. [pp.450–451]
8. ___ Flavoprotein may be a central component. [p.451]
9. ___ Tendrils are sometimes involved. [p.451]
10. ___ Generally, statoliths are involved. [p.450]

27.10. BIOLOGICAL CLOCKS AND THEIR EFFECTS [pp.452–453]

Selected Words: Pfr [p.452], Pr [p.452], *circadian* [p.452], *short-day* plants [p.452], *long-day* plants [p.452], *day-neutral* plants [p.452]

Boldfaced, Page-Referenced Terms

[p.452] biological clocks _____

_____

[p.452] phytochrome _____

_____

[p.452] circadian rhythm _____

_____

[p.452] photoperiodism _____

_____

Matching

Choose the most appropriate answer for each term.

1. ___photoperiodism [p.452]
2. ___Pr [p.452]
3. ___day-neutral plants [p.452]
4. ___phytochrome activation [p.452]
5. ___"long-day" plants [p.452]
6. ___circadian rhythms [p.452]
7. ___Pfr [p.452]
8. ___biological clocks [p.452]
9. ___rhythmic leaf movements [p.452]
10. ___phytochrome [p.452]

A. Biological activities that recur in cycles of twenty-four hours or so
B. Flower in spring when daylength exceeds a critical value
C. Internal time-measuring mechanisms with roles in adjusting daily activities
D. Any biological response to a change in the relative length of daylight and darkness in the 24-hour cycle; active Pfr may be an alarm button for this process
E. A blue-green pigment that absorbs red or far-red wavelengths, with different results
F. An example of a circadian rhythm
G. Active form of phytochrome
H. Flower when mature enough to do so
I. Inactive form of phytochrome
J. May induce plant cells to take up free calcium ions or induce certain plant cell organelles to release them

Labeling

Identify each numbered part of the accompanying illustration.

11. _____ [p.452]
12. _____ [p.452]
13. _____ [p.452]
14. _____ [p.452]
15. _____ [p.452]

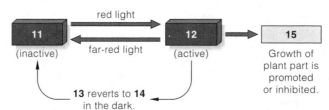

Fill-in-the-Blanks

(16) _____ [p.452] is any biological response to change in the relative length of daylight and darkness in the cycle of twenty-four hours. (17) _____ [p.452], phytochrome's active form, may be a switching mechanism for the process; it might trigger the synthesis of specific (18) _____ [p.452] in specific types of plant cells. "Long-day plants" flower in spring when day length becomes (19) [choose one] ❑ shorter ❑ longer [p.452] than some critical value. "Short-day plants" flower in late summer or early autumn when daylength becomes (20) [choose one] ❑ shorter ❑ longer [p.452] than some critical value. "Day-neutral plants" flower whenever they become (21) _____ [p.452] enough to do so without regard to daylength. Spinach is a (22) _____ - _____ [p.452] plant because it will not flower and produce seeds unless it is exposed to fourteen hours of light every day for two weeks. Cocklebur is termed a (23) _____ - _____ [p.452] plant because it flowers after a single night that is longer than 8-1/2 hours.

27.11. LIFE CYCLES END, AND TURN AGAIN [p.454]

Boldfaced, Page-Referenced Terms

[p.454] abscission _____

_____

[p.454] senescence _____

_____

[p.454] dormancy _____

_____

Choice

For questions 1–10, choose from the following:

      a. senescence   b. abscission   c. entering dormancy   d. breaking dormancy

1. ___ Dropping of flowers, leaves, fruits, and other plant parts [p.454]

2. ___ A process at work between fall and spring; temperatures become milder, and rain and nutrients become available again. [p.454]

3. ___ Strong cues are short days, long, cold nights, and dry, nitrogen-deficient soil. [p.454]

4. ___ The zone consists of thin-walled parenchyma cells at the base of a petiole or some other part about to drop from the plant. [p.454]

5. ___ The sum total of processes leading to the death of a plant or some of its parts. [p.454]

6. ___ A recurring cue for this process is a decrease in daylength. [p.454]

7. ___ When a plant stops growing under conditions that appear quite suitable for growth [p.454]

8. ___ The forming of ethylene in cells near the breakpoints may trigger the process. [p.454]

9. ___ Interrupting nutrient diversion to reproductive parts blocks this process in stems, leaves, and roots. [p.454]

10. ___ Gardeners can postpone this process by removing flower buds from plants to maintain vegetative growth. [p.454]

Self-Quiz

___ 1. A stamen is _____. [p.440]
    a. composed of a stigma
    b. the mature male gametophyte
    c. the site where microspores are produced
    d. part of the vegetative phase of an angiosperm

___ 2. The portion of the carpel that contains an ovule is the _____. [p.440]
    a. stigma
    b. anther
    c. style
    d. ovary

___ 3. The phase in the life cycle of plants that gives rise to spores is known as the _____. [p.438]
a. gametophyte
b. embryo
c. sporophyte
d. seed

___ 4. An immature fruit is a(n) _____ and an immature seed is a(n) _____. [p.442]
a. ovary; megaspore
b. ovary; ovule
c. megaspore; ovule
d. ovule; ovary

___ 5. In flowering plants, one sperm nucleus fuses with that of an egg, and a zygote forms that develops into an embryo. Another sperm fuses with _____. [p.441]
a. a primary endosperm cell to produce three cells, each with one nucleus
b. a primary endosperm cell to produce one cell with one triploid nucleus
c. both nuclei of the endosperm mother cell, forming a primary endosperm cell with a single triploid nucleus
d. one of the smaller megaspores to produce what will eventually become the seed coat

___ 6. "Simple, aggregate, multiple, and accessory" refer to types of _____. [p.442]
a. carpels
b. seeds
c. fruits
d. ovaries

___ 7. "When a leaf falls or is torn away from a jade plant, a new plant can develop from the leaf, from meristematic tissue." This statement refers to _____. [p.444]
a. parthenogenesis
b. runners
c. tissue culture propagation
d. vegetative propagation

___ 8. Auxins _____. [p.448]
a. cause flowering
b. promote stomatal closure
c. promote cell division
d. promote cell elongation in coleoptiles and stems

___ 9. _____ is demonstrated by a germinating seed whose first root always curves down while the stem always curves up. [p.450]
a. Phototropism
b. Photoperiodism
c. Gravitropism
d. Thigmotropism

___ 10. Light of _____ wavelengths is the main stimulus for phototropism. [p.451]
a. blue
b. yellow
c. red
d. green

___ 11. 2,4-D, a potent dicot weed killer, is a synthetic _____. [p.448]
a. auxin
b. gibberellin
c. cytokinin
d. phytochrome

___ 12. All the processes that lead to the death of a plant or any of its organs are called _____. [p.454]
a. dormancy
b. vernalization
c. abscission
d. senescence

___ 13. Phytochrome is converted to an active form, _____, at sunrise and reverts to an inactive form, _____, at sunset, night, or in the shade. [p.452]
a. Pr; Pfr
b. Pfr; Pfr
c. Pr; Pr
d. Pfr; Pr

___ 14. When a perennial or biennial plant stops growing under conditions suitable for growth, it has entered a state of _____. [p.454]
a. senescence
b. vernalization
c. dormancy
d. abscission

Chapter Objectives/Review Questions

1. _____ refers to two or more species jointly evolving as an outcome of close ecological interactions. [p.436]
2. Describe the role of a pollinator. [p.436]
3. Be able to distinguish between sporophytes and gametophytes. [p.436]
4. Be able to identify the various parts of a typical flower and state their functions. [pp.438–439]
5. Walled microspores form in pollen sacs and develop into _____ _____. [p.439]
6. Distinguish between a flower that is *perfect* and one that is *imperfect*. [p.439]
7. Describe the condition called *allergic rhinitis*. [p.439]
8. Relate the sequence of events and structures involved that give rise to microspores and megaspores. [p.440]
9. _____ is the transfer of pollen grains to a receptive stigma. [p.440]
10. What structures represent the male gametophyte and female gametophyte in flowering plants? List the contents of each. [pp.440–441]
11. The endosperm mother cell in the embryo sac is composed of two haploid _____. [p.440]
12. Describe the *double fertilization* that occurs uniquely in the flowering plant life cycle. [p.441]
13. After pollination and double fertilization, a(n) _____ and nutritive tissue form in the _____, which becomes a seed. [p.441]
14. How is endosperm formed? What is the function of endosperm? [p.441]
15. Describe the formation of the embryo sporophyte; give the origin and formation of seeds and fruits. [p.442]
16. Review the general types of fruits produced by flowering. [p.442]
17. Seeds and fruits are structurally adapted for _____ by air currents, water currents, and many kinds of animals. [p.443]
18. Distinguish between parthenogenesis, vegetative propagation, and tissue culture propagation; cite an example for each. [p.444]
19. Be able to list representative plant examples of a runner, a rhizome, a corm, a tuber, and a bulb. [p.442]
20. _____ is a resumption of growth after a time of arrested embryonic development. [p.446]
21. Define *imbibition*, and describe its role in germination. [p.446]
22. List the environmental factors that influence germination. [p.446]
23. The primary _____ breaks through the seed coat first. [p.446]
24. The basic patterns of growth and development are heritable, dictated by the plant's _____. [p.446]
25. Compare and contrast the major features of the growth and development of a monocot plant and a dicot plant. [p.447]
26. _____ are signaling molecules secreted by some cells that travel to other, target cells and stimulate or inhibit gene activity. [p.448]
27. _____ of a multicelled organism means the number, size, and volume of cells increase. [p.448]
28. Explain why plant growth is measured in quantitative terms and plant development in qualitative terms. [p.448]
29. Be able to state the known functions of the following plant hormones: auxins, cytokinins, abscissic acid, and ethylene.
30. Synthetic auxins are used as _____, which, at suitable concentrations, kill some plants but not others. [p.449]
31. Describe the form of growth inhibition known as apical dominance. [p.449]
32. Define phototropism, gravitropism, and thigmotropism and cite examples of each. [pp.450–451]
33. Explain how statoliths form the basis for gravity-sensing mechanisms. [p.450]
34. Plants make the strongest phototropic response to light of _____ wavelengths; _____ is a yellow pigment molecule that absorbs blue wavelengths. [p.451]
35. Provide an example of how mechanical stress can affect plants. [p.451]
36. Plants have internal time-measuring mechanisms called biological _____. [p.452]
37. The alarm button for some biological clocks in plants is the blue-green pigment molecule called _____. [p.452]
38. What are circadian rhythms? Give an example. [p.452]
39. Phytochrome is converted to an active form, _____, at sunrise, when red wavelengths dominate the sky. It reverts to an inactive form, _____, at sunset, at night, or even in shade, where far-red wavelengths predominate. [p.452]

40. _____ is a biological response to a change in the relative length of daylight and darkness in a 24-hour cycle. [p.452]
41. _____ serves as a switching mechanism in the biological clock governing flowering responses. [p.452]
42. Describe the photoperiodic responses of "long-day," "short-day," and "day-neutral" plants. [p.452]
43. _____ is the dropping of leaves, flowers, fruits, or other plant parts. [p.454]
44. Describe the events that signal plant senescence. [p.454]
45. List environmental cues that send a plant into dormancy. [p.454]
46. What conditions are instrumental in the dormancy-breaking process? [p.454]

Integrating And Applying Key Concepts

1. In terms of botanical morphology, a flower is interpreted as "a reproductive shoot bearing organs." Try to list structural evidence, not discussed in the chapter, that botanists might have discovered that led them to this view.
2. An oak tree has grown up in the middle of a forest. A lumber company has just cut down all of the surrounding trees except for a narrow strip of woods that includes the oak. How will the oak be likely to respond as it adjusts to its changed environment? To what new stresses will it be exposed? Which hormones will most probably be involved in the adjustment?

28

TISSUES, ORGAN SYSTEMS, AND HOMEOSTASIS

Meerkats, Humans, It's All the Same

EPITHELIAL TISSUE
    General Characteristics
    Cell-to-Cell Contacts
    Glandular Epithelium and Glands

CONNECTIVE TISSUE
    Soft Connective Tissues
    Specialized Connective Tissues

MUSCLE TISSUE

NERVOUS TISSUE

Focus on Science: **FRONTIERS IN TISSUE RESEARCH**

ORGAN SYSTEMS
    Overview of the Major Organ Systems
    Tissue and Organ Formation

HOMEOSTASIS AND SYSTEMS CONTROL
    Concerning the Internal Environment
    Mechanisms of Homeostasis

Interactive Exercises

Meerkats, Humans, It's All the Same [pp.458–459]

28.1. EPITHELIAL TISSUE [pp.460–461]

Selected Words: "meerkats" [p.458], interstitial fluid [p.458], anatomy [p.458], physiology [p.458], *simple* epithelium [p.460], *stratified* epithelium [p.460], basement membrane [p.460], squamous [p.460], cuboidal [p.460], columnar [p.460], "secretion" [p.461], "excretion" [p.461], hormones [p.461]

Boldfaced, Page-Referenced Terms

[p.458] internal environment _____

_____

[p.458] homeostasis _____

_____

[p.458] tissue _____

_____

[p.458] organ _____

_____

[p.458] organ system _____

_____

[p.459] division of labor _____

_____

[p.460] epithelium (plural, epithelia) _____

_____

[p.460] tight junctions _____

_____

[p.460] adhering junctions _____

_____

[p.460] gap junctions _____

_____

[p.461] gland cells _____

_____

[p.461] exocrine glands _____

_____

[p.461] endocrine glands _____

_____

True–False

If the statement is true, write a T in the blank. If the statement is false, correct it by changing the underlined word(s) and writing the correct word(s) in the answer blank.

_____1. Physiology is the study of the way body parts are arranged in an organism. [p.458]

_____2. Groups of like cells that work together to perform a task are known as an organ. [p.458]

_____3. Most animals are constructed of only four types of tissue: epithelial, connective, nervous, and muscle tissues. [p.458]

_____4. Mammalian skin contains squamous epithelium and other tissues. [p.460]

_____5. The more tight junctions there are in a tissue, the more permeable the tissue will be. [p.461]

_____6. Endocrine glands secrete their products through ducts that empty onto an epithelial surface. [p.461]

_____7. Endocrine-cell products include digestive enzymes, saliva, and mucus. [p.461]

Fill-in-the-Blanks

Groups of like cells and intercellular substances that work together to perform a task are known as a(n)

(8) _____ [p.458]. Groups of *different* types of tissues that interact to carry out a task are known as a(n)

(9) _____ [p.458]. Each cell engages in basic (10) _____ [p.459] activities that assure its own

survival. The combined contributions of cells, tissues, organs, and organ systems help maintain a stable

(11) _____ _____ [p.459] that is required for individual cell survival.

Whereas a specific kind of tissue (for example, simple squamous epithelium) is composed of cells that

look very similar and do similar jobs, a specific (12) _____ [p.458] (for example, a kidney) is composed

of different tissues that cooperate to do a specific job (in this case, to remove waste products from blood,

produce urine, and maintain the composition of body fluids). Kidneys, a urinary bladder, and various tubes

are grouped together into a(n) (13) _____ _____ [pp.458–459], which adds the functions of urine storage and elimination to the jobs that the kidneys do. A multicellular (14) _____ [recall p.4 of text] is most often composed of organ systems that cooperate to keep activities running smoothly in a coordinated fashion; this maintenance of stable operating conditions in the internal environment is known as (15) _____ [p.458].

All the diverse body parts found in different animals can be assembled from a few (16) _____ [p.458] types through variations in the way they are combined and arranged. Somatic cells compose the physical structure of the animal body; they become differentiated into four main types of specialized tissue: (17) _____ [p.459], (18) _____ [p.459], muscle, and nervous tissues.

(19) _____ [p.460] tissues have a(n) (20) _____ [p.460] surface, which is exposed to either a body fluid or the outside environment. Between the other surface of (19) and the connective tissue on which each rests is a(n) (21) _____ _____ [p.460]. When (19) tissues exist as a single layer, they generally serve as a(n) (22) _____ [p.460] for body cavities, ducts, and tubes (for example, the (22) of blood vessels). When (19) tissues exist as two or more layers (stratified), they generally serve as a(n) (23) _____ [p.460] barrier (for example, the outer region of your skin). Various kinds of (24) _____ [p.460] hold cells in a tissue together, prevent leakage through the tissue, or help cells communicate with each other by connecting their cytoplasmic regions. (25) _____ [p.461] cells of (19) tissues may absorb, produce, and secrete a variety of substances. Examples of substances secreted by (26) _____ [p.461] glands are oil, mucus, saliva, milk, digestive enzymes, sweat and ear wax; all such (26) secretions are released onto the free (19) surface through ducts or tubes. By contrast, (27) _____ [p.461] produced by (28) _____ [p.461] glands are secreted directly into the bloodstream or fluid bathing the gland (interstitial fluid).

Identification

29. Identify the three types of simple epithelia pictured below, and fill in the blanks with the word(s) describing where these epithelia are generally found.

simple (a.) _____ epithelium located in (b.) _____ _____ of lungs, walls of blood vessels [p.460]

simple (c.) _____ epithelium located in tubular parts of (d.) _____ in kidneys, various glands [p.460]

simple (e.) _____ epithelium located in part of lining of (f.) _____ and respiratory tract [p.460]

28.2. CONNECTIVE TISSUE [pp.462–463]

28.3. MUSCLE TISSUE [p.464]

28.4. NERVOUS TISSUE [p.465]

28.5. *Focus on Science*: **FRONTIERS IN TISSUE RESEARCH** [p.465]

Selected Words: collagen [p.462], elastin [p.462], "ground substance" [p.462], fibroblasts [p.462], tendons [p.462], ligaments [p.462], platelet [p.463], *plasma* [p.463], *contract* [p.464], *striated* [p.464], "involuntary" [p.464], *laboratory-grown epidermis* [p.465], *designer organs* [p.465], type I *diabetes mellitus* [p.465]

Boldfaced, Page-Referenced Terms

[p.462] loose connective tissue _____

_____

[p.462] dense, irregular connective tissue _____

_____

[p.462] dense, regular connective tissue _____

_____

[p.462] cartilage _____

_____

[p.463] bone tissue _____

_____

[p.463] adipose tissue _____

_____

[p.463] blood _____

_____

[p.464] skeletal muscle tissue _____

_____

[p.464] smooth muscle tissue _____

_____

[p.464] cardiac muscle tissue _____

_____

[p.465] nervous tissue _____

_____

[p.465] neuroglia _____

_____

[p.465] neurons _____

_____

True–False

If the statement is true, write a T in the blank. If the statement is false, correct it by changing the underlined word(s) and writing the correct word(s) in the answer blank.

_____1. Muscle bundles are identical to <u>individual</u> skeletal muscle cells. [p.464]

_____2. Both skeletal and cardiac muscle tissues are <u>striated</u>. [p.464]

_____3. Cardiac muscle cells are fused, end-to-end, but each cell contracts <u>independently of</u> other cardiac muscle cells. [p.464]

_____4. Smooth muscle tissue is involuntary and not <u>striated</u>. [p.464]

_____5. Smooth muscle tissue is located in the walls of the <u>intestine</u>. [p.464]

_____6. Neurons conduct messages to other <u>neurons</u> or to muscles or glands. [p.465]

Fill-in-the-Blanks

Cells of connective tissues are scattered throughout an extensive extracellular (7) _____ _____ [p.462]. (8) _____ [p.462] connective tissue contains a weblike scattering of strong, flexible protein fibers and a few highly elastic protein fibers arranged in a semifluid ground substance and serves as a support framework for epithelium and internal organs. (9) _____ [p.462] and (10) _____ [p.462] are examples of dense, regular connective tissue that help connect elements of the skeletal and muscular systems.

(11) _____ - _____ _____ [p.462] is a product surgeons can place over a wound so that cells in the sheet interact biochemically and make attachments with underlying cells, and damaged or missing tissue can be regenerated. People who have type I diabetes mellitus would be very interested in having a(n) (12) _____ _____ [p.462] implanted that would secrete insulin; biotechnologists are working out the ways to assemble such a package of cells.

Matching

Choose the most appropriate answer for each term. A letter may be used more than once and matched with more than one numbered item.

13. ___adipose tissue [p.463]

14. ___blood [p.463]

15. ___bone [p.463]

16. ___cardiac muscle [p.464]

17. ___cartilage [p.462]

18. ___neuroglia [p.465]

19. ___neurons [p.465]

20. ___skeletal muscle [p.464]

21. ___smooth muscle [p.464]

A. Involuntary contractile tissue
B. Rubber-like tissue that serves as a structural model for various bones
C. Storage region of calcium salts; may be involved in blood cell production also
D. Make up more than 50% of the volume of nervous tissue
E. Its fluid matrix is called plasma; involved in transporting various cells and substances
F. Storage region for fat
G. The communication units of most nervous systems
H. Striated

Label–Match

Identify each of the illustrations below by labeling it with one of the following: connective, epithelial, muscle, nervous, or gametes. Complete the exercise by matching and entering *all* appropriate letters and numbers from each group below in the parentheses following each label. Three of these have no matching letters. All have at least one matching number.

22. _____ (   ) [p.462]

23. _____ (   ) [recall p.460]

24. _____ (   ) [p.464]

25. _____ (   ) [p.464]

26. _____ (   ) [p.462]

27. _____ (   ) [recall p.460]

28. _____ (   ) [p.463]

29. _____ (   ) [recall p.460]

30. _____ (   ) [p.463]

31. _____ (   ) [p.465]

32. _____ (   ) [p.464]

33. _____ (   ) [recall p.460]

34. _____ (   ) [p.463]

A. Adipose
B. Bone
C. Cardiac
D. Dense, regular
E. Loose
F. Simple columnar
G. Simple cuboidal
H. Simple squamous
I. Smooth
J. Skeletal

1. Absorption
2. Maintain diploid number of chromosomes in sexually reproducing populations
3. Communication by means of electrochemical signals
4. Energy reserve
5. Contraction for voluntary movements
6. Diffusion
7. Padding
8. Contract to propel substances along internal passageways; not striated
9. Attaches muscle to bone and bone to bone
10. In vertebrates, provides the strongest internal framework of the organism
11. Elasticity
12. Secretion
13. Pumps circulatory fluid; striated
14. Insulation
15. Transport of nutrients and waste products to and from body cells
16. Blood cells are produced in some of these.

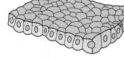
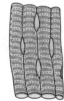

22.   23.

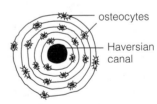

24.   25.   26.   27.   28.

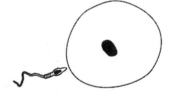
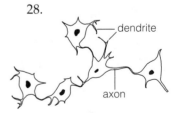

29.   30.   31.

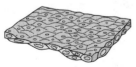
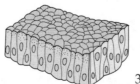
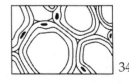

32.   33.   34.

28.6. ORGAN SYSTEMS [pp.466–467]

Selected Words: *midsagittal* plane [p.466], *dorsal* [p.466], *ventral* [p.466], *anterior* [p.467], *posterior* [p.467], superior [p.467], inferior [p.467], germ cells [p.467], "somatic" [p.467]

Boldfaced, Page-Referenced Terms

[p.466] integumentary system _____

_____

[p.466] muscular system _____

_____

[p.466] skeletal system _____

_____

[p.466] nervous system _____

_____

[p.466] endocrine system _____

_____

[p.466] circulatory system _____

_____

[p.467] lymphatic system _____

_____

[p.467] respiratory system _____

_____

[p.467] digestive system _____

_____

[p.467] urinary system _____

_____

[p.467] reproductive system _____

_____

[p.467] ectoderm _____

_____

[p.467] mesoderm _____

_____

[p.467] endoderm _____

_____

Fill-in-the-Blanks

In (1) _____ _____ [p.467] (immature reproductive cells that later develop into (2) _____ [p.467]), a special form of cell division known as (3) _____ [p.467] occurs. In all other body tissues that consist of (4) _____ [p.467] cells, the usual form of cell division, (5) _____ [p.467], occurs. The life of almost any animal begins with two gametes merging to form a fertilized egg, which undergoes reorganization and then divides by mitosis to form first undifferentiated cells, then three types of primary (6) _____ [p.467] (groups of similar cells that perform similar activities) in the early embryo. Eventually, the outer layer of skin and the tissues of the nervous system are formed from the relatively unspecialized embryonic tissue known as (7) _____ [p.467] located on the embryo's surface. The inner lining of the gut and the major organs formed from the embryonic gut develop from the internal embryonic tissue known as (8) _____ [p.467]. Most of the internal skeleton, muscle, the circulatory, reproductive, and urinary systems, and the connective tissue layers of the gut and body covering are formed from (9) _____ [p.467], the embryonic tissue composed of cells that can move about like amoebae.

Complete the Table

Supply the name of the "primary" tissue of the embryo that does the job indicated by becoming specialized in particular ways.

Primary Tissue	Functions
[p.467] 10.	Forms internal skeleton and muscle, circulatory, reproductive, and urinary systems
[p.467] 11.	Forms inner lining of gut and linings of major organs formed from the embryonic gut
[p.467] 12.	Forms outer layer of skin and the tissues of the nervous system

Labeling

Label each organ system described.

13. _____ Picks up nutrients absorbed from gut and transports them to cells throughout body [p.466]

14. _____ Helps cells use nutrients by supplying them with oxygen and relieving them of CO_2 wastes [p.467]

15. _____ Helps maintain the volume and composition of body fluids that bathe the body's cells [p.467]

16. _____ Provides basic framework for the animal and supports other organs of the body [p.466]

17. _____ Uses chemical messengers to control and guide body functions [p.466]

18. _____ Protects the body from viruses, bacteria, and other foreign agents; collects and returns some interstitial (tissue) fluid to the bloodstream [p.467]

19. _____ Produces younger, temporarily smaller versions of the animal [p.467]

20. _____ Breaks down larger food molecules into smaller nutrient molecules that can be absorbed by body fluids and transported to body cells [p.467]

21. _____ Consists of contractile parts that move the body through the environment and propel substances about in the animal [p.466]

22. _____ Serves as an electrochemical communications system in the animal's body [p.466]

23. _____ In the meerkat, serves as a heat catcher in the morning and protective insulation at night [p.466]

Matching

Match the most appropriate function with each system shown on the next page.

24. ___Male: production and transfer of sperm to the female. Female: production of eggs; provision of a protected nutritive environment for developing embryo and fetus. Both systems have hormonal influences on other organ systems. [p.467]

25. ___Ingestion of food, water; preparation of food molecules for absorption; elimination of food residues from the body. [p.467]

26. ___Movement of internal body parts; movement of whole body; maintenance of posture; heat production. [p.466]

27. ___Detection of external and internal stimuli; control and coordination of responses to stimuli; integration of activities of all organ systems. [p.466]

28. ___Protection from injury and dehydration; body temperature control; excretion of some wastes; reception of external stimuli; defense against microbes. [p.466]

29. ___Provisioning of cells with oxygen; removal of carbon dioxide wastes produced by cells; pH regulation. [p.467]

30. ___Support, protection of body parts; sites for muscle attachment, blood cell production, and calcium and phosphate storage. [p.466]

31. ___Hormonal control of body functioning; works with nervous system in integrative tasks. [p.466]

32. ___Maintenance of the volume and composition of extracellular fluid. [p.467]

33. ___Rapid internal transport of many materials to and from cells; helps stabilize internal temperature and pH. [p.466]

34. ___Return of some extracellular fluid to blood; roles in immunity (defense against specific invaders of the body). [p.467]

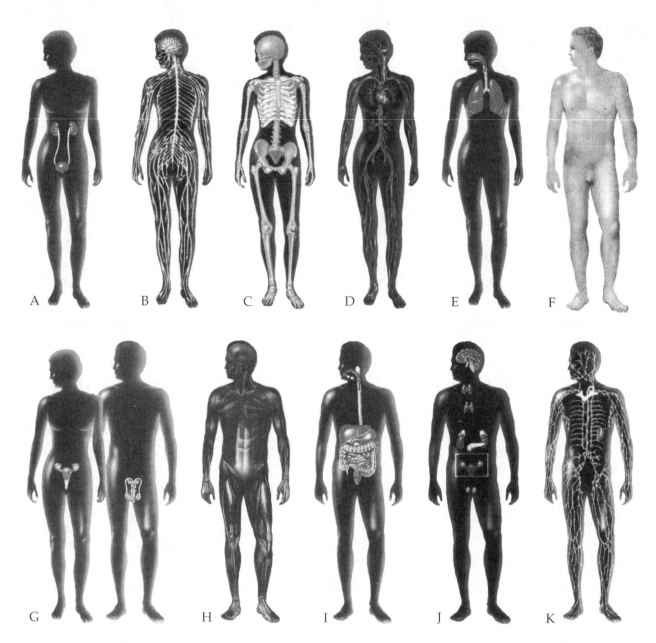

A   B   C   D   E   F

G   H   I   J   K

Fill-in-the-Blanks

There are four major body cavities in humans. Lungs are located in the (35) _____ [p.466] cavity, the brain is in the (36) _____ [p.466] cavity, and, in the female, ovaries and the urinary bladder are in the (37) _____ [p.466] cavity. In humans, the (38) _____ [p.467] plane divides the body into right and left halves. The (39) _____ [p.467] plane divides the body into (40) _____ [p.467] (front) and posterior (back) parts. In humans, the (41) _____ [p.467] plane divides it into superior (upper) and (42) _____ [p.467] (lower) parts. The urinary system of an animal maintains the volume and composition of the internal environment and is responsible for the disposal of (43) _____ [p.467] wastes; fecal material is not considered in the category. The endocrine system is generally responsible for internal (44) _____ [p.466] control; together with the (45) _____ [p.466] system, it integrates and governs physiological processes.

28.7. HOMEOSTASIS AND SYSTEMS CONTROL [pp.468–469]

Selected Words: *interstitial* fluid [p.468], *plasma* [p.468], "set points" [p.468], homeostatic control mechanism [p.469]

Boldfaced, Page-Referenced Terms

[p.468] extracellular fluid _____

_____

[p.468] sensory receptors _____

_____

[p.468] stimulus _____

_____

[p.468] integrator _____

_____

[p.468] effectors _____

_____

[p.468] negative feedback mechanism _____

_____

[p.469] positive feedback mechanism _____

_____

True–False

If the statement is true, write a T in the blank. If false, make it true by changing the underlined word(s) and writing the correct word(s) in the blank.

_____1. The process of childbirth is an example of a <u>negative</u> feedback mechanism. [p.469]

_____2. The human body's <u>effectors</u> are, for the most part, muscles and glands. [p.468]

_____3. An integrator is constructed in such a way that it is informed of specific energy changes in the environment and relays messages about them to a <u>receptor</u>. [p.468]

_____4. <u>Plasma</u> is the fluid portion of blood located in blood vessels. <u>Interstitial</u> <u>fluid</u> is located outside of blood vessels, lymphatic vessels, and body cells, but within the body's outer boundaries. [p.468]

Fill-in-the-Blanks

In a(n) (5) _____ [p.469] feedback mechanism, a chain of events is set in motion that intensifies the original condition before returning to a set point; sexual arousal and childbirth are two examples. Generally, physiological controls work by means of (6) _____ [p.469] feedback, in which an activity changes some condition in the internal environment, and the change causes the condition to be reversed; the maintenance of body temperature close to a "set point" is an example. Your brain is a(n) (7) _____ [p.469], a control point where different bits of information are pulled together in the selection of a response. Muscles and (8) _____ [p.469] are examples of effectors. Internal temperature control of the husky body is achieved by (9) _____ [p.469] in the skin and elsewhere sensing a temperature change at the body's surface and relaying the neural information to an integrator. In this example, the (10) _____ [p.469] in the brain (see text Figure 28.14 on p.469) compares neural input against a set point. This part of the brain then sends output signals to (11) _____ [p.469]. Examples of these are the (12) _____ [p.469] glands, which evaporate water from the tongue.

Self-Quiz

_____ 1. Which of the following is *not* included in connective tissues? [p.464]
   a. bone
   b. blood
   c. cartilage
   d. skeletal muscle

_____ 2. A surrounding material within which some-thing originates, develops, or is contained is known as a _____. [p.462]
   a. lamella
   b. ground substance
   c. plasma
   d. lymph

_____ 3. Blood is considered to be a(n) _____ tissue. [p.463]
   a. epithelial
   b. muscular
   c. connective
   d. none of these

_____ 4. _____ are abundant in tissues of the heart and liver where they promote diffu-sion of ions and small molecules from cell to cell. [p.460]
   a. Adhesion junctions
   b. Filter junctions
   c. Gap junctions
   d. Tight junctions

_____ 5. Muscle that is not striped and is involuntary is _____. [p.464]
   a. cardiac
   b. skeletal
   c. striated
   d. smooth

_____ 6. Chemical and structural bridges link groups or layers of like cells, uniting them in struc-ture and function as a cohesive _____. [p.460]
   a. organ
   b. organ system
   c. tissue
   d. cuticle

_____ 7. A fish embryo was accidentally stabbed by a graduate student in developmental biology. Later, the embryo developed into a creature that could not move and had no supportive or circulatory systems. Which embryonic tis-sue had suffered the damage? [p.467]
   a. ectoderm
   b. endoderm
   c. mesoderm
   d. protoderm

___ 8. A tissue whose cells are striated and fused at the ends by cell junctions so that the cells contract as a unit is called _____ tissue. [p.464]
   a. smooth muscle
   b. dense fibrous connective
   c. supportive connective
   d. cardiac muscle

___ 9. The secretion of tears, milk, sweat, and oil are functions of _____ tissues. [p.461]
   a. epithelial
   b. loose connective
   c. lymphoid
   d. nervous

___ 10. Memory, decision making, and issuing commands to effectors are functions of _____ tissue. [p.465]
   a. connective
   b. epithelial
   c. muscle
   d. nervous

___ 11. An animal that feels heated from the sun moves to an environment that tends to cool its body. This is an example of _____. [p.468]
   a. intensifying an original condition
   b. a positive feedback mechanism
   c. a positive phototropic response
   d. a negative feedback mechanism

___ 12. Which group is arranged correctly from smallest structure to largest? [p.464]
   a. muscle cells, muscle bundle, muscle
   b. muscle cells, muscle, muscle bundle
   c. muscle bundle, muscle cells, muscle
   d. none of the above

Matching

Choose the most appropriate answer for each term.

13. ___circulatory system [p.466]

14. ___digestive system [p.467]

15. ___endocrine system [p.466]

16. ___lymphatic system [p.467]

17. ___integumentary system [p.466]

18. ___muscular system [p.466]

19. ___nervous system [p.466]

20. ___reproductive system [p.467]

21. ___respiratory system [p.467]

22. ___skeletal system [p.466]

23. ___urinary system [p.467]

A. Picks up nutrients absorbed from gut and transports them to cells throughout body
B. Helps cells use nutrients by supplying them with oxygen and relieving them of CO_2 wastes
C. Helps maintain the volume and composition of body fluids that bathe the body's cells
D. Provides basic framework for the animal and supports other organs of the body
E. Uses chemical messengers to control and guide body functions
F. Protects the body from viruses, bacteria, and other foreign agents; collects and returns extracellular fluid to the bloodstream
G. Produces younger, temporarily smaller versions of the animal
H. Breaks down larger food molecules into smaller nutrient molecules that can be absorbed by body fluids and transported to body cells
I. Consists of contractile parts that move the body through the environment and propel substances about in the animal
J. Serves as an electrochemical communications system in the animal's body
K. In the meerkat, serves as a heat catcher in the morning and protective insulation at night

Chapter Objectives/Review Questions

1. Explain how the meerkat maintains a rather constant internal environment in spite of changing external conditions. [pp.458–459]
2. Cells are the basic units of life; in a multicellular animal, like cells are grouped into a(n) _____. [p.458]
3. Explain how, if each cell can perform all its basic activities, organ systems contribute to cell survival. [pp.459,466]
4. _____ tissues cover the body surface of all animals and line internal organs from gut cavities to vertebrate lungs; this tissue always has one _____ surface; the opposite surface adheres to a(n) _____ _____. [pp.460–461]
5. Know the characteristics of the various types of tissues. Know the types of cells that compose each tissue type and be able to cite some examples of organs that contain significant amounts of each tissue type. [pp.460–465]
6. List the functions carried out by epithelial tissue, and state the general location of each type. [p.460]
7. Explain the nature of three different cell-to-cell junctions, and state the types of tissues in which these junctions occur. [pp.460–461]
8. Explain the meaning of the term *gland*, cite three examples of glands, and state the extracellular products secreted by each. [p.461]
9. Connective tissue cells and fibers are surrounded by a(n) _____ _____. [p.462]
10. Describe the basic features of connective tissue, and explain how they enable connective tissue to carry out its various tasks. [pp.462–463]
11. List three functions of blood. [p.463]
12. Distinguish among skeletal, cardiac, and smooth muscle tissues in terms of location, structure, and function. [p.464]
13. Muscle tissues contain specialized cells that can _____. [p.464]
14. Neurons are organized as lines of _____. [p.465]
15. List each of the eleven principal organ systems in humans and match each to its main task. [pp.466–467]
16. Describe the ways by which extracellular fluid helps cells survive. [p.468]
17. Draw a diagram that illustrates the mechanism of homeostatic control. [pp.468–469]
18. Describe the relationships among receptors, integrators, and effectors in a negative feedback system. [pp.468–469]

Integrating and Applying Key Concepts

Explain why, of all places in the body, marrow is located on the interior of long bones. Explain why your bones are remodeled after you reach maturity. Why does your body not keep the same mature skeleton throughout life?

29

INTEGRATION AND CONTROL: NERVOUS SYSTEMS

Interactive Exercises

Why Crack the System? [pp.472-473]

29.1. NEURONS—THE COMMUNICATION SPECIALISTS [pp.474-475]

29.2. A CLOSER LOOK AT ACTION POTENTIALS [pp.476-477]

Selected Words: crack [p.472], receptors [p.473], integrators [p.473], effectors [p.473], *input* zones [p.474], *trigger* zone [p.474], *conducting* zone [p.474], *output* zone [p.474], *graded* signals [p.476], *local* signals [p.476], threshold level [p.476], *all-or-nothing* event [p.476], spike [p.476]

Boldfaced, Page-Referenced Terms

[p.473] nervous system _____

_____

[p.473] sensory neurons _____

_____

[p.473] stimulus _____

_____

[p.473] interneurons _____

[p.473] motor neurons _____

_____

[p.474] dendrites _____

_____

[p.474] axon _____

_____

[p.474] resting membrane potential _____

_____

[p.474] action potential _____

_____

[p.475] sodium-potassium pumps _____

_____

[p.476] positive feedback _____

_____

Fill-in-the-Blanks

Nerve cells that conduct messages are called (1) _____ [p.473]. (2) _____ [p.473] cells, which
support and nurture the activities of neurons, make up less than half the volume of the nervous system.
(3) _____ [p.473] neurons respond to specific kinds of environmental stimuli, (4) _____ [p.473]
connect different neurons in the spinal cord and brain, and (5) _____ [p.473] neurons are linked with
muscles or glands. All neurons have a(n) (6) _____ _____ [p.474] that contains the nucleus and the
metabolic means to carry out protein synthesis. (7) _____ [p.474] are short, slender extensions of
(6), and together these two neuronal parts are the neurons' "input zone" for receiving (8) _____ [p.474].
The (9) _____ [p.474] is a single long cylindrical extension away from the (10) _____ _____
[p.474]; in motor neurons, the (11) _____ has finely branched (12) _____ [p.474] that terminate on
muscle or gland cells and are "output zones," where messages are sent on to other cells.

A neuron at rest establishes unequal electric charges across its plasma membrane, and a(n)
(13) _____ _____ [p.474] is maintained. Another name for (13) is the (14) _____ _____
_____ [p.474]; it represents an ability for the membrane to be disturbed. Weak disturbances of the
neuronal membrane might set off only slight changes across a small patch, but strong disturbances can
cause a(n) (15) _____ _____ [p.474], which is an abrupt, short-lived reversal in the polarity of
charge across the plasma membrane of the neuron. For a fraction of a second, the cytoplasmic side of a bit

of membrane becomes positive with respect to the outside. The (16) _____ [p.474] that travels along the neural membrane is nothing more than short-lived changes in the membrane potential.

How is the resting membrane potential established, and what restores it between action potentials? The concentrations of (17) _____ [p.474] ions (K+), sodium ions (18) (_____+) [p.474], and other charged substances are not the same on the inside and outside of the neuronal membrane. (19) _____ [p.475] proteins that span the membrane affect the diffusion of specific types of ions across it. (20) _____ [p.475] proteins that span the membrane pump sodium and potassium ions against their concentration gradients across it by using energy stored in ATP. A neuronal membrane has many more positively charged (21) _____ [p.475] ions inside than out and many more positively charged (22) _____ [p.475] ions outside than in. There are about (23) _____ [p.475] times more potassium ions on the cytoplasmic side as outside, and there are about (24) _____ [p.475] times more sodium ions outside as inside. Some channel proteins leak ions through them all the time; others have (25) _____ [p.475] that open only when stimulated. Transport proteins called (26) _____-_____ _____ [p.475] counter the leakage of ions across the neuronal membrane and maintain the resting membrane potential. In all neurons, stimulation at an input zone produces (27) _____ [p.476] signals that do not spread very far (half a millimeter or less). (28) _____ [p.476] means that signals can vary in magnitude—small or large—depending on the intensity and (29) _____ [p.476] of the stimulus. When stimulation is intense or prolonged, graded signals can spread into an adjacent (30) _____ _____ [p.476] of the membrane—the site where action potentials can be initiated.

A(n) (31) _____ _____ [p.474] is an abrupt, brief reversal in voltage difference; that moves along a neuron. Once an action potential has been achieved, it is a(n) (32) _____-_____-_____ [p.476] event; its amplitude will not change even if the strength of the stimulus changes. The minimum change in membrane potential needed to achieve an action potential is the (33) _____ [p.476] value.

The cytoplasm next to the plasma membrane of a neuron at rest is more (34) [choose one] ❏ positive ❏ negative [p.477] than the interstitial fluid just outside the membrane. During an action potential, the inside of a disturbed patch of membrane becomes more (35) [choose one] ❏ positive ❏ negative than the outside. After an action potential, (36) _____ [p.477] conditions are restored at the membrane patch.

Labeling

Identify the parts of the neuron illustrated below.

37. _____ _____ [p.474]

38. _____ _____ [p.475]

39. _____ [p.475]

40. _____-_____ _____ [p.475]

41. _____ _____ [p.475]

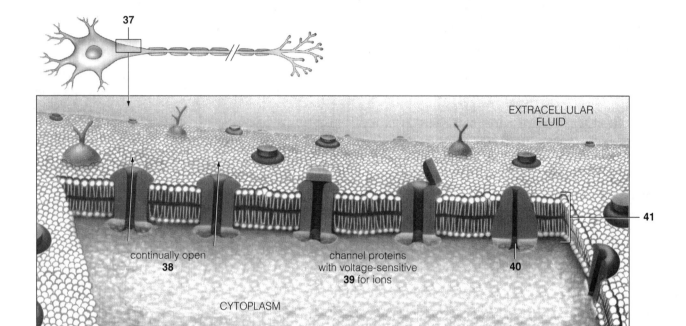

EXTRACELLULAR
FLUID

continually open
38

channel proteins
with voltage-sensitive
39 for ions

41

40

CYTOPLASM

Identify the numbered parts of the illustration at right.

42. _____ _____ [p.477]

43. _____ [p.477]

44. _____ _____ _____ [p.477]

45. _____ [p.477]

46. _____ [p.477]

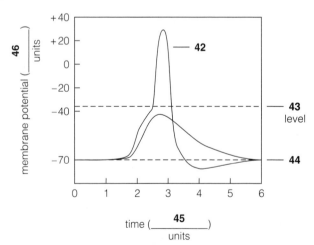

29.3. CHEMICAL SYNAPSES [pp.478-479]
29.4. PATHS OF INFORMATION FLOW [pp.480-481]
29.5. *Focus on Health:* SKEWED INFORMATION FLOW [p.482]

*Selected Words: pre*synaptic cell [p.478], *post*synaptic cell [p.478], *excitatory* effect [p.478], *inhibitory* effect [p.478], neuromuscular junction [p.478], *neuromodulators* [p.478], EPSPs [p.479], *depolarizing* effect [p.479], IPSPs [p.479], *hyperpolarizing* effect [p.479], summation [p.479], *divergent, convergent,* and *reverberating* circuits [p.480], Schwann cells [p.480], *multiple sclerosis* [p.480], stretch reflex [p.481], *botulism* [p.482], *tetanus* [p.482]

Boldfaced, Page-Referenced Terms

[p.478] neurotransmitters _____

_____

[p.478] chemical synapse _____

_____

[p.478] acetylcholine (ACh) _____

_____

[p.479] synaptic integration _____

_____

[p.480] nerves _____

_____

[p.480] myelin sheath _____

_____

[p.480] reflexes _____

_____

Fill-in-the-Blanks

The junction specialized for transmission between a neuron and another cell is called a(n) (1) _____

_____ [p.478]. Usually, the signal being sent to the receiving cell is carried by chemical messengers

called (2) _____ [p.478]. (3) _____ [p.478] is an example of this type of chemical messenger that

diffuses across the synaptic cleft, combines with protein receptor molecules on the muscle cell membrane,

and soon thereafter is rapidly broken down by enzymes. At a(n) (4) _____ [p.479] synapse, the

membrane potential is driven toward the threshold value and increases the likelihood that depolarization

and an action potential will occur. At a(n) (5) _____ [p.479] synapse, the membrane potential is driven

away from the threshold value, and the receiving neuron becomes hyperpolarized and is less likely to

achieve an action potential. A specific transmitter substance can have either excitatory or inhibitory effects

depending on which type of protein channel it opens up in the (6) _____ [p.478] membrane.

(7) _____ [p.478] are neuromodulators that inhibit perceptions of pain and may have roles in

memory and learning, emotional states, temperature regulation, and sexual behavior. (8) _____

_____ [p.479] at the cellular level is the moment-by-moment tallying of all excitatory and inhibitory signals acting on a neuron. Incoming information is (9) _____ [p.479] by cell bodies, and the charge differences across the membranes are either enhanced or inhibited.

Some narrow-diameter neurons are wrapped in lipid-rich (10) _____ [p.480] produced by specialized neuroglial cells called Schwann cells; each of these is separated from the next by a(n) (11) _____ _____ [p.480]—a small gap where the axon is exposed to extracellular fluid.

A(n) (12) _____ [p.480] is an involuntary sequence of events elicited by a stimulus. During a (13) _____ _____ [p.481], a muscle contracts involuntarily whenever conditions cause a stretch in length; many of these help you maintain an upright posture despite small shifts in balance.

Imbalances can occur at chemical synapses; a neurotoxin produced by *Clostridium tetani* blocks the release of GABA and glycine, thus freeing the motor neurons from (14) _____ [p.482] control, which may cause tetanus—a prolonged, spastic paralysis that can lead to death.

Labeling

Label the parts of the nerve illustrated at the right and below.

15. _____ [p.480]

16. _____ _____ [p.480]

17. _____ _____ [p.480]

18. _____ [p.480]

19. _____ _____ [p.480]

20. _____ _____ [p.480]

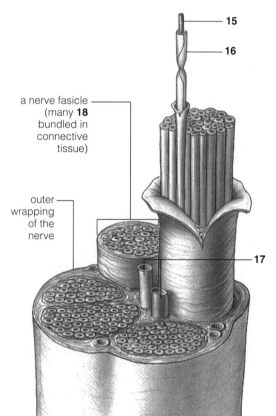

a nerve fasicle
(many **18**
bundled in
connective
tissue)

outer
wrapping
of the
nerve

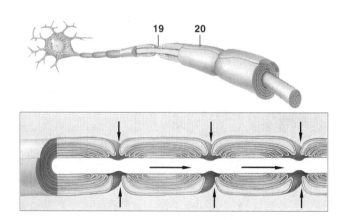

Labeling

Label the parts of neurons and types of neurons in the illustration.

21. _____ _____ [p.481]

22. _____ [p.481]

23. _____ _____ [p.481]

24. _____ [p.481]

25. _____ _____ [p.481]

26. _____ [p.481]

27. _____ _____ [p.481]

28. _____ [p.481]

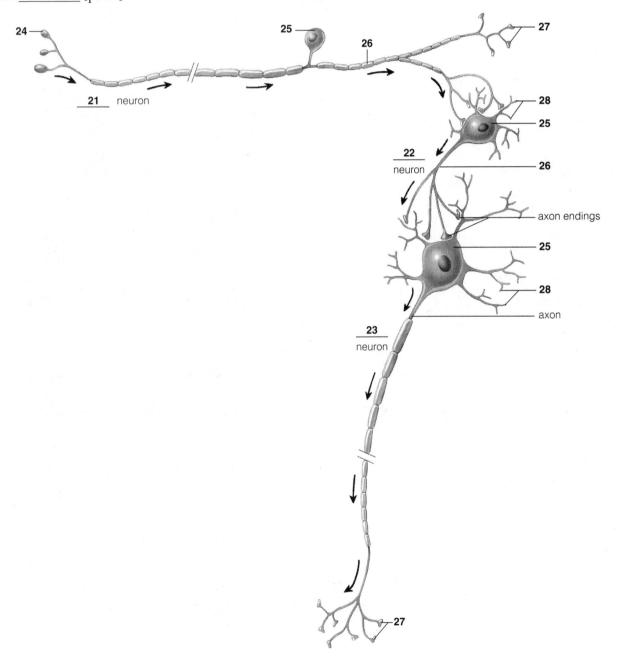

Matching

Match the choices below with the correct number in the diagram. (Two of the numbers match with two lettered choices.) [All from p.481]

29. ___

30. ___

31. ___

32. ___

33. ___

34. ___

35. ___

36. ___

37. ___

A. Response
B. Action potentials generated in motor neuron and propagated along its axon toward muscle
C. Motor neuron synapses with muscle cells
D. Muscle cells contract
E. Local signals in receptor endings of sensory neuron
F. Muscle spindle stretches
G. Action potentials generated in all muscle cells innervated by motor neuron
H. Stimulus
I. Axon endings synapse with motor neuron
J. Spinal cord
K. Action potential propagated along sensory neuron toward spinal cord

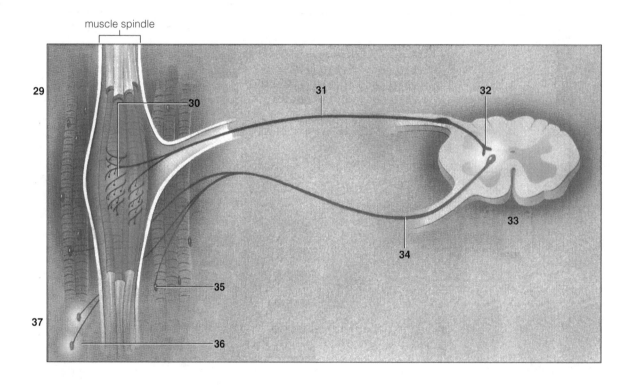

29.6. INVERTEBRATE NERVOUS SYSTEMS [pp.482-483]

29.7. VERTEBRATE NERVOUS SYSTEMS—AN OVERVIEW [pp.484-485]

29.8. THE MAJOR EXPRESSWAYS [pp.486-487]

Selected Words: radial symmetry [p.482], *bilateral* symmetry [p.482], nerve net [p.483], cephalization [p.483], *afferent* [p.485], *efferent* [p.485], *spinal* nerves [p.486], *cranial* nerves [p.486], *fight-flight* response [p.487], "rebound effect" [p.487], *meningitis* [p.487], white matter [p.487], gray matter [p.487]

Boldfaced, Page-Referenced Terms

[p.482] nerve net _____

_____

[p.484] neural tube _____

_____

[p.485] central nervous system _____

_____

[p.485] peripheral nervous system _____

_____

[p.486] somatic nerves _____

_____

[p.486] autonomic nerves _____

_____

[p.486] parasympathetic nerves _____

_____

[p.487] sympathetic nerves _____

_____

[p.487] spinal cord _____

_____

[p.487] meninges _____

_____

Fill-in-the-Blanks

Animals such as jellyfishes and sea anemones are invertebrates with (1) _____ [p.482] symmetry; their body parts are arranged, like spokes of a bike wheel, outward from a central axis. Cnidarians such as these have a(n) (2) _____ _____ [p.482] design for their nervous systems; even though signals travel slowly in all directions, the network is equally responsive to food or potential (3) _____ [p.482] coming from any direction.

(4) _____ [p.483] are the simplest animals with bilateral symmetry; bilaterally symmetrical animals tend to concentrate sensory cells at the body's leading end, a trend called (5) _____ [p.483] (the formation of a head).

All motor-nerves-to-skeletal muscle pathways and all sensory pathways make up the (6) _____ [p.486] nervous system. The remaining nerve tissue, which generally is not under conscious control, is collectively known as the (7) _____ [p.486] nervous system; it is subdivided into two parts: (8) _____ [pp.486-487] nerves, which respond to emergency situations, and (9) _____ [p.486] nerves, which oversee the restoration of normal body functioning. The (10) _____ _____ _____ [p.485] consists of the brain and spinal cord. A skull encloses the brain, and the (11) _____ _____ [p.487] encloses and protects the spinal cord in vertebrates. In humans, thirty-one pairs of spinal nerves connect with the spinal cord and are grouped by anatomical region; twelve pairs of (12) _____ [p.486] nerves connect parts of the head and neck with brain centers.

Labeling

Identify the divisions of the nervous system in the posterior view at the right.

13. _____ _____ [p.485]

14. _____ _____ [p.485]

15. _____ _____ [p.485]

16. _____ _____ [p.485]

17. _____ _____ [p.485]

18. _____ _____ [p.485]

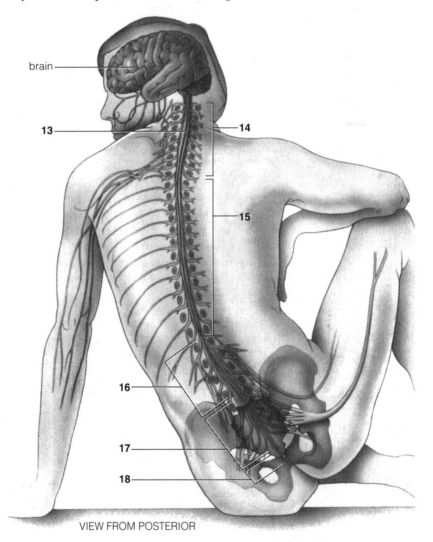

VIEW FROM POSTERIOR

Labeling

Label each numbered part of the accompanying illustration.

19. _____ [p.486]

20. _____ [p.486]

21. _____ [p.486]

22. _____ [p.486]

23. _____ [p.486]

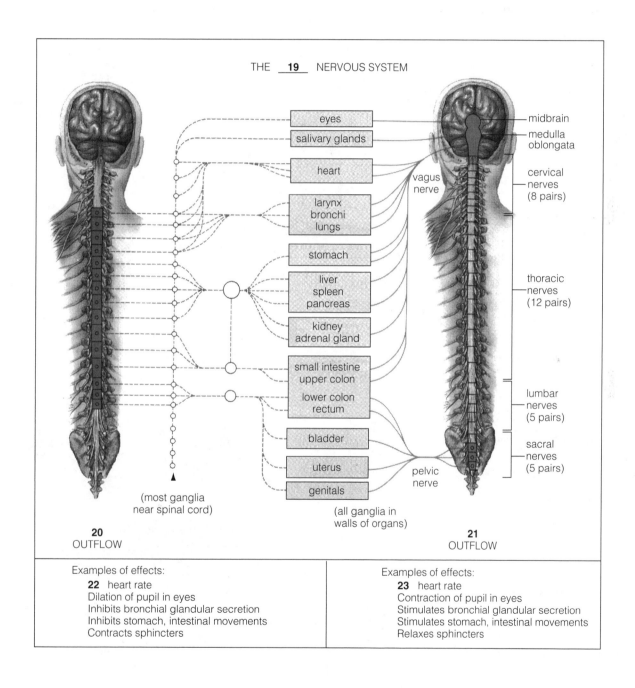

THE __19__ NERVOUS SYSTEM

eyes

salivary glands

heart

larynx
bronchi
lungs

stomach

liver
spleen
pancreas

kidney
adrenal gland

small intestine
upper colon

lower colon
rectum

bladder

uterus

genitals

vagus
nerve

pelvic
nerve

midbrain

medulla
oblongata

cervical
nerves
(8 pairs)

thoracic
nerves
(12 pairs)

lumbar
nerves
(5 pairs)

sacral
nerves
(5 pairs)

(most ganglia
near spinal cord)

(all ganglia in
walls of organs)

20
OUTFLOW

21
OUTFLOW

Examples of effects:
22 heart rate
Dilation of pupil in eyes
Inhibits bronchial glandular secretion
Inhibits stomach, intestinal movements
Contracts sphincters

Examples of effects:
23 heart rate
Contraction of pupil in eyes
Stimulates bronchial glandular secretion
Stimulates stomach, intestinal movements
Relaxes sphincters

Fill-in-the-Blanks

The (24) _____ _____ [p.487] is a region of local integration and reflex connections with nerve pathways leading to and from the brain; its (25) _____ _____ [p.487], which contains myelinated sensory and motor axons, is the expressway that connects the brain with the peripheral nervous system.

The (26) _____ _____ [p.487] includes neuronal cell bodies, dendrites, nonmyelinated axon terminals, and neuroglial cells; this is the integrative zone that deals mainly with (27) _____ [p.487] for limb movements (such as walking) and organ activity (such as bladder emptying).

Labeling

Identify the numbered parts of the accompanying illustrations.

28. _____ _____ [p.487]

29. _____ [p.487]

30. _____ [p.487]

31. _____ _____ [p.487]

32. _____ [p.487]

33. _____ _____ [p.487]

34. _____ _____ [p.487]

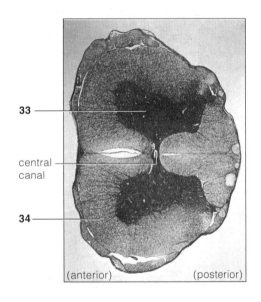

central canal

(anterior)   (posterior)

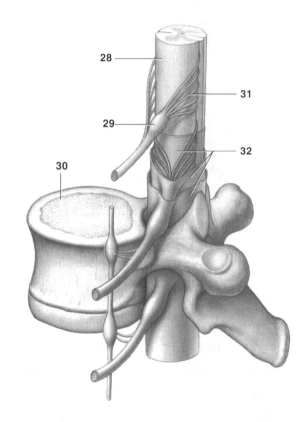

29.9. THE VERTEBRATE BRAIN [pp.488-489]

29.10. MEMORY [p.490]

29.11. *Focus on Health:* DRUGGING THE BRAIN [pp.490-491]

Selected Words: medulla oblongata [p.488], cerebellum [p.488], pons [p.488], optic lobes [p.488], olfactory lobes [p.488], cerebrum [p.488], thalamus [p.488], hypothalamus [p.488], amygdala [p.489], hippocampus [p.489], *EEG,* electroencephalogram [p.489], cerebrospinal fluid [p.489], *short-term* storage [p.490], *long-term* storage [p.490], *amnesia* [p.490], *Parkinson's disease* [p.490], *Alzheimer's disease* [p.490], stimulants [p.490], *amphetamines* [p.491], depressants [p.491], cirrhosis [p.491], *analgesics* [p.491], *heroin* [p.491], psychedelics [p.491], *LSD* [p.491]

Boldfaced, Page-Referenced Terms

[p.488] brain _____

_____

[p.488] brain stem _____

_____

[p.488] hindbrain _____

_____

[p.488] midbrain _____

_____

[p.488] forebrain _____

_____

[p.488] cerebral cortex _____

_____

[p.488] reticular formation _____

_____

[p.489] limbic system _____

_____

[p.489] blood-brain barrier _____

_____

[p.490] memory _____

_____

[p.490] drug addiction _____

_____

Fill-in-the-Blanks

The hindbrain is an extension and enlargement of the upper spinal cord; it consists of the (1) _____

_____ [p.488], which contains the control centers for the heartbeat rate, blood pressure, and breathing

reflexes. The hindbrain also includes the (2) _____ [p.488], which helps coordinate motor responses

associated with refined limb movements, maintenance of posture, and spatial orientation. The (3) _____

[p.488] is a major routing station for nerve tracts passing between the cerebellum and cerebral cortex.

The (4) _____ [p.488] evolved as a center that coordinates reflex responses to images and sounds.

The midbrain, pons, and medulla oblongata make up the brain stem; within its core, the (5) _____

_____ [p.488] is a major network of interneurons that extends its entire length and helps govern an

organism's level of nervous system functioning.

The (6) _____ [p.488] contains two cerebral hemispheres, the surfaces of which are composed of

gray matter. Underlying the cerebral hemispheres is the (7) _____ [p.488], a region that monitors

internal organs and influences hunger, thirst, and (8) _____ [p.488] behaviors, and the (9) _____ [p.488], the coordinating center where some motor pathways converge and relay signals to and from the cerebrum. Both the spinal cord and the brain are bathed in (10) _____ _____ [p.489].

The storage of individual bits of information somewhere in the brain is called (11) _____ [p.490]. Experiments suggest that at least two stages are involved in its formation. One is a(n) (12) _____-_____ _____ [p.490] period, lasting only a few hours; information then becomes spatially and temporally organized in neural pathways. The other is a(n) (13) _____-_____ _____ [p.490]; information then is put in a different neural representation and is permanently filed in the brain.

(14) _____ [p.491] are analgesics produced by the brain that inhibit regions concerned with our emotions and perception of (15) _____ [p.491]. Imbalances in (16) _____ [pp.478,491] can produce emotional disturbances.

Matching

Match the named part with the letter that describes its function.

17. ___cerebellum [p.488]

18. ___corpus callosum [pp.488-489]

19. ___hypothalamus [p.488]

20. ___limbic system [p.489]

21. ___medulla oblongata [p.488]

22. ___occipital lobe [p.489]

23. ___olfactory lobes [p.488]

24. ___prefrontal cortex [p.490]

25. ___primary motor cortex [p.489]

26. ___primary somatic sensory cortex [p.489]

27. ___temporal lobe [p.489]

28. ___thalamus [p.488]

A. Monitors activities of internal organs; influences behaviors related to thirst, hunger, reproductive cycles, and temperature control
B. Coordinates muscles required for speech; houses many banks of fact memories
C. Receives inputs from cochleas of inner ears
D. Issues commands to muscles
E. Relays and coordinates sensory signals to the cerebrum
F. Receives inputs from receptors in nasal epithelium
G. Receives and processes input from body-feeling areas
H. Broad channel of white matter that keeps the two cerebral hemispheres communicating with each other
I. Coordinates nerve signals for maintaining balance, posture, and refined limb movements
J. Receives inputs from retinas of eyeballs
K. Connects pons and spinal cord; contains reflex centers involved in respiration, stomach secretion, and cardiovascular function
L. Contains brain centers that coordinate activities underlying emotional expression and memory

Labeling

Identify each numbered part of the accompanying illustration.

29. _____ [p.488]

30. _____ _____ [p.488]

31. _____ [p.488]

32. _____ [p.488]

33. _____ [p.488]

34. _____ _____ [p.488]

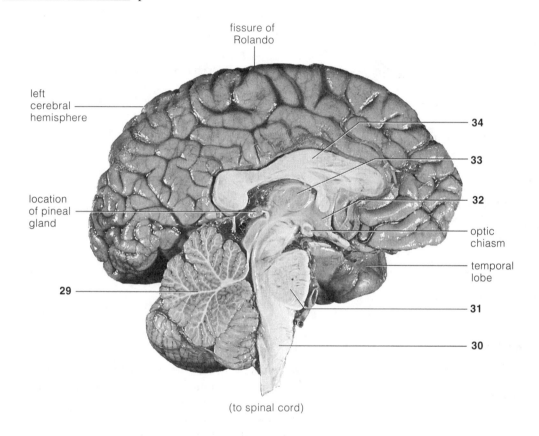

fissure of Rolando

left cerebral hemisphere

location of pineal gland

29

optic chiasm

temporal lobe

31

30

(to spinal cord)

Matching

Choose the most appropriate category for each drug.

35. ___amphetamines [p.491]

36. ___barbiturates [p.491]

37. ___caffeine [p.490]

38. ___cocaine [p.490]

39. ___ethyl alcohol [p.491]

40. ___heroin [p.491]

41. ___LSD [p.491]

42. ___marijuana [p.491]

43. ___nicotine [p.490]

A. Depressant or hypnotic drug
B. Narcotic analgesic drug
C. Psychedelic or hallucinogenic drug
D. Stimulant

Self-Quiz

___ 1. Which of the following is *not* true of an action potential? [pp.476-477]
   a. It is a short-range signal that can vary in size along a neuron.
   b. It is an all-or-none brief reversal in membrane potential.
   c. Its strength doesn't diminish as it passes along the axon.
   d. It is self-propagating.

___ 2. The resting membrane potential _____. [pp.474-475]
   a. exists as long as a charge difference sufficient to do work exists across a membrane
   b. occurs because there are more potassium ions outside the neuronal membrane than there are inside
   c. occurs because of the unique distribution of receptor proteins located on the dendrite exterior
   d. is brought about by a local change in membrane permeability caused by a greater-than-threshold stimulus

___ 3. An action potential is brought about by _____. [p.476]
   a. a sudden membrane impermeability
   b. the movement of negatively charged proteins through the neuronal membrane
   c. the movement of lipoproteins to the outer membrane
   d. a local change in membrane permeability caused by a greater-than-threshold stimulus

___ 4. The _____ are the protective coverings of the brain. [p.487]
   a. ventricles
   b. meninges
   c. tectums
   d. olfactory bulbs
   e. pineal glands

___ 5. The right hemisphere of the brain is generally responsible for _____. [p.488]
   a. music
   b. mathematics
   c. speech
   d. analytical skills
   e. reading ability

Matching

Choose the most appropriate statement for each term.

6. ___amphetamine [p.491]

7. ___axon [p.474]

8. ___cell body [p.474]

9. ___cerebellum [p.488]

10. ___cerebrum [p.488]

11. ___endorphin [p.491]

12. ___hypothalamus [p.488]

13. ___interneuron [p.473]

14. ___limbic system [p.489]

15. ___medulla oblongata [p.488]

16. ___parasympathetic [pp.486-487]

17. ___reticular formation [p.488]

18. ___sympathetic [p.487]

A. The part of the brain that controls the basic responses necessary to maintain life processes (breathing, heartbeat) is the _____.
B. Governs aspects of memory archiving and emotional behavior
C. Natural analgesic
D. The input zone of a neuron includes the _____.
E. _____ nerves generally dominate internal events when "fight or flight" situations occur.
F. The center for balance and coordination in the human brain is the _____.
G. The conducting zone of a neuron is the _____.
H. Connects motor centers of medulla oblongata and spinal cord with the cerebral cortex; determines activity level of cerebrum.
I. Monitors internal organs, acts as gatekeeper to the limbic system, helps the reasoning centers of the brain to dampen rage and hatred and governs hunger, thirst, and sex drives.
J. _____ are responsible for integration in the nervous system.
K. The center of consciousness and intelligence is the _____.
L. _____ nerves generally dominate internal events when environmental conditions permit normal body functioning.
M. _____ is a stimulant.

Chapter Objectives/Review Questions

1. Outline some of the ways by which information flow is regulated and integrated in the human body. [pp.473,478-482]
2. Summarize current best guesses about how the evolution of animal nervous systems may have occurred. [pp.482-484]
3. Draw a neuron and label it according to its four general zones, its specific structures, and the specific function(s) of each structure. [pp.474-475]
4. Define *resting membrane potential*; explain what establishes it and how it is used by the cell neuron. [p.474]
5. Describe the distribution of the invisible array of large proteins, ions, and other molecules in a neuron, both at rest and as a neuron experiences a change in potential. [pp.474-475]
6. Define *action potential* by stating its three main characteristics. [p.474]
7. Define *sodium-potassium pump* and state how it helps maintain the resting membrane potential. [p.475]
8. Explain the chemical basis of the action potential. Look at text Figure 29.7 on p.477 and determine which part of the curve represents the following:
   a. the point at which the stimulus was applied;
   b. the events prior to achievement of the threshold value;
   c. the opening of the ion gates and the diffusing of the ions;
   d. the change from net negative charge inside the neuron to net positive charge and back again to net negative charge; and
   e. the active transport of sodium ions out of and potassium ions into the neuron. [pp.476-477]
9. Explain how graded signals differ from action potentials. [p.476]
10. Understand how a nerve impulse is received by a neuron, conducted along a neuron, and transmitted across a synapse to a neighboring neuron, muscle, or gland. [pp.478-479]
11. Explain what a reflex is by drawing and labeling a diagram and telling how it functions. [pp.480-481]
12. Contrast the central and peripheral nervous systems. [p.485]
13. Explain how parasympathetic nerve activity balances sympathetic nerve activity. [pp.486-487]
14. Describe the basic structural and functional organization of the spinal cord. In your answer, distinguish spinal cord from vertebral column. [p.487]
15. For each part of the brain, state how the behavior of a normal person would change if he or she suffered a stroke in that part of the brain. [pp.488-489]
16. Describe how the cerebral hemispheres are related to the other parts of the forebrain. [pp.488-489]
17. List the major classes of psychoactive drugs, and provide an example of each class. [pp.490-491]

Integrating and Applying Key Concepts

Suppose that anger is eventually determined to be caused by excessive amounts of specific transmitter substances in the brains of angry people. Also suppose that an inexpensive antidote to anger that neutralizes these anger-producing transmitter substances is readily available. Can violent murderers now argue that they have been wrongfully punished because they were victimized by their brain's transmitter substances and could not have acted in any other way? Suppose an antidote is prescribed to curb violent tempers in an easily angered person. Suppose also that the person forgets to take the pill and subsequently murders a family member. Can the murderer still claim to be victimized by transmitter substances?

30

SENSORY RECEPTION

Interactive Exercises

Different Strokes for Different Folks [pp.494–495]

30.1. OVERVIEW OF SENSORY PATHWAYS [p.496]

30.2. SOMATIC SENSATIONS [p.497]

30.3. SENSES OF HEARING AND BALANCE [pp.498–499]

Selected Words: "ultrasounds" [p.494], stimulus (pl. stimuli) [p.494], nociceptors [p.495], *amplitude* [p.496], *frequency* [p.496], *somatic* pain [p.497], *visceral* pain [p.497], *equilibrium* position [p.498], *motion sickness* [p.498], oval window [p.499], organ of Corti [p.499], basilar membrane [p.499], tectorial membrane [p.499], cochlea [p.499]

Boldfaced, Page-Referenced Terms

[p.494] echolocation _____

_____

[p.494] sensory systems _____

_____

[p.494] sensation _____

_____

[p.494] perception _____

_____

[p.495] mechanoreceptors _____

_____

[p.495] thermoreceptors _____

_____

[p.495] pain receptors _____

_____

[p.495] chemoreceptors _____

_____

[p.495] osmoreceptors _____

_____

[p.495] photoreceptors _____

_____

[p.496] sensory adaption _____

_____

[p.496] somatic sensations _____

_____

[p.496] special senses _____

_____

[p.497] somatosensory cortex _____

_____

[p.497] pain _____

_____

[p.498] inner ear _____

_____

[p.498] vestibular apparatus _____

_____

[p.498] hearing _____

_____

[p.499] middle ear _____

_____

[p.499] external ear _____

_____

[p.499] hair cells _____

_____

Fill-in-the-Blanks

Finely branched peripheral endings of sensory neurons that detect specific kinds of stimuli are
(1) _____ [p.495]. A(n) (2) _____ [p.495] is any form of energy change in the environment that the
body actually detects. (3) _____ [p.495] detect substances dissolved in water or air; (4) _____
[p.495] detect pressure, stretching, and vibrational changes; (5) _____ [p.495] detect the energy of
visible and ultraviolet light; (6) _____ [p.495] detect radiant energy associated with temperature
changes. A(n) (7) _____ [pp.494–495] is a conscious awareness of change in internal or external
conditions; this is not to be confused with (8) _____ [p.494], which is an understanding of what a
sensation means. A sensory system consists of sensory receptors for specific stimuli, (9) _____
_____ [p.494] that conduct information from those receptors to the brain, and (10) _____
_____ [p.494] where information is evaluated.

Matching

Select the best match for each item below.

11. ___Vision is associated with _____. [p.495]
12. ___Pain is associated with _____. [p.495]
13. ___Hearing is detected by _____. [p.498]
14. ___CO$_2$ concentration in the blood is detected by _____.
    [p.495, by definition]
15. ___Environmental temperature is detected by _____. [p.495]
16. ___Internal body temperature is detected by _____. [p.495]
17. ___Touch is detected by _____. [p.495]
18. ___Hair cells in the ear's organ of Corti are _____. [p.499]
19. ___Pacinian corpuscles in the skin are _____. [pp.496–497]
20. ___Any stimulus that causes tissue damage is a _____. [p.495]
21. ___The movement of fluid in the inner ear is associated with _____. [pp.495,499]

A. chemoreceptors
B. mechanoreceptors
C. nociceptors
D. photoreceptors
E. thermoreceptors

Fill-in-the-Blanks

The somatic sensations (awareness of (22) _____ [p.497], pressure, heat, (23) _____ [p.497], and
pain) start with receptor endings that are embedded in (24) _____ [p.497] and other tissues at the
body's surfaces, in (25) _____ [p.497] muscles, and in the walls of internal organs. All skin
(26) _____ [pp.495,497] are easily deformed by pressure on the skin's surface; these make you aware of
touch, vibrations, and pressure. (27) _____ [p.497] nerve endings serve as "heat" receptors, and their
firing of action potentials increases with increases in temperature. (28) _____ [p.497] is the perception
of injury to some body region. (29) _____ [p.497] in skeletal muscle, joints, tendons, ligaments, and
(30) _____ [p.497] are responsible for awareness of the body's position in space and of limb movements.
The (31) _____ _____ [p.498] in each of your inner ears contains organs of equilibrium that help
keep the body balanced in relation to gravity, velocity, acceleration, and other forces that influence its
position and movement.

Label–Match

Identify each numbered part in the illustrations to the right. Complete the exercise by entering the appropriate letter in the parentheses that follow the labels.

32. _____ _____ _____ ( ) [p.497]

33. _____ _____ ( ) [p.497]

34. _____ _____ ( ) [p.497]

35. _____ [p.497]

36. _____ [p.497]

37. _____ _____ ( ) [p.497]

A. React continually to ongoing stimuli
B. React to fine textures, touch, and vibrations
C. Involved in sensing heat, light pressure, and pain
D. Adapt very slowly to low-frequency vibrations

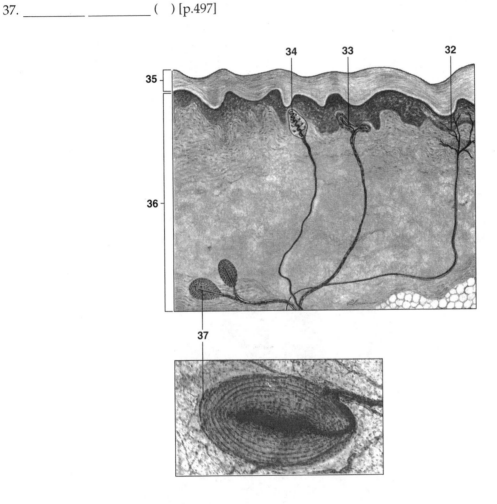

Fill-in-the-Blanks

The (38) _____ [p.496] senses a stimulus based on which nerve pathways carry the signals, the (39) _____ [p.496] of signals from each axon of that pathway, and the (40) _____ [p.496] of axons carrying the signals to the (38). The special senses of (41) _____ [pp.498–499] and balance involve variously-shaped tubes that contain fluid and hair cells. The (42) _____ [p.498] (perceived loudness) of sound depends on the height (and depth) of the sound wave. The (43) _____ [p.498] (perceived pitch) of sound depends on how many wave cycles per second occur. The faster the vibrations, the (44) [choose one] ❑ higher ❑ lower [p.498] the sound. Hair cells are (45) [choose one] ❑ nociceptors ❑ mechanoreceptors

❏ thermoreceptors [pp.495,499] that detect vibrations. The hammer, anvil, and stirrup are located in the (46) [choose one] ❏ inner ear ❏ middle ear [p.498]. The (47) _____ [p.498] is a coiled tube that resembles a snail shell and contains the (48) _____ _____ _____ [p.499]—the organ that changes vibrations into electrochemical impulses. Structures that detect rotational acceleration in humans compose the (49) _____ _____ [p.498].

Labeling

Identify each numbered part of the accompanying illustrations.

50. _____ _____ [p.498]
51. _____ [p.498]
52. _____ _____ [p.498]
53. _____ _____ [p.498]
54. _____ _____ [p.499]
55. _____ _____ [p.499]
56. _____ _____ [p.499]

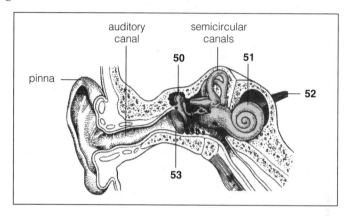

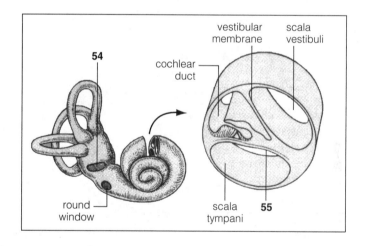

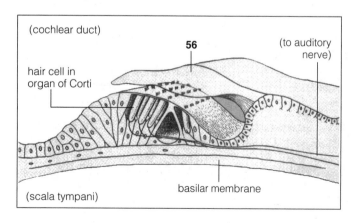

30.4. SENSE OF VISION [pp.500–501]

30.5. VISUAL PERCEPTION [p.502]

30.6. SENSES OF TASTE AND SMELL [p.503]

Selected Words: *simple* eyes [p.500], *compound* eyes [p.500], photoreceptor [p.500], iris [p.500], pupil [p.500], ciliary muscle [p.501], *astigmatism* [p.501], *nearsightedness* [p.501], *farsightedness* [p.501], rhodopsin [p.502], fovea [p.502], receptive field [p.502], visual cortex [p.502], olfactory bulb [p.503], limbic system [p.503]

Boldfaced, Page-Referenced Terms

[p.500] vision _____

_____

[p.500] eyes _____

_____

[p.500] visual field _____

_____

[p.500] lens _____

_____

[p.500] cornea _____

_____

[p.500] camera eyes _____

_____

[p.500] retina _____

_____

[p.501] visual accommodation _____

_____

[p.502] rod cells _____

_____

[p.502] cone cells _____

_____

[p.503] taste receptors _____

_____

[p.503] olfactory receptors _____

_____

[p.503] pheromones _____

_____

Fill-in-the-Blanks

(1) _____ [pp.500,502] is a stream of photons—discrete energy packets. A(n) (2) _____ [p.502] is a cell in which photons are absorbed by pigment molecules and photon energy is transformed into the electrochemical energy of a nerve signal. (3) _____ [p.500] requires precise light focusing onto a layer of photoreceptive cells that are dense enough to sample details of the light stimulus, followed by image formation in the brain. Crustaceans and insects have (4) _____ _____ [p.500], each of which is an array of units that contain a lens and a bundle of photoreceptors; the units all contribute to form a mosaic image. (5) _____ [p.500] are well-developed photoreceptor organs that allow at least some degree of image formation. The (6) _____ [p.500] is a transparent cover of the lens area, and the (7) _____ [p.502] consists of tissue containing densely packed photoreceptors. In the vertebrate eye, lens adjustments ensure that the (8) _____ _____ [p.501] for a specific group of light rays lands on the retina. (9) _____ _____ [p.501] refers to the lens adjustments that bring about precise focusing onto the retina. (10) _____ [p.501] people focus light from nearby objects posterior to the retina. (11) _____ [p.502] cells are concerned with daytime vision and, usually, color perception. The (12) _____ [p.502] is a funnel-shaped pit on the retina that provides the greatest visual acuity.

Labeling

Identify each numbered part of the accompanying illustration.

13. _____ _____ [p.501]

14. _____ [p.501]

15. _____ [p.501]

16. _____ [p.501]

17. _____ _____ [p.501]

18. _____ _____ [p.501]

19. _____ [p.501]

20. _____ [p.501]

21. _____ _____ [p.501]

22. _____ _____ [p.501]

23. _____ [p.501]

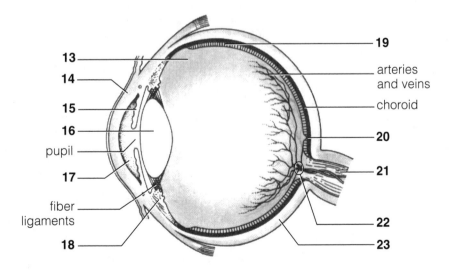

Self-Quiz

_____ 1. The correctly matched pair is _____.
[p.497]
   a. bulb of Krause—temperatures below 20°C
   b. Meissner's corpuscle—temperatures above 45°C
   c. Pacinian corpuscle—detection of pain
   d. Ruffini endings—painful freezing sensations caused by temperatures below 10°C

_____ 2. The principal place in the human ear where sound waves are amplified is _____.
[p.499]
   a. the pinna
   b. the ear canal
   c. the middle ear
   d. the organ of Corti
   e. none of the above

_____ 3. The place where vibrations are translated into patterns of nerve impulses is _____.
[p.499]
   a. the pinna
   b. the ear canal
   c. the middle ear
   d. the organ of Corti
   e. none of the above

_____ 4. Accommodation involves the ability to _____. [p.501]
   a. change the sensitivity of the rods and cones by means of transmitters
   b. change the width of the lens by relaxing or contracting certain muscles
   c. change the curvature of the cornea
   d. adapt to large changes in light intensity
   e. all of the above

_____ 5. Nearsightedness is caused by _____.
[p.501]
   a. eye structure that focuses an image in front of the retina
   b. uneven curvature of the lens
   c. eye structure that focuses an image posterior to the retina
   d. uneven curvature of the cornea
   e. none of the above

_____ 6. Olfactory receptors _____. [p.503]
   a. detect substances involved in taste perception
   b. detect stimuli involved in determining equilibrium position
   c. in vomeronasal organs detect pheromones
   d. generally send signals to the olfactory bulb (lobe) and limbic system
   e. both c and d

Matching

7. _____ciliary muscle [p.501]

8. _____cornea [pp.500–501]

9. _____fovea [p.502]

10. _____iris [p.500]

11. _____retina [pp.500,502]

12. _____rhodopsin [p.502]

13. _____sclera [p.500]

14. _____visual cortex [p.502]

A. The sensations of sight are produced here.
B. The adjustable ring of contractile and connective tissues that controls the amount of light entering the eye is the _____.
C. Rods and cones are located in the _____.
D. The white protective fibrous tissue of the eye is the _____.
E. The outer transparent protective covering part of the eyeball is the _____.
F. The highest concentration of cones is in the _____.
G. The shape of the lens is changed by action of this
H. Absorber of photon energy corresponding to blue to green wavelengths

Chapter Objectives/Review Questions

1. List the six kinds of receptors, and identify the type of stimulus energy that each type detects. [p.495]
2. Distinguish the types of stimuli detected by tactile and stretch receptors from those detected by hearing and equilibrium receptors. [pp.497–499]
3. Follow a sound wave from the pinna to the organ of Corti; mention the name of each structure it passes and state where the sound wave is amplified and where the pattern of pressure waves is translated into electrochemical impulses. [pp.498–499]
4. Explain what a visual system is, and list four of the five aspects of a visual stimulus that are detected by different components of a visual system. [p.500]
5. Contrast the structure of compound eyes with the structures of invertebrate pigment cups and of the human eye. [pp.500–502]
6. Define *nearsightedness* and *farsightedness,* and relate each to eyeball structure. [p.501]
7. Explain the general principles that affect how light is detected by photoreceptors and changed into electrochemical messages. [pp.501–502]
8. Distinguish the sense of taste from the sense of smell. Describe how the receptors work, tell where they are located, and state where both senses are perceived and interpreted. [p.503]

31

ENDOCRINE CONTROL

Interactive Exercises

Hormone Jamboree [pp.506–507]

31.1. THE ENDOCRINE SYSTEM [pp.508–509]

Selected Words: target cells [p.508], vomeronasal organ [p.506], *endon* [p.509], *krinein* [p.509]

Boldfaced, Page-Referenced Terms

[p.508] hormones _____

_____

[p.508] neurotransmitters _____

_____

[p.508] local signaling molecules _____

_____

[p.508] pheromones _____

_____

[p.509] endocrine system _____

_____

Matching

Choose the most appropriate answer for each term.

1. ___hormones [p.508]
2. ___neurotransmitters [p.508]
3. ___vomeronasal organ [p.508]
4. ___target cells [p.508]
5. ___local signaling molecules [p.508]
6. ___pheromones [p.508]

A. Signaling molecules released from axon endings of neurons that act swiftly on target cells
B. A pheromone detector discovered in humans
C. Released by many types of body cells and alter conditions within localized regions of tissues
D. Nearly odorless secretions of particular exocrine gland; common signaling molecules that act on cells of other animals of the same species and help integrate social behavior
E. Secretions from endocrine glands, endocrine cells, and some neurons that the bloodstream distributes to nonadjacent target cells
F. Cells that have receptors for any given type of signaling molecule

Complete the Table

7. Complete the table below by identifying the numbered components of the endocrine system shown in the illustration on p.391 as well as the hormones produced by each.

Gland Name	Number	Hormones Produced
[p.509] a. Hypothalamus		
[p.509] b. Pituitary, anterior lobe		
[p.509] c. Pituitary, posterior lobe		
[p.509] d. Adrenal glands (cortex)		
[p.509] e. Adrenal glands (medulla)		
[p.509] f. Ovaries (two)		
[p.509] g. Testes (two)		
[p.509] h. Pineal		
[p.509] i. Thyroid		
[p.509] j. Parathyroids (four)		
[p.509] k. Thymus		
[p.509] l. Pancreatic islets		

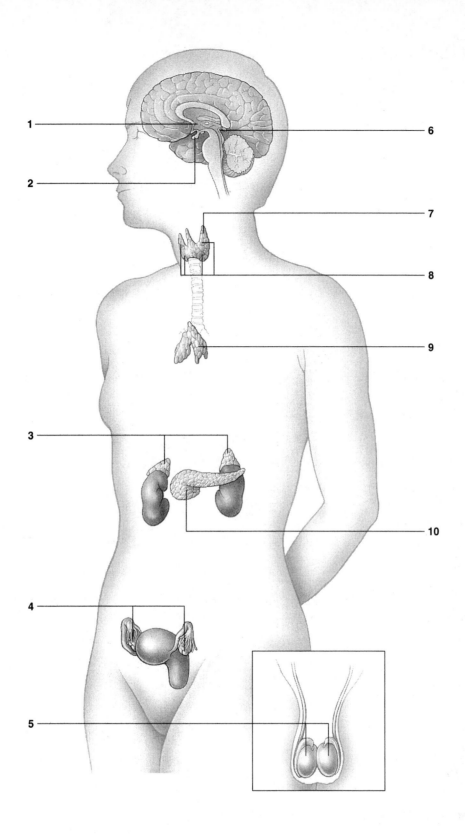

31.2. SIGNALING MECHANISMS [pp.510–511]

Selected Words: *testicular feminization syndrome* [p.510]

Boldfaced, Page-Referenced Terms

[p.510] steroid hormones _____

_____

[p.511] peptide hormones _____

_____

[p.511] second messenger _____

_____

Choice

For questions 1–10, choose from the following:

a. steroid hormones   b. peptide hormones

1. ___ Lipid-soluble molecules derived from cholesterol; can diffuse directly across the lipid bilayer of a target cell's plasma membrane [p.510]

2. ___ Various peptides, polypeptides, and glycoproteins [p.511]

3. ___ One example involves testosterone, defective receptors, and a condition called testicular feminization syndrome. [p.510]

4. ___ Hormones that often require assistance from second messengers [p.511]

5. ___ Hormones that bind to receptors at the plasma membrane of a cell; the receptor then activates specific membrane-bound enzyme systems, which in turn initiate reactions leading to the cellular response [p.511]

6. ___ Lipid-soluble molecules that move through the target cell's plasma membrane to the nucleus where it binds to some type of protein receptor. The hormone-receptor complex moves into the nucleus and interacts with specific DNA regions to stimulate or inhibit transcription of mRNA. [p.510]

7. ___ Water-soluble signaling molecules that may incorporate anywhere from 3 to 180 amino acids [p.511]

8. ___ Involves molecules such as cyclic AMP that activate many enzymes in cytoplasm that, in turn, cause alteration in some cell activity [p.511]

9. ___ Glucagon is an example. [p.511]

10. ___ Cyclic AMP relays a signal into the cells interior to activate protein kinase A. [p.511]

31.3. THE HYPOTHALAMUS AND PITUITARY GLAND [pp.512–513]

Selected Words: *posterior* lobe [p.512], *anterior* lobe [p.512], *second* capillary bed [p.512]

Boldfaced, Page-Referenced Terms

[p.512] hypothalamus _____

_____

[p.512] pituitary gland _____

_____

[p.513] releasers _____

_____

[p.513] inhibitors _____

_____

Choice–Match

Label each hormone given below with an "A" if it is secreted by the anterior lobe of the pituitary, a "P" if it is released from the posterior pituitary, or an "I" if it is released from intermediate tissue. Complete the exercise by entering the letter of the corresponding action in the parentheses following each label.

1. ___( ) ACTH [p.512]

2. ___( ) ADH [p.512]

3. ___( ) FSH [p.512]

4. ___( ) STH (GH) [p.512]

5. ___( ) LH [p.512]

6. ___( ) MSH [p.512]

7. ___( ) OCT [p.512]

8. ___( ) PRL [p.512]

9. ___( ) TSH [p.512]

A. Stimulates egg and sperm formation in ovaries and testes
B. Targets are pigmented cells in skin and other surface coverings; induces color changes in response to external stimuli and affects some behaviors
C. Stimulates and sustains milk production in mammary glands
D. Stimulates progesterone secretion, ovulation, and corpus luteum formation in females; promotes testosterone secretion and sperm release in males
E. Induces uterine contractions and milk movement into secretory ducts of the mammary glands
F. Stimulates release of thyroid hormones from the thyroid gland
G. Acts on the kidneys to conserve water required in control of extracellular fluid volume
H. Stimulates release of adrenal steroid hormones from the adrenal cortex
I. Promotes growth in young; induces protein synthesis and cell division; roles in adult glucose and protein metabolism

Dichotomous Choice

Circle one of two possible answers given between parentheses in each statement.

10. The (hypothalamus/pituitary gland) region of the brain monitors internal organs and activities related to their functioning, such as eating and sexual behavior; it also secretes some hormones. [p.512]
11. The (posterior/anterior) lobe of the pituitary stores and secretes two hormones, ADH and OCT, that are produced by the hypothalamus. [p.512]
12. The (posterior/anterior) lobe of the pituitary produces and secretes its own hormones that govern the release of hormones from other endocrine glands. [p.512]
13. Humans lack the (posterior/intermediate) lobe of the pituitary gland, one that is possessed by many other vertebrates. [p.512]
14. Most hypothalamic hormones acting in the anterior pituitary lobe are (releasers/inhibitors) and cause target cells there to secrete hormones of their own. [p.513]
15. Some hypothalamic hormones slow down secretion from their targets in the anterior pituitary; these are classed as (releasers/inhibitors). [p.512]

31.4. EXAMPLES OF ABNORMAL PITUITARY OUTPUT [p.514]

Selected Words: *gigantism* [p.514], *pituitary dwarfism,* [p.514], *acromegaly* [p.514], *diabetes insipidus* [p.514]

Complete the Table

1. Complete the table below to summarize examples of abnormal pituitary output.

Condition	Hormone/Abnormality	Characteristics
[p.514] a.	Excessive somatotropin produced during childhood	Affected adults are proportionally similar to a normal person but larger
[p.514] b.	Insufficient somatotropin produced during childhood	Affected adults are proportionally similar to a normal person but much smaller
[p.514] c.	Diminished ADH secretion by a damaged posterior pituitary lobe	Large volumes of dilute urine are secreted, causing life-threatening dehydration
[p.514] d.	Excessive somatotropin output during adulthood when long bones can no longer lengthen	Abnormal thickening of bone, cartilage, and other connective tissues in the hands, feet, and jaws

31.5. SOURCES AND EFFECTS OF OTHER HORMONES [p.515]

Complete the Table

1. Complete the table below by matching the gland/organ and the hormone(s) produced by it to the descriptions of hormone action. Refer to text Table 31.3, p.515.

Gland/Organ

A. adrenal cortex
B. adrenal medulla
C. thyroid
D. parathyroids
E. testes
F. ovaries
G. pancreas (alpha cells)
H. pancreas (beta cells)
I. pancreas (delta cells)
J. thymus
K. pineal

Hormones

(a) thyroxine and triiodothyronine
(b) glucagon
(c) PTH
(d) androgens (includes testosterone)
(e) somatostatin
(f) thymosins
(g) glucocorticoids
(h) estrogens, general (includes progesterone)
(i) epinephrine
(j) melatonin
(k) insulin
(l) progesterone
(m) mineralocorticoids (including aldosterone)
(n) calcitonin
(o) norepinephrine

Gland/Organ	Hormone	Hormone Action
[p.515] a.		Elevates calcium levels in blood
[p.515] b.		Influences carbohydrate metabolism of insulin-secreting cells
[p.515] c.		Required in egg maturation and release; preparation of uterine lining for pregnancy and its maintenance in pregnancy; influences growth and development; genital development; maintains sexual traits
[p.515] d.		Promote protein breakdown and conversion to glucose in most cells
[p.515] e.		Lowers blood sugar level in muscle and adipose tissue
[p.515] f.		In general, required in sperm formation, genital development, and maintenance of sexual traits; influences growth and development
[p.515] g.		In most cells, regulates metabolism, plays roles in growth and development
[p.515] h.		In the gonads, influences daily biorhythms and influences gonad development and reproductive cycles
[p.515] i.		In liver, muscle, and adipose tissues; raises blood sugar level, fatty acids; increases heart rate and force of contraction
[p.515] j.		Targets lymphocytes, has roles in immunity
[p.515] k.		Raises blood sugar level
[p.515] l.		Prepares, maintains uterine lining for pregnancy; stimulates breast development
[p.515] m.		In kidneys, promotes sodium reabsorption and control of salt-water balance
[p.515] n.		In smooth muscle cells of blood vessels, promotes constriction or dilation of blood vessels
[p.515] o.		In bone, lowers calcium levels in blood

31.6. FEEDBACK CONTROL OF HORMONAL SECRETIONS [pp.516–517]

Selected Words: inhibit [p.516], *stimulate* [p.516], *fight-flight* response [p.516], *goiter* [p.517], *hypothyroidism* [p.517], *hyperthyroidism* [p.517], *primary* reproductive organs [p.517]

Boldfaced, Page-Referenced Terms

[p.516] negative feedback _____

[p.516] positive feedback _____

_____

[p.516] adrenal cortex _____

_____

[p.516] adrenal medulla _____

_____

[p.516] thyroid gland _____

_____

[p.517] gonads _____

_____

Matching

Choose the most appropriate answer for each term.

1. ___adrenal medulla [p.516]
2. ___ACTH [p.516]
3. ___nervous system [p.516]
4. ___glucocorticoids [p.516]
5. ___CRH [p.516]
6. ___cortisol [p.516]
7. ___fight-flight response [p.516]
8. ___cortisol-like drugs [p.516]
9. ___adrenal cortex [p.516]
10. ___feedback mechanisms [p.516]

A. Results from actions of epinephrine and norepinephrine in times of excitement or stress
B. Outer portion of each adrenal gland; some of its cells secrete hormones such as glucocorticoids
C. Negative ultimately inhibits hormone secretion, positive stimulates further hormone secretion
D. Initiates a stress response during chronic stress, injury, or illness; cortisol helps to suppress inflammation
E. Inner portion of the adrenal gland; neurons located here release epinephrine and norepinephrine
F. Stimulates the adrenal cortex to secrete cortisol; this helps raise the level of glucose by preventing muscle cells from taking up more blood glucose
G. Used to counter asthma and other chronic inflammatory disorders
H. Help maintain blood glucose concentration and help suppress inflammatory responses
I. Secreted by the hypothalamus in response to falling glucose blood level; stimulates the anterior pituitary to secrete ACTH
J. Secreted by the adrenal cortex; blocks the uptake and use of blood glucose by muscle cells; also stimulates liver cells to form glucose from amino acids

Fill-in-the-Blanks

Thyroxine and triiodothyronine are the main hormones secreted by the human (11) _____ gland [p.516]. They are critical for normal development of many tissues, and they control overall (12) _____ [pp.516–517] rates in humans and other warm-blooded animals. The synthesis of thyroid hormones requires (13) _____ [p.517], which is obtained from food. When iodine is absorbed from the gut, iodine is converted to (14) _____ [p.517]. In the absence of that form of the element, blood levels of thyroid hormones decrease. The anterior pituitary responds by secreting (15) _____ [p.517]. When thyroid hormones cannot be synthesized, the feedback signal continues—and so does TSH secretion. TSH secretion continues and overstimulates the thyroid gland and causes (16) _____ [p.517], an enlargement of the thyroid gland.

(17) _____ [p.517] results from insufficient blood level concentrations of thyroid hormones. Such (18) _____ [p.517] adults are often overweight, sluggish, dry-skinned, intolerant of cold, and sometimes confused and depressed. (19) _____ [p.517] results from excess concentrations of thyroid hormones. Affected adults show an increased heart rate, heat intolerance, elevated blood pressure, profuse sweating, and weight loss even when caloric intake increases. Affected individuals typically are nervous and agitated, and have trouble sleeping. The primary reproductive organs are known as (20) _____ [p.517]. These organs produce and secrete sex (21) _____ [p.517] essential to reproduction. These organs are known as the (22) _____ [p.517] in human males and (23) _____ [p.517] in females. Testes secrete (24) _____ [p.517]. Ovaries secrete estrogens and progesterone. All these hormones influence (25) _____ sexual traits [p.517]. Both types of organs produce (26) _____, or sex cells [p.517].

31.7. RESPONSES TO LOCAL CHEMICAL CHANGES [pp.518–519]

Selected Words: *rickets* [p.518], *exocrine* cells [p.518], *endocrine* cells [p.518], *alpha* cells [p.518], *glucagon* [p.518], *beta* cells [p.518], *insulin* [p.518], *delta* cells [p.518], *diabetes mellitus* [p.518], "type 1 diabetes" [p.519], "type 2 diabetes" [p.519]

Boldfaced, Page-Referenced Terms

[p.518] parathyroid glands _____

_____

[p.518] pancreatic islet _____

_____

Fill-in-the-Blanks

Humans have four (1) _____ [p.518] glands positioned next to the posterior (or back) of the human thyroid; they secrete (2) _____ [p.518] in response to a low (3) _____ [p.518] level in blood. This hormone induces living bone cells to secrete enzymes that digest bone tissue and thereby release (4) _____ [p.518] and other minerals to interstitial fluid, then to the blood. It enhances calcium (5) _____ [p.518] from the filtrate flowing from the nephrons of the kidneys. PTH induces some kidney cells to secrete enzymes that act on blood-borne precursors of the active form of vitamin (6) _____ [p.518], a hormone. This hormone stimulates (7) _____ [p.518] cells to increase calcium absorption from the gut lumen. In a child with vitamin D deficiency, too little calcium and phosphorus are absorbed and so rapidly growing bones develop improperly. The resulting bone disorder is called (8) _____ [p.518] and is characterized by bowed legs, a malformed pelvis, and in many cases a malformed skull and rib cage.

Complete the Table

9. Complete the table below to summarize function of the pancreatic islets.

Pancreatic Islet Cells	Hormone Secreted	Hormone Action
[p.518] a. Alpha cells		
[p.518] b. Beta cells		
[p.518] c. Delta cells		

Dichotomous Choice

Circle one of two possible answers given between parentheses in each statement.

10. Insulin deficiency can lead to diabetes mellitus, a disorder in which the glucose level (rises/decreases) in the blood, then in the urine. [p.518]
11. In a person with diabetes mellitus, urination becomes (reduced/excessive), so the body's water-solute balance becomes disrupted; people become abnormally dehydrated and thirsty. [pp.518–519]
12. Lacking a steady glucose supply, body cells of persons with diabetes mellitus begin breaking down their own fats and proteins for (energy/water). [p.519]
13. Weight loss occurs and (amino acids/ketones) accumulate in blood and urine; this promotes excessive water loss with a life-threatening disruption of brain function. [p.519]
14. After a meal, blood glucose rises; pancreatic beta cells secrete (glucagon/insulin); targets use glucose or store it as glycogen. [p.519]
15. Blood glucose levels decrease between meals. (Glucagon/Insulin) is secreted by stimulated pancreas alpha cells; targets convert glycogen back to glucose, which then enters the blood. [p.519]
16. In (type 1 diabetes/type 2 diabetes) the body mistakenly mounts an autoimmune response against its own insulin-secreting beta cells and destroys them. [p.519]
17. Juvenile-onset diabetes is also known as (type 1 diabetes/type 2 diabetes); these patients survive with insulin injections. [p.519]
18. In (type 1 diabetes/type 2 diabetes), insulin levels are close to or above   normal, but target cells fail to respond to insulin. [p.519]
19. (Type 1 diabetes/Type 2 diabetes) usually is manifested during middle age and is less dramatically dangerous than the other type; beta cells produce less insulin as a person ages. [p.519]

31.8. HORMONAL RESPONSES TO ENVIRONMENTAL CUES [pp.520–521]

Selected Words: melatonin [p.520], "jet lag" [p.520], *winter blues* [p.520], ecdysone [p.521]

Boldfaced, Page-Referenced Terms

[p.520] pineal gland _____

_____

[p.520] puberty _____

_____

[p.521] molting _____

_____

Choice

For questions 1–10, choose from the following:

a. melatonin   b. ecdysone

1. ___ The hormone that controls molting [p.521]
2. ___ Hormone secreted by the pineal gland [p.520]
3. ___ High blood levels of this hormone in winter (long nights) suppresses sexual activity in hamsters. [p.520]
4. ___ Winter blues [p.520]
5. ___ Chemical interactions that cause an old cuticle to detach from the epidermis and muscles [p.520]
6. ___ Triggers puberty [p.520]
7. ___ Suppresses growth of a bird's gonads in fall and winter [p.520]
8. ___ Jet lag [p.520]
9. ___ Waking up at sunrise [p.520]
10. ___ Produced and stored in molting glands by insects and crustaceans [p.520]

Self-Quiz

___ 1. The _____ governs the release of hormones from other endocrine glands; it is controlled by the _____.
  a. pituitary, hypothalamus
  b. pancreas, hypothalamus
  c. thyroid, parathyroid glands
  d. hypothalamus, pituitary
  e. pituitary, thalamus

___ 2. Neurons of the _____ produce ADH and oxytocin that are stored within axon endings of the _____. [p.512]
  a. anterior pituitary, posterior pituitary
  b. adrenal cortex, adrenal medulla
  c. posterior pituitary, hypothalamus
  d. posterior pituitary, thyroid
  e. hypothalamus, posterior pituitary

___ 3. If you were lost in the desert and had no fresh water to drink, the level of _____ in your blood would increase as a means to conserve water. [p.512]
  a. insulin
  b. corticotropin
  c. oxytocin
  d. antidiuretic hormone
  e. salt

For questions 4–6, choose from the following answers:
  a. estrogen
  b. PTH
  c. FSH
  d. somatotropin
  e. prolactin

___ 4. _____ stimulates bone cells to release calcium and phosphate and the kidneys to conserve it. [p.518]

___ 5. _____ stimulates and sustains milk production in mammary glands. [p.512]

___ 6. _____ is the hormone associated with pituitary dwarfism, gigantism, and acromegaly. [p.514]

For questions 7–9, choose from the following answers:
  a. adrenal medulla
  b. adrenal cortex
  c. thyroid
  d. anterior pituitary
  e. posterior pituitary

___ 7. The _____ produces glucocorticoids that help increase the level of glucose in blood. [p.516]

___ 8. The gland that is most closely associated with emergency situations is the _____. [p.516]

___ 9. The overall metabolic rates of warm-blooded animals, including humans, depend on hormones secreted by the _____ gland. [pp.516–517]

___10. If all sources of calcium were eliminated from your diet, your body would secrete more _____ in an effort to release calcium stored in your body and send it to the tissues that require it. [p.518]
  a. parathyroid hormone
  b. aldosterone
  c. calcitonin
  d. mineralocorticoids
  e. none of the above

Matching

Choose the most appropriate answer for each term.

11. ___ADH

12. ___ACTH and TSH

13. ___FSH and LH

14. ___GnRH

15. ___STH

16. ___glucocorticoids

17. ___cortisol

18. ___epinephrine and norepinephrine

19. ___thyrosine and triiodothyronine

20. ___estrogens and progesterone

21. ___PTH

22. ___glucagon

23. ___testosterone

24. ___insulin

25. ___pheromones

26. ___somatostatin

27. ___melatonin

28. ___ecdysone

A. In times of excitement or stress, these adrenal medulla hormones help adjust blood circulation and fat and carbohydrate metabolism

B. Hormones secreted by the ovaries; influence secondary sexual traits

C. Hormone secreted by beta pancreatic cells; lowers blood glucose level

D. Abnormal amount of this anterior pituitary lobe hormone has different effects on human growth during childhood and adulthood

E. Anterior pituitary lobe hormones that orchestrate secretions from the adrenal gland and thyroid gland, respectively

F. Adrenal cortex hormones that help increase the level of glucose in blood

G. Hormone secreted by alpha pancreatic cells; raises the blood glucose level

H. The hormone that largely controls molting in insects and crustaceans

I. Major thyroid hormones having widespread effects such as controlling the overall metabolic rates of warm-blooded animals

J. A glucocorticoid that comes into play when the body is under stress, as when the blood glucose level declines below a set point

K. Hypothalmic hormone that brings about secretion of FSH and LH, which are gonadotropins

L. Hormone secreted by the testes; influences secondary sexual traits

M. Hormone secreted by the pineal gland; influences the growth and development of gonads

N. Antidiuretic hormone produced by the posterior pituitary lobe; promotes water reabsorption when the body must conserve water

O. Hormone secreted by delta pancreatic cells; helps control digestion and absorption of nutrients; can also block secretion of insulin and glucagon

P. Nearly odorless hormone-like secretions of certain exocrine glands; they diffuse through water or air to cellular targets outside the animal body

Q. Anterior pituitary lobe hormones that act through the gonads to influence gamete formation and secretion of the sex hormones required in sexual reproduction

R. Hormone secreted by the parathyroid glands in response to low blood calcium levels

Chapter Objectives/Review Questions

1. Hormones, neurotransmitters, local signaling molecules, and pheromones are all known as _____ molecules that carry out integration. [p.508]
2. _____ are the secretory products of endocrine glands, endocrine cells, and some neurons. [p.508]
3. Define *neurotransmitters, local signaling molecules,* and *pheromones.* [p.508]
4. Collectively, the body's sources of hormones came to be called the _____ system. [p.509]
5. Be able to locate and name the components of the human endocrine systems on a diagram such as text Figure 31.2b. [p.509]
6. Target cells have _____ receptors that hormones and other signaling molecules interact with and have diverse effects on physiological processes. [p.510]
7. Contrast the proposed mechanisms of hormonal action on target cell activities by (a) steroid hormones and (b) peptide hormones that are proteins or are derived from proteins. [pp.510–511]
8. The _____ and the pituitary gland interact closely as a major neural-endocrine control center. [p.512]
9. Explain how, even though the anterior and posterior lobes of the pituitary are compounded as one gland, the tissues of each part differ in character. [pp.512–513]
10. Identify the hormones released from the posterior lobe of the pituitary, and state their target tissues. [p.512]
11. Identify the hormones produced by the anterior lobe of the pituitary and tell which target tissues or organs each acts on. [pp.512–513]
12. Most hypothalamic hormones acting in the anterior lobe are _____; they cause target cells to secrete hormones of their own but others are _____ that slow down secretion from their targets. [p.513]
13. Pituitary dwarfism, gigantism, and acromegaly are all associated with abnormal secretion of _____ by the pituitary gland. [p.514]
14. One cause of _____ _____ is damage to the pituitary's posterior lobe and the diminished secretion or lack of ADH. [p.514]
15. Be familiar with the major human hormone sources, their secretions, main targets, and primary actions as shown on text Table 31.3. [p.515]
16. With _____ feedback, an increase or decrease in the concentration of a secreted hormone triggers events that inhibit further secretion. [p.516]
17. With _____ feedback, an increase in the concentration of a secreted hormone triggers events that stimulate further secretion. [p.516]
18. The adrenal _____ secretes glucocorticoids. [p.516]
19. The _____ system initiates the stress response. [p.516]
20. Describe the role of cortisol in a stress response. [p.516]
21. The adrenal _____ contains neurons that secrete epinephrine and norepinephrine. [p.516]
22. List the features of the fight-flight response. [p.516]
23. Over time, a sustained response is made to the low blood level of TSH produced by the thyroid gland; this causes an enlargement known as a form of _____. [p.517]
24. Describe the characteristics of hypothyroidism and hyperthyroidism. [p.517]
25. The _____ are primary reproductive organs that produce and secrete hormones with essential roles in reproduction. [p.517]
26. Name the glands that secrete PTH, and state the function of this hormone. [p.518]
27. Describe an ailment called rickets, and state its cause. [p.518]
28. Give two examples that illustrate the effects of *local signaling molecules.* [p.518]
29. Be able to name the hormones secreted by alpha, beta, and delta pancreatic cells; list the effect of each. [p.518]
30. Describe the symptoms of diabetes mellitus, and distinguish between type 1 and type 2 diabetes. [pp.518–519]
31. The pineal gland secretes the hormone _____; relate two examples of the action of this hormone. [p.520]
32. Explain the cause of "jet lag" and "winter blues." [p.520]
33. The invertebrate hormone _____ is related to the control of the phenomenon known as _____, which occurs among crustaceans and insects. [p.521]

Integrating and Applying Key Concepts

Suppose you suddenly quadruple your already high daily consumption of calcium. State which body organs would be affected, and tell how they would be affected. Name two hormones whose levels would most probably be affected, and tell whether your body's production of them would increase or decrease. Suppose you continue this high rate of calcium consumption for ten years. Can you predict the organs that would be subject to the most stress as a result?

32

PROTECTION, SUPPORT, AND MOVEMENT

Interactive Exercises

Of Men, Women, and Polar Huskies [pp.524-525]

32.1. VERTEBRATE SKIN [p.526]

32.2. *Focus on Health:* SUNLIGHT AND SKIN [p.527]

Selected Words: cuticle [p.525], keratin [p.526], melanin [p.526], *albinos* [p.526], hemoglobin [p.526], carotene [p.526], *acne* [p.526], *cold sores* [p.527], *Herpes simplex* [p.527]

Boldfaced, Page-Referenced Terms

[p.525] integument _____

_____

[p.525] skin _____

_____

[p.526] epidermis _____

_____

[p.526] dermis _____

_____

[p.526] keratinocytes _____

_____

[p.526] melanocytes _____

_____

[p.526] hair _____

_____

[p.527] vitamin D _____

_____

[p.527] Langerhans cells _____

_____

[p.527] Granstein cells _____

_____

Fill-in-the-Blanks

Human skin is an organ system that consists of two layers: the outermost (1) _____ [p.526], which
contains mostly dead cells, and the (2) _____ [p.526], which contains hair follicles, nerves, tiny muscles
associated with the hairs, and various types of glands. The (3) _____ [p.526] layer, with its loose
connective tissue and store of fat in (4) _____ [p.526] tissue, lies beneath the skin.

Labeling

Label the numbered parts of the accompanying illustration. [All answers from text Figure 32.3, p.526]

5. _____

6. _____ _____ _____

7. _____ _____

8. _____ _____

9. _____ _____

10. _____ _____

11. _____ _____

12. _____

13. _____

14. _____

Choice

Match the following proteins (or protein derivatives) with the particular ability that each substance lends to the skin. For questions 15-21, choose from these letters:

a. collagen   b. elastin   c. keratin   d. melanin   e. hemoglobin

15. ___ Fibers that run through the dermis and strengthen it [p.526]

16. ___ Protects against loss of moisture [p.526]

17. ___ Protects against ultraviolet radiation and sunburn; contributes to skin color [p.526]

18. ___ Helps skin to be flexible, yet return to its previous shape when stretched [p.526]

19. ___ Beaks, hooves, hair, fingernails and claws contain a lot of this. [p.526]

20. ___ Helps ward off bacterial attack by making the skin surface rather impermeable [p.526]

21. ___ Located in red blood cells; binds with O_2; contributes to skin color [p.526]

32.3. SKELETAL SYSTEMS [pp.528-529]

32.4. CHARACTERISTICS OF BONE [pp.530-531]

Selected Words: appendicular [skeleton] [p.528], *axial* [skeleton] [p.528], *herniated disks* [p.529], *compact* bone [p.530], *spongy* bone [p.530], osteocytes [p.530], *osteoporosis* [p.531], osteoclast [p.531], *fibrous, cartilaginous,* and *synovial joints* [p.531], *strain* [p.531], *sprain* [p.531], *osteoarthritis* [p.531], *rheumatoid arthritis* [p.531]

Boldfaced, Page-Referenced Terms

[p.528] hydrostatic skeleton _____

_____

[p.528] exoskeleton _____

_____

[p.528] endoskeleton _____

_____

[p.528] vertebrae (singular, vertebra) _____

_____

[p.529] intervertebral disks _____

_____

[p.530] bones _____

_____

[p.530] red marrow _____

_____

[p.530] yellow marrow _____

_____

[p.531] joint _____

_____

[p.531] ligaments _____

_____

Fill-in-the-Blanks

All motor systems are based on muscle cells that are able to (1) _____ [p.528] and (2) _____ [p.528], and on the presence of a medium against which the (3) _____ [p.528] force can be applied. Longitudinal and (4) _____ [p.528] muscle layers work as an antagonistic muscle system, in which the action of one motor element opposes the action of another. A membrane filled with fluid resists compression and can act as a(n) (5) _____ [p.528] skeleton. Arthropods have antagonistic muscles attached to an (6) [choose one] ❒ endoskeleton ❒ exoskeleton [p.528]. (7) _____ [p.530] are hard compartments that enclose and protect the brain, spinal cord, heart, lungs, and other vital organs of vertebrates. Bones support and anchor (8) _____ [p.530] and soft organs, such as eyes. (9) _____ [p.530] systems are found in the long bones of mammals and contain living bone cells that receive their nutrients from the blood. (10) _____ _____ [p.530] is a major site of blood cell formation. Bone tissue serves as a "bank" for (11) _____ [p.530], (12) _____ [p.530], and other mineral ions; depending on metabolic needs, the body deposits ions into and withdraws ions from this "bank."

Bones develop from osteoblasts secreting material inside the shaft and on the surface of the cartilage model. Bone can also give ions back to interstitial fluid as (13) _____ [p.531] dissolve out component minerals and remodel bone in response to (14) _____ [p.531] hormone deficiencies, lack of exercise, and lack of calcium in the diet. Extreme decreases in bone density result in (15) _____ [p.531], particularly among older women.

The (16) _____ [p.528] skeleton includes the skull, vertebral column, ribs, and breastbone; the (17) _____ [p.528] skeleton includes the pectoral and pelvic girdles and the forelimbs and hindlimbs, when they exist in vertebrates.

(18) _____ [p.531] joints are freely movable and are lubricated by a fluid secreted into the capsule of dense connective tissue that surrounds the bones of the joint. Bones are often tipped with (19) _____ [p.531]; as a person ages, the cartilage at (20) _____ [p.531] joints may simply wear away, a condition called (21) _____ [p.531]. By contrast, in (22) _____ _____ [p.531], the synovial membrane becomes inflamed, cartilage degenerates, and bone becomes deposited in the joint.

Labeling

Identify each numbered part of the illustrations below and on the next page.

23. _____ _____ [p.530]
24. _____ _____ [p.530]
25. _____ _____ [p.530]
26. _____ _____ [p.530]
27. _____ _____ _____ [p.530]

28. _____ _____ [p.530]
29. _____ _____ [p.530]
30. _____ _____ [p.530]
31. _____ [p.530]

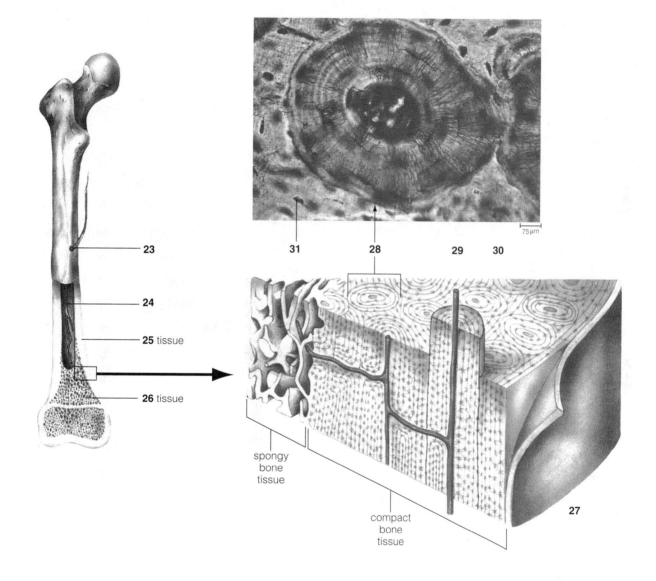

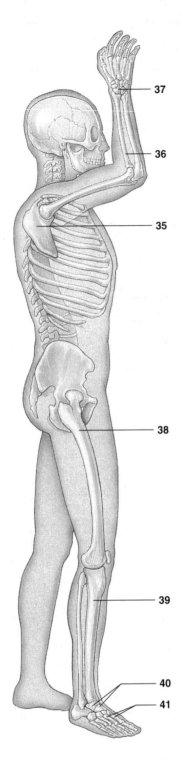

32. _____ [p.528]

33. _____ [p.528]

34. _____ [p.528]

35. _____ [p.528]

36. _____ [p.528]

37. _____ _____ [p.528]

38. _____ [p.528]

39. _____ [p.528]

40. _____ _____ [p.528]

41. _____ _____ [p.528]

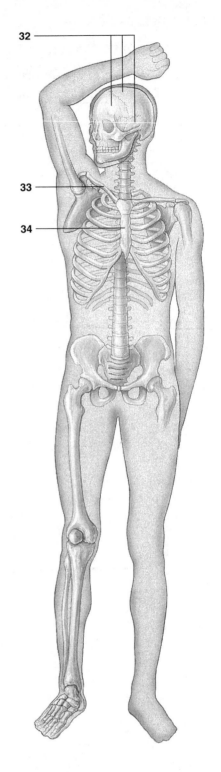

32.5. SKELETAL-MUSCULAR SYSTEMS [pp.532-533]

32.6. A CLOSER LOOK AT MUSCLES [pp.534-535]

Selected Words: contract [p.532], triceps brachii [p.533], biceps brachii [p.533], pectoralis major [p.533], rectus abdominis [p.533], quadriceps femoris [p.533], tibialis anterior [p.533], gastrocnemius [p.533], gluteus maximus [p.533], latissimus dorsi [p.533], Z lines [p.534], myofibril [p.534], glycogen [p.535], cross-bridge [p.535]

Boldfaced, Page-Referenced Terms

[p.532] skeletal muscles _____

_____

[p.532] tendons _____

_____

[p.534] sarcomeres _____

_____

[p.534] actin _____

_____

[p.534] myosin _____

_____

[p.534] sliding-filament model _____

_____

[p.535] creatine phosphate _____

_____

[p.535] oxygen debt _____

_____

Fill-in-the-Blanks

(1) _____ [p.532] muscle is striated, largely voluntary, and generally attached to (2) _____ [p.532] or cartilage by means of (3) _____ [p.532]. (4) _____ [p.532] muscle interacts with the skeleton to bring positional changes of body parts and to move the animal through its environment.

Together, the skeleton and its attached muscles are like a system of levers in which rigid rods,

(5) _____ [p.532], move about at fixed points, called (6) _____ [p.532]. Most attachments are close to joints, so a muscle has to shorten only a small distance to produce a large movement of some body part. When the (7) _____ [p.532] contracts, the elbow joint bends (flexes). As it relaxes and as its partner, the (8) _____ [p.532], contracts, the forelimb extends and straightens. As skeletal muscles contract, they transmit force to (9) _____ [p.532] and make them move. (10) _____ [p.532] attach skeletal muscles to bones.

Sequence

Arrange in order of decreasing size.

11. ___   14. ___   A. Muscle fiber (muscle cell) [p.534]   D. Muscle [p.534]
                        B. Myosin filament [p.534]   E. Myofibril [p.534]
12. ___   15. ___   C. Muscle bundle [p.534]   F. Actin filament [p.534]
13. ___   16. ___

Labeling

Identify each numbered part of the accompanying illustration. [All are from p.533]

17. _____ _____

18. _____ _____

19. _____ _____

20. _____ _____

21. _____ _____

22. _____ _____

23. _____ _____

24. two parts of the _____

25. _____

26. _____ _____

27. _____ _____

28. _____ _____

29. _____ _____

30. _____ _____

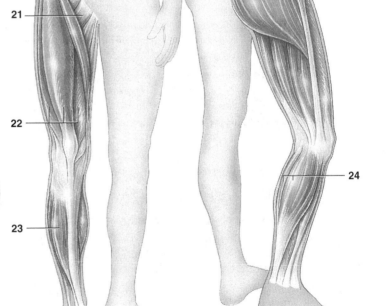

Labeling

Label the numbered parts of the accompanying illustrations. [All are from p.534]

31. _____ bundle

32. _____ _____

33. _____

34. _____

35. _____ _____

36. _____ _____

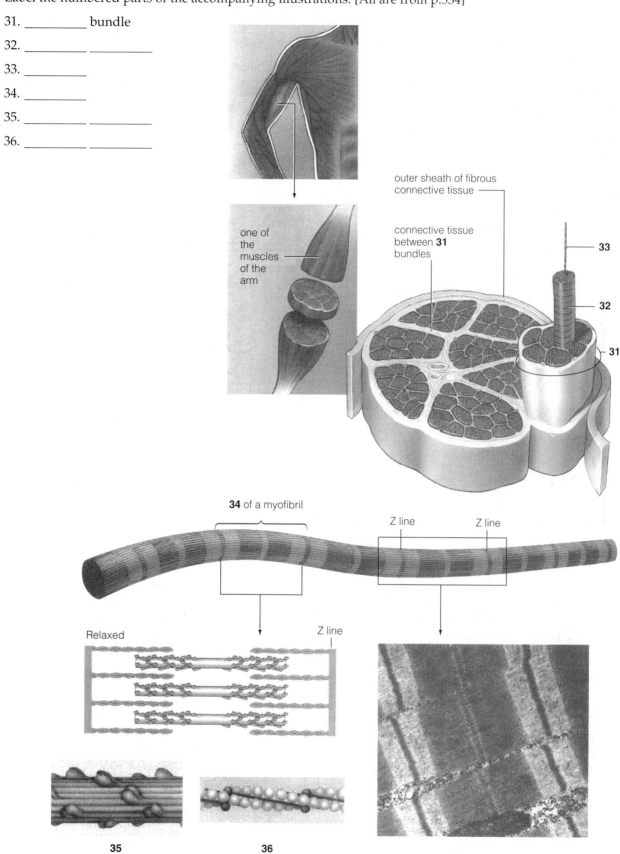

one of the muscles of the arm

outer sheath of fibrous connective tissue

connective tissue between **31** bundles

33

32

31

34 of a myofibril

Z line   Z line

Relaxed   Z line

35   36

Fill-in-the-Blanks

Each muscle cell contains (37) _____ [p.534]: threadlike structures packed together in parallel array. Every (37) is functionally divided into (38) _____ [p.534], which appear to be striped and are arranged one after another along its length. Each myofibril contains (39) _____ [p.534] filaments, which have cross-bridges and (40) _____ [p.534] filaments, which are thin and lack cross-bridges.

According to the (41) _____-_____ [p.534] model, (42) _____ [p.534] filaments physically slide along actin filaments and pull them toward the center of a(n) (43) _____ [p.534] during a contraction.

Fill-in-the-Blanks

[All are from p.535]

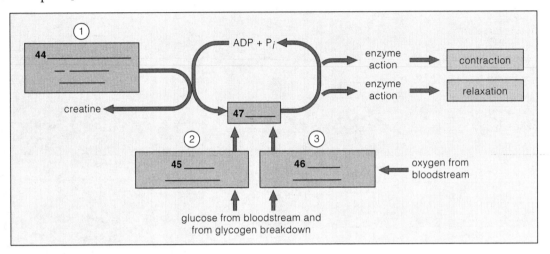

Complete the Table

For each item in the table below, state its specific role in muscle contraction.

Aerobic respiration [p.535]	48.
ATP [p.534]	49.
Creatine phosphate [p.535]	50.
Glycogen [p.535]	51.
Glycolysis + lactate fermentation [p.535]	52.
Myosin heads [p.535]	53.
Sarcomere [p.534]	54.

32.7. CONTROL OF MUSCLE CONTRACTION [p.536]
32.8. PROPERTIES OF WHOLE MUSCLES [p.537]

Selected Words: acetylcholine (ACh) [p.536], *excitable* [p.536], tropomyosin and troponin [p.536], *isometric, isotonic,* and *lengthening* contractions [p.537], *muscle fatigue* [p.537], *aerobic exercise* [p.537], *strength training* [p.537]

Boldfaced, Page-Referenced Terms

[p.536] action potential _____

_____

[p.536] sarcoplasmic reticulum _____

_____

[p.537] muscle tension _____

_____

[p.537] motor unit _____

_____

[p.537] muscle twitch _____

_____

[p.537] tetanus _____

_____

[p.537] exercise _____

_____

Fill-in-the-Blanks

Muscle cells have three properties in common: contractility, elasticity, and (1) _____ [p.536]. Like a(n) (2) _____ [p.536], a muscle cell is "excited" when a wave of electrical disturbance, a(n) (3) _____ _____ [p.536], travels along its cell membrane. Neurons that send signals to muscles are (4) _____ [p.536] neurons.

The energy that drives the forming and breaking of the cross-bridges comes immediately from (5) _____ [recall p.535], which obtained its phosphate group from (6) _____ _____ [recall p.535], which is stored in muscle cells.

(7) _____ [recall p.534] of an entire muscle is brought about by the combined decreases in length of the individual sarcomeres that make up the myofibrils of the muscle cells. The parallel orientation of muscle's component parts directs the force of contraction toward a(n) (8) _____ [recall p.534] that must be pulled in some direction.

When excited by incoming signals from motor neurons, muscle cell membranes cause (9) _____ [p.536] ions to be released from the (10) _____ _____ [p.536]. These ions remove all obstacles that might interfere with myosin heads binding to the sites along (11) _____ [p.536] filaments. When muscle cells relax/rest, calcium ions are sent back to the sarcoplasmic reticulum by (12) _____

_____ [p.536]. By controlling the (13) _____ _____ [p.536] that reach the (14) _____ _____ [p.536] in the first place, the nervous system controls muscle contraction by controlling calcium ion levels in muscle tissue.

The larger the (15) _____ [p.537] of a muscle, the greater its strength. A(n) (16) _____ [p.537] neuron and the muscle cells under its control are called a(n) (17) _____ _____ [p.537]. A(n) (18) _____ _____ [p.537] is a response in which a muscle contracts briefly when reacting to a single, brief stimulus and then relaxes. The strength of the muscular contraction depends on how far the (19) _____ [p.537] response has proceeded by the time another signal arrives. (20) _____ [p.537] is the state of contraction in which a motor unit that is being stimulated repeatedly is maintained. In a(n) (21) _____ [p.537] contraction, the nervous system activates only a small number of motor units; in a stronger contraction, a larger number are activated at a high (22) _____ [p.537] of stimulation.

Labeling

Identify the numbered parts of the accompanying illustrations.

23. _____ [p.536]

24. _____ _____ [p.536]

25. _____ _____ [p.536]

26. _____ _____ _____ [p.536]

27. _____ [p.537]

28. _____ _____ [p.537]

29. _____ _____ [p.537]

30. _____ (_____)[p.537]

31. _____ [p.537]

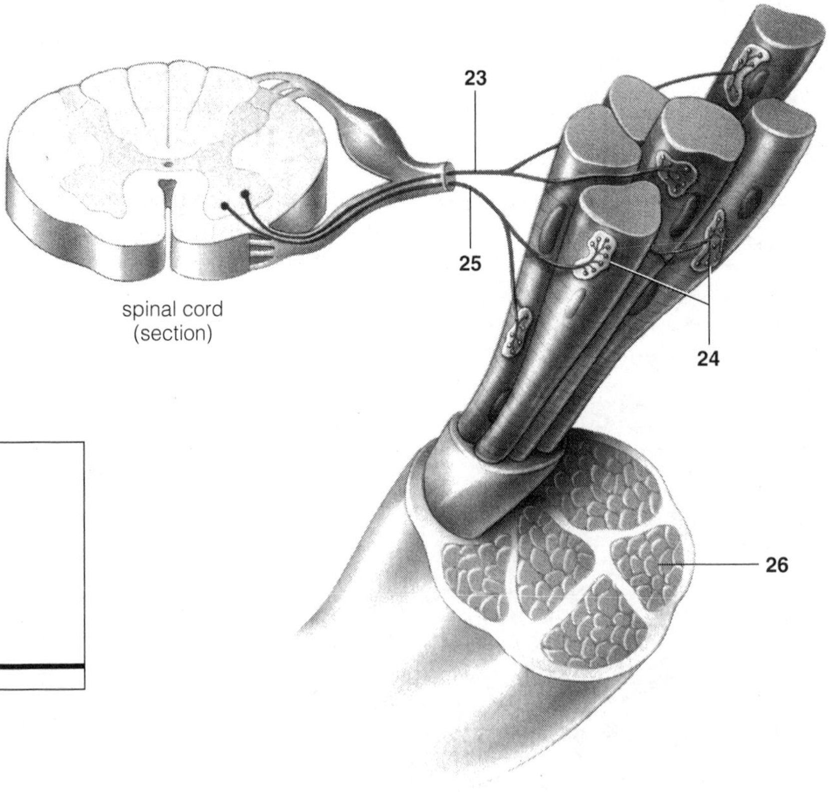

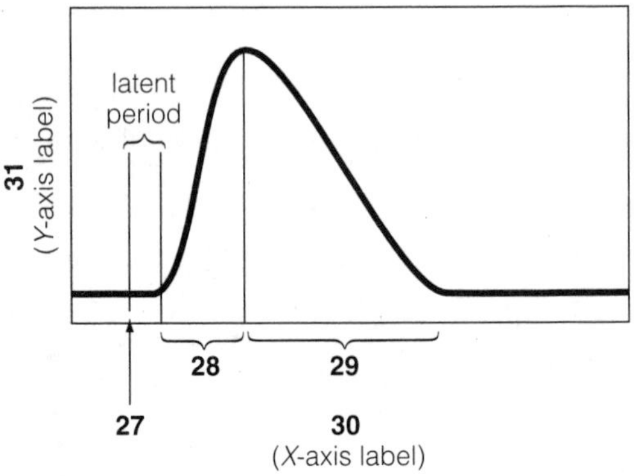

spinal cord
(section)

23

25

24

26

latent
period

31
(Y-axis label)

28

29

27

30
(X-axis label)

Complete the Table

For each item in the table below, state its specific role in muscle contraction.

Calcium ions [p.536]	32.
Motor neuron [p.536]	33.
Sarcoplasmic reticulum [p.536]	34.

Fill-in-the-Blanks

(35) _____ _____ [p.539] are synthetic hormones that mimic the effects of testosterone in building greater (36) _____ [p.539] mass in both men and women. It is illegal for competitive athletes to use them because of unfair advantage and because of the side effects. In men, (37) _____ [p.539], baldness, shrinking (38) _____ [p.539], and infertility are the first signs of damage. Aside from these physical side effects, some men experience uncontrollable (39) _____ [p.539], delusions, and wildly manic behavior. In women, (40) _____ [p.539] hair becomes more noticeable, (41) _____ _____ [p.539] become irregular, breasts may shrink, and the (42) _____ [p.539] may become deeper.

Self-Quiz

For questions 1-7, choose from the following answers:

    a. bone
    b. cartilage
    c. epidermis
    d. dermis
    e. hypodermis

___ 1. Fat cells in adipose tissue are most likely to be located in this. [p.526]

___ 2. Keratinized squamous cells are most likely to be located in this. [p.526]

___ 3. Melanin in melanocytes is most likely to be here. [p.526]

___ 4. Smooth muscles attached to hairs are probably here. [p.526]

___ 5. This makes up the original "model" of the skeletal framework. [p.530]

___ 6. The receiving ends of sensory receptors are most likely here. [p.526]

___ 7. This serves as a "bank" for withdrawing and depositing calcium and phosphate ions. [p.530]

For questions 8-11, choose from the following answers:

    a. Ligaments
    b. Osteoclasts
    c. Osteocytes
    d. Red marrow
    e. Tendons

___ 8. _____ secrete bone-building enzymes. [p.530]

___ 9. Major site of blood cell formation. [p.530]

___10. _____ remove Ca^{++} and $PO_4^{\equiv}$ ions from bones and put them into the bloodstream. [p.531]

___11. _____ attach muscles to bone. [p.532]

For questions 12-16, choose from the following answers:

    a. An action potential
    b. Cross-bridge formation
    c. The sliding-filament model
    d. Tension
    e. Tetanus

___12. _____ is a mechanical force that causes muscle cells to shorten if it is not exceeded by opposing forces. [p.537]

___13. A wave of electrical disturbance that moves along a neuron or muscle cell in response to a threshold stimulus. [p.536]

___14. _____ is a large contraction caused by repeated stimulation of motor units that are not allowed to relax. [p.537]

___15. _____ is assisted by calcium ions and ATP. [p.536]

___16. _____ explains how myosin filaments move to the centers of sarcomeres and back. [p.534]

For questions 17-20, choose from the following answers:

    a. actin
    b. myofibril
    c. myosin
    d. sarcomere
    e. sarcoplasmic reticulum

___17. A(n) _____ contains many repetitive units of muscle contraction. [p.534]

___18. The repetitive unit of muscle contraction. [p.534]

___19. Thin filaments that depend upon calcium ions to clear their binding sites so that they can attach to parts of thick filaments. [p.534]

___20. _____ stores calcium ions and releases them in response to an action potential. [p.536]

Chapter Objectives/Review Questions

1. Name four functions of human skin. [p.525]
2. Describe the two-layered structure of human skin and identify the items located in each layer. [p.526]
3. Compare invertebrate and vertebrate motor systems in terms of skeletal and muscular components and their interactions. [pp.528-529]
4. Identify human bones by name and location. [pp.528-529]
5. Explain the various roles of osteoclasts, osteocytes, cartilage models, and long bones in the development of human bones. [pp.530-531]
6. Refer to text Figure 32.14 on p.533 and indicate (a) a muscle used in sit-ups, (b) another used in dorsally flexing and inverting the foot, and (c) another used in flexing the elbow joint. [p.533]
7. Describe the fine structure of a muscle fiber; use terms such as *myofibril*, *sarcomere*, *Z lines*, *actin*, and *myosin*. [pp.534-535]
8. Explain in detail the structure of muscles, from the molecular level to the organ systems level. Then explain how biochemical events occur in muscle contractions and how antagonistic muscle action refines movements. [pp.534-536]
9. List, in sequence, the biochemical and fine structural events that occur during the contraction of a skeletal muscle fiber and explain how the fiber relaxes. [pp.534-536]
10. Distinguish twitch contractions from tetanic contractions. [p.537]

Integrating and Applying Key Concepts

If humans had an exoskeleton rather than an endoskeleton, would they move differently from the way they do now? Name any advantages or disadvantages that having an exoskeleton instead of an endoskeleton would present in human locomotion.

33

CIRCULATION

Interactive Exercises

Heartworks [pp.540–541]

33.1. CIRCULATORY SYSTEMS—AN OVERVIEW [pp.542–543]

Selected Words: electrocardiogram, ECG [p.540], *closed* and *open* circulatory systems [p.542], blood *volume* and *velocity* [p.542]

Boldfaced, Page-Referenced Terms

[p.542] circulatory system _____

_____

[p.542] interstitial fluid _____

_____

[p.542] blood _____

_____

[p.542] heart _____

_____

[p.542] capillary beds _____

_____

[p.542] capillaries _____

_____

[p.543] pulmonary circuit _____

_____

[p.543] systemic circuit _____

_____

[p.543] lymphatic system _____

_____

Fill-in-the-Blanks

Cells survive by taking in from their surroundings what they need, (1) _____ [p.541], and giving back
to their surroundings materials which they don't need: (2) _____ [p.541]. In most animals, substances
move rapidly to and from living cells by way of a(n) (3) _____ [p.541] circulatory system.
(4) _____ [p.541], a fluid connective tissue within the (5) _____ [p.541] and blood vessels, is the
transport medium.

Most of the cells of animals are bathed in a(n) (6) _____ _____ [p.542]; blood is constantly
delivering nutrients and removing wastes from that fluid. The (7) _____ [p.542] generates the pressure
that keeps blood flowing. Blood flows (8) [choose one] ❏ rapidly ❏ slowly [p.542] through large diameter
vessels to and from the heart, but where the exchange of nutrients and wastes occurs, in the (9) _____
[p.542] beds, the blood is divided up into vast numbers of smaller-diameter vessels with tremendous surface
area that enables the exchange to occur by diffusion. An elaborate network of drainage vessels attracts
excess interstitial fluid and reclaimable (10) _____ [p.543] and returns them to the circulatory system.
This network is part of the (11) _____ [p.543] system, which also helps clean the blood of disease
agents.

Nutrients are absorbed into the blood from the (12) _____ [p.541] and (13) _____ [p.541]
systems. Carbon dioxide is given to the (14) _____ [p.541] system for elimination, and excess water,
solutes, and wastes are eliminated by the (15) _____ [p.541] system.

The heart is a pumping station for two major blood transport routes: the (16) _____ [p.543]
circulation to and from the lungs, and the (17) _____ [p.543] circulation to and from the rest of the
body.

In the pulmonary circuit, the heart pumps (18) _____ [p.543]-poor blood to the lungs; then the
(19) _____ [p.543]-enriched blood flows back to the (20) _____ [p.543]. In fishes, blood flows in
(21) _____ [p.543] circuit(s) away from and back to the heart.

Labeling

Label the numbered parts in the illustrations shown at the right.

22. _____ [p.542]

23. _____ _____ [p.542]

24. _____ [p.542]

Describe the kind of circulatory system in:

25. Creature A. _____

_____ [p.542]

26. Creature B. _____

_____ [p.542]

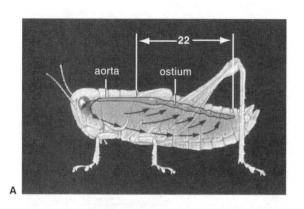

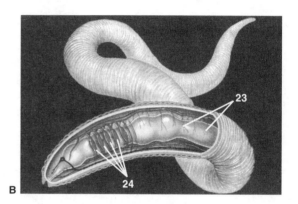

33.2. CHARACTERISTICS OF BLOOD [pp.544–545]

33.3. *Focus on Health:* BLOOD DISORDERS [p.546]

33.4. BLOOD TRANSFUSION AND TYPING [pp.546–547]

Selected Words: erythrocytes [p.544], leukocytes [p.545], neutrophils, eosinophils, basophils, monocytes, and lymphocytes [p.545], macrophages [p.545], megakaryocytes [p.545], *hemorrhagic, chronic, hemolytic, iron deficiency, B_{12} deficiency,* and *sickle-cell* anemias [p.546], *thalassemia* [p.546], *polycythemia* [p.546], *infectious mononucleosis* [p.546], *leukemias* [p.546], Rh+ and Rh- markers [p.547], *erythroblastosis fetalis* [p.547]

Boldfaced, Page-Referenced Terms

[p.544] plasma _____

_____

[p.544] red blood cells _____

_____

[p.544] white blood cells _____

_____

[p.544] platelets _____

_____

[p.545] stem cells _____

_____

[p.545] cell count _____

_____

[p.546] anemias _____

_____

[p.546] blood transfusions _____

_____

[p.546] agglutination _____

_____

[p.547] ABO blood typing _____

_____

[p.547] Rh blood typing _____

_____

Fill-in-the-Blanks

Blood is a highly specialized fluid (1) _____ [p.544] tissue that helps stabilize internal (2) _____
[p.544] and equalize internal temperature throughout an animal's body. In addition to supplying cells with
oxygen and other nutrients, removing carbon dioxide, waste products, and cellular secretions, blood serves
as a highway for (3) _____ [p.544] cells that fight infections and scavenge debris in tissues. Blood
volume for average-size adult humans amounts to approximately (4) [choose one] ❒ 2–3 ❒ 4–5 ❒ 6–7 ❒ 8–10
❒ 11–15 [p.544] quarts. Oxygen binds with the (5) _____ [p.545] atom in a hemoglobin molecule. The
red blood (6) _____ _____ [p.545] in males is about 5.4 million cells per microliter of blood; in
females, it is 4.8 million per microliter. The plasma portion constitutes approximately (7) _____
to _____ [p.544] percent of the total blood volume. Erythrocytes are produced in the (8) _____
_____ [p.545]. (9) _____ _____ [p.545] are immature cells not yet fully differentiated.
(10) _____ [p.545] and monocytes are highly mobile and phagocytic; they chemically detect, ingest, and
destroy bacteria, foreign matter, and dead cells. (11) _____ [p.545] (thrombocytes) are cell fragments
that aid in forming blood clots.

In humans, red blood cells lack their (12) _____ [p.545], but they contain enough materials to
sustain them for about (13) _____ [p.545] months. Platelets also have no (14) _____ [p.545], but
they last a maximum of (15) _____ [p.545] days in the human bloodstream.

If you are blood type (16) _____ [p.547], you have no antibodies against A or B markers in your
plasma. If you are type (17) _____ [p.547], you have antibodies against A and B markers in your
plasma. (18) [choose one] ❒ Women ❒ men [p.547] who are (19) [choose one] ❒ Rh$^+$ ❒ Rh$^-$ [p.547] have to be
careful so that their fetuses don't develop erythroblastosis fetalis.

Complete the Table

20. Complete the following table, which describes the components of blood. [All are from p.544]

Components	Relative Amounts	Functions
Plasma portion (50%–60% of total volume):		
Water	91%–92% of plasma volume	Solvent
a. (albumin, globulins, fibrinogen, etc.)	7%–8%	Defense, clotting, lipid transport, roles in extracellular fluid volume, etc.
Ions, sugars, lipids, amino acids, hormones, vitamins, dissolved gases	1%–2%	Roles in extracellular fluid volume, pH, etc.
Cellular portion (40%–50% of total volume):		
b.	4,800,000–5,400,000 per microliter	O_2, CO_2 transport
White blood cells:		
c.	3,000–6,750	Phagocytosis
d.	1,000–2,700	Immunity
Monocytes (macrophages)	150–720	Phagocytosis
Eosinophils	100–360	Roles in inflammatory response, immunity
Basophils	25–90	Roles in inflammatory response, anticlotting
e.	250,000–300,000	Roles in clotting

Label–Match

Identify the numbered cell types in the illustration below. Complete the exercise by matching and entering the letter of the appropriate function in the parentheses following the given cell types. A letter may be used more than once. Cell types 21 and 27 have no matching letters.

21. _____ [p.545]

22. _____ _____ (   ) [p.545]

23. _____ (   ) [p.545]

24. _____ (   ) [p.545]

25. _____ (   ) [p.545]

26. _____ (   ) [p.545]

27. _____ [p.545]

28. _____ (   ) [p.545]

A. Phagocytosis
B. Plays a role in the inflammatory response
C. Plays a role in clotting
D. Immunity
E. O$_2$, CO$_2$ transport

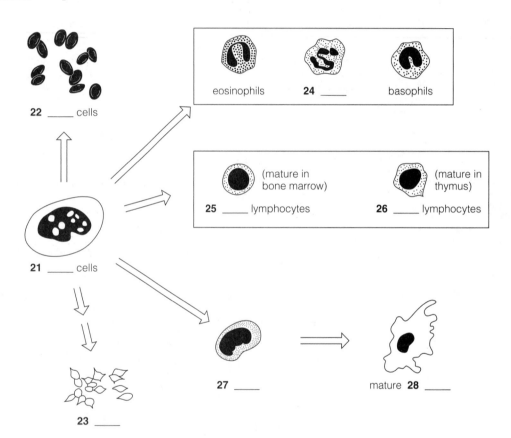

Complete the Table

Complete the following table. [All are from p.546]

	Red Blood Cell Disorders	White Blood Cell Disorders
29.	Potential hazard for strict vegetarians and alcoholics	
30.	Results from sudden blood loss, as from a severe wound	
31.	Results from insufficiency of a particular metal ion in the diet	
32.	Results from a gene mutation that creates an abnormal form of hemoglobin; causes distorted red blood cell shape	
33.	Results from far too many red blood cells, could be caused by cancer of the bone marrow	
34.		Too many monocytes and lymphocytes caused by infection by the Epstein-Barr virus
35.		A malignant increase in the number of white blood cells.

33.5. HUMAN CARDIOVASCULAR SYSTEM [pp.548–549]

33.6. THE HEART IS A LONELY PUMPER [pp.550–551]

Selected Words: "cardiovascular" [p.548], *pulmonary* circuit [p.548], *systemic* circuit [p.548], capillary bed [p.549], *coronary* circulation [p.550], endothelium [p.550], atrium [p.550], ventricle [p.550]

Boldfaced, Page-Referenced Terms

[p.548] arteries _____

_____

[p.548] arterioles _____

_____

[p.548] venules _____

_____

[p.548] veins _____

_____

[p.548] aorta _____

_____

[p.550] cardiac cycle _____

_____

[p.551] cardiac conduction system _____

_____

[p.551] cardiac pacemaker _____

_____

Fill-in-the-Blanks

The heart is a pumping station for two major blood transport routes: the (1) _____ [p.548] circulation to and from the lungs, and the (2) _____ [p.548] circulation to and from the rest of the body.

A receiving zone of a vertebrate heart is called a(n) (3) _____ [p.550]; a departure zone is called a(n) (4) _____ [p.550]. Each contraction period is called (5) _____ [p.550]; each relaxation period is (6) _____ [p.550].

During a cardiac cycle, contraction of the (7) _____ [p.551] is the driving force for blood circulation; (8) _____ [p.550] contraction helps fill the ventricles.

Labeling

Look at text Figure 33.11 on p.549 and use a red pen to redden all tubes in this illustration (except the pulmonary artery) that are indicated by "aorta" or "artery." Memorize the names, then fill in all of the blanks. Also, redden the parts of the heart that contain oxygen-rich blood.

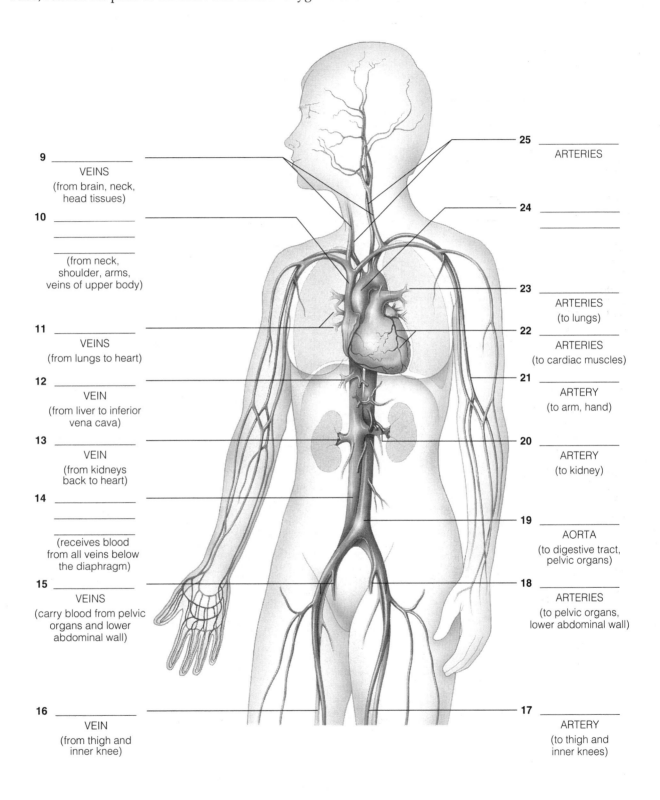

9 _____ _____
VEINS
(from brain, neck,
head tissues)

10 _____

_____

_____
(from neck,
shoulder, arms,
veins of upper body)

11 _____
VEINS
(from lungs to heart)

12 _____
VEIN
(from liver to inferior
vena cava)

13 _____
VEIN
(from kidneys
back to heart)

14 _____

_____

_____
(receives blood
from all veins below
the diaphragm)

15 _____
VEINS
(carry blood from pelvic
organs and lower
abdominal wall)

16 _____
VEIN
(from thigh and
inner knee)

25 _____
ARTERIES

24 _____

_____

23 _____
ARTERIES
(to lungs)

22 _____
ARTERIES
(to cardiac muscles)

21 _____
ARTERY
(to arm, hand)

20 _____
ARTERY
(to kidney)

19 _____
AORTA
(to digestive tract,
pelvic organs)

18 _____
ARTERIES
(to pelvic organs,
lower abdominal wall)

17 _____
ARTERY
(to thigh and
inner knees)

Labeling

Identify each numbered part of the accompanying illustration.

26. _____ [p.550]

27. _____ _____ _____ [p.550]

28. _____ _____ [p.550]

29. _____ _____ [p.550]

30. _____ _____ _____ [p.550]

31. _____ _____ [p.550]

32. _____ _____ _____ [p.550]

33. _____ _____ _____ [p.550]

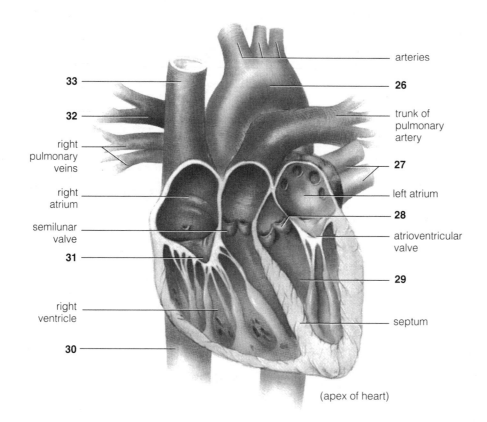

33.7. BLOOD PRESSURE IN THE CARDIOVASCULAR SYSTEM [pp.552–553]
33.8. FROM CAPILLARY BEDS BACK TO THE HEART [pp.554–555]
33.9. *Focus on Health:* CARDIOVASCULAR DISORDERS [pp.556–557]
33.10. HEMOSTASIS [p.558]

Selected Words: *systolic* pressure [p.552], *diastolic* pressure [p.552], *pulse* pressure [p.552], medulla oblongata [p.553], diffusion zones [p.554], *edema* [p.554], *elephantiasis* [p.554], bulk flow [p.555], *hypertension* [p.556], *atherosclerosis* [p.556], *heart attacks* [p.556], *strokes* [p.556], *arteriosclerosis* [p.556], *atherosclerotic plaque* [p.557], *thrombus* [p.557], *embolus* [p.557], *angina pectoris* [p.557], *stress electrocardiograms* [p.557], *angiography* [p.557], *coronary bypass surgery* [p.557], *laser* and *balloon angioplasty* [p.557], *arrhythmias* [p.557], *bradycardia* [p.557], *tachycardia* [p.557]

Boldfaced, Page-Referenced Terms

[p.552] blood pressure _____

_____

[p.553] vasodilation _____

_____

[p.553] vasoconstriction _____

_____

[p.553] baroreceptor reflex _____

_____

[p.554] ultrafiltration _____

_____

[p.554] reabsorption _____

_____

[p.557] LDLs, low-density lipoproteins _____

_____

[p.557] HDLs, high-density lipoproteins _____

_____

[p.558] hemostasis _____

_____

Labeling

Identify each numbered part of the accompanying illustrations.

1. _____ [p.552]

2. _____ [p.552]

3. _____ [p.552]

4. _____ [p.552]

5. _____ _____, _____ _____ [p.552]

6. _____ [p.552]

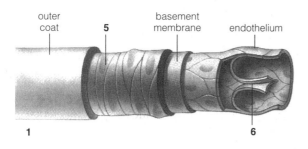

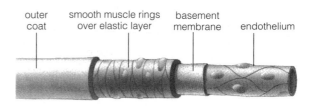

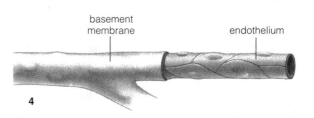

Fill-in-the-Blanks

A(n) (7) _____ [p.552] carries blood away from the heart. A(n) (8) _____ [p.554] is a blood vessel with such a small diameter that red blood cells must flow through it single-file; its wall consists of no more than a single layer of (9) _____ [p.554] cells resting on a basement membrane. In each (10) _____ _____ [p.554], small molecules move between the bloodstream and the (11) _____ [p.554] fluid. (12) _____ [pp.552, 554] are in the walls of veins and prevent backflow. Both (13) _____ [p.555] and (14) _____ [p.555] serve as temporary reservoirs for blood volume.

(15) _____ [p.552] are pressure reservoirs that keep blood flowing smoothly away from the heart while the (16) _____ [recall pp.550–551] are relaxing. (17) _____ [p.553] are control points where adjustments can be made in the volume of blood flow to be delivered to different capillary beds. They offer great resistance to flow, so there is a major drop in (18) _____ [p.552] in these tubes.

One cause of (19) _____ [p.556] is the rupture of one or more blood vessels in the brain. A (20) _____ _____ [p.557] blocks a coronary artery. (21) _____ _____ [p.557] is a term for a formation that can include cholesterol, calcium salts, and fibrous tissue. It is not healthful to have a high concentration of (22) _____ [p.557]-density lipoproteins in the bloodstream. A clot that stays in place is a(n) (23) _____ [p.557], but a clot that travels in the bloodstream is an embolus. Bleeding is stopped by several mechanisms that are referred to as (24) _____ [p.558]; the mechanisms include blood vessel spasm, (25) _____ _____ _____ [p.558], and blood (26) _____ [p.558]. Reactions cause rod-shaped proteins to assemble into long (27) _____ [p.558] fibers, which adhere to exposed (28) _____ [p.558] fibers in damaged vessel walls. These trap blood cells, platelets, and components of plasma. Under normal conditions, a clot eventually forms at the damaged site.

True–False

If the statement is true, write a T in the blank. If the statement is false, correct it by changing the underlined word(s) and writing the correct word(s) in the answer blank.

_____29. The pulse rate is the difference between the systolic and the diastolic pressure readings. [p.552]

_____30. The total volume of blood flowing in blood vessels remains more or less constant in the human body; blood pressure also remains constant throughout the circuit. [p.552]

More Fill-in-the-Blanks

Blood pressure is normally high in the (31) _____ [recall pp.550, 552] immediately after leaving the heart, but then it drops as the fluid passes along the circuit through different kinds of blood vessels. As blood flows through a vessel, it rubs against the vessel's inner wall and generates friction which causes some energy in the form of (32) _____ [p.552] to be lost. This explains why (32) drops as blood makes its way back to the heart. The rate of blood flow in a blood vessel is inversely proportional to the vessel's resistance to flow. In other words, the rate of blood flow decreases as resistance to blood flow (33) [choose one] ❑ increases ❑ decreases [p.552]. Arterial walls are thick, muscular, and (34) _____ [p.552] and have large diameters. Arteries present [choose one] (35) ❑ much ❑ little [p.552] resistance to blood flow, so pressure [choose one] (36) ❑ drops a lot ❑ does not drop much [p.552] in the arterial portion of the systemic and pulmonary circuits.

The greatest drop in pressure occurs at [choose one] (37) ❑ capillaries ❑ arterioles ❑ veins [p.552]. With this slowdown, blood flow can now be allotted in different amounts to different regions of the body in response to signals from the (38) _____ [p.553] system and endocrine system or even changes in local chemical conditions.

When a person is resting, blood pressure is influenced most by reflex centers in the (39) _____ _____ [p.553]. When the mean arterial pressure increases, reflex centers command the heart to [choose one] (40) ❑ beat more slowly ❑ beat faster [p.553], and command smooth muscle cells in arteriole walls to [choose one] (41) ❑ contract ❑ relax [p.553], which results in [choose one] (42) ❑ vasodilation ❑ vasoconstriction [p.552], which brings more oxygen and nutrients to the tissue.

33.11. LYMPHATIC SYSTEM [pp.558–559]

Selected Words: lymph [p.558], lymph vascular system [p.558], lymph capillaries [p.558], lymph vessels [p.558], tonsils [p.559], right lymphatic duct [p.559], thoracic duct [p.559], bone marrow [p.59], "valves" [p.559]

Boldfaced, Page-Referenced Terms

[p.558] lymphoid organs and tissues _____

_____

[p.559] lymph nodes _____

_____

[p.559] spleen _____

_____

[p.559] thymus gland _____

_____

Labeling

Identify each numbered part of the accompanying illustrations.

1. _____ [p.559]
2. _____ _____ _____ [p.559]
3. _____ _____ [p.559]
4. _____ _____ [p.559]
5. _____ [p.559]
6. _____ _____ [p.559]

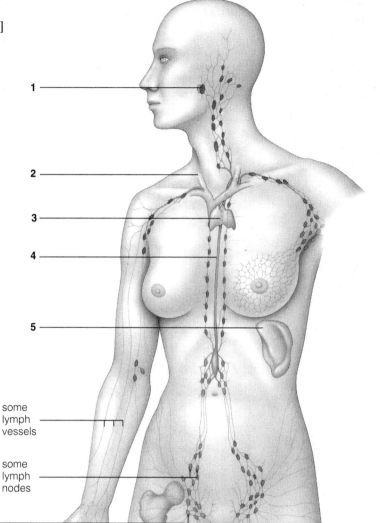

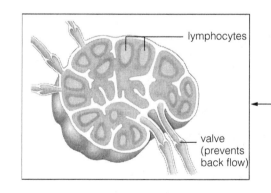

some lymph vessels

some lymph nodes

lymphocytes

valve (prevents back flow)

Fill-in-the-Blanks

(7) _____ [p.559] vessels reclaim fluid lost from the bloodstream, purify the blood of microorganisms, and transport (8) _____ [p.558] from the (9) _____ _____ [p.558] to the bloodstream.

Self-Quiz

___ 1. Most of the oxygen in human blood is transported by _____. [pp.544–545]
   a. plasma
   b. serum
   c. platelets
   d. hemoglobin
   e. leukocytes

___ 2. Of all the different kinds of white blood cells, two classes of _____ are the ones that respond to *specific* invaders and confer *immunity* to a variety of disorders. [pp.544–545]
   a. basophils
   b. eosinophils
   c. monocytes
   d. neutrophils
   e. lymphocytes

___ 3. Open circulatory systems generally lack _____. [p.542]
   a. a heart
   b. arterioles
   c. capillaries
   d. veins
   e. arteries

___ 4. Red blood cells originate in the _____. [p.545]
   a. liver
   b. spleen
   c. yellow bone marrow
   d. thymus gland
   e. red bone marrow

___ 5. Hemoglobin contains _____. [p.545]
   a. copper
   b. magnesium
   c. sodium
   d. calcium
   e. iron

___ 6. The pacemaker of the human heart is the _____. [p.551]
   a. sinoatrial node
   b. semilunar valve
   c. inferior vena cava
   d. superior vena cava
   e. atrioventricular node

___ 7. During systole, _____. [p.550]
   a. oxygen-rich blood is pumped to the lungs
   b. the heart muscle tissues contract
   c. the atrioventricular valves suddenly open
   d. oxygen-poor blood from all parts of the human body, except the lungs, flows toward the right atrium
   e. none of the above

___ 8. _____ are reservoirs of blood pressure in which resistance to flow is low. [p.552]
   a. Arteries
   b. Arterioles
   c. Capillaries
   d. Venules
   e. Veins

___ 9. Begin with a red blood cell located in the superior vena cava and travel with it in proper sequence as it goes through the following structures. Which will be *last* in sequence? [pp.548–549]
   a. aorta
   b. left atrium
   c. pulmonary artery
   d. right atrium
   e. right ventricle

___10. The lymphatic system is the principal avenue in the human body for transporting _____. [p.558]
   a. fats
   b. wastes
   c. carbon dioxide
   d. amino acids
   e. interstitial fluids

Chapter Objectives/Review Questions

1. Describe how the circulatory, respiratory, digestive, and urinary systems cooperate to help a multicellular animal survive. [p.541]
2. Distinguish between open and closed circulatory systems. Provide examples of animals with one or the other. [p.542]
3. Describe how vertebrate circulatory systems have evolved from the fish model to the mammalian model. [p.543]
4. Describe the composition of human blood, using percentages of total volume. [p.544]
5. State where erythrocytes, leukocytes, and platelets are produced. [p.545]
6. Describe how blood is typed for the ABO blood group and for the Rh factor. [pp.546–547]
7. List the factors that cause blood to leave the heart and the factors that cooperate to return blood to the heart. [pp.548–551]
8. Explain what causes a heart to beat. Then describe how the rate of heartbeat can be slowed down or speeded up. [p.551]
9. Explain what causes high pressure and low pressure in the human circulatory system. Then show where major drops in blood pressure occur in humans. [pp.552–555]
10. Describe how the structures of arteries, capillaries, and veins differ. [pp.552–553]
11. Describe the exchanges that occur in the capillary bed regions and the mechanisms that cause the exchanges. [p.554]
12. Explain how veins and venules can act as reservoirs of blood volume. [pp.554–555]
13. Distinguish a stroke from a coronary artery blockage, or occlusion. [pp.556–557]
14. Describe how hypertension develops, how it is detected, and whether it can be corrected. [p.556]
15. State the significance of high- and low-density lipoproteins to cardiovascular disorders. [p.557]
16. Describe the events that occur in the production of a blood clot. [p.558]
17. Describe the composition and function of the lymphatic system. [pp.558–559]

Integrating and Applying Key Concepts

You observe that some people appear as though fluid had accumulated in their lower legs and feet. Their lower extremities resemble those of elephants. You inquire about what is wrong and are told that the condition is caused by the bite of a mosquito that is active at night. Construct a testable hypothesis that would explain (1) why the fluid was not being returned to the torso, as normal, and (2) what the mosquito did to its victims.

34

IMMUNITY

Interactive Exercises

Russian Roulette, Immunological Style [pp.562–563]

34.1. THREE LINES OF DEFENSE [p.564]

34.2. COMPLEMENT PROTEINS [p.565]

34.3. INFLAMMATION [pp.566–567]

Selected Words: smallpox [p.562], pasteurization [p.563], anthrax [p.563], *athlete's foot* [p.564], *nonspecific* responses [p.564], "immune" response [p.564], edema [p.567], *chemotaxins* [p.567], *lactoferrin* [p.567], *endogenous pyrogen* [p.567], "set point" [p.567], *interleukin-1* [p.567]

Boldfaced, Page-Referenced Terms

[p.563] vaccination _____

[p.564] pathogens _____

_____

[p.564] lysozyme _____

_____

[p.565] complement system _____

_____

[p.565] lysis _____

_____

[p.566] neutrophils _____

_____

[p.566] eosinophils _____

_____

[p.566] basophils _____

_____

[p.566] macrophages _____

_____

[p.566] acute inflammation _____

_____

[p.566] mast cells _____

_____

[p.566] histamine _____

_____

[p.567] fever _____

_____

Fill-in-the-Blanks

Several barriers prevent pathogens from crossing the boundaries of your body. Intact skin and

(1) _____ [p.564] membranes are effective barriers. (2) _____ [p.564] is an enzyme that destroys the

cell wall of many bacteria. (3) _____ [p.564] fluid destroys many food-borne pathogens in the gut.

Normal (4) _____ [p.564] residents of the skin, gut, and vagina outcompete pathogens for resources and

help keep their numbers under control.

When a sharp object cuts through the skin and foreign microbes enter, some plasma proteins come to

the rescue and seal the wound with a(n) (5) _____ [p.564] mechanism. (6) _____ [p.564] white

blood cells engulf bacteria soon thereafter. Also, there are about twenty plasma proteins (collectively

referred to as the (7) _____ _____ [p.565]) that are activated one after another in a "cascade" of

reactions to help destroy invading microorganisms. (8) _____ [p.565] are Y-shaped proteins that lock

onto specific foreign targets and thereby tag them for destruction by phagocytes or by activating the complement system. The inflammatory response engages in battle both specific and nonspecific invaders. When the complement system is activated, circulating basophils and mast cells in tissues release (9) _____ [p.566], which dilates (10) _____ [p.566] and makes them "leaky," so fluid seeps out and causes the inflamed area to become swollen and warm.

Complete the Table
Complete the table by providing the specific functions carried out by each of the four different kinds of white blood cells listed below.

Basophils [p.566]	11.
Eosinophils [p.566]	12.
Neutrophils [p.566]	13.
Macrophages [p.566]	14.

34.4. THE IMMUNE SYSTEM [pp.568–569]
34.5. LYMPHOCYTE BATTLEGROUNDS [p.570]
34.6. CELL-MEDIATED RESPONSES [pp.570–571]

Selected Words: *immunological specificity* [p.568], *immunological memory* [p.568], *self* markers [p.568], *nonself* markers [p.568], *effector* cells [p.568], *memory* cells [p.568], "touch killing" [p.568], *primary* (immune) response [p.569], *secondary* (immune) response [p.569], *antibody-mediated* responses [p.569], *cell-mediated* response [p.570], virgin T cell [p.570], clone [p.570], interleukins [p.570], *perforins* [p.570]

Boldfaced, Page-Referenced Terms

[p.568] B lymphocytes (B cells) _____

_____

[p.568] T lymphocytes (Tcells) _____

_____

[p.568] immune system _____

_____

[p.568] antigen _____

_____

[p.568] MHC markers _____

_____

[p.568] antigen-MHC complexes _____

_____

[p.568] antigen-presenting cell _____

_____

[p.568] helper T cells _____

_____

[p.568] cytotoxic T cells _____

_____

[p.569] antibodies _____

_____

[p.570] TCRs (T-Cell Receptors) _____

_____

[p.570] apoptosis _____

_____

[p.571] natural killer cells (NK cells) _____

_____

Fill-in-the-Blanks

If the (1) _____ [recall p.564] defenses (fast-acting white blood cells such as neutrophils, eosinophils,

and basophils, slower-acting macrophages, and the plasma proteins involved in clotting and complement)

fail to repel the microbial invaders, then the body calls on its (2) _____ _____ [p.568], which

identifies *specific* targets to kill and *remembers* the identities of its targets. Your own unique (3) _____

_____ [p.568] patterns identify your cells as "self" cells. Any other surface pattern is, by definition,

(4) _____ [p.568], and doesn't belong in your body.

The principal actors of the immune system are (5) _____ [p.568] descended from stem cells [consult

text Figure 33.7 on p.545] in the bone marrow, which have two different strategies of action to deal with

their different kinds of enemies. (6) _____ _____ [p.569] clones secrete antibodies and act

principally against the extracellular (ones that stay outside the body cells) enemies that are pathogens in

blood or on the cell surfaces of body tissues. (7) _____ _____ [p.570] clones descend from

lymphocytes that matured in the (8) _____ [p.570] where they acquired specific markers on their cell

surfaces; they defend principally against intracellular pathogens such as (9) _____ [p.570], and against

any (10) _____ [p.570] cells and grafts of foreign tissue) that are perceived as abnormal or foreign.

Labeling

Identify each numbered part in the accompanying illustration. [All are from p.569]

11. _____ - _____ _____

12. _____ _____ _____ _____

13. _____ _____ _____

14. _____

15. _____ - _____ _____ _____

16. _____ _____ _____

17. _____

18. _____

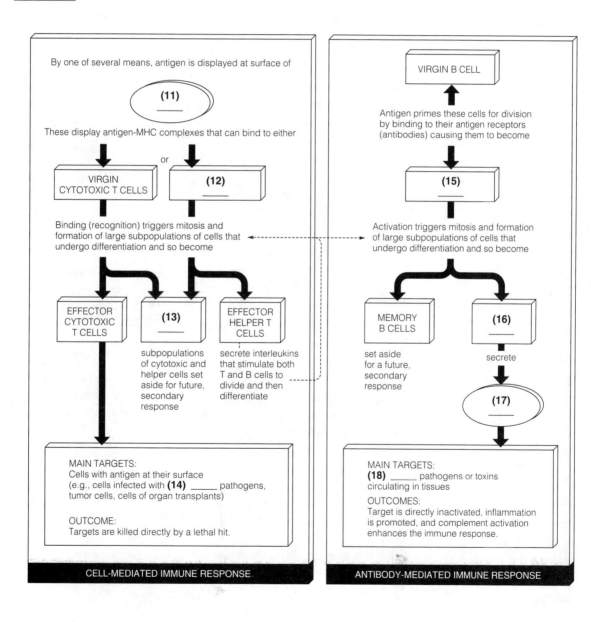

By one of several means, antigen is displayed at surface of

(11) _____

These display antigen-MHC complexes that can bind to either

VIRGIN CYTOTOXIC T CELLS   or   **(12)**

Binding (recognition) triggers mitosis and formation of large subpopulations of cells that undergo differentiation and so become

EFFECTOR CYTOTOXIC T CELLS   **(13)** _____   EFFECTOR HELPER T CELLS

(13) subpopulations of cytotoxic and helper cells set aside for future, secondary response

EFFECTOR HELPER T CELLS secrete interleukins that stimulate both T and B cells to divide and then differentiate

MAIN TARGETS:
Cells with antigen at their surface (e.g., cells infected with **(14)** _____ pathogens, tumor cells, cells of organ transplants)

OUTCOME:
Targets are killed directly by a lethal hit.

CELL-MEDIATED IMMUNE RESPONSE

VIRGIN B CELL

Antigen primes these cells for division by binding to their antigen receptors (antibodies) causing them to become

(15)

Activation triggers mitosis and formation of large subpopulations of cells that undergo differentiation and so become

MEMORY B CELLS   **(16)** _____

MEMORY B CELLS set aside for a future, secondary response

(16) secrete

(17)

MAIN TARGETS:
(18) _____ pathogens or toxins circulating in tissues

OUTCOMES:
Target is directly inactivated, inflammation is promoted, and complement activation enhances the immune response.

ANTIBODY-MEDIATED IMMUNE RESPONSE

Choice

For functions 19–23, choose the appropriate white blood cell type:

> a. effector cytotoxic T cells   b. effector helper T cells   c. macrophages
> d. memory cells   e. effector B cells

19. ___ Lymphocytes that directly destroy body cells already infected by certain viruses or by intracellular pathogens [p.570]

20. ___ A portion of B and T cell populations that were set aside as a result of a first encounter, now circulate freely and respond rapidly to any later attacks by the same type of invader [p.569]

21. ___ Lymphocytes and their progeny that produce antibodies [p.569]

22. ___ Lymphocytes that serve as master switches of the immune system; stimulate the rapid division of B cells and cytotoxic T cells [p.569]

23. ___ Nonlymphocytic white blood cells that develop from monocytes, engulf anything perceived as foreign, and alert helper T cells to the presence of specific foreign agents [p.566]

Labeling

Identify each numbered part in the accompanying illustration and its caption. [All are from p.571]

24. _____

25. _____ - _____ _____

26. _____ _____

27. _____ _____

This illustration depicts activated

(27) _____ _____

cells carrying out a(n)

(28) _____ -mediated

immune response.

Note: If the same number is used more than once, that is because the label is the same in all such situations.

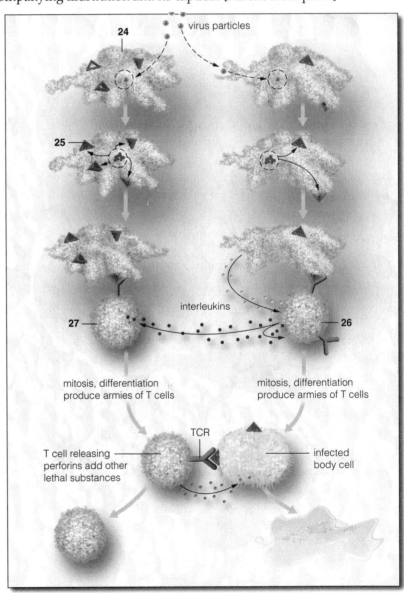

Fill-in-the-Blanks

The (29) _____ _____ _____ [p.569] is the route taken during a first-time contact with an antigen; the secondary immune response is more rapid because patrolling battalions of (30) _____ _____ [p.568] are in the bloodstream on the lookout for enemies they have conquered before. When these meet up with recognizable (31) _____ [p.569], they divide at once, producing large clones of B or T cells within 2–3 days.

In addition to effector B cells and cytotoxic T cells (executioner lymphocytes that mature in the thymus), (32) _____ _____ [p.571] cells mature in other lymphoid tissues, do not bind to antigen-MHC complexes but search out any cell that is either coated with complement proteins or antibodies, *or* bears any foreign molecular pattern. When cytotoxic T cells find foreign cells, they secrete (33) _____ [p.570] and other toxic substances to poison and lyse the offenders.

34.7. ANTIBODY-MEDIATED RESPONSES [pp.572–573]
34.8. *Focus on Health:* CANCER AND IMMUNOTHERAPY [p.573]
34.9. IMMUNE SPECIFICITY AND MEMORY [pp.574–575]

Selected Words: antigen-presenting cell [p.572], *IgM, IgD, IgG, IgA, IgE* [p.573], *carcinomas, sarcomas, leukemia* [p.573], *immunotherapy* [p.573], *monoclonal antibodies* [p.573], *LAK cells* [p.573], *therapeutic vaccines* [p.573], *dendritic* cells [p.573], *variable* regions [p.574], *clonal selection* hypothesis [p.575], "immunological memory" [p.575], *secondary* immune response [p.575] *immunological specificity* [p.575]

Boldfaced, Page-Referenced Terms

[p.573] immunoglobulins (Igs) _____

_____

Fill-in-the-Blanks

Antibodies are plasma proteins that are part of the (1) _____ [p.573] group of proteins. Some of these circulate in blood; others are present in other body fluids or bound to B cells. Although all antibodies are proteins, each kind has (2) _____ _____ [p.572] that only match up with a particular shape of (3) _____ [p.572]. When freely-circulating antibody molecules bind antigen, they tag an invader for destruction by (4) _____ [p.573] and complement activation. Antibody-mediated responses generally target (5) _____ _____ [p.573] and toxins, which are freely circulating in tissues or in body fluids. (6) _____ [p.573] can't bind to pathogens or toxins hidden in a host cell.

Identify each numbered part in the accompanying illustration and its caption. [All are from p.572]

7. _____ _____

8. _____ - _____ _____

9. _____ _____ _____

10. _____

11. _____ _____

12. _____ _____

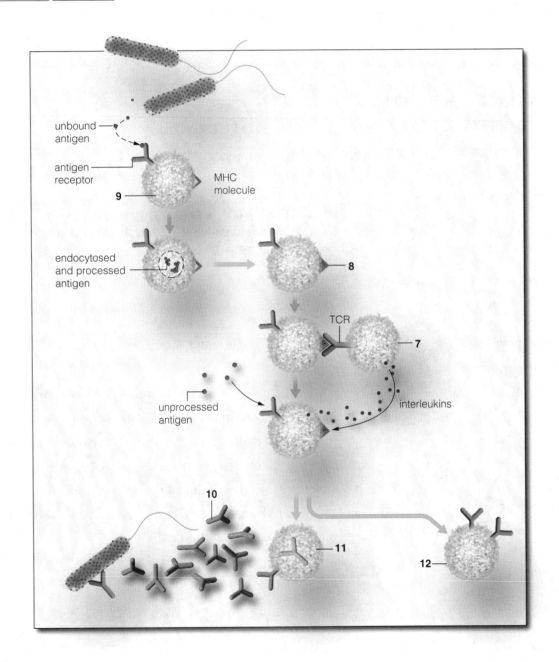

Choice

For questions 13–17, choose from the following:

a. IgA   b. IgE   c. IgG   d. IgM   e. Monoclonal antibodies

13. ___ Antibodies that activate complement proteins and neutralize many toxins; long-lasting; the only antibodies to cross the placenta during pregnancy and protect a fetus with the mother's acquired immunities; also in early mother's milk [p.573]

14. ___ Y-shaped molecules produced by clones of a hybrid cell formed by fusing an effector B cell with a cell from a B cell tumor [p.573]

15. ___ Y-shaped molecules that enter mucus-coated surfaces and neutralize infectious agents, also in mother's milk [p.573]

16. ___ Antibodies that start the inflammatory response if parasitic worms attack the body; attach to basophils and mast cells that can generate an allergic response [p.573]

17. ___ The first antibodies to be secreted during immune responses; trigger the complement cascade after binding antigen; also can bind targets together in clumps for removal by phagocytes [p.573]

Labeling

Identify each numbered part in the illustration below and its caption.

18. _____ [p.575]

19. _____ [p.575]

20. _____ _____ [p.572]

21. _____ _____ [p.575]

22. _____ _____ [p.575]

This illustration depicts selection of a(n) (20) _____ _____ [p.575] cell, the descendants of which produced the specific (18) _____ [p.575] that can combine with a specific antigen.

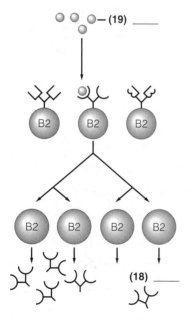

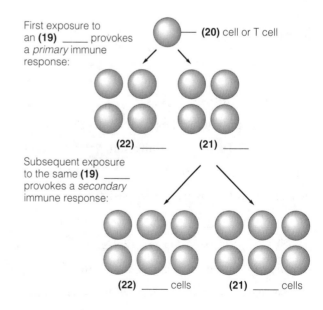

Fill-in-the-Blanks

The (23) _____ _____ _____ [p.575] explains how an individual has immunological memory, which is the basis of a secondary immune response; the theory also explains in part how *self* cells are distinguished from (24) _____ [recall p.568] cells in the vertebrate immune response.

(25) _____ [p.574] of gene segments drawn at random from receptor-encoding regions of DNA helps give each T or B cell a different gene sequence that will dictate one protein shape out of a (26) _____ [p.574] possible antigen receptor shapes.

(27) _____ [p.573] refers to cells that have lost control over cell division. Milstein and Kohler developed a means of producing large amounts of (28) _____ _____.

34.10. DEFENSES ENHANCED, MISDIRECTED, OR COMPROMISED [pp.576–577]
34.11. *Focus on Health:* AIDS—THE IMMUNE SYSTEM COMPROMISED [pp.578–579]

Selected Words: *active* and *passive* immunization [p.576], *asthma* [p.576], *hay fever* [p.576], *anaphylactic shock* [p.576], antihistamines [p.576], *Grave's disorder* [p.577], *myasthenia gravis* [p.577], *rheumatoid arthritis* [p.577], *severe combined immune deficiencies* (SCIDs) [p.577], AIDS [p.578], HIV [p.578], *Pneumocystis carinii* [p.578]

Boldfaced, Page-Referenced Terms

[p.576] immunization _____

_____

[p.576] vaccine _____

_____

[p.576] allergens _____

_____

[p.576] allergy _____

_____

[p.577] autoimmune response _____

_____

Fill-in-the-Blanks

Deliberately provoking the production of memory lymphocytes is known as (1) _____ [p.576]. In a(n) (2) _____ [p.576] immunization, a vaccine containing antigens is injected into the body or taken orally. The first one elicits a(n) (3) _____ _____ _____ [p.576], and the second one (a booster shot) elicits a secondary immune response, which causes the body to produce more effector cells and (4) _____ _____ [p.576] to provide long-lasting protection.

(5) _____ [p.576] is an altered secondary response to a normally harmless substance that may actually cause injury to tissues. In a(n) (6) _____ _____ [p.577] the body mobilizes its forces against self-antigens associated with its own tissues. (7) _____ _____ [p.577] is an example of this

kind of disorder in which antibodies tag acetylcholine receptors on skeletal muscle cells and cause progressive weakness. (8) _____ _____ [p.577] is a similar kind of disorder in which skeletal joints are chronically inflamed. AIDS is a constellation of disorders that follow infection by the (9) _____ _____ _____ [p.578]. In the United States, transmission has occurred most often among intravenous drug abusers who share needles and among (10) _____ _____ [p.578]. HIV is one of the (11) _____ _____ [p.578]; its genetic material is RNA rather than DNA, and it has several copies of an enzyme (12) _____ _____ [p.578], which uses the viral RNA as a template for making DNA, which is then inserted into a host chromosome. In 1997, an estimated (13) _____ [p.578] thousand people in North America were HIV-infected. Worldwide, an estimated (14) _____ [p.578] were HIV-infected; of these more than (15) _____ [p.578] million people were from sub-Saharan Africa.

Complete the Table

Indicate with a check (√) the age(s) of vaccination. [All are from p.576]

Age Vaccination Is Administered

Disease	(a) 0–4 months after birth	(b) 6–18 months after birth	(c) 4–6 years after birth	(d) 11–12 years after birth
16. Diphtheria				
17. *Hemophilus influenzae*				
18. Hepatitis B				
19. Measles				
20. Mumps				
21. Polio				
22. Rubella				
23. Tetanus				
24. Whooping cough				

Self-Quiz

___ 1. All the body's phagocytes are derived from stem cells in the _____. [p.566]
   a. spleen
   b. liver
   c. thymus
   d. bone marrow
   e. thyroid

___ 2. The plasma proteins that are activated when they contact a bacterial cell are collectively known as the _____ system. [p.565]
   a. shield
   b. complement
   c. Ig G
   d. MHC
   e. HIV

___ 3. _____ are divided into two groups: T cells and B cells. [p.568]
   a. Macrophages
   b. Lymphocytes
   c. Platelets
   d. Complement cells
   e. Cancer cells

___ 4. _____ produce and secrete antibodies that set up bacterial invaders for subsequent destruction by macrophages. [pp.572–573]
   a. B cells
   b. Phagocytes
   c. T cells
   d. Bacteriophages
   e. Thymus cells

___ 5. Antibodies are shaped like the letter _____. [p.572]
   a. Y
   b. W
   c. Z
   d. H
   e. E

___ 6. The markers for every cell in the human body are referred to by the letters _____. [p.568]
   a. HIV
   b. MBC
   c. RNA
   d. DNA
   e. MHC

___ 7. Effector B cells _____. [pp.569,572–573]
   a. fight against extracellular bacteria, viruses, some fungal parasites, and some protozoans
   b. develop from B cells
   c. manufacture and secrete antibodies
   d. do not divide and form clones
   e. all of the above

___ 8. Clones of B or T cells are _____. [pp.568–569]
   a. being produced continually
   b. sometimes known as memory cells if they keep circulating in the bloodstream
   c. only produced when their surface proteins recognize other specific proteins previously encountered
   d. produced and mature in the bone marrow
   e. both (b) and (c)

___ 9. Whenever the body is reexposed to a specific sensitizing agent, IgE antibodies cause _____. [p.573]
   a. histamine and other substances associated with inflammation to be produced
   b. clonal cells to be produced
   c. macrophages to be released
   d. the immune response to be suppressed
   e. none of the above

___10. The clonal selection theory explains _____. [p.575]
   a. how the human body can "remember" a first encounter with an antigen and form a secondary immune response in any later encounter with the same kind of antigen
   b. how B cells differ from T cells
   c. how so many different kinds of antigen-specific receptors can be produced by lymphocytes
   d. how memory cells are set aside from effector cells
   e. how antigens differ from antibodies

Matching

Choose the most appropriate description for each term.

11. ___allergy [p.576]
12. ___antibody [pp.572–573]
13. ___antigen [p.568]
14. ___macrophage [p.566]
15. ___clone [p.570]
16. ___complement [p.565]
17. ___histamine [p.566]
18. ___MHC marker [p.568]
19. ___effector B cell [p.569]
20. ___T cell [p.571]

A. Begins its development in bone marrow, but matures in the thymus gland
B. Cells that have directly or indirectly descended from the same parent cell
C. A potent chemical that causes blood vessels to dilate and let protein pass through the vessel walls
D. Y-shaped immunoglobulin
E. Nonself marker that can trigger formation of lymphocyte armies
F. The progeny of activated B cells signaled by helper T cells
G. A group of about twenty proteins that participate in the inflammatory response
H. An altered secondary immune response to a substance that is normally harmless to other people
I. The basis for self-recognition at the cell surface
J. Principal perpetrator of phagocytosis

Chapter Objectives/Review Questions

1. Describe typical external barriers that organisms present to invading organisms. [p.564]
2. List and discuss four nonspecific defense responses that serve to exclude microbes from the body. [p.564]
3. Explain how the complement system is related to an inflammatory response. [pp.565–567]
4. Distinguish between the antibody-mediated response pattern and the cell-mediated response pattern. [pp.568–573]
5. Explain what is meant by primary immune pathway as contrasted with secondary immune pathway. [p.569]
6. Explain what monoclonal antibodies are and tell how they are currently being used in passive immunization and cancer treatment. [p.573]
7. Describe the clonal selection theory and tell what it helps to explain. [p.575]
8. Describe two ways that people can be immunized against specific diseases. [p.576]
9. Distinguish allergy from autoimmune disorder. [pp.576–577]
10. Describe how AIDS specifically interferes with the human immune system. [pp.577–579]

Integrating and Applying Key Concepts

Suppose you wanted to get rid of forty-seven warts that you have on your hands by treating them with monoclonal antibodies. Outline the steps you would have to take.

35

RESPIRATION

Interactive Exercises

Conquering Chomolungma [pp.582–583]

35.1. THE NATURE OF RESPIRATION [p.584]
35.2. INVERTEBRATE RESPIRATION [p.585]
35.3. VERTEBRATE RESPIRATION [pp.586–587]

Selected Words: hypoxia [p.582], integument [p.585], book lungs [p.585], *external* gill [p.586], *internal* gill [p.586], gill filament [p.586], airways [p.587], larynx [p.587]

Boldfaced, Page-Referenced Terms

[p.583] respiration _____

_____

[p.583] respiratory system _____

_____

[p.584] pressure gradients _____

_____

[p.584] partial pressure _____

_____

[p.584] respiratory surface _____

_____

[p.584] Fick's law _____

_____

[p.584] hemoglobin _____

_____

[p.585] integumentary exchange _____

_____

[p.585] gills _____

_____

[p.585] tracheal respiration _____

_____

[p.586] countercurrent flow _____

_____

[p.586] lungs _____

_____

[p.587] vocal cords _____

_____

[p.587] glottis _____

_____

Fill-in-the-Blanks

A(n) (1) _____ [p.586] is an outfolded, thin, moist membrane endowed with blood vessels; it may (as in fish) be protected by bony covering or (as in aquatic insects) it may be naked. Gas transfer is enhanced by (2) _____ _____ [p.586], in which water flows past the bloodstream in the opposite direction. Insects have (3) _____ [p.585] (chitin-lined air tubes leading from the body surface to the interior). At sea level, atmospheric pressure is approximately 760 mm Hg, and oxygen represents about (4) _____ [p.584] percent of the total volume.

The energy to drive animal activities comes mainly from (5) _____ _____ [p.583], which uses (6) _____ [p.583] and produces (7) _____ _____ [p.583] wastes. In a process called (8) _____ [p.583], animals move (6) into their internal environment and give up (7) to the external environment.

All respiratory systems make use of the tendency of any gas to diffuse down its (9) _____ _____ [p.584]. Such a (9) exists between (6) in the atmosphere where pressure is (10) ❏ high ❏ low [p.584] and the metabolically active cells in body tissues. Pressure is (11) ❏ highest ❏ lowest [p.584] where (6) is used up rapidly. Another (9) exists between (7) in body tissues where (12) [choose one] ❏ high ❏ low [p.584] pressure exists and the atmosphere, with its (13) ❏ higher ❏ lower [p.584] amount of (7).

The more extensive the (14) _____ _____ [p.584] of a respiratory surface membrane and the larger the differences in (15) _____ _____ [p.584] across it, the faster a gas diffuses across the membrane. (16) _____ [p.584] is an important transport pigment, each molecule of which can bind loosely with as many as four O_2 molecules in the lungs.

A(n) (17) _____ [p.586] is an internal respiratory surface in the shape of a cavity or sac. In all lungs, (18) _____ [p.587] carry gas to and from one side of the respiratory surface, and (19) _____ [p.587] in blood vessels carries gas to and from the other side.

(20) _____ [p.582] is the medical name for oxygen deficiency; it is characterized by faster breathing, faster heart rate, and anxiety at altitudes of 8,000 feet above sea level.

Labeling

Identify the numbered parts of the accompanying illustration, which shows the respiratory system of many fishes.

21. _____ _____ [p.586]

22. _____ _____ [p.586]

23. _____ - _____ _____ [p.586]

24. _____ - _____ _____ [p.586]

25. _____ [p.586]

26. _____ [p.586]

water in

21

gill arch

22

23

24

26

25

direction of **25** flow (gray arrow) and **26** flow (black arrow)

35.4. HUMAN RESPIRATORY SYSTEM [pp.588–589]

Selected Words: pulmonary capillaries [p.588], pleural membrane [p.588], intercostal muscles [p.588], diaphragm [p.588], *laryngitis* [p.589], *pleurisy* [p.589], "bronchial tree" [p.589]

Boldfaced, Page-Referenced Terms

[p.588] alveolus (plural, alveoli) _____

_____

[p.589] pharynx _____

_____

[p.589] larynx _____

_____

[p.589] epiglottis _____

_____

[p.589] trachea _____

_____

[p.589] bronchus (plural, bronchi) _____

_____

[p.589] diaphragm _____

_____

[p.589] bronchioles _____

_____

Fill-in-the-Blanks

Every time you take a breath, you are (1) _____ [p.588] the respiratory surfaces of your lungs. Each lung has approximately 300 million tiny thin-walled outpouching air sacs called (2) _____ [p.588]. Inhaled air dead-ends in these (2), each of which is surrounded by a network of pulmonary (3) _____ [p.588] into which oxygen from air diffuses and is exchanged for (4) _____ _____ [p.588] waste gases diffusing from the bloodstream into the alveolar cavities.

The (5) _____ _____ [p.589] surrounds each lung. In succession, air passes through the nasal cavities, pharynx, and (6) _____ [p.589], past the epiglottis into the (7) _____ [p.589] (the space between the true vocal cords), into the trachea, and then to the (8) _____ [p.589], (9) _____ [p.589], and alveolar ducts. Exchange of gases occurs across the epithelium of the (10) _____ [p.589].

In addition to gas exchange, breathing has other functions such as ridding the body of excess heat and water vapor, adjusting the body's acid-base balance, and producing (11) _____ [p.588] to communicate with other creatures.

Labeling

Identify each numbered part of the accompanying illustration.

12. _____ _____ [p.588]

13. _____ [p.588]

14. _____ [p.588]

15. _____ [p.588]

16. _____ _____ [p.588]

17. _____ [p.588]

18. _____ [p.588]

19. _____ [p.588]

20. _____ _____ [p.588]

21. _____ _____ [p.588]

22. _____ _____ [p.588]

23. _____ [p.588]

24. _____ (singular), _____ (plural) [p.588]

25. _____ [p.588]

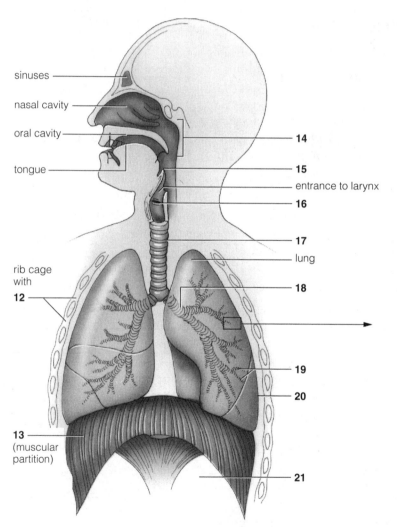

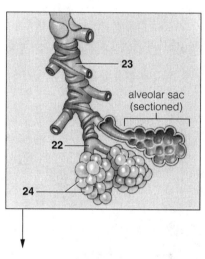

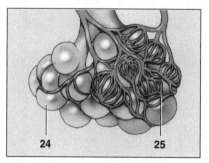

Complete the Table

26. Complete the following table with the structures that carry out the functions listed.

Structure	Function
a. [p.588]	Thin-walled sacs where O_2 diffuses into body fluids and CO_2 diffuses out
b. [p.589]	Increasingly branched airways that connect the trachea and alveoli
c. [p.588]	Muscle sheet that separates the chest (thoracic) cavity from the abdominal cavity
d. [p.588]	Airway where breathing is blocked while swallowing and where sound is produced
e. [p.588]	Airway that enhances speech sounds; connects nasal cavity with larynx

35.5. BREATHING—CYCLIC REVERSALS IN AIR PRESSURE GRADIENTS [pp.590–591]

35.6. GAS EXCHANGE AND TRANSPORT [pp.592–593]

35.7. *Focus on Health:* WHEN THE LUNGS BREAK DOWN [pp.594–595]

Selected Words: inhalation [p.590], *exhalation* [p.590], *atmospheric* pressure [p.590], *intrapulmonary* pressure [p.590], *intrapleural* pressure [p.590], hypoxia [p.591], *carbon monoxide poisoning* [p.592], medulla oblongata [p.593], *apnea* [p.593], *sudden infant death syndrome* (SIDS) [p.593], *bronchitis* [p.594], *emphysema* [p.594], *secondhand smoke* [p.594], "smoker's cough" [p.595], *pot* [p.595]

Boldfaced, Page-Referenced Terms

[p.590] respiratory cycle _____

_____

[p.591] acclimatization _____

_____

[p.591] erythropoietin _____

_____

[p.592] heme groups _____

_____

[p.592] oxyhemoglobin, HbO_2 _____

_____

[p.592] carbaminohemoglobin, $HbCO_2$ _____

_____

[p.592] carbonic anhydrase _____

_____

Fill-in-the-Blanks

During inhalation, the (1) _____ [p.590] moves downward and flattens, and the (2) _____ _____ [p.590] moves outward and upward. When these things happen, the chest cavity volume (3) [choose one] ❐ increases ❐ decreases [p.590] and the internal pressure (4) [choose one] ❐ rises ❐ drops ❐ stays the same [p.590].

When the thoracic (chest) cavity expands during inhalation, the lungs must expand too because water molecules in the (5) _____ [p.590] fluid are cohesive and behave like a "glue," binding the lungs to the thoracic wall. With *active* expiration (breathing out), the muscles in the abdominal wall contract, pressure is exerted upwardly on the (6) _____ [p.591], which decreases the volume of the chest cavity. The internal (7) _____ [p.591] muscles also contract, pulling the chest wall downward and inward and further decreasing the volume of the chest cavity. Internal pressure (8) [choose one] ❐ rises ❐ drops ❐ stays the same and air flows up the airways and outward because pressure in the alveoli is (9) [choose one] ❐ greater ❐ less [infer from p.590] than the atmospheric pressure.

Quiet exhalation is (10) _____ [p.590]; the muscles relax and the lungs recoil without any further energy outlays after an inhalation.

·(11) _____ [p.591] is the major hormone that governs red blood cell production. Secreted by specific kidney cells, (11) stimulates (12) _____ _____ [p.591] in bone marrow to divide repeatedly, yielding many more than the usual two to three million that are released each second in normal adult humans. Increased red blood cell production increases the blood's (13) _____-_____ [p.591] capacity, which is important for aerobic animals living at high altitudes.

Oxygen is said to exert a(n) (14) _____ _____ [recall p.584] of 760/21 or 160 mm Hg. (15) _____ [p.592] alone moves oxygen from the alveoli into the bloodstream, and it is enough to move (16) _____ _____ [p.592] in the reverse direction as each follows its own pressure gradient. (17) _____ [p.593] in red blood cells boosts oxygen transport from the lungs by 70 times and boosts carbon dioxide transport away from tissues by 17 times. About 60 percent of the carbon dioxide in the blood is transported as (18) _____ [p.592]. When oxygen-rich blood reaches a(n) (19) _____ [p.592] tissue capillary bed, oxygen diffuses outward, and carbon dioxide moves from tissues into the capillaries. Red blood cells contain the enzyme (20) _____ _____ [p.592], which converts CO_2 not bound to hemoglobin to carbonic acid (H_2CO_3) and promotes the diffusion of CO_2 from interstitial fluid into the bloodstream. When the (21) _____ _____ [p.592] of carbon dioxide is lower in the alveoli than in the neighboring blood capillaries, carbonic acid dissociates to form water and carbon dioxide. The rate of breathing is governed by clusters of cells that make up a respiratory center in the (22) _____ _____ [p.593], which monitors and coordinates signals coming in from arterial walls, from blood vessels, and other brain regions. The respiratory center regulates contractions of the diaphragm and intercostal muscles associated with inhalation and exhalation.

(23) _____ [p.594] is the distension of lungs and the loss of gas exchange efficiency such that running, walking, and even exhaling are painful experiences. At least 90 percent of all (24) _____ _____ [p.595] deaths are the result of cigarette smoking; only about 10 percent of afflicted individuals will survive.

Self-Quiz

___ 1. Most forms of life depend on _____ to obtain oxygen and eliminate carbon dioxide. [p.583]
   a. active transport
   b. bulk flow
   c. diffusion
   d. osmosis
   e. muscular contractions

___ 2. _____ is the most abundant gas in Earth's atmosphere. [p.584]
   a. Water vapor
   b. Oxygen
   c. Carbon dioxide
   d. Hydrogen
   e. Nitrogen

___ 3. With respect to respiratory systems, counter-current flow is a mechanism that explains how _____. [p.586]
   a. oxygen uptake by blood capillaries in the gill filaments of fish gills occurs
   b. ventilation occurs
   c. intrapleural pressure is established
   d. sounds originating in the vocal cords of the larynx are formed
   e. all of the above

___ 4. _____ have the most efficient respiratory system. [p.587]
   a. Amphibians
   b. Reptiles
   c. Birds
   d. Mammals
   e. Humans

___ 5. Immediately before reaching the alveoli, air passes through the _____. [p.588]
   a. bronchioles
   b. glottis
   c. larynx
   d. alveolar ducts
   e. trachea

___ 6. During inhalation, _____. [p.590]
   a. the pressure in all alveolar sacs is lower than the atmospheric pressure
   b. the pressure in all alveolar sacs is greater than the atmospheric pressure

   c. the diaphragm moves upward and becomes more curved
   d. the thoracic cavity volume decreases
   e. all of the above

___ 7. Oxyhemoglobin _____. [p.592]
   a. releases oxygen more readily in tissues with high rates of cellular respiration
   b. tends to release oxygen in places where the temperature is lower
   c. tends to hold on to oxygen when the pH of the blood drops
   d. tends to give up oxygen in regions where partial pressure of oxygen exceeds that in the lungs
   e. all of the above

___ 8. Oxygen moves from alveoli to the bloodstream _____. [p.592]
   a. whenever the concentration of oxygen is greater in alveoli than in the blood
   b. by means of active transport
   c. by using the assistance of carbamino hemoglobin
   d. principally due to the activity of carbonic anhydrase in the red blood cells
   e. by all of the above

___ 9. Which of the following paired items is incorrectly associated in the process that matches air flow with blood flow? [p.593]
   a. aortic bodies—low partial pressure of oxygen
   b. hypothalamus—controls rhythmic pattern of breathing
   c. apneustic center—lengthens inhalation
   d. pneumotaxic center—shortens inhalation
   e. carotid bodies—low partial pressure of oxygen

___10. Nonsmokers live an average of _____ longer than people in their mid-twenties who smoke two packs of cigarettes each day. [p.595]
   a. 6 months
   b. 1–2 years
   c. 3–5 years
   d. 7–9 years
   e. over 12 years

Matching

11. ___bronchioles [p.588]

12. ___bronchitis [p.594]

13. ___carbonic anhydrase [p.592]

14. ___emphysema [p.594]

15. ___glottis [p.587]

16. ___hypoxia [p.582]

17. ___intercostal muscles [pp.590–591]

18. ___larynx [p.588]

19. ___oxyhemoglobin [p.592]

20. ___pharynx [p.588]

21. ___pleurisy [p.589]

22. ___tracheal respiration [p.585]

23. ___ventilation [p.584]

24. ___integumentary exchange [p.585]

A. Membrane that encloses human lung becomes inflamed and swollen; painful breathing generally results

B. HbO_2

C. Occurs in terrestrial insects

D. Throat passageway that connects to <u>both</u> the respiratory tract below <u>and</u> the digestive tract

E. Inflammation of the two principal passageways that lead air into the human lungs

F. One set contracts when air is leaving the lungs; another set contracts when lungs are filling with air

G. The opening into the "voicebox"

H. Finer and finer branchings that lead to alveoli

I. Occurs in flatworms, earthworms, and many other invertebrates; gases diffuse directly across the body surface covering

J. An enzyme that increases the rate of production of H_2CO_3 from CO_2 and H_2O

K. Lungs have become distended and inelastic so that walking, running, and even exhaling are difficult

L. Where sound is produced by vocal cords

M. Movements that keep air or water moving across a respiratory surface

N. Too little oxygen is being distributed in the body's tissues

Chapter Objectives/Review Questions

1. List some of the ways that respiratory systems are adapted to unusual environments. [pp.582–583,585–587, 591]

2. Understand how the human respiratory system is related to the circulatory system, to cellular respiration, and to the nervous system. [pp.583, 588]

3. Understand the behavior of gases and the types of respiratory surfaces that participate in gas exchange. [pp.584–587]

4. Describe how incoming oxygen is distributed to the tissues of insects and contrast this process with the process that occurs in mammals. [pp.585, 588–589]

5. Define *countercurrent flow*, and explain how it works. State where such a mechanism is found. [p.586]

6. List all the principal parts of the human respiratory system, and explain how each structure contributes to transporting oxygen from the external world to the bloodstream. [pp.588–589]

7. Describe the relationship of the human lung to the pleural sac and to the chest (thoracic) cavity. [p.589]

8. Explain why oxygen diffuses from alveolar air spaces, through interstitial fluid, and across capillary epithelium. Then explain why carbon dioxide diffuses in the reverse direction. [p.592]

9. Explain why oxygen diffuses from the bloodstream into the tissues far from the lungs. Then explain why carbon dioxide diffuses into the bloodstream from the same tissues. [p.592]

10. Describe what happens to carbon dioxide when it dissolves in water under conditions normally present in the human body. [p.592]

11. List the structures involved in detecting carbon dioxide levels in the blood and in regulating the rate of breathing. Name the location of each structure. [pp.592–593]

12. Distinguish bronchitis from emphysema. Then explain how lung cancer differs from emphysema. [pp.594–595]

Integrating and Applying Key Concepts

Consider the amphibians—animals that generally have aquatic larval forms (tadpoles) and terrestrial adults. Outline the respiratory changes that you think might occur as an aquatic tadpole metamorphoses into a land-going juvenile.

36

DIGESTION AND HUMAN NUTRITION

Interactive Exercises

Lose It—And It Finds Its Way Back [pp.598–599]

36.1. THE NATURE OF DIGESTIVE SYSTEMS [pp.600–601]
36.2. OVERVIEW OF THE HUMAN DIGESTIVE SYSTEM [p.602]

Selected Words: anorexia nervosa [p.598], bulimia [p.598], crop [p.600], gizzard [p.600], lumen [p.602]

Boldfaced, Page-Referenced Terms

[p.599] nutrition _____

_____

[p.599] digestive system _____

_____

[p.600] incomplete digestive system _____

_____

[p.600] complete digestive system _____

_____

[p.600] mechanical processing _____

_____

[p.600] motility _____

_____

[p.600] secretion _____

_____

[p.600] digestion _____

_____

[p.600] absorption _____

_____

[p.600] elimination _____

_____

[p.600] ruminants _____

_____

[p.600] gut _____

_____

Fill-in-the-Blanks

(1) _____ [p.599] is a large concept that encompasses processes by which food is ingested, digested, absorbed, and later converted to the body's own (2) _____ [p.599], lipids, proteins, and nucleic acids. A digestive system is some form of body cavity or tube in which food is reduced first to (3) _____ [p.599] and then to small (4) _____ [p.599]. Digested nutrients are then (5) _____ [p.599] into the internal environment. A(n) (6) _____ [p.600] digestive system has only one opening, two-way traffic, and a highly branched gut cavity that serves both digestive and (7) _____ [p.600] functions. A(n) (8) _____ [p.600] digestive system has a tube or cavity with regional specializations and a(n) (9) _____ [p.600] at each end. (10) _____ [p.600] involves the muscular movement of the gut wall, but (11) _____ [p.600] is the release into the space inside the tube of enzyme fluids and other substances required to carry out digestive functions.

The human digestive system is a tube, 21–30 feet long in an adult, that has regions specialized for different aspects of digestion and absorption. They are, in order: the mouth, pharynx, esophagus, (12) _____ [p.602], (13) _____ _____ [p.602], large intestine, rectum, and (14) _____ [p.602]. Various (15) _____ [p.602] structures secrete enzymes and other substances that are also essential to the breakdown and absorption of nutrients; these include the salivary glands, liver, gallbladder, and (16) _____ [p.602]. The (17) _____ [p.599] system distributes nutrients to cells throughout the body. The (18) _____ [p.599] system supplies oxygen to the cells so that they can oxidize the carbon atoms of food molecules, thereby changing them to the waste product, (19) _____ _____ [p.599], which is eliminated by the same system. If excess water, salts, and wastes accumulate in the blood, the (20) _____ [p.599] system and skin will maintain the volume and composition of blood and other body fluids.

Labeling

Identify each numbered structure in the accompanying illustration.

21. _____ _____ [p.602]

22. _____ [p.602]

23. _____ [p.602]

24. _____ [p.602]

25. _____ [p.602]

26. _____ _____ [p.602]

27. _____ _____ [p.602]

28. _____ [p.602]

29. _____ [p.602]

30. _____ [p.602]

31. _____ [p.602]

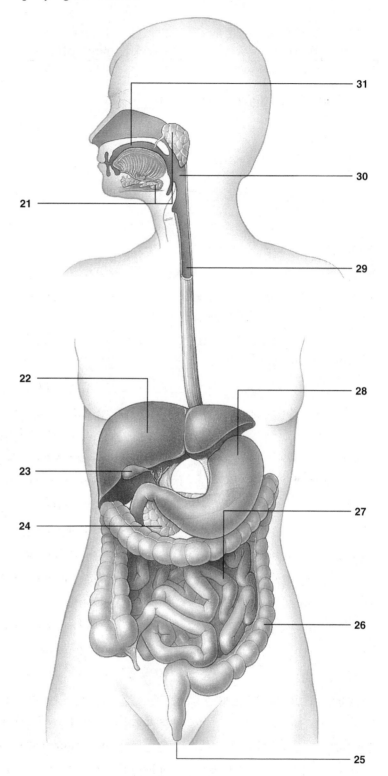

36.3. INTO THE MOUTH, DOWN THE TUBE [p.603]

36.4. DIGESTION IN THE STOMACH AND SMALL INTESTINE [pp.604–605]

Selected Words: caries [p.603], gingivitis [p.603], periodontal disease [p.603], heartburn [p.604], peptic ulcer [p.604], *Helicobacter pylori* [p.604], small intestine [p.604], "emulsion" [p.605], CCK (cholecystokinin) [p.605], GIP (glucose insulinotropic peptide) [p.605]

Boldfaced, Page-Referenced Terms

[p.603] tooth _____

_____

[p.603] saliva _____

_____

[p.603] pharynx _____

_____

[p.603] esophagus _____

_____

[p.603] sphincter _____

_____

[p.604] stomach _____

_____

[p.604] gastric fluid _____

_____

[p.604] chyme _____

_____

[p.604] pancreas _____

_____

[p.604] liver _____

_____

[p.604] gallbladder _____

_____

[p.605] bile _____

_____

[p.605] emulsification _____

_____

Complete the Table

1. Complete the following table by naming the organs described.

Organ | *Main Functions*

Organ	Main Functions
[p.603] a.	Mechanically breaks down food, mixes it with saliva
[p.603] b.	Moistens food; starts polysaccharide breakdown; buffers acidic foods in mouth
[p.604] c.	Stores, mixes, dissolves food; kills many microorganisms; starts protein breakdown; empties in a controlled way
[pp.604–605] d.	Digests and absorbs most nutrients
[pp.604–605] e.	Produces enzymes that break down all major food molecules; produces buffers against hydrochloric acid from stomach; secretes bicarbonate
[p.605, recall p.602] f.	Secretes bile for fat emulsification; plays roles in carbohydrate, fat, and protein metabolism
[p.605, recall p.602] g.	Stores, concentrates bile from liver
[recall p.602] h.	Stores, concentrates undigested matter by absorbing water and salts
[recall pp.600,602] i.	Controls expulsion of undigested and unabsorbed residues from end of gut

Fill-in-the-Blanks

Saliva contains an enzyme (2) _____ _____ [p.603] that breaks down starch. Contractions force the larynx against a cartilaginous flap called the (3) _____ [p.603], which closes off the trachea. The (4) _____ [p.603] is a muscular tube that propels food to the stomach. Any alternating progression of contracting and relaxing muscle movements along the length of a tube is known as (5) _____ [p.604]. (6) _____ [p.604] is an enzyme that works in the stomach to begin to break down proteins.

   Carbohydrates include sugars and (7) _____ [pp.603–604], the name commonly given to polysaccharides. Rice, cereal, pasta, bread, and white potatoes are composed of many polysaccharide molecules that are too large to be absorbed into the internal environment. If these foods are chewed thoroughly, (8) _____ _____ [p.604] in the mouth digests them to the (9) _____ [p.604] (double sugar) level. Because no carbohydrate digestion occurs in the (10) _____ [p.604], if you gulped down your food, starch digestion would again begin in the (11) _____ _____ [p.604] where (12) _____ [p.604] produced by the pancreas would do what should have been done in the mouth. Digestion of disaccharides to monosaccharides (simple sugars) also occurs in the (13) _____ _____ [p.604]. The enzymes responsible are (14) _____ [p.604] with names such as sucrase, lactase, and maltase.

   Proteins are digested to protein fragments, beginning in the (15) _____ [p.604] by (16) _____ [p.604] secreted by the lining of the stomach. Protein fragments are subsequently digested to smaller protein fragments in the (17) _____ _____ [p.604] by enzymes known as trypsin and chymotrypsin produced by the (18) _____ [p.604]. Eventually the smaller protein fragments are digested to (19) _____ _____ [p.604] by means of carboxypeptidase produced by the pancreas and by aminopeptidase produced by glands in the intestinal lining.

Another name for **fat** is triglycerides. (20) _____ [p.604], produced by the pancreas but acting in the (21) _____ _____ [p.604], breaks down one triglyceride molecule into (22) _____ _____ [p.604] and monoglyceride molecules. (23) _____ [p.605], which is made by the liver, stored in the (24) _____ [p.605], and does not contain digestive enzymes emulsifies the fat droplet (converts it into small droplets coated with bile salts) thereby increasing the surface area of the substrate upon which (25) _____ [p.604] can act.

Nutrients are also mostly digested and absorbed in the (26) _____ _____ [p.604]. (23) is made by the liver, is stored in the gallbladder, and works in the (27) _____ _____ [p.604]. (28) _____ [p.604] is an example of an enzyme that is made by the pancreas but works in the small intestine to convert protein fragments to amino acids. (29) _____ _____ [p.604] are made in the pancreas, but convert DNA and RNA into nucleotides in the small intestine.

True False

If the statement is true, write a T in the blank. If the statement is false, correct it by changing the underlined word(s) and writing the correct word(s) in the answer blank.

_____30. Amylase digests starch, lipase digests lipids, and <u>proteases</u> break peptide bonds. [p.604]

_____31. ATP is the end product of <u>digestion</u>. [Recall Section 7.6; look at p.604]

36.5. ABSORPTION IN THE SMALL INTESTINE [pp.606–607]
36.6. DISPOSITION OF ABSORBED ORGANIC COMPOUNDS [p.608]
36.7. THE LARGE INTESTINE [p.609]

Selected Words: surface-to-volume ratio [p.606], "brush border" cell [p.606], absorption [p.606], chylomicrons [p.606], feces [p.609], cecum [p.609], rectum [p.609], anus [p.609], gastrin [p.609], *constipation* [p.609], *appendicitis* [p.609], *colon cancer* [p.609]

Boldfaced, Page-Referenced Terms

[p.606] villi (singular, villus) _____

_____

[p.606] microvilli, (singular, microvillus) _____

_____

[p.607] segmentation _____

_____

[p.607] micelle formation _____

_____

[p.609] colon _____

_____

[p.609] bulk _____

_____

[p.609] appendix _____

_____

Short Answer

1. What is the pool of amino acids used for in the human body? [See text Figure 36.12 on p.608]

_____

2. Which breakdown products result from carbohydrate and fat digestion? [See text Table 36.1 on p.604]

_____

3. Monosaccharides, free fatty acids, and monoglycerides all have three uses; identify them. [p.608]

_____

Fill-in-the-Blanks

Immediately after a meal, the body's cells take up the (4) _____ [p.608] being absorbed from the small intestine and use it as an energy source. Excess amounts of (4) and other organic compounds are converted mainly to (5) _____ [p.608], which get stored in adipose tissue. Some of the excess is converted to (6) _____ [p.608], which is stored mainly in the (7) _____ [p.608] and in muscle tissue. Absorbed amino acids are synthesized into structural (8) _____ [p.608] and enzymes, as well as into some nitrogen-containing (9) _____ [p.608], which act as chemical messengers, and (10) _____ [p.608] which are assembled into DNA, RNA, and ATP. Between meals, the body uses the (11) _____ [p.608] reservoirs as the main energy source; energy from them is transferred to ATP during cellular respiration. Amino acid conversions in the liver form (12) _____ [p.608], which is potentially toxic to cells; the liver immediately converts this substance to (13) _____ [p.608], a much less toxic waste product that is expelled by the urinary system from the body.

True–False

If the statement is true, write a T in the blank. If the statement is false, correct it by changing the underlined word(s) and writing the correct word(s) in the answer blank.

_____14. The appendix has no known underlined digestive functions. [p.609]

_____15. Water and sodium ions are absorbed into the bloodstream from the lumen of the large intestine. [p.609]

_____16. Fatty acids and monoglycerides recombine into fats inside epithelial cells lining the colon. [p.607]

36.8. HUMAN NUTRITIONAL REQUIREMENTS [pp.610–611]

36.9. VITAMINS AND MINERALS [pp.612–613]

36.10. *Focus on Science:* WEIGHTY QUESTIONS, TANTALIZING ANSWERS [pp.614–615]

Selected Words: the zone diet [p.610], *complete* protein [p.611], *incomplete* protein [p.611], sucrose polyester [p.611], NPU (net protein utilization) [p.611], *organic* and *inorganic* substances [p.612], free radical [p.613], body mass index (BMI) [p.614], *identical twins* [p.615], "set point" [p.615]

Boldfaced, Page-Referenced Terms

[p.610] kilocalories _____

_____

[p.610] food pyramids _____

_____

[p.611] essential fatty acids _____

_____

[p.611] essential amino acids _____

_____

[p.612] vitamins _____

_____

[p.612] minerals _____

_____

[p.614] obesity _____

_____

[p.615] *ob* gene _____

_____

[p.615] leptin _____

_____

Fill-in-the-Blanks

Use the (1) _____ _____ [p.610] diagram at the right, as revised in 1992, to devise a well-balanced diet for yourself. Group 1, the trapezoidal base, represents the group of complex (2) _____ [p.610], which includes rice, pasta, cereal and (3) _____ [p.610]. (4) From this group [choose one] ❐ 0 ❐ 2–3 ❐ 2–4 ❐ 3–5 ❐ 6–11 [p.610] servings every day are needed to supply energy and fiber. Group 2 represents the (5) _____ [p.610] group. Use the choices in (4) to indicate the number of servings (6) _____ [p.610] that are needed from this group each day. Group 3 is the (7) _____ [p.610] group, from which (8) _____ [p.610] servings are needed each day. Choices include mango, oranges, and (9) _____ [p.610], cantaloupe, pineapple, or 1 cup of fresh (10) _____ [p.610]. Group 4 includes foods that are a source of nitrogen: nuts, poultry, fish, legumes, and (11) _____ [p.610]. From this group, the body's (12) _____ [p.610] and nucleic acids are constructed. (13) _____ [p.610] servings are required every day because the human body cannot synthesize eight of the 20 essential (14) _____ _____ [p.611] that are used to construct proteins, and must get them in their food supplies. The foods in Group 5, the (15) _____ [p.610], yogurt, and cheese group, supply calcium, vitamins A, D, B_2, and B_{12}. You need (16) _____ [p.610] servings every day. The foods in Group 6 provide extra calories but few vitamins and minerals; (17) _____ [p.610] servings are needed every day.

Group 6 ——
Group 5 ——
Group 4 ——
Group 3 ——
Group 2 ——
Group 1 ——

Complete the Table

18. Complete the following table by determining how many kilocalories the people described should take in daily, given the stated exercise level, in order to *maintain* their weight. [Consult p.614 of the text.]

Height	Age	Sex	Level of Physical Activity	Present Weight (lb.)	Number of Kilocalories/Day
5'6"	25	Female	Moderately active	138	a.
5'10"	18	Male	Very active	145	b.
5'8"	53	Female	Not very active	143	c.

Problem

19. Compute the body mass index (BMI) for each of the people described in #18, and state whether any are above (+) or below (–) the ideal weight [text Figure 36.16, p.614]. By how much?

The BMI of (a) is _____. _____ the ideal weight.

The BMI of (b) is _____. _____ the ideal weight.

The BMI of (c) is _____. _____ the ideal weight.

Fill-in-the-Blanks

(20) _____ _____ [p.610] are the body's main sources of energy; they should make up

(21) _____ to _____ [p.610] percent of the human male's (average size) daily caloric intake.

(22) _____ [p.611] and cholesterol are components of animal cell membranes.

Fat deposits are used primarily as (23) _____ _____ [p.611], but they also cushion many organs and provide insulation. Lipids should constitute less than (24) _____ [p.611] percent of the human diet. One teaspoon a day of polyunsaturated oil supplies all (25) _____ _____ _____ [p.611] that the body cannot synthesize. (26) _____ [p.611] are digested to twenty common amino acids, of which eight are (27) _____ [p.611], cannot be synthesized, and must be supplied by the diet. Animal proteins such as (28) _____ [p.611] and (29) _____ [p.611] contain high amounts of essential amino acids (that is, they are complete). (30) _____ [p.612] are organic substances needed in small amounts in order to build enzymes or help them catalyze metabolic reactions. (31) _____ [p.612] are inorganic substances needed for a variety of uses.

Related Problems

You are a 19-year-old male, very sedentary (TV, sleep, and computers), with a 6'1" medium frame and you weigh 195 pounds.

32. Use text Figure 36.16, p.614, to calculate the number of calories required to sustain your present weight. Are you underweight, overweight, or just right? _____

33. How many kilocalories are you allowed to ingest every day to reach your desired weight? _____

Complete the Table

Use the new, improved food pyramid to construct a one-day diet that would eventually allow the 19-year-old male in the preceding example to reach his ideal desired weight if he ate a similar diet every day. Place your choices in the table below. [All are from p.610]

How many servings from each group below is he allowed to have daily?	What, specifically, could he choose to eat?
34a. complex carbohydrates	34b.
35a. fruits	35b.
36a. vegetables	36b.
37a. dairy group	37b.
38a. assorted proteins	38b.
39a. The "sin" group at the top	39b.

Self-Quiz

___ 1. The process that moves nutrients into the blood or lymph is _____. [p.600]
a. ingestion
b. absorption
c. assimilation
d. digestion
e. none of the above

___ 2. The enzymatic digestion of proteins begins in the _____. [p.604]
a. mouth
b. stomach
c. liver
d. pancreas
e. small intestine

___ 3. The enzymatic digestion of starches begins in the _____. [p.604]
a. mouth
b. stomach
c. liver
d. pancreas
e. small intestine

___ 4. The greatest amount of absorption of digested nutrients occurs in the _____. [p.606]
a. stomach
b. pancreas
c. liver
d. colon
e. small intestine

___ 5. Glucose moves through the membranes of the small intestine mainly by _____. [p.607]
a. peristalsis
b. osmosis
c. diffusion
d. active transport
e. bulk flow

___ 6. Which of the following is *not* found in bile? [p.605]
a. lecithin
b. salts
c. digestive enzymes
d. cholesterol
e. pigments

___ 7. The average American consumes approximately _____ pounds of sugar per year. [p.610]
   a. 25
   b. 50
   c. 75
   d. 104
   e. 125

___ 8. Of the following, _____ has (have) the highest net protein utilization. [p.611]
   a. peanuts
   b. eggs
   c. broccoli
   d. beans
   e. bread

___ 9. Which of the following acts to slow emptying of the stomach so that food is not moved faster than it can be processed? [p.605]
   a. fear, depression, and other emotional upsets
   b. gastrin secreted from the lining of the stomach triggers the secretion of stomach acid
   c. glucose and fat in the small intestine call for insulin secretion, which helps cells absorb glucose
   d. secretion prods the pancreas to secrete bicarbonate
   e. cholecystokinin (CCK) enhances secretin's action and causes the gallbladder to contract

___ 10. Which of the following structures and/or activities are *not* associated with the small intestine? [p.609]
   a. micelle formation
   b. maximum compaction of bulk
   c. bile salts assist the absorption of fatty acids and monoglycerides
   d. triglycerides and proteins combine, forming chylomicrons
   e. substances cross the free surface of each villus by active transport, osmosis, and diffusion

___ 11. The element needed by humans for blood clotting, nerve impulse transmission, and bone and tooth formation is _____. [p.613]
   a. magnesium
   b. iron
   c. calcium
   d. iodine
   e. zinc

Matching

Choose the most appropriate answer for each term.

12. ___anorexia nervosa [p.598]

13. ___brush border cell [p.606]

14. ___bulimia [p.598]

15. ___complex carbohydrates [p.610]

16. ___essential amino acids [p.611]

17. ___essential fatty acids [p.611]

18. ___mineral [p.612]

19. ___rickets [p.612]

20. ___scurvy [p.612]

21. ___vitamin [p.612]

A. Linoleic acid is one example
B. Phenylalanine, lysine, and methionine are 3 of 8
C. Bears approximately 1,700 microvilli; involved in absorption of nutrients from small intestine
D. Obsessive dieting + skewed perception of body weight
E. Vitamin C deficiency
F. Vitamin D deficiency in young children
G. Organic substances that help enzymes to do their jobs; required in small amounts for good health
H. Feasting followed by vomiting or taking laxatives
I. Inorganic substances required for good health
J. Long chains of simple sugars; in cereal grains, legumes, and white potatoes

Chapter Objectives/Review Questions

1. Describe the specific tasks that your digestive system does for you every day. Then state how it interacts with the respiratory, circulatory, and urinary systems to supply your cells with raw materials and eliminate wastes. [p.599]
2. Distinguish between incomplete and complete digestive systems, and tell which is characterized by (a) specialized regions and (b) two-way traffic. [p.600]
3. Define and distinguish among motility, secretion, digestion, and absorption. [p.600]
4. Explain how, during digestion, food is mechanically broken down. Then explain how it is chemically broken down. [pp.602–604]
5. List all parts (in order) of the human digestive system through which food actually passes. Then list the auxiliary organs that contribute one or more substances to the digestive process. [pp.602–605]
6. Tell which foods undergo digestion in each of the following parts of the human digestive system, and state what the food is broken into: oral cavity, stomach, small intestine, large intestine. [pp.602–605]
7. List the enzyme(s) that act in (a) the oral cavity, (b) the stomach, and (c) the small intestine. Then tell where each enzyme was originally made. [pp.603–605]
8. Describe how the digestion and absorption of fats differ from the digestion and absorption of carbohydrates and proteins. [pp.604–607]
9. List the items that leave the digestive system and enter the circulatory system during the process of absorption. [pp.604,607]
10. Describe the cross-sectional structure of the small intestine, and explain how its structure is related to its function. [pp.605–607]
11. Explain how the human body manages to meet the energy and nutritional needs of the various body parts even though the person may be feasting sometimes and fasting at other times. [p.608]
12. Describe events that occur in the human colon. Describe three disorders of the large intestine. [p.609]
13. Reproduce from memory the food pyramid diagram as revised in 1992. Identify each of the six components, list the numerical range of servings permitted from each group, and also list some of the choices available. [p.610]
14. Compare the contributions of carbohydrates, proteins, and fats to human nutrition with the contributions of vitamins and minerals. [pp.610–613]
15. Summarize the daily nutritional requirements of a 25-year-old man who is 5'11" and works at a desk job and exercises very little. State what he needs in energy, carbohydrates, proteins, and lipids, and name at least six vitamins and six minerals that he needs to include in his diet every day. [pp.610–614]
16. Construct an ideal diet for yourself for one 24-hour period. Calculate the number of calories necessary to maintain your weight [see p.614] and then use the food pyramid [p.610] to choose exactly what to eat and how much. [pp.610,614]
17. State what is meant by net protein utilization, and distinguish vitamins from minerals. [pp.611–613]
18. Name five minerals that are important in human nutrition, and state the specific role of each. [p.613]
19. List the factors that influence us to gain weight, and describe precisely how our normal system of weight control can fail. [pp.614–615]

Integrating and Applying Key Concepts

Suppose you could not eat solid food for two weeks and you had only water to drink. List in correct sequential order the measures your body would take to try to preserve your life. Mention the command signals that are given as one after another critical point is reached, and tell which parts of the body are the first and the last to make up for the deficit.

37

THE INTERNAL ENVIRONMENT

Interactive Exercises

Tale of the Desert Rat [pp.618–619]

37.1. URINARY SYSTEM [pp.620–621]

Selected Words: internal environment [p.618], "metabolic water" [p.618], solutes [p.619], ammonia [p.620], uric acid [p.620], *glomerular* capillaries [p.621], *peritubular* capillaries [p.621]

Boldfaced, Page-Referenced Terms

[p.619] interstitial fluid _____

_____

[p.619] blood _____

_____

[p.619] extracellular fluid _____

_____

[p.620] urinary excretion _____

_____

[p.620] urea _____

_____

[p.621] urinary system _____

_____

[p.621] kidneys _____

_____

[p.621] urine _____

_____

[p.621] ureter _____

_____

[p.621] urinary bladder _____

_____

[p.621] urethra _____

_____

[p.621] nephrons _____

_____

[p.621] Bowman's capsule _____

_____

[p.621] glomerulus _____

_____

[p.621] proximal tubule _____

_____

[p.621] loop of Henle _____

_____

[p.621] distal tubule _____

_____

[p.621] collecting duct _____

_____

Fill-in-the-Blanks

The body gains water by absorbing water from the slurry in the lumen of the small intestine and from

(1) _____ [p.620] during condensation reactions. The mammalian body loses water mostly by excretion

of (2) _____ [p.620], evaporation through the skin and (3) _____ [p.620], elimination of feces from

the gut, and (4) _____ [p.620] as the body is cooled. (5) _____ [p.620] behavior, in which the brain

compels the individual to seek liquids, influences the gain of water.

The body gains solutes by absorption of substances from the gut, by the secretion of hormones and

other substances, and by (6) _____ [p.620], which produces CO_2 and other waste products of

degradative reactions. Besides CO_2, there are several major metabolic wastes that must be eliminated:

(7) _____ [p.620], formed when amino groups are detached from amino acids; (8) _____ [p.620],

which is produced in the liver during reactions that link two ammonia molecules to CO_2 and release a

molecule of water, and (9) _____ _____ [p.620], which is formed in reactions that break down

nucleic acids.

Labeling

Identify each numbered part of the accompanying illustrations.

10. _____ [p.620]

11. _____ [p.620]

12. _____ _____ [p.620]

13. _____ [p.620]

14. _____ [p.621]

15. _____ [p.621]

16. _____ [p.621]

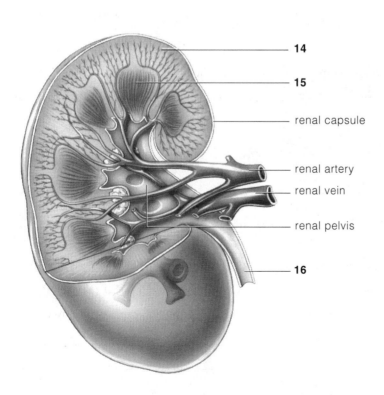

Labeling

Identify each numbered part of the accompanying illustrations. [All are from p.621]

17. _____ _____

18. _____ _____

19. _____ _____

20. _____ _____

21. _____ _____

22. _____ _____ _____

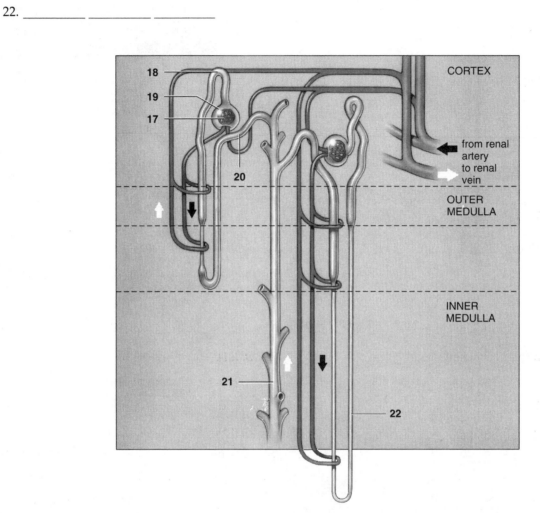

37.2. URINE FORMATION [pp.622–623]

37.3. *Focus on Health:* **WHEN THE KIDNEYS BREAK DOWN** [p.624]

Selected Words: *uremic toxicity* [p.624], *glomerulonephritis* [p.624], *kidney stones* [p.624], *kidney dialysis machine* [p.624], *"dialysis"* [p.624], *hemodialysis* [p.624], *peritoneal dialysis* [p.624]

Boldfaced, Page-Referenced Terms

[p.622] filtration _____

_____

[p.622] tubular reabsorption _____

_____

[p.622] tubular secretion _____

_____

[p.623] ADH (antidiuretic hormone) _____

_____

[p.623] aldosterone _____

_____

[p.623] angiotensin II _____

_____

[p.623] thirst center _____

_____

[p.624] renal failure _____

_____

True–False

If the statement is true, write a T in the blank. If the statement is false, correct it by changing the underlined word(s) and writing the correct word(s) in the answer blank.

_____1. When the body rids itself of excess water, urine becomes more dilute. [p.623]

_____2. Water reabsorption into peritubular capillaries is achieved by osmosis and active transport. [p.623]

Fill-in-the-Blanks

In mammals, urine formation occurs in a pair of (3) _____ [p.621]. Each contains about a million

tubelike blood-filtering units called (4) _____ [p.621]. The function of (3) depends on intimate links

between the (4) and the (5) _____ [p.622]. In every (4), water and reclaimable substances dissolved in

blood flow from a(n) (6) _____ [p.622] into a set of capillaries inside the (7) _____ _____

[p.621], then into a second set of capillaries that thread around the tubular parts of the nephron, then back to

the bloodstream, leaving the kidney. Urine composition and volume depend on three processes: *filtration* of

blood at the (8) _____ [p.622] of a nephron, with (9) _____ _____ [p.622] providing the force

for filtration; *reabsorption*, in which water and (10) _____ [p.622] move out of tubular parts of the

nephron and back into adjacent (11) _____ _____ [p.622]; and (12) _____ _____ [p.622],

in which excess ions and a few foreign substances move out of those capillaries and back into the distal

tubule of the nephron so that they are disposed of in the urine. (13) _____ [p.621] carry urine away

from the kidney to the (14) _____ _____ [p.621], where it is stored until it is released via a tube

called the (15) _____ [p.621], which carries urine to the outside.

Two hormones, ADH and (16) _____ [p.623], adjust the reabsorption of water and (17) _____

[p.623] along the distal tubules and collecting ducts. An increase in the secretion of aldosterone causes

(18) [choose one] ❏ more ❏ less [p.623] sodium to be excreted in the urine. If a person eats an unnecessarily large amount of table salt (sodium chloride) due to taste preferences, the excess sodium must be excreted or it will cause the body to retain excess water, which leads to a rise in (19) _____ _____ [recall Section 33.9, p.556]. Chronically high (19) is called hypertension; it can damage the kidneys, the vascular system, and the brain. When the nephrons of both kidneys can no longer perform their excretory and regulatory functions, (20) _____ _____ [p.624] results. (20) can be caused by (21) _____ _____ [p.624], by continued high doses of aspirin and other drugs, by ingestion of lead, arsenic, pesticides and other toxins, or by abnormal retention of metabolic wastes. Increased secretion of (22) _____ [p.623] enhances water reabsorption at distal tubules and collecting ducts when the body must conserve water. When excess water must be excreted, ADH secretion is (23) [choose one] ❏ stimulated ❏ inhibited [p.623].

By adjusting the blood's (24) _____ [p.623] and composition, kidneys help maintain conditions in the extracellular fluid.

37.4. THE ACID–BASE BALANCE [p.624]
37.5. ON FISH, FROGS, AND KANGAROO RATS [p.625]
37.6. MAINTAINING BODY TEMPERATURE [pp.626–627]

Selected Words: *metabolic acidosis* [p.624], *bicarbonate–carbon dioxide* buffer system [p.624], *behavioral temperature regulation* [p.626], *hypothermia* [p.627], *frostbite* [p.627], *"panting"* [p.627], *hyperthermia* [p.627], *fever* [p.627]

Boldfaced, Page-Referenced Terms

[p.624] acid–base balance _____

_____

[p.626] core temperature _____

_____

[p.626] radiation _____

_____

[p.626] conduction _____

_____

[p.626] convection _____

_____

[p.626] evaporation _____

_____

[p.626] ectotherms _____

_____

[p.626] endotherms _____

_____

[p.626] heterotherms _____

_____

[p.627] peripheral vasoconstriction _____

_____

[p.627] pilomotor response _____

_____

[p.627] shivering response _____

_____

[p.627] nonshivering heat production _____

_____

[p.627] peripheral vasodilation _____

_____

[p.627] evaporative heat loss _____

_____

Fill-in-the-Blanks

The (1) _____ [p.624] control the acid–base balance of body fluids by controlling the levels of dissolved

ions, especially (2) _____ [p.624] ions. The extracellular pH of humans must be maintained between

7.37 and (3) _____ [p.624]. (4) [choose one] ❒ Acids ❒ Bases [recall Section 2.6, p.30] lower the pH and

(5) [choose one] ❒ acids ❒ bases [recall Section 2.6, p.30] raise it. If you were to drink a gallon of orange

juice that contains citric acid and ascorbic acid (Vitamin C) the pH would be (6) [choose one] ❒ raised

❒ lowered [p.624], but the effect is minimized when excess (7) _____ [p.624] ions are neutralized by

(8) _____ [p.624] ions in the bicarbonate-carbon dioxide buffer system. Only the (9) _____ [p.624]

system eliminates excess H^+ and restores buffers. Desert-dwelling kangaroo rats have very long

(10) _____ _____ _____ [p.625] so that nearly all (11) _____ [p.625] that reaches their

very long collecting ducts is reabsorbed. In freshwater ecosystems, bony fishes and amphibians tend to gain

(12) _____ [p.625] and lose (13) _____ [p.625]; they produce (14) ❒ very dilute ❒ very concentrated

[p.625] urine.

   In the brain of mammals, the (15) _____ [p.626] is the seat of temperature control. Thermoreceptors

located deep in the body are called (16) _____ [p.627] thermoreceptors. The (17) _____ _____

[p.627] contains smooth muscles that erect hairs or feathers and create an insulative layer of still air that

helps prevent heat loss. (18) _____ _____ [p.627] is a response to cold stress in which the

bloodstream's convective delivery of heat to the body's surface is reduced. A drop in body temperature

below tolerance levels is referred to as (19) _____ [p.627].

True–False

If the statement is true, write a T in the blank. If the statement is false, correct it by changing the underlined word(s) and writing the correct word(s) in the answer blank.

_____20. Jackrabbits are underlined{endotherms}. [p.626]

_____21. When the core temperature of the human body falls underlined{a few degrees}, consciousness is lost and heart muscle action becomes irregular. [p.627]

_____22. When the core temperature of the human body causes underlined{hypothermia} to occur, ventricular fibrillation may set in and death soon follows. [p.627]

Matching

Choose the most appropriate answer for each term.

23. ___conduction [p.626]

24. ___convection [p.626]

25. ___ectotherm [p.626]

26. ___endotherm [p.626]

27. ___evaporation [p.626]

28. ___heterotherm [p.626]

29. ___radiation [p.626]

A. Body temperature determined more by heat exchange with the environment than by metabolic heat
B. Heat transfer by air or water heat-bearing currents away from or toward a body
C. Body temperature determined largely by metabolic activity and by precise controls over heat produced and heat lost
D. Direct transfer of heat energy between two objects in direct contact with each other
E. The emission of energy in the form of infrared or other wavelengths that are converted to heat by the absorbing body
F. Body temperature fluctuating at some times and heat balance controlled at other times
G. In changing from the liquid state to the gaseous state, the energy required is supplied by the heat content of the liquid

Self-Quiz

___ 1. _____ forms as amino groups are removed from amino acids. [p.620]
   a. Water
   b. Uric acid
   c. Urea
   d. Ammonia
   e. Carbon dioxide

___ 2. An entire subunit of a kidney that purifies blood and restores solute and water balance is called a _____. [p.621]
   a. glomerulus
   b. loop of Henle
   c. nephron
   d. ureter
   e. none of the above

___ 3. In humans, the thirst center is located in the _____. [p.623]
   a. adrenal cortex
   b. thymus
   c. heart
   d. adrenal medulla
   e. hypothalamus

___ 4. The longer the _____, the greater an animal's capacity to conserve water and to concentrate solutes to be excreted in the urine. [p.625]
   a. loop of Henle
   b. proximal tubule
   c. ureter
   d. Bowman's capsule
   e. collecting tubule

___ 5. During reabsorption, sodium ions cross the proximal tubule walls into the interstitial fluid principally by means of _____. [p.623]
a. phagocytosis
b. countercurrent multiplication
c. bulk flow
d. active transport
e. all of the above

___ 6. Filtration of the blood in the kidney takes place in the _____. [p.622]
a. loop of Henle
b. proximal tubule
c. distal tubule
d. glomerulus
e. all of the above

___ 7. _____ primarily controls the concentration of solutes in urine. [p.623]
a. Insulin
b. Glucagon
c. Antidiuretic hormone
d. Aldosterone
e. Epinephrine

___ 8. Hormonal control over excretion primarily affects _____. [p.623]
a. Bowman's capsules
b. distal tubules
c. proximal tubules
d. the urinary bladder
e. loops of Henle

___ 9. The last portion of the excretory system passed by urine before it is eliminated from the body is the _____. [p.621]
a. renal pelvis
b. bladder
c. ureter
d. collecting ducts
e. urethra

___10. Which of the following processes cause an animal to lose body heat as a liquid is converted to a gaseous form of the same substance? [p.626]
a. conduction
b. convection
c. evaporation
d. peripheral vasoconstriction
e. radiation

___11. Normally, the extracellular pH of the human body must be maintained between _____ and _____; only the urinary system eliminates excess _____ and restores _____. [p.626]
a. 6.45–7.30; NH_4^+; urea
b. 7.37–7.45; H^+; buffers
c. 7.50–7.85; H^+; glucose
d. 7.90–8.30; NH_4^+; urea
e. 8.15–8.35; OH^-; glucose

Chapter Objectives/Review Questions

1. List some of the factors that can change the composition and volume of body fluids. [pp.618–620,622–623]
2. List three soluble by-products of animal metabolism that are potentially toxic. [p.620]
3. List successively the parts of the human urinary system that constitute the path of urine formation and excretion. [pp.620–621]
4. Locate the processes of filtration, tubular reabsorption, and tubular secretion along a nephron, and tell what makes each process happen. [pp.622–623]
5. State explicitly how the hypothalamus, posterior pituitary, adrenal cortex, sensory receptors in the heart and blood vessels, gland cells in the arteriole wall near the glomeralus, and distal tubules of the nephrons are interrelated in regulating water and solute levels in body fluids. [pp.622–623]
6. List two kidney disorders and explain what can be done if kidneys become too diseased to work properly. [p.624]
7. Describe the role of the kidney in maintaining the pH of the extracellular fluids between 7.35 and 7.45. [p.624]
8. Describe how the details of vertebrate urinary systems differ as they adapt to (a) life in freshwater habitats, (b) fish life in the open ocean, and (c) life in the desert. [p.625]
9. State the different physiological strategies employed by ectotherms, endotherms, and heterotherms in their regulation of bodily heat gain and heat loss as their body temperature remains within a normal range. [pp.626–627]
10. State how each of the three types in #9 are able to cope with (a) cold stress and (b) heat stress. [pp.626–627]

Integrating and Applying Key Concepts

The hemodialysis machine used in hospitals is expensive and time-consuming. Currently, there are no artificial kidneys capable of allowing people who have nonfunctional kidneys to purify their blood by themselves, without having to go to a hospital or clinic. Which aspects of the hemodialysis procedure do you think have presented the most problems in development of a method of home self-care? If you had an unlimited budget and were appointed head of a team to develop such a procedure and its instrumentation, what strategy would you pursue?

38

REPRODUCTION AND DEVELOPMENT

Interactive Exercises

From Frog to Frog and Other Mysteries [pp.630–631]

38.1. THE BEGINNING: REPRODUCTIVE MODES [pp.632–633]

38.2. STAGES OF DEVELOPMENT—AN OVERVIEW [pp.634–635]

Selected Words: *Rana pipiens* [p.630], *reproductive timing* [p.632], *outermost* primary tissue layer [p.634], *innermost* primary tissue layer [p.634], *intermediate* primary tissue layer [p.634]

Boldfaced, Page-Referenced Terms

[p.630] zygotes _____

_____

[p.632] sexual reproduction _____

_____

[p.632] asexual reproduction _____

_____

[p.633] yolk _____

_____

[p.634] embryos _____

_____

[p.634] gamete formation _____

_____

[p.634] fertilization _____

_____

[p.634] cleavage _____

_____

[p.634] blastomeres _____

_____

[p.634] gastrulation _____

_____

[p.634] ectoderm _____

_____

[p.634] endoderm _____

_____

[p.634] mesoderm _____

_____

[p.634] organ formation _____

_____

[p.634] growth and tissue specialization _____

_____

Fill-in-the-Blanks

New sponges budding from parent sponges and a flatworm dividing into two flatworms represent examples of (1) _____ [p.632] reproduction. This type of reproduction is useful when gene-encoded traits are strongly adapted to a limited set of (2) _____ [p.632] conditions. Separation into male and female sexes requires special reproductive structures, control mechanisms, and behaviors; this cost is offset by a selective advantage: (3) _____ [p.632] in traits among the offspring. Often, mating requires special forms of behavior, such as (4) _____ [p.632], that can promote fertilization. Mating also requires built-in controls that can synchronize the timing of (5) _____ [p.632] formation, sexual readiness, even parental behavior in two individuals. The maturation of sperm in one individual exactly when the eggs mature in another individual is a question of (6) _____ [p.632] timing. Finding and actually recognizing a(n) (7) _____ [p.633] mate of the same species is another challenge. Assuring the survival of (8) _____ [p.633] is also costly. Many invertebrates, bony fishes, and frogs simply release eggs and motile sperm into the (9) _____ [p.633] surroundings. Such species invest energy in producing numerous (10) _____ [p.633], often thousands of them. Nearly all land animals rely on (11) _____ [p.633] fertilization, the union of sperm and egg inside the female's body. They invest energy to construct elaborate (12) _____ [p.633] organs, such as a penis and a uterus.

Finally, animals set aside energy in forms that can (13) _____ [p.633] the developing individual until it is developed enough to feed itself. Nearly all animal eggs contain (14) _____ [p.633], which is a protein-rich, lipid-rich substance that nourishes embryonic stages. The eggs of some species have much more yolk than others. Sea urchins produce (15) _____ [p.633] eggs with very little yolk but produce large numbers of them. Mother (16) _____ [p.633] lay truly yolky eggs. The yolk nourished the bird (17) _____ [p.633] through an extended period of development. Human eggs have almost no yolk.

Sequence

Arrange the following events in correct chronological sequence. Write the letter of the first event next to 18, the letter of the second event next to 19, and so on.

18. ___   A. Gastrulation [p.634]

19. ___   B. Fertilization [p.634]

20. ___   C. Cleavage [p.634]

21. ___   D. Growth, tissue specialization [p.634]

22. ___   E. Organ formation [p.634]

23. ___   F. Gamete formation [p.634]

Complete the Table

24. Complete the table below by entering the correct germ layer (ectoderm, mesoderm, or endoderm) that forms the tissues and organs listed.

Tissues/Organs	Germ Layer
Muscle, circulatory organs	[p.634] a.
Nervous system tissues	[p.634] b.
Inner lining of the gut	[p.634] c.
Circulatory organs (blood vessels, heart)	[p.634] d.
Outer layer of the integument	[p.634] e.
Reproductive and excretory organs	[p.634] f.
Organs derived from the gut	[p.634] g.
Most of the skeleton	[p.634] h.
Connective tissues of the gut and integument	[p.634] i.

38.3. EARLY MARCHING ORDERS [pp.636–637]

38.4. HOW DO SPECIALIZED TISSUES AND ORGANS FORM? [pp.638–639]

Selected Words: gray crescent [p.636], *vegetal* pole [p.637], *animal* pole [p.637], *radial* cleavage [p.637], animal-vegetal axis [p.637], blastocoel [p.637], morula [p.637], *incomplete* cleavage [p.637], *rotational* cleavage [p.637], *identical twins* [p.637], *fraternal twins* [p.637], *programmed cell death* [p.639], apoptosis [p.639]

Boldfaced, Page-Referenced Terms

[p.636] oocyte _____

_____

[p.636] sperm _____

_____

[p.637] cytoplasmic localization _____

_____

[p.637] blastula _____

_____

[p.637] blastocyst _____

_____

[p.638] cell differentiation _____

_____

Matching

Choose the most appropriate answer for each term.

1. ___oocyte [p.636]

2. ___sperm [p.636]

3. ___gray crescent [p.636]

4. ___cytoplasmic localization [p.637]

5. ___vegetal pole [p.637]

6. ___animal pole [p.637]

7. ___radial [p.637]

8. ___blastula [p.637]

9. ___incomplete [p.637]

10. ___rotational [p.637]

11. ___blastocyst [p.637]

12. ___identical twins [p.637]

13. ___fraternal twins [p.637]

A. A type of blastula resulting from rotational cleavage; outer cells become organized as a thin layer giving rise to the placenta and the inner cells mass on one side of a fluid-filled cavity to give rise to the embryo.

B. Arises from a splitting of the first two blastomeres of the inner cell mass; possess identical genes

C. In vertebrate eggs with polarity, the end closest to the nucleus; a frog egg is a good example.

D. Embryonic stage characterized by blastomeres and a fluid-filled cavity, the blastocoel

E. Consists of paternal DNA and cellular equipment enabling this cell to reach and penetrate an egg

F. An area of intermediate pigmentation and maternal messages in some animal eggs

G. Type of cleavage occurring in mammal eggs

H. By virtue of where they form, blastomeres end up with different maternal messages; helps seal the developmental fate of each cell's descendants.

I. Type of cleavage where the furrow runs horizontally and vertically with the animal-vegetal axis; typical of frog and sea urchin eggs

J. Term for an immature egg; contains regionally localized aspects known as "maternal messages"

K. Type of cleavage occurring in eggs with a large yolk volume; early divisions are restricted to a small caplike region near the animal pole; typical of reptiles, birds, and most fishes

L. Arise from two oocytes that matured and were fertilized during the same menstrual cycle

M. The yolk-rich end of an egg exhibiting polarity

Choice

For questions 14–27, choose from the following:

a. cell differentiation   b. morphogenesis

14. ___ Cells divide, grow, migrate, and change in size. [p.639]

15. ___ Migrating Schwann cells stick to adhesion proteins on the surface of axons but no blood vessels. [p.639]

16. ___ A cell selectively activates genes and synthesizes proteins not found in other cell types. [p.638]

17. ___ Cells "know" where to move because of responses to adhesive cues. [p.639]

18. ___ Apoptosis occurring in a developing body part [p.639]

19. ___ Transparent fibers in the human eye lens incorporate crystallin proteins produced by cells uniquely capable of activating the genes to produce them. [p.638]

20. ___ Refers to a program of orderly changes in an embryo's size, shape, and proportions [p.639]

21. __ Programmed cell death that helps sculpt body parts. [p.639]

22. __ Cells become specialized without loss of genetic information. [p.638]

23. __ In an African clawed frog, an intestinal cell nucleus transplanted into an enucleated unfertilized egg directs the development of a complete frog. [pp.638–639]

24. __ Cells send out and use pseudopods that move them along prescribed migration routes. [p.639]

25. __ From gastrulation onward, certain groups of genes are activated in some cells but not others. [p.638]

26. __ Through controlled, localized events, the size, shape, and proportion of body parts emerge. [p.639]

27. __ Elongation of ectodermal cells along an embryo's midline elongate to form a neural plate; microfilaments act to change cell shapes that result in neural tube formation. [p.639]

38.5. PATTERN FORMATION [pp.640–641]

Selected Words: *Drosophila* [p.640], *maternal effect* genes [p.640], *gap* genes [p.640], *pair-rule* genes [p.640], *segment polarity* genes [p.640], *homeotic* genes [p.640], apical ectodermal ridge [p.641], *physical* constraints [p.641], *architectural* constraints [p.641], *phyletic* constraints [p.641]

Boldfaced, Page-Referenced Terms

[p.640] theory of pattern formation _____

_____

[p.640] fate map _____

_____

[p.641] embryonic induction _____

_____

Fill-in-the-Blanks

Based upon studies of gene (1) _____ [p.640] in *Drosophila* and other organisms, researchers constructed a theory of (2) _____ _____ [p.640]. Here are its key points: during development, classes of (3) _____ [p.640] genes are activated in orderly sequence, at prescribed times. Interactions among the (3) genes are guided by (4) _____ [p.640] proteins. They result in the appearance of different gene products that are spatially organized relative to one another in the (5) _____ [p.640]. Different genes are activated and (6) _____ [p.640] along the embryo's anterior-posterior axis and dorsal-ventral axis. Certain protein products of this (7) _____ [p.640] gene expression diffuse through the embryo and create chemical (8) _____ [p.640] that help identify each cell's identity. (9) _____ [p.640] genes are a class of (10) _____ [p.640] genes that specify the development of specific body parts.

Matching

In *Drosophila*, specific types of genes act as master organizers that establish the location, shapes, and sizes of various body parts in the animal during pattern formation. Match each gene with its function.

11. ___gap genes [p.640]

12. ___homeotic genes [p.640]

13. ___maternal effect genes [p.640]

14. ___pair-rule genes [p.640]

15. ___segment polarity genes [p.640]

A. Genes for regulatory proteins that are transcribed, translated, or both; these products become localized in different parts of the egg cytoplasm and are activated in the zygote, where they activate or inhibit gap genes.

B. These genes are activated by differences in concentrations of gap gene products; products of these genes accumulate in bands.

C. Divide the embryo into segment-size units

D. Map out broad body regions

E. Control major developmental pathways; collectively govern the developmental fate of each body segment; activated by interactions of gap, pair-rule, and segment polarity genes

Fill-in-the-Blanks

Exposure to a gene product released from one tissue can change the developmental fate of an adjacent embryonic tissue. This effect is known as embryonic (16) _____ [p.641]. Experiments on chick wings demonstrated an interaction between (17) _____ [p.641] and (18) _____ [p.641] where a chick wing forms. In this case, (19) _____ [p.641] induces overlying ectodermal cells to elongate and form a narrow ridge, the (20) _____ _____ [p.641] ridge, or AER. The AER is a population of self-sustaining cells that do not mingle with cells around them. If AER is removed, development of limbs (21) _____ [p.641]. If extra AER is grafted above the forelimb mesoderm, (22) _____ [p.641] wing parts develop. If mesoderm from a leg bud is grafted under the normal wing AER, a leg forms. If nonwing mesoderm is grafted under the normal wing AER, AER regresses and the limb stops (23) _____ [p.641].

On the nature of induction, slowly degradable proteins known as (24) _____ [p.641] create key concentration gradients as they diffuse from an inducing tissue into adjoining tissues. The signals from (24) are strongest at the beginning of the gradients and weaken with (25) _____ [p.641]. Thus, cells at different positions along the gradient are exposed to different chemical information—which guides the (26) _____ _____ [p.641] of different genes in different parts of the embryo.

Morphogens and other inducers switch on blocks of genes in (27) _____ [p.641]. Their targets include (28) _____ [p.641] genes, which belong to animals as evolutionarily distant as roundworms and vertebrates. As organs are first forming, the gene products interact with (29) _____ [p.641] elements. Together, they activate and inhibit blocks of genes in similar ways among major animal groups. They direct the (30) _____ [p.641] that map out the overall body plan, including its major axes. If they fail to do so when organs are forming, major disasters follow—as when a heart gets constructed at the wrong location. Why do we find no more than a few dozen body plans among all animals? Apparently, the first stages of (31) _____ [p.641] are open to evolutionary change but early (32) _____ [p.641] formation is not. In addition to the known physical and architectural constraints, there are also (33) _____ [p.641] constraints on change. These are imposed on each lineage by master (34) _____ [p.641] genes that operate when organs first form and that govern induction of the basic body plan. Once a body part has been formed owing to their interactions, it's hard to start over again.

38.6. REPRODUCTIVE SYSTEM OF HUMAN MALES [pp.642–643]

38.7. MALE REPRODUCTIVE FUNCTION [pp.644–645]

Selected Words: *secondary* sexual traits [p.642], epididymis [p.642], vasa deferentia [p.642], prostaglandins [p.643], *prostate cancer* [p.643], *testicular cancer* [p.643], spermatogonia [p.644], *incomplete* cytoplasmic divisions [p.644], Leydig cells [p.644], Sertoli cells [p.645]

Boldfaced, Page-Referenced Terms

[p.642] testes (singular, testis) _____

_____

[p.642] seminiferous tubules _____

_____

[p.644] testosterone _____

_____

[p.645] LH _____

_____

[p.645] FSH _____

_____

Labeling

The numbered items in the following illustrations represent missing information; complete the blanks in the following narrative to supply that information. Some illustrated structures are numbered more than once to aid identification.

Within each testis and following repeated (1) _____ [p.645] divisions of undifferentiated diploid cells

just inside the (2) _____ [p.644] tubule walls, (3) _____ [p.645] occurs to form haploid, mature

(4) _____ [p.645]. Males produce sperm continuously from puberty onward. Sperm leaving a testis

enter a long coiled duct, the (5) _____ [p.644]; the sperm are stored in the last portion of this organ.

When a male is sexually aroused, muscle contractions quickly propel the sperm through a thick-walled tube,

the (6) _____ _____ [p.644], then to ejaculatory ducts and finally the (7) _____ [p.644], which

opens at the tip of the penis. During the trip to the urethra, glandular secretions become mixed with the

sperm to form semen. (8) _____ _____ [p.644] secrete fructose to nourish the sperm and

prostaglandins to induce contractions in the female reproductive tract. (9) _____ _____ [p.644]

secretions help neutralize vaginal acids. (10) _____ [p.644] glands secrete mucus to lubricate the penis,

aid vaginal penetration, and improve sperm motility.

head (DNA inside an enzyme-rich cap)   midpiece (with mitochondria)   tail (contains microtubules)

4

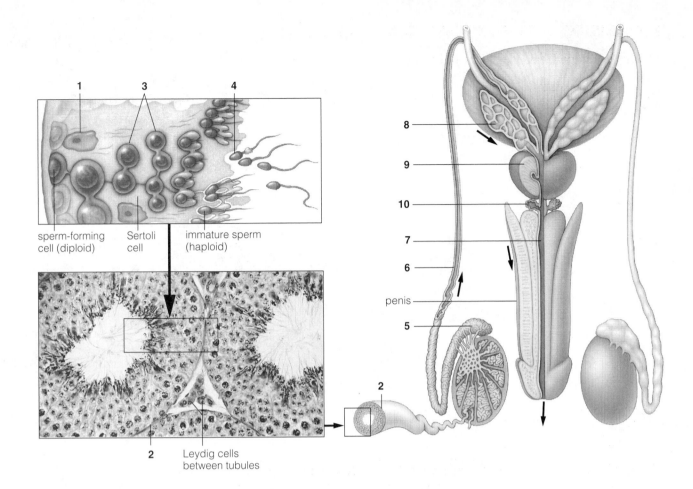

sperm-forming cell (diploid)

Sertoli cell

immature sperm (haploid)

penis

Leydig cells between tubules

Dichotomous Choice

Circle one of two possible answers given between parentheses in each statement.

11. Testosterone is secreted by (Leydig/hypothalamus) cells. [p.644]

12. (Testosterone/FSH) governs the growth, form, and functions of the male reproductive tract. [p.644]

13. Sexual behavior, aggressive behavior, and secondary sexual traits are associated with (LH/testosterone). [p.644]

14. LH and FSH are secreted by the (anterior/posterior) lobe of the pituitary gland. [p.645]

15. The (testes/hypothalamus) governs sperm production by controlling secretion of testosterone, LH, and FSH. [p.645]

16. When blood levels of testosterone (increase/decrease), the hypothalamus secretes GnRH that stimulates the pituitary to step up the release of LH and FSH, which travel the bloodstream to targets in the testes. [p.645]

17. Within the testes, (LH/FSH) acts on Leydig cells; in response, they secrete testosterone, which helps stimulate sperm formation and development. [p.645]

18. (Sertoli/Leydig) cells have receptors for FSH, which is necessary to initiate spermatogenesis at puberty. [p.645]

19. Feedback loops to the hypothalamus lead to (increased/decreased) testosterone secretion and sperm formation. [p.645]

20. A(n) (elevated/depressed) testosterone level in blood slows down the release of GnRH. [p.645]

21. (Leydig/Sertoli) cells release inhibin. This protein hormone acts on the hypothalamus and pituitary to cut back on the release of GnRH and FSH. [p.645]

Labeling

Identify each numbered part of the illustration at right.

22. _____ [p.645]

23. _____ _____ [p.645]

24. _____ _____ [p.645]

25. _____ _____ [p.645]

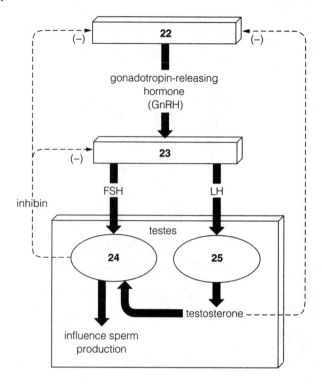

38.8. REPRODUCTIVE SYSTEM OF HUMAN FEMALES [pp.646–647]

38.9. FEMALE REPRODUCTIVE FUNCTION [pp.648–649]

38.10. VISUAL SUMMARY OF THE MENSTRUAL CYCLE [p.650]

Selected Words: primary reproductive organs [p.646], *follicular* phase [p.646], *luteal* phase [p.646], *menopause* [p.647], *endometriosis* [p.647]

Boldfaced, Page-Referenced Terms

[p.646] uterus _____

_____

[p.646] endometrium _____

_____

[p.646] menstrual cycle _____

_____

[p.646] ovulation _____

_____

[p.647] estrogens _____

_____

[p.647] progesterone _____

_____

[p.649] corpus luteum _____

_____

Fill-in-the-Blanks

The numbered items in the illustrations below represent missing information; complete the blanks in the following narrative to supply that information. Some illustrated structures are numbered more than once to aid identification.

An immature egg (oocyte) is released from one (1) _____ [pp.647–648] of a pair. From each ovary, a(n) (2) _____ [pp.647–648] forms a channel for transport of the immature egg to the (3) _____ [pp.647–648], a hollow, pear-shaped organ where the embryo grows and develops. The lower narrowed part of the uterus is the (4) _____ [p.647]. The uterus has a thick layer of smooth muscle, the (5) _____ [p.647], lined inside with connective tissue, glands, and blood vessels; this lining is called the (6) _____ [p.647]. The (7) _____ [pp.647–648], a muscular tube, extends from the cervix to the body surface; this tube receives sperm and functions as part of the birth canal. At the body surface are external genitals

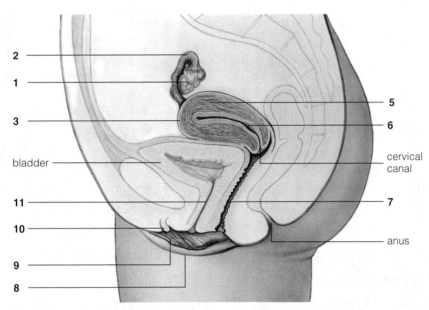

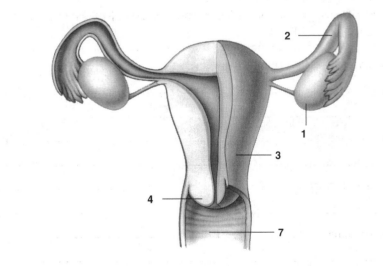

(vulva) that include organs for sexual stimulation. Outermost is a pair of fat-padded skin folds, the (8) _____ _____ [p.647]. Those folds enclose a smaller pair of skin folds, the (9) _____ _____ [p.647]. The smaller folds partly enclose the (10) _____ [p.647], an organ sensitive to stimulation. The location of the (11) _____ [p.647] is about midway between the clitoris and the vaginal opening.

Fill-in-the-Blanks

(12) _____ [p.648] occurs in the ovaries so that a normal female infant has about 2 million primary

oocytes with the division process halted in the (13) [choose one] ❏ I ❏ II [p.648] stage. By age seven, only

about (14) _____ [p.648] remain. A primary oocyte surrounded by a nourishing layer of granulosa cells

is called a(n) (15) _____ [p.648]. When a female enters puberty, at the start of a menstrual cycle, the

(16) _____ [p.648] secretes a hormone (GnRH) in amounts that cause the (17) _____ [p.648]

pituitary to step up its secretion of follicle-stimulating hormone (FSH) and luteinizing hormone (LH). The

blood concentration of these two hormones increases and is carried by the blood to all parts of the body.

That increase causes the (18) _____ [p.648] to grow. The (19) _____ [p.648] begins to increase in

size and more layers of cells form around it. (20) _____ [p.648] deposits accumulate between the oocyte

and the layers. In time, all the deposits form the (21) _____ _____ [p.648], a noncellular coating

around the oocyte.

(22) _____ [p.648] and (23) _____ [p.648] stimulate cells outside the zona pellucida to secrete

(24) _____ [p.648]. A(n) (24)-containing fluid accumulates in the follicle, and estrogen levels in

(25) _____ [p.648] begin to increase. About eight to ten hours before being released from the ovary, the

(26) _____ [p.648] completes meiosis I. And then its (27) _____ [p.648] divides, forming two cells.

One cell, the (28) _____ _____ [p.648], ends up with nearly all the cytoplasm. The other cell is

the first of three (29) _____ _____ [p.648]. The meiotic parceling of the chromosomes among all

four cells gives the secondary oocyte a (30) _____ [p.648] chromosome number—which is the exact

number required for (31) _____ [p.648] and for (32) _____ [p.648] reproduction.

About halfway through the menstrual cycle, the (33) _____ [p.648] gland detects the rise in the

blood level of estrogens. It responds with a brief outpouring of (34) _____ [p.648]. This (34) surge

causes rapid vascular changes that make the (35) _____ [p.648] swell quickly. The surge also induces

(36) _____ [p.648] to digest the bulging follicle wall. The weakened wall ruptures. Fluid escapes and

carries the (37) _____ _____ [p.648] with it. Thus, the midcycle surge of LH triggers

(38) _____ [p.648]—the release of a secondary oocyte from the ovary.

The estrogens released early in the menstrual cycle also help pave the way for a possible (39) _____

[p.649]. Estrogens stimulate growth of the (40) _____ [p.649] and its glands. Just before the midcycle

surge of LH, follicle cells start to secrete some (41) _____ [p.649] as well as estrogens. Blood vessels

grow rapidly in the thickened (42) _____ [p.649]. At (43) _____ [p.649], the estrogens act on tissue

around the (44) _____ [p.649] canal. The cervix starts to secrete large quantities of a thin, clear

(45) _____ [p.649], which is an ideal medium for sperm to swim through.

The midcycle surge of LH that triggers ovulation also induces the formation of a(n) (46) _____

_____ [p.649] in the ovary. This yellowish, glandular structure forms by cell differentiation among

(47) _____ [p.649] cells left behind in the follicle. Secretions from the (46) influence the rest of the cycle.

The corpus luteum secretes (48) _____ [p.649] and some estrogen. (48) prepares the reproductive

tract for the arrival of a(n) (49) _____ [p.649]. This hormone makes cervical (50) _____ [p.649] turn thick and sticky; it may keep normal bacterial inhabitants of the (51) _____ [p.649] out of the uterus. Progesterone also will maintain the (52) _____ [p.649] during pregnancy.

A corpus luteum persists for about (53) _____ [p.649] days. All the while, the (54) _____ [p.649] is calling for minimal FSH secretion, which stops other (55) _____ from developing. If a(n) (56) _____ [p.649] does not burrow into the endometrium, the corpus luteum will self-destruct during the last days of the cycle. It will secrete certain (57) _____ [p.649] that apparently can disrupt its own functioning.

Following this, (58) _____ [p.649] and (59) _____ [p.649] levels in blood decline rapidly, and so the (60) _____ [p.649] starts to break down. Blood as well as the sloughed endometrial tissues make up a(n) (61) _____ [p.649] flow, which continues for three to six days. Thus, the cycle begins anew. By (62) _____ [p.649], the supply of oocytes is dwindling, hormone secretions slow down, and in time menstrual cycles—and fertility—will be over.

38.11. PREGNANCY HAPPENS [p.651]

38.12. FORMATION OF THE EARLY EMBRYO [pp.652–653]

38.13. EMERGENCE OF THE VERTEBRATE BODY PLAN [p.654]

38.14. WHY IS THE PLACENTA SO IMPORTANT? [p.655]

38.15. EMERGENCE OF DISTINCTLY HUMAN FEATURES [pp.656–657]

38.16. *Focus on Health:* MOTHER AS PROVIDER, PROTECTOR, POTENTIAL THREAT [pp.658–659]

Selected Words: orgasm [p.651], *embryonic* period [p.652], *fetal* period [p.652], blastocyst [p.652], at-home *pregnancy tests* [p.653], *first, second, third* trimester [p.652], lanugo [p.657], *teratogens* [p.658], *rubella* [p.659], *thalidomide* [p.659], *anti-acne drugs* [p.659], *fetal alcohol syndrome* [p.659], *secondhand smoke* [p.659]

Boldfaced, Page-Referenced Terms

[p.651] ovum (plural, ova) _____

_____

[p.652] fetus _____

_____

[p.652] fetal period _____

_____

[p.652] implantation _____

_____

[p.653] amnion _____

_____

[p.653] yolk sac _____

_____

[p.653] chorion _____

_____

[p.653] allantois _____

_____

[p.654] somites _____

_____

[p.655] placenta _____

_____

Fill-in-the-Blanks

Following sexual intercourse, (1) _____ [p.651] are in the vagina. Fertilization most often takes place in the upper portion of the (2) _____ [p.651]. When living sperm contact an oocyte, they release (3) _____ [p.651] that clear a path through the zona pellucida. Usually only one sperm fuses with the secondary oocyte; only its nucleus and centrioles do not degenerate in the oocyte's cytoplasm. Sperm penetration induces the secondary oocyte and the first polar body to finish meiosis (4) _____ [p.651]. There are now three polar bodies and a mature egg, or (5) _____ [p.651]. As the sperm and egg nuclei fuse, their chromosomes restore the (6) _____ [p.651] number for a brand new zygote.

Matching

Choose the most appropriate answer for each term.

7. ___embryonic period [p.652]

8. ___fetus [p.652]

9. ___fetal period [p.652]

10. ___first trimester [p.652]

11. ___second trimester [p.652]

12. ___third trimester [p.652]

13. ___human blastocyst [p.652]

14. ___implantation [p.652]

A. Extends from the seventh month until birth
B. Extends from the start of the fourth month to the end of the sixth month
C. The blastocyst adheres to the uterine lining and sends out projections that invade the mother's tissues.
D. The time span from the third to the end of the eighth week of pregnancy
E. Defined by a trophoblast, a blastocoel, and an inner cell mass
F. The first three months of pregnancy
G. Term for the new, distinctly human individual at the end of the embryonic period
H. From the start of the ninth week until birth, organs enlarge and become specialized.

Sequence

Arrange the following human developmental stages of early human embryo formation in the proper chronological sequence (day 1 to day 14). Write the letter of the first stage next to 15. The letter of the final process is written next to 23.

15. ___   A. A blastocyst develops from the morula. [pp.652–653]
16. ___   B. The four-cell stage is completed by the second cleavage division
17. ___   to produce four cells. [p.652]
18. ___   C. Cleavage begins about 24 hours after fertilization. [p.652]
19. ___   D. A connecting stalk has formed between the embryonic disk and
20. ___   chorion; chorionic villi begin to form. [p.653]
21. ___   E. Blood-filled spaces begin to form in maternal tissue; the
22. ___   chorionic cavity starts to form. [p.653]
23. ___   F. Cleavage divisions have produced a ball of six to twelve cells. [p.652]
          G. The yolk sac, embryonic disk, and amniotic cavity have started to form. [p.653]
          H. A morula, a ball of sixteen to thirty-two cells is produced; the morula gives rise to the embryo
             and attached membranes. [p.652]
          I. The blastocyst begins to burrow into the endometrium. [pp.652–653]

Matching

Link each extra embryonic membrane with its function.

24. ___ allantois [p.653]   A. Outermost membrane; will become part of the spongy blood-engorged
                                  placenta
25. ___ amnion [p.653]   B. Directly encloses the embryo and cradles it in a buoyant protective fluid
26. ___ chorion [p.653]   C. In humans, some of this membrane becomes a site for blood cell formation,
                                  some will give rise to germ cells, the forerunners of gametes.
27. ___ yolk sac [p.653]   D. In humans, the urinary bladder as well as blood vessels form from it

Sequence

Arrange the following human developmental stages in the emerging vertebrate body plan in the proper chronological sequence (day 15 to days 24–25). Write the letter of the first stage next to 28, and so on.

28. ___ [p.654]   A. Morphogenesis occurs; the neural tube and somites form; mesoderm somites give rise
                     to the axial skeleton, skeletal muscles, and much of the dermis.

29. ___ [p.654]   B. Pharyngeal arches form to contribute to formation of the face, neck, mouth, nasal
                     cavities, larynx, and pharynx.

30. ___ [p.654]   C. The primitive streak forms to mark the onset of gastrulation in vertebrate embryos.

Labeling

Identify each numbered placental component in the illustration below.

31. _____ [p.655]

32. _____ _____ [p.655]

33. _____ [p.655]

34. _____ [p.655]

35. _____ [p.655]

36. _____ [p.655]

37. _____ _____ [p.655]

38. _____ [p.655]

39. _____ _____ [p.655]

40. _____ [p.655]

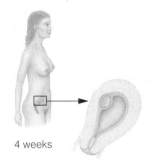

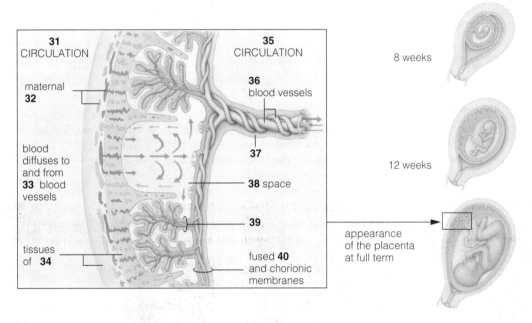

Sequence

Arrange the following human developmental stages demonstrating the emergence of distinctly human features in the proper chronological sequence (four weeks to full term or completion of the fetal period). Write the letter of the first stage next to 41. The letter of the final process is written next to 45.

41. ___ [p.656]   A. Movements begin as nerves make functional connections with developing muscles; legs kick, arms wave, fingers grasp, and the mouth puckers.

42. ___ [p.656]   B. The length of the fetus increases from 16 centimeters to 50 centimeters, and weight increases from about 7 ounces to 7.5 pounds.

43. ___ [p.656]   C. The human embryo has a tail and pharyngeal arches, and limbs; fingers and toes are sculpted from embryonic paddles; the circulatory system becomes more complex, and the head develops.

44. ___ [pp.656–657] D. The human embryo is distinctly human as compared to other vertebrate embryos; upper and lower limbs are well formed; fingers and then toes have separated; primordial tissues of all internal and external structures have developed; the tail has become stubby.

45. ___ [p.657]   E. Human features are visible at the boundary of the embryonic and fetal periods; the embryo floats in amniotic fluid and the chorion covers the amnion.

Labeling

Identify each numbered part of the accompanying illustrations.

46. _____ _____ [p.654]

47. _____ _____ [p.654]

48. _____ _____ [p.654]

49. _____ _____ [p.654]

50. _____ _____ [p.654]

51. _____ _____ [p.654]

52. _____ _____ [p.654]

53. _____ [p.654]

54. _____ _____ [p.654]

55. _____ [p.656]

56. _____ _____ [p.656]

57. _____ [p.656]

58. _____ [p.656]

59. _____ [p.656]

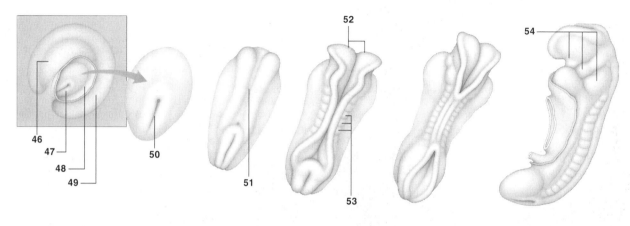

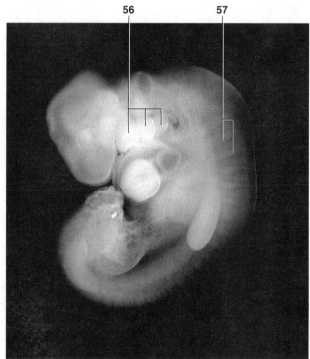

A human embryo at (**55**) weeks after conception.

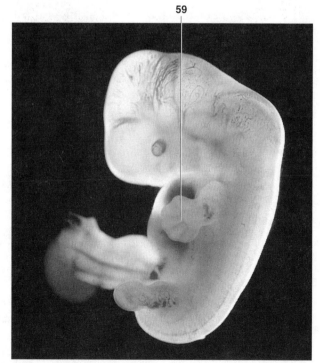

A human embryo at (**58**) weeks after conception.

Choice

For questions 60–70, choose from the following threats to human development:

a. nutrition   b. infections   c. prescription drugs   d. alcohol   e. cocaine
f. cigarette smoke   g. all of the preceding threats

60. ___ About 60–70 percent of newborns of these women have FAS. [p.659]

61. ___ In the case of rubella, there is a 50 percent chance some organs won't form properly. [pp.658–659]

62. ___ In cases of exposure to the secondhand type, children were smaller, died of more postdelivery complications, and had twice as many heart abnormalities. [p.659]

63. ___ Increasing B-complex intake of the mother before conception and during early pregnancy reduces the risk that an embryo will develop severe neural tube defects. [p.658]

64. ___ The "crack" type disrupts nervous system development of a child. [p.659]

65. ___ Teratogens [pp.658–659]

66. ___ Women who used the tranquilizer thalidomide during the first trimester gave birth to infants with missing or severely deformed arms and legs. [p.659]

67. ___ A pregnant woman must eat enough so that her body weight increases by 20–25 pounds, on average. [p.658]

68. ___ Facial deformities, poor coordination, and sometimes, heart defects. [p.659]

69. ___ As birth approaches, a fetus makes greater demands of this type from the mother that will influence the remaining developmental events. [p.658]

70. ___ Vaccination before pregnancy can prevent it. [pp.658–659]

38.17. FROM BIRTH ONWARD [pp.660–661]

38.18. CONTROL OF HUMAN FERTILITY [pp.662–663]

Selected Words: labor [p.660], relaxin [p.660], *breech birth* [p.660], *afterbaby blues* [p.660], *lactation* [p.660], *breast cancer* [p.661], *prenatal* [p.661], *postnatal* [p.661], *Werner's syndrome* [p.661], *Alzheimer's disease* [p.661], *abstinence* [p.662], *rhythm method* [p.662], *withdrawal* [p.662], *douching* [p.662], *vasectomy* [p.662], *tubal ligation* [p.662], *spermicidal foam* [p.663], *spermicidal jelly* [p.663], *IUDs* [p.663], *diaphragm* [p.663], *condoms* [p.663], *birth control pill* [p.663], *Depo-Provera* injection [p.663], *Norplant* [p.663], *morning-after pills* [p.663]

Boldfaced, Page-Referenced Terms

[p.661] aging _____

_____

Matching

Choose the most appropriate answer for each term.

1. ___labor [p.660]
2. ___breech birth [p.660]
3. ___afterbaby blues [p.660]
4. ___lactation [p.660]
5. ___breast cancer [p.661]
6. ___prenatal [p.661]
7. ___postnatal [p.661]
8. ___aging [p.661]
9. ___Werner's syndrome [p.661]
10. ___Alzheimer's disease [p.661]

A. The portion of an individual's lifespan before birth
B. An aging disorder correlated with three mutated genes; normal gene products are a part of a toxic waste removal system of the brain.
C. When any part of the fetus other than the head becomes positioned near the birth canal first
D. Result of collective processes of the breakdown of cell structure and function; leads to structural changes and gradual loss of body functions
E. Term for the birth process or delivery
F. Milk production occurring in the mammary glands in the mother's breasts
G. The portion of an individual's lifespan after birth
H. An aging disorder caused by a mutation of the gene coding for helicase, an enzyme effective in DNA replication and repair
I. Postpartum depression of new mothers
J. Obesity, high cholesterol, and high estrogen levels contribute to this undesirable cellular transformation.

Matching

Choose the most appropriate answer for each term.

11. ___abstinence [p.662]
12. ___rhythm method [p.662]
13. ___withdrawal [p.662]
14. ___douching [p.662]
15. ___vasectomy [p.662]
16. ___tubal ligation [p.662]
17. ___spermicidal foam and jelly [p.663]
18. ___diaphragm [p.663]
19. ___condoms [p.663]
20. ___birth control pill [p.663]
21. ___Depo-Provera and Norplant [p.663]
22. ___RU-486 [p.663]

A. A flexible, dome-shaped device inserted into the vagina and positioned over the cervix before intercourse; 84 percent effective
B. Progestin injections or implants that inhibit ovulation; 95–96 percent effective
C. A woman's oviducts are cauterized or cut and tied off; extremely effective
D. Rinsing the vagina with a chemical right after intercourse; next to useless
E. Thin, tight-fitting sheaths worn over the penis during intercourse; 85–93 percent reliable
F. Avoiding intercourse during a woman's fertile period; 74 percent effective
G. The morning-after pill that interferes with hormones that control events between ovulation and implantation; use is still controversial in the United States
H. Cutting and tying off each vas deferens of a man; extremely effective
I. An oral contraceptive made of synthetic estrogens and progestins that suppress oocyte maturation and ovulation; 94 percent effective
J. No sexual intercourse; foolproof
K. Removing the penis from the vagina before ejaculation; an ineffective method
L. Chemicals toxic to sperm are transferred from an applicator into the vagina just before intercourse; 82–83 percent effective

38.19. *Focus on Health:* SEXUALLY TRANSMITTED DISEASES [pp.664–665]

38.20. *Focus on Bioethics:* TO SEEK OR END PREGNANCY [p.666]

Selected Words: type II *Herpes* virus [p.664], *AIDS* [p.664], *behavioral* controls [p.664], *gonorrhea* [p.664], *syphilis* [p.665], *pelvic inflammatory disease* [p.665], *genital herpes* [p.665], type II *Herpes simplex* virus [p.665], *genital warts* [p.665], chlamydial nongonococcal urethritis [p.665], *in vitro fertilization* [p.666], *abortion* [p.666]

Boldfaced, Page-Referenced Terms

[p.664] sexually transmitted diseases, or STDs _____

_____

Matching

Match each of the following with *all* applicable diseases.

1. _____ Can severely damage the brain and spinal cord in ways leading to various forms of insanity and paralysis [p.665]

2. _____ Has no cure [pp.664–665]

3. _____ Can cause severe cramps, fever, vomiting, and sterility due to scarring and blocking of the oviducts [pp.664–665]

4. _____ Caused by a motile, corkscrew-shaped bacterium, *Treponema pallidum* [p.665]

5. _____ Infected women typically have miscarriages, stillbirths, or sickly infants [p.665]

6. _____ Caused by a bacterium with pili, *Neisseria gonorrhoeae* [pp.664–665]

7. _____ Caused by direct contact with the viral agent; about 25 million people in the United States infected by it [p.665]

8. _____ Produces a chancre (localized ulcer) 1–8 weeks following infection [p.665]

9. _____ Chronic infections by this can lead to cervical cancer [p.665]

10. _____ Can lead to lesions in the eyes that cause blindness in babies born to mothers with this disease [p.665]

11. _____ Acyclovir decreases the healing time and may also decrease the pain and viral shedding from the blisters [p.665]

12. _____ Eight weeks following infection, a flattened, painless chancre harbors many motile bacteria [p.665]

13. _____ May be cured by antibiotics but can be infected again [pp.664–665]

14. _____ Can be treated with tetracycline and sulfonamides [p.665]

15. _____ Generally preventable by correct condom usage [pp.664–665]

A. AIDS
B. Chlamydial infection
C. Genital herpes
D. Gonorrhea
E. Pelvic inflammatory disease
F. Syphilis

Fill-in-the-Blanks

(16) _____ _____ [p.666] fertilization means conception outside the body; this is an expensive, sometimes controversial method of assisting childless couples to have children. (17) _____ [p.666] is the dislodging and removal of the blastocyst, embryo, or fetus from the uterus. This is a very controversial method of dealing with unwanted pregnancies.

Self-Quiz

___ 1. The process of cleavage most commonly produces a(n) _____. [p.637]
   a. zygote
   b. blastula
   c. gastrula
   d. third germ layer
   e. organ

___ 2. The formation of three germ (embryonic) tissue layers occurs during _____. [p.634]
   a. gastrulation
   b. cleavage
   c. pattern formation
   d. morphogenesis
   e. neural plate formation

___ 3. Muscles differentiate from _____ tissue. [p.634]
   a. ectoderm
   b. mesoderm
   c. endoderm
   d. parthenogenetic
   e. yolky

___ 4. The gray crescent is _____. [p.636]
   a. formed where the sperm penetrates the egg
   b. part of only one blastomere after the first cleavage
   c. the yolky region of the egg
   d. where the first mitotic division begins
   e. formed opposite from where the sperm enters the egg

___ 5. The type of cleavage observed in mammal eggs is _____. [p.637]
   a. incomplete
   b. animal-vegetal axis
   c. radial
   d. rotational
   e. cytoplasmic localization

___ 6. Of the following, _____ is *not* an aspect of morphogenesis. [p.639]
   a. active cell migration
   b. expansion and folding of whole sheets of cells
   c. apoptosis
   d. shaping the proportion of body parts
   e. initiating production of a protein by some cells and not others

___ 7. The differentiation of a body part in response to signals from an adjacent body part is _____. [p.641]
   a. contact inhibition
   b. ooplasmic localization
   c. embryonic induction
   d. pattern formation
   e. none of the above

___ 8. A homeotic mutation _____. [p.640]
   a. may cause a leg to develop on the head where an antenna should grow
   b. may affect pattern formation
   c. affects morphogenesis
   d. may alter the path of development
   e. all of the above

For questions 9–12, choose from the following answers:
   a. Leydig cells
   b. seminiferous tubules
   c. vas deferens
   d. epididymis
   e. prostate

___ 9. The _____ connects a structure on the surface of the testis with the ejaculatory duct. [p.643]

___ 10. Testosterone is produced by the _____. [p.644]

___ 11. Meiosis occurs in the _____. [p.642]

___ 12. Sperm mature in the _____. [p.642]

For questions 13–15, choose from the following answers:

    a. blastocyst   d. oviduct
    b. allantois   e. cervix
    c. yolk sac

___13. The _____ lies between the uterus and the vagina. [p.647]

___14. The _____ is a pathway from the ovary to the uterus. [p.647]

___15. The _____ results from the process known as cleavage. [p.637]

For questions 16–20, choose from the following answers:

    a. AIDS   d. Gonorrhea
    b. Chlamydial infection   e. Syphilis
    c. Genital herpes

___16. _____ is a disease caused by a spherical bacterium (*Neisseria*), it is curable by prompt diagnosis and treatment. [pp.664–665]

___17. _____ is a disease caused by a spiral bacterium (*Treponema*) that produces a localized ulcer (a chancre). [p.665]

___18. _____ is an incurable disease caused by the human immunodeficiency virus. [p.664]

___19. _____ is a disease caused by an intracellular parasite that lives in the genital and urinary tracts; also causes NGU. [p.665]

___20. _____ is an extremely contagious viral infection that causes sores on the mouth area and genitals; acyclovir may be effective. [p.665]

Chapter Objectives/Review Questions

1. Understand how asexual reproduction differs from sexual reproduction. Know the advantages and problems associated with having separate sexes. [p.632]
2. Explain why evolutionary trends in many groups of organisms tend toward developing more complex, sexual strategies rather than retaining simpler, asexual strategies. [pp.632–633]
3. Describe early embryonic development and distinguish among the following: gamete formation, fertilization, cleavage, gastrulation, organ formation, and growth and tissue specialization. [pp.634–635]
4. Name each of the three embryonic tissue layers and the organs formed from each. [p.634]
5. Explain what causes polarity to occur during oocyte maturation in the mother, and state how polarity influences later development. [p.636]
6. By virtue of where they form, blastomeres end up with different maternal messages; this outcome of cleavage is known as _____ _____. [pp.636–637]
7. Cleavage patterns of major animal groups correlates with the amount and distribution of _____ in their eggs. [p.637]
8. Distinguish the animal pole from the vegetal pole of animal eggs. [p.637]
9. Describe *radial* cleavage, *incomplete* cleavage, and *rotational* cleavage; cite animal groups that exhibit each type of cleavage. [p.637]
10. The successive cuts of cleavage produce the embryonic stage known as a(n) _____. [p.637]
11. Tell how the origin of identical twins differs from that of fraternal twins. [p.637]
12. Define *differentiation*, and give one example of cells in a multicellular organism that have undergone differentiation. [p.638]
13. _____ refers to a program of orderly changes in an embryo's size, shape, and proportions. [p.639]
14. Be able to list three major activities occurring as morphogenesis occurs. [p.639]
15. Describe the mechanisms involved in the model of pattern formation utilizing fate maps, maternal effect genes, gap genes, pair-rule genes, segment polarity genes, and homeotic genes. [p.640]
16. Exposure to a gene product released from one tissue can change the developmental fate of an adjacent embryonic tissue; this effect is termed _____ _____. [p.641]
17. Name the primary male reproductive organs; list secondary sexual traits determined by these organs. [p.642]
18. Follow the path of a mature sperm from the seminiferous tubules to the urethral exit. List every structure encountered along the path, and state the contribution to the nurturing of the sperm. [pp.642–643]
19. Describe how a man examines himself for testicular cancer. [p.643]

20. Name the four hormones that directly or indirectly control male reproductive function. Diagram the negative feedback mechanisms that link the hypothalamus, anterior pituitary, and testes in controlling gonadal function. [pp.644–645]
21. Diagram the structure of a sperm, label its components, and state the function of each. [p.645]
22. Trace the path of a sperm from the urethral exit to the place where fertilization normally occurs. Mention in correct sequence all major structures of the female reproductive tract that are passed along the way, and state the principal function of each structure. [pp.644–647]
23. Distinguish the follicular phase of the menstrual cycle from the luteal phase, and explain how the two cycles are synchronized by hormones from the anterior pituitary, hypothalamus, and ovaries. [pp.646–650]
24. State which hormonal event brings about ovulation and which other hormonal events bring about the onset and finish of menstruation. [pp.648–649]
25. List the physiological factors that bring about erection of the penis during sexual stimulation and the factors that bring about ejaculation. [p.651]
26. List the similar events that occur in both male and female orgasm. [p.651]
27. Distinguish between the embryonic period and the fetal period. [p.652]
28. Describe the events that occur during the first month of human development; include the developed structures of the blastocyst as seen at about 15 days. [pp.652–653]
29. List and describe the four extraembryonic membranes and their functions. [p.653]
30. Characterize the human developmental events occurring from day 15 to day 25. [p.654]
31. The _____ is a blood-engorged organ of endometrial and extraembryonic membrane. [p.653]
32. Describe the characteristics of the developing human fetus during the second trimester. [pp.656–657]
33. The embryonic period of human development ends after the _____ week is over; the embryo is now clearly a human _____. [p.656]
34. Explain why the mother must be particularly careful of her diet, health habits, and life-style during the first trimester after fertilization (especially during the first six weeks). [pp.658–659]
35. Generally describe the process of labor or delivery. [p.660]
36. Milk production, or _____, occurs in mammary glands in the mother's breasts. [p.660]
37. An animal lifespan is divided into _____ (before birth) and _____ (after birth) stages. [p.661]
38. List the factors that bring about aging of an animal. [p.661]
39. Distinguish between Werner's syndrome and Alzheimer's disease. [p.661]
40. Describe two different types of sterilization. [p.662]
41. State which birth control offers protection against sexually transmitted diseases. [p.663]
42. Identify the three most effective birth control methods used in the United States and the four least effective birth control methods. [p.663]
43. For each STD described in *Focus on Health*, know the causative organism and the symptoms of the disease. [pp.664–665]
44. State the physiological circumstances that would prompt a couple to try in vitro fertilization. [p.666]

Integrating and Applying Key Concepts

1. If embryonic induction did not occur in a human embryo, how would the eye region appear? What would happen to the forebrain and epidermis? If controlled cell death did not happen in a human embryo, how would its hands appear? its face?
2. What rewards do you think a society should give a woman who has at most two children during her lifetime? In the absence of rewards or punishments, how can a society encourage women not to have abortions and yet ensure that the human birth rate does not continue to increase?

39

POPULATION ECOLOGY

Interactive Exercises

Tales of Nightmare Numbers [pp.670–671]

39.1. CHARACTERISTICS OF POPULATIONS [p.672]

39.2. *Focus on Science:* ELUSIVE HEADS TO COUNT [p.673]

39.3. POPULATION SIZE AND EXPONENTIAL GROWTH [pp.674–675]

Selected Words: pre-reproductive, reproductive, and *post-reproductive* ages [p.672], *habitat* [p.672], *crude* density [p.672], *Larrea* [p.672]

Boldfaced, Page-Referenced Terms

[p.671] ecology _____

_____

[p.672] demographics _____

[p.672] population size _____

_____

[p.672] age structure _____

_____

[p.672] reproductive base _____

_____

[p.672] population density _____

_____

[p.672] population distribution _____

_____

[p.673] capture-recapture method _____

_____

[p.674] immigration _____

_____

[p.674] emigration _____

_____

[p.674] migrations _____

_____

[p.674] zero population growth _____

_____

[p.674] per capita _____

_____

[p.674] net reproduction per individual per unit time, or r _____

_____

[p.675] exponential growth _____

_____

[p.675] doubling time _____

_____

[p.675] biotic potential _____

_____

Matching

Choose the most appropriate answer for each term.

1. ___demographics [p.672]
2. ___population size [p.672]
3. ___population density [p.672]
4. ___habitat [p.672]
5. ___population distribution [p.672]
6. ___age structure [p.672]
7. ___reproductive base [p.672]
8. ___crude density [p.672]
9. ___pre-reproductive, reproductive, and post-reproductive [p.672]
10. ___clumped dispersion [p.672]
11. ___nearly uniform dispersion [p.672]
12. ___random dispersion [p.672]

A. Includes pre-reproductive and reproductive age categories
B. The general pattern in which the individuals of the population are dispersed through a specified area
C. When individuals of a population are more evenly spaced than they would be by chance alone
D. The number of individuals in some specified area or volume of a habitat
E. Occurs only when individuals of a population neither attract nor avoid one another when conditions are fairly uniform through the habitat, and when resources are available all the time
F. The number of individuals in each of several to many age categories
G. The measured number of individuals in a specified area
H. The number of individuals that contribute to a population's gene pool
I. The type of place where a species normally lives
J. Categories of a population's age structure
K. The vital statistics of a population
L. Individuals of a population form aggregations at specific habitat sites; most common dispersion pattern

Fill-in-the-Blanks

Population size increases as a result of births and (13) _____ [p.674]—the arrival of new residents from other populations of the species. Population size decreases as a result of deaths and (14) _____ [p.674], whereby individuals permanently move out of the population. Population size also changes on a predictable basis as a result of daily or seasonal events called (15) _____ [p.674]. However, this is a recurring round trip between two areas, so these transient effects need not be initially considered. If it is assumed that immigration is balancing emigration over time, the effects of both on population size may be ignored. This allows the definition of (16) _____ _____ [p.674] growth as an interval during which the number of births is balanced out by the number of deaths. The population size has no overall increase or decrease.

Births, deaths, and other variables that might affect population size can be measured in terms of (17) _____ _____ [p.674] rates, or rates per individual. *Capita* means (18) _____ [p.674], as in head counts. Visualize 5,000 mice living in a large cornfield. If the female mice collectively give birth to 2000 mice per month, the birth rate would be (19) _____ [p.674] per mouse, per month. If 500 of the 5,000 die during the month, the death rate would be (20) _____ [p.674] per mouse per month.

If it is assumed the birth rate and death rate remain constant, both can be combined into a single variable—the net (21) _____ [p.674] per individual per unit (22) _____ [p.674], or *r* for short. For the mouse population in the example above, the value of *r* is (23) _____ [p.674] per mouse per month. This example provides a method of representing population growth by the equation, (24) _____ [p.674].

If the monthly increases in the individuals in a population are plotted against time, one finds a graph line in the shape of a(n) (25) _____ [p.675]. Then one knows that the (26) _____ [p.675] growth is

being tracked. The length of time it takes for a population to double in size is its (27) _____ [p.675]
time. When a population displays its (28) _____ _____ [p.674] in terms of growth, it is the
maximum rate of increase per individual under ideal conditions.

Problems

For questions 29–33, consider the equation $G = rN$, where G = the population growth rate per unit time, r = the
net population growth rate per individual per unit time, and N = the number of individuals in the population.

29. Assume that r remains constant at 0.2.

    a. As the value of G increases, what happens to the value of N? [pp.674–675]

    _____

    b. If the value of G decreases, what happens to the value of N? [pp.674–675]

    _____

    c. If the net reproduction per individual stays the same and the population grows faster, then what
       must happen to the number of individuals in the population? [pp.674–675]

    _____

30. If a society decides it is necessary to lower its value of N through reproductive means because
    supportive resources are dwindling, it must lower either its net reproduction per individual per unit or
    its [pp.674–675] _____.

31. The equation $G = rN$ expresses a direct relationship between G and $r \times N$. If G remains constant and N
    increases, what must the value of r do? (In this situation, r varies inversely with N.) [pp.674–675]

    _____

32. Look at line (a) in the graph at the right. After seven hours have elapsed, approximately how many
    individuals are in the population? [pp.674–675]

    _____

33. Look at line (b) in the same graph.

    a. After 24 hours have elapsed, approximately how many
       individuals are in the population? [pp.674–675]

    _____

    b. After 28 hours have elapsed, approximately how many
       individuals are in the population? [pp.674–675]

    _____

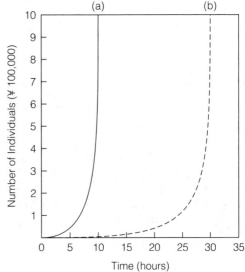

39.4. LIMITS ON THE GROWTH OF POPULATIONS [pp.676–677]

39.5. LIFE HISTORY PATTERNS [pp.678–679]

39.6. *Focus on Science:* NATURAL SELECTION AND THE GUPPIES OF TRINIDAD [p.679]

Selected Words: *sustainable* supply of resources [p.676], *bubonic plague* [p.677], *pneumonic plague* [p.677], *Yersinia pestis* [p.677], cohort [p.678], *Type I* curves [p.678], *Type II* curves [p.678], *Type III* curves [p.678], pike-cichlid [p.679]

Boldfaced, Page-Referenced Terms

[p.676] limiting factor _____

_____

[p.676] carrying capacity _____

_____

[p.676] logistic growth _____

_____

[p.677] density-dependent control _____

_____

[p.677] density-independent factors _____

_____

[p.678] life history pattern _____

_____

[p.678] survivorship curve _____

_____

Fill-in-the-Blanks

If (1) _____ [p.676] factors (essential resources in short supply) act on a population, population growth tapers off. (2) _____ _____ [p.676] refers to the maximum number of individuals of a population that can be sustained indefinitely by the environment. S-shaped growth curves are characteristic of (3) _____ [p.676] population growth. The plot of (3) growth levels off once the (4) _____ _____ [p.676] is reached. In the equation $G = r_{max}N[(K - N)/K]$, as the value of N approaches the value of K, and K and r_{max} remain constant, the value of G (5) [choose one] ❑ increases ❑ decreases ❑ cannot be determined by humans, even if they know algebra [pp.676–677]. In an overcrowded population, predators, parasites, and disease agents serve as (7) _____-_____ [p.677] controls. When an event such as a freak summer snowstorm in the Colorado Rockies causes more deaths or fewer births in a butterfly population (with no regard to crowding or dispersion patterns), the controls are said to be (8) _____-_____ [p.677] factors. Food availability is a density-(9) _____ [p.677] factor that works to cut back population size when it approaches the environment's (10) _____ _____ [p.677]. Environmental disruptions such as forest fires and floods are density-(11) _____ [p.677] factors that may push a population above or below its tolerance range for a given variable.

The term (12) _____ _____ [p.678] pattern refers to the individuals of a species in a population that exhibit particular morphological, physiological, and behavioral traits that are adaptive to different conditions at different times in the life cycle. (13) Life and health _____ [p.678] companies use information summarized from life tables, which show trends in mortality and life expectancy. A(n) (14) _____ [p.678] is a group of individuals that is tracked from the time of birth until the last one dies. A(n) (15) _____ [p.678] table lists the completed data on a population's age-specific death schedule. Such tables can be converted to more cheery (16) "_____" schedules, which list the number of individuals that reach some specified age (x). (17) _____ [p.678] curves are the graph lines that emerge when ecologists plot the age-specific survival of a cohort in a particular habitat. Type (18) _____ [p.678] survivorship curves illustrate populations having low survivorship early in life. Organisms whose life cycles reflect high survivorship until fairly late in life with a later large increase in deaths have a type (19) _____ [p.678] survivorship curve. Organisms such as lizards, small mammals, and some songbirds, are just as likely to be killed or die of disease at any age and reflect the type (20) _____ [p.678] survivorship curve. Provided people have access to good health care, human populations typically show a type (21) _____ [p.678] survivorship curve.

Matching

Choose the most appropriate answer(s) for each phrase. A letter may be used more than once, and a blank may contain more than one letter.

22. _____ cohort example [p.678]

23. _____ example(s) of density-independent controls [p.677]

24. _____ example(s) of density-dependent controls [p.677]

25. _____ selective effects of predation by killifish and pike cichlids [p.679]

A. Drought, floods, earthquakes
B. Bubonic plague and pneumonic plague
C. Food availability
D. Adrenal enlargement in wild rabbits in response to crowding
E. Heavy applications of pesticides in your backyard
F. All of the 1987 human babies of New York City
G. Natural selection in the life history patterns of Trinidadian guppies
H. Freak snowstorms in the Rocky Mountains

39.7. HUMAN POPULATION GROWTH [pp.680–681]

39.8. CONTROL THROUGH FAMILY PLANNING [pp.682–683]

39.9. POPULATION GROWTH AND ECONOMIC DEVELOPMENT [pp.684–685]

39.10. SOCIAL IMPACT OF NO GROWTH [p.685]

Selected Words: Homo habilis [p.681], *Vibrio cholerae* [p.681], *baby-boomers* [p.683], *preindustrial* stage [p.684], *transitional* stage [p.684], *industrial* stage [p.684], *postindustrial* stage [p.684], *postpone* [p.685]

Boldfaced, Page-Referenced Terms

[p.682] total fertility rate _____

_____

[p.684] demographic transition model _____

_____

Graph Construction

1. Construct a graph of the following data in the space provided on the next page.

Year	Estimated World Population
1650	500,000,000
1850	1,000,000,000
1930	2,000,000,000
1975	4,000,000,000
1986	5,000,000,000
1993	5,500,000,000
1995	5,700,000,000
1997	5,800,000,000
2050 (projected)	9,000,000,000+

a. Estimate the year that the world contained 3 billion humans. _____(Consult your graph.)
b. Estimate the year that Earth will house 8 billion humans. _____ (Consult your graph.)
c. Do you expect Earth to house 8 billion humans within your lifetime? _____ (Consult your graph.)

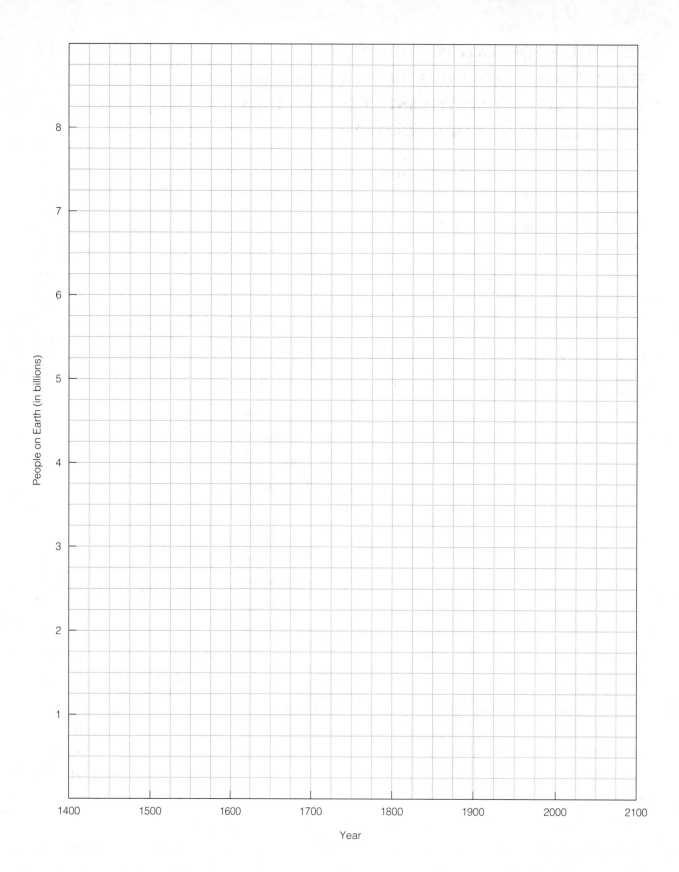

True–False

If the statement is true, write a T in the blank. If the statement is false, make it correct by changing the underlined word(s) and writing the correct word(s) in the answer blank.

_____2. In 1998, the human population approached 8 billion. [p.680]

_____3. Even if we could double our present food supply, death from starvation could still reach 20 million to 40 million people a year. [p.680]

_____4. Compared with the geographic spread of other organisms, it has taken the human population an extremely long period of time to expand into diverse environments. [p.680]

_____5. Managing food supplies through agriculture has had the effect of increasing the carrying capacity for the human population. [p.680]

_____6. By bringing many disease agents under control and by tapping into concentrated existing forms of energy, humans utilized factors that had previously limited their population growth. [p.681]

_____7. The stupendously accelerated growth of the human population continues and can be sustained indefinitely. [p.681]

Fill-in-the-Blanks

If the annual current average population growth rate of 1.47 percent is maintained for the human

population, there may be more than (8) _____ [p.682] billion people by the year 2050. Large efforts will

have to be made to increase the amount of (9) _____ [p.682] to sustain that many people. Such gross

manipulation of resources is likely to intensify (10) _____ [p.682], which almost certainly will adversely

affect our water supplies, the atmosphere, and productivity on land and in the seas. (11) _____

_____ [p.682] programs educate individuals about choosing how many children they will have, and

when. Carefully developed and administered programs may bring about a long-term decline in

(12) _____ [p.682] rates. To arrive at zero population growth, the average "replacement rate" would

have to be slightly higher than (13) _____ [p.682] children per couple, for some (14) _____ [p.682]

children die before reaching reproductive age. A more useful measure of global population trends is the

total (15) _____ [p.682] rate. This is defined as the average number of children born to women during

their reproductive years, as estimated on the basis of current age-specific birth rates. In 1997, the average

rate was (16) _____ [p.682] children per woman, an impressive decline from 1950, when the rate was

(17) _____ [p.682]. In the United States, a cohort of 78 million (18) _____-_____ [p.682] is

being tracked since it formed in 1946 when soldiers returned home from World War II. Although the

United States has a narrow reproductive base, more than (19) _____-_____ [p.683] of the world

population falls in the broad pre-reproductive base. One way to slow growth is for women to bear children

during their early (20) _____ [p.683], rather than in the mid-teens or early twenties. (21) _____

[p.683] is the country described to have the world's most extensive family planning program. Family

planning programs on a global scale are designed to help (22) _____ [p.683] the size of the human

population. Even if the human population reaches a level of zero growth, it will continue to grow for sixty

years, for its (23) _____ [p.683] base already consists of a staggering number of individuals.

Matching

Choose the most appropriate country examples to match with each population age structure.

24. ___zero growth [p.683]

25. ___rapid growth (more than a third of the world population) [p.683]

26. ___negative growth [p.683]

27. ___slow growth [p.683]

A. United States, Canada
B. Germany, Hungary
C. Kenya, Nigeria, Saudi Arabia
D. Denmark, Italy

Sequence

Arrange the following stages of the demographic transition model in correct chronological sequence. Write the letter of the first step next to 28, the letter of the second step next to 29, and so on.

28. ___   A. Industrial stage: population growth slows and industrialization is in full swing [p.684]

29. ___   B. Preindustrial stage: harsh living conditions, high birth rates and low death rates, slow population growth [p.684]

30. ___   C. Postindustrial stage: zero population growth is reached; birth rate falls below death rate, and population size slowly decreases [p.684]

31. ___   D. Transitional stage: industrialization begins, food production rises, and health care improves; death rates drop, birth rates remain high to give rapid population growth [p.684]

Fill-in-the-Blanks

Today, the United States, Canada, Australia, Japan, nations of the former Soviet Union, and most countries of western Europe are in the (32) _____ stage [p.684]. Their growth rate is slowly (33) [choose one] ❐ increasing ❐ decreasing [p.684]. In Germany, Bulgaria, Hungary, and some other countries, the death rates exceed the birth rates, and the populations are getting (34) _____ [p.684]. Mexico and other less developed countries are in the (35) _____ [p.684] stage. In many countries, population growth outpaces economic growth, death rates will increase, and they may be stalled in the (36) _____ stage [p.684]. Enormous disparities in (37) _____ [p.684] development are driving forces for immigration and emigration. Claiming that population growth affects economic health, many governments restrict (38) _____ [p.684].

India has a whopping (39) _____ [p.685] percent of the human population. By comparison, the Unites States has merely (40) _____ [p.685] percent. The highly industrialized United States produces (41) _____ [p.685] percent of all the world's goods and services. It uses (42) _____ [p.685] percent of the world's processed minerals and available, nonrenewable sources of energy. Its people consume (43) _____ [p.685] times more goods and services than the average person in India. The United States generates at least (44) _____ [p.685] percent of the global pollution and trash. By contrast, India produces about (45) _____ [p.685] percent of all goods and services and uses (46) _____ [p.685] percent of the available minerals and nonrenewable energy resources. India generates only about (47) _____ [p.685] percent of the pollution and trash. Tyler Miller, Jr., estimated that it would take (48) _____ [p.685] billion impoverished individuals in India to have as much environmental impact as (49) _____ [p.685] million Americans.

For all species on our planet, the biological implications of extremely rapid (50) _____ [p.685] are truly staggering. Yet so are the (51) _____ [p.685] implications of what will happen when and if the human population declines to the point of zero population growth—and stays there. If the population ever does reach and maintain zero growth over time, a larger proportion of the individuals will end up in the (52) _____ [p.685] age brackets. To care for older, nonproductive citizens, fewer productive ones must carry more and more of the (53) _____ [p.685] burden. Through our special abilities to undergo rapid cultural evolution, we have (54) _____ [p.685] the action of most of the factors that limit growth. No amount of (55) _____ [p.685] intervention can repeal the ultimate laws governing population growth, as imposed by the (56) _____ _____ [p.685] of the environment.

Self-Quiz

___ 1. The number of individuals that contribute to a population's gene pool is _____. [p.672]
   a. the population density
   b. the population growth
   c. the population birth rate
   d. the population size

___ 2. The number of individuals in a given area or volume of a habitat is _____. [p.672]
   a. the population density
   b. the population growth
   c. the population birth rate
   d. the population size

___ 3. A population that is growing exponentially in the absence of limiting factors can be illustrated accurately by a(n) _____. [p.675]
   a. S-shaped curve
   b. J-shaped curve
   c. curve that terminates in a plateau phase
   d. tolerance curve

___ 4. How are the individuals in a population most often dispersed? [p.672]
   a. clumped
   b. very uniform
   c. nearly uniform
   d. random

___ 5. Assuming immigration is balancing emigration over time, _____ may be defined as an interval in which the number of births is balanced by the number of deaths. [p.674]
   a. the lack of a limiting factor
   b. exponential growth
   c. saturation
   d. zero population growth

___ 6. Assuming the birth and death rate remain constant, both can be combined into a single variable, r, or _____. [p.674]
   a. the per capita rate
   b. the minus migration factor
   c. exponential growth
   d. the net reproduction per individual per unit time

___ 7. _____ is a way to express the growth rate of a given population. [p.675]
   a. Doubling time
   b. Population density
   c. Population size
   d. Carrying capacity

___ 8. Any population that is not restricted in some way will grow exponentially _____. [p.675]
   a. except in the case of bacteria
   b. irrespective of doubling time
   c. if the death rate is even slightly greater than the birth rate
   d. all of the above

___ 9. The maximum number of individuals of a population (or species) that a given environment can sustain indefinitely defines _____. [p.676]
   a. the carrying capacity of the environment
   b. exponential growth
   c. the doubling time of a population
   d. density-independent factors

___10. In natural communities, some feedback mechanisms operate whenever populations change in size; they are _____. [p.677]
   a. density-dependent factors
   b. density-independent factors
   c. always intrinsic to the individuals of the community
   d. always extrinsic to the individuals of the community

___11. _____ are any essential resources that are in short supply for a population. [p.676]
   a. Density-independent factors
   b. Extrinsic factors
   c. Limiting factors
   d. Intrinsic factors

___12. Which of the following is *not* characteristic of logistic growth? [p.676]
   a. S-shaped curve
   b. leveling off of growth as carrying capacity is reached
   c. unrestricted growth
   d. slow growth of a low-density population followed by rapid growth

___13. The population growth rate (G) is equal to the _____ net population growth rate per individual (r) and number of individuals (N). [p.674]
   a. sum of
   b. product of
   c. doubling of
   d. difference between

___14. $G = r_{max} N (K - N/K)$ represents _____. [p.676]
   a. exponential growth
   b. population density
   c. population size
   d. logistic growth

___15. The beginning of industrialization, a rise in food production, improvement of health care, rising birth rates, and declining death rates describes the _____ stage of the demographic transition model. [p.684]
   a. preindustrial
   b. transitional
   c. industrial
   d. postindustrial

___16. A group of individuals that is typically tracked from the time of birth until the last species dies is a _____. [p.678]
   a. cohort
   b. species
   c. type I curve group
   d. density-independent group

___17. The survivorship curve typical of industrialized human populations is type _____. [p.678]
   a. I
   b. II
   c. III
   d. none of the above types

Chapter Objectives/Review Questions

1. Be able to define: *ecology, demographics, population, habitat, population size, population density, population distribution, age structure,* and *reproductive base.* [pp.671,672]
2. In a population, the pre-reproductive ages and the reproductive ages together are counted as the _____ _____. [p.672]
3. List and describe the three patterns of dispersion illustrated by populations in a habitat. [p.672]
4. Distinguish immigration from emigration; define *migration.* [p.674]
5. Define *zero population growth* and describe how achieving it would affect population size. [p.674]
6. In the equation $G = rN$, as long as r holds constant, any population will show _____ growth. [p.675]

7. Calculate a population growth rate (*G*); use values for birth, death, and number of individuals (*N*) that seem appropriate. [p.674]
8. State how increasing the death rate of a population affects its doubling time. [p.675]
9. _____ _____ is the maximum rate of increase per individual under ideal conditions. [p.675]
10. The _____ rate of increase in population growth depends on the age at which each individual reproduces, and how many offspring are produced. [p.675]
11. List several examples of limiting factors and explain how they influence population curves. [p.676]
12. Explain the phrase "exceeding the carrying capacity," as applied to a population of organisms in an environment. [p.676]
13. Understand the meaning of the logistic growth equation and know how to calculate values of *G* by using the logistic growth equation. Understand the meaning of r_{max} and *K*. [pp.676–677]
14. Contrast the conditions that promote J-shaped growth curves with those that promote S-shaped curves in populations. [pp.675–676]
15. Define *density-dependent growth* of populations; cite one example. [p.677]
16. Define *density-independent factors* and list two examples; indicate how such factors affect populations. [p.677]
17. Most species have a(n) _____ _____ pattern in that its individuals exhibit particular morphological, physiological, and behavioral traits that are adaptive to different conditions at different times in the life cycle. [p.678]
18. Life insurance companies and ecologists track a(n) _____, a group of individuals from the time of birth until the last one dies. [p.678]
19. Understand the significance and use of life tables; be able to list and interpret the three survivorship curves. [p.678]
20. Explain how the construction of life tables and survivorship curves can be useful to humans in managing the distribution of scarce resources. [p.678]
21. Guppy populations targeted by killifish tend to be larger, less streamlined, and more brightly colored and guppy populations targeted by pike-cichlids tend to be smaller, more streamlined, and duller in color patterning. Other life history pattern differences exist between the two groups. After consideration of the research results obtained by Reznick and Endler, provide an explanation for these differences. [p.679]
22. Be able to list three possible reasons that growth of the human population is out of control. [p.680]
23. Most governments are trying to lower birth rates by _____ _____ programs. [p.682]
24. Define *total fertility rate*. [p.682]
25. What is the significance of the cohort of 78 million baby boomers? [p.683]
26. List and describe the four stages of the demographic transition model. [p.684]
27. Differences in population growth among countries correlate with levels of _____ development. [p.684]
28. Be able to generally compare the implications of the resource consumptions of India and the United States. [p.685]
29. In the final analysis, no amount of _____ intervention can repeal the ultimate laws governing population growth, as imposed by the _____ _____ of the environment. [p.685]

Integrating and Applying Key Concepts

Assume that the world has reached zero population growth. The year is 2110, and there are 10.5 billion individuals of *Homo pollutans* on Earth. You have seen stories on the community television screen about how people used to live 120 years ago. List the ways that life has changed, and comment on the events that no longer happen because of the enormous human population.

40

COMMUNITY INTERACTIONS

Interactive Exercises

No Pigeon Is an Island [pp.688–689]

40.1. WHICH FACTORS SHAPE COMMUNITY STRUCTURE? [p.690]

40.2. MUTUALISM [p.691]

Selected Words: *partition* the fruit supply [p.689], *fundamental* niche [p.690], *realized* niche [p.690], *neutral* relationship [p.690], *obligatory* mutualism [p.691]

Boldfaced, Page-Referenced Terms

[p.690] habitat _____

_____

[p.690] community _____

_____

[p.690] niche _____

_____

[p.690] commensalism _____

_____

[p.690] mutualism _____

_____

[p.690] interspecific competition _____

_____

[p.690] predation _____

_____

[p.690] parasitism _____

_____

[p.690] symbiosis _____

_____

Fill-in-the-Blanks

The type of place where you will normally find an organism is its (1) _____ [p.690]. The populations of all species in a habitat associate with one another as a(n) (2) _____ [p.690]. The (3) _____ [p.690] of an organism is defined by its "profession" in a community, in the sum of activities and relationships in which it engages to secure and use the resources necessary for its survival and reproduction. The (4) _____ [p.690] niche is the one that could prevail in the absence of competition and other factors that could constrain its acquisition and use of resources. The (5) _____ [p.690] niche is one that is more constrained and that shifts in large and small ways over time, as individuals of the species respond to a mosaic of changes. Each species has a(n) (6) _____ [p.690] relationship with most species in its habitat; that is, they do not affect each other (7) _____ [p.690]. Tree-roosting birds are (8) _____ [p.690] with trees; the trees get nothing but are not harmed. With (9) _____ [p.690], benefits flow both ways between the interacting species. The flowering plants and their pollinators form (10) _____ [p.690] relationships from which both participating populations benefit (two-way exploitation). With (11) _____ [p.690] competition, disadvantages flow both ways between species. (12) _____ [p.690] and (13) _____ [p.690] are interactions that directly benefit one species and directly hurt the other. Commensalism, mutualism, and parasitism are all forms of (14) _____ [p.690]; it means "living together."

Complete the Table

15. Complete the following table to describe how each of the organisms listed is intimately dependent on the other for survival and reproduction in an obligatory mutualistic symbiotic interaction.

Organism	Dependency
[p.691] a. Yucca moth	
[p.691] b. Yucca plant	

40.3. COMPETITIVE INTERACTIONS [pp.692–693]
40.4. PREDATION AND PARASITISM [pp.694–695]
40.5. *Focus on the Environment:* THE COEVOLUTIONARY ARMS RACE [pp.696–697]

Selected Words: *intraspecific* competition [p.692], *interspecific* competition [p.692], *Paramecium* [p.692], *Pisaster* [p.693], *Mytilus* [p.693], *Littorina littorea* [p.693], *carrying capacity* [p.694], *three-level* interaction [p.695], *ecto*parasites [p.695], *endo*parasites [p.695], *biological controls* [p.695], *Lithops* [p.696], *Dendrobates* [p.696], *Ranunculus* [p.697]

Boldfaced, Page-Referenced Terms

[p.692] competitive exclusion _____

_____

[p.692] keystone species _____

_____

[p.693] resource partitioning _____

_____

[p.694] predators _____

_____

[p.694] prey _____

_____

[p.694] parasites _____

_____

[p.694] hosts _____

_____

[p.694] coevolution _____

_____

[p.695] microparasites _____

_____

[p.695] macroparasites _____

_____

[p.695] social parasites _____

_____

[p.696] camouflage _____

_____

[p.696] warning coloration _____

_____

[p.696] mimicry _____

_____

[p.696] moment-of-truth defenses _____

_____

Fill-in-the-Blanks

(1) _____ [p.692] competition means individuals of the same species compete with one another.

(2) _____ [p.692] competition occurs between populations of different species. (3) _____ [p.692] competition is usually the least intense of the two types. According to the concept of (4) _____ _____ [p.692], two species that require identical resources cannot coexist indefinitely. A(n) (5) _____ [p.692] species is a dominant species that dictates community structure. *Pisaster*, a sea star, is a(n) (6) _____ [p.693] species that controls abundances of mussels, limpets, chitons, and barnacles. If sea stars are removed from experimental plots, (7) _____ [p.693], the main prey of sea stars, become the strongest competitors; they crowd out seven other species of invertebrates. When sea stars are removed, the community shrinks from fifteen species to eight.

In (8) _____ _____ [p.693], competing species coexist by subdividing a category of similar resources. A good example is the nine species of fruit-eating (9) _____ [p.693] in the same New Guinea forest. They all require the same resource: fruit. The pigeons overlap only slightly in their use of the resource, because each species specializes in fruits of a particular (10) _____ [p.693]. Another example is of three species of annual plants, each adapted to exploiting a different portion of the habitat. Each has a(n) (11) _____ [p.693] system that grows to a different depth in the soil than the others.

Matching

Choose the most appropriate answer for each term.

12. ___ a cycle turning on plants, herbivores, and carnivores [p.694]

13. ___ ectoparasites [p.695]

14. ___ endoparasites [p.695]

15. ___ microparasites [p.695]

16. ___ macroparasites [p.695]

17. ___ social parasites [p.695]

18. ___ biological controls [p.695]

A. A sometimes risky alternative to chemical pesticides
B. Bacteria, viruses, protozoans, and sporozoans; microscopically small, reproduce rapidly
C. Live on a host's surface
D. Complete their life cycle by drawing on social behaviors of another species; the cowbird is a good example
E. Canadian lynx and snowshoe hare
F. Large, includes many flatworms, roundworms (nematodes), and arthropods such as fleas and ticks
G. Live inside a host's body; some species live on or in one or more hosts for their entire life cycle

Fill-in-the-Blanks

Organisms that exhibit (19) _____ [p.696] have adaptations in form, patterning, color, and behavior that help them blend in with their surroundings and escape detection. Toxic types of prey often have (20) _____ _____ [p.696], or conspicuous patterns and colors that predators learn to recognize as "avoid me" signals. (21) _____ [p.696] is a term applied to prey organisms that bear close resemblance to dangerous, unpalatable, or hard-to-catch species. (22) _____-of-_____ [p.696] defenses refers to last-ditch tricks used by prey species for their survival as they are cornered or under attack. Predators may counter prey defenses with their own marvelous (23) _____ [p.697].

Matching

Match each of the following descriptions with the most appropriate answer. The same letter may be used more than once. Use only one letter per blank.

24. ___Cornered earwigs, skunks, and stink beetles producing awful odors [p.696]

25. ___Pisaster, a keystone species, controls abundances of mussels, limpets, chitons, and barnacles [p.693]

26. ___Different species of New Guinea pigeons coexisting in the same environment by eating fruits of different sizes [p.693]

27. ___Canadian lynx and snowshoe hare [p.694]

28. ___Grasshopper mice plunging the noxious chemical-spraying tail end of their beetle prey into the ground to feast on the head end [p.697]

29. ___Resemblance of *Lithops*, a desert plant, to a small rock [p.696]

30. ___An inedible butterfly is a model for the edible *Dismorphia* [p.697]

31. ___In the same area, drought-tolerant foxtail grasses have a shallow, fibrous root system, mallow plants have a deeper taproot system, and smartweed has a taproot system that branches in topsoil and in soil below the roots of other species [p.693]

32. ___Aggressively stinging yellow jackets are the likely model for nonstinging edible wasps [p.697]

33. ___Two species of *Paramecium* requiring identical resources cannot coexist indefinitely [p.692]

34. ___Baboon on the run turns to give canine tooth display to a pursuing leopard [p.696]

35. ___Least bittern with coloration similar to surrounding withered reeds [p.696]

A. Parasitism
B. Predator–prey interactions
C. Camouflage
D. Warning coloration
E. Mimicry
F. Moment-of-truth defenses
G. Competitive exclusion
H. Resource partitioning
I. Predator responses to prey

40.6. FORCES CONTRIBUTING TO COMMUNITY STABILITY [pp.698–699]

40.7. COMMUNITY INSTABILITY [p.700]

40.8. *Focus on the Environment:* EXOTIC AND ENDANGERED SPECIES [pp.700–701]

Selected Words: *facilitate* their own replacement [p.698], *natural* and *active* restoration of the climax community [p.699], *marshes* [p.699], *riparian zones* [p.699], *mangrove swamps* [p.699], *jump* dispersal [p.700], *Cryphonectria* [p.700], *Solenopsis* [p.700], *Oryctolagus* [p.700], *myxamatosis* [p.701], *Pueraria* [p.701]

Boldfaced, Page-Referenced Terms

[p.698] ecological succession _____

_____

[p.698] pioneer species _____

_____

[p.698] climax community _____

_____

[p.698] primary succession _____

_____

[p.698] secondary succession _____

_____

[p.698] climax-pattern model _____

_____

[p.699] wetlands _____

_____

[p.700] geographic dispersal _____

_____

[p.700] exotic species _____

_____

[p.700] endangered species _____

_____

Fill-in-the-Blanks

When a community develops in a predictable sequence from pioneers to a stable end array of species that remain in equilibrium over some region, this is known as ecological (1) _____ [p.698]. (2) _____ [p.698] species are opportunistic colonizers of vacant habitats that are noted for their high dispersal rates and rapid growth. As time passes, the (3) _____ [p.698] are replaced by species that are more competitive; these species are themselves then replaced until the array of species stabilizes under the prevailing habitat conditions. This persistent array of species is the (4) _____ [p.698] community. When pioneer species begin to colonize a barren habitat such as a volcanic island, it is known as (5) _____ [p.698] succession. Once established, the pioneers improve conditions for other species and often set the stage for their own (6) _____ [p.698].

When an area within a community is disturbed and then recovers to move again toward the stable, self-perpetuating climax community, it is known as (7) _____ [p.698] succession. Some scientists hypothesize that the colonizers (8) _____ [p.698] their own replacement while other scientists subscribe to the idea that the sequence of (9) _____ [p.698] depends on what species arrive first to compete successfully with species that could replace them. According to the (10) _____-_____ [p.698] model, a community is adapted to a total pattern of environmental factors—including climate, soil, topography, wind, species interactions, and recurring disturbances such as fires and chance events. Small-scale changes contribute to the internal dynamics of the (11) _____ [p.698] as a whole. For example, a tree falling in a tropical forest opens a gap in the forest canopy that allows more light to reach that gap. Here the growth of previously suppressed small trees and the germination of (12) _____ [p.699] or shade-intolerant species is encouraged. Giant sequoia trees in climax communities of the California Sierra Nevada are best maintained by modest (13) _____ [p.699], which eliminate trees and shrubs that compete with young sequoias but do not damage the older sequoias. Thus, recurring small-scale changes are built into the overall workings of many communities. The example of secondary succession regrowth following Mount St. Helen's violent eruption in 1980 is classified as (14) _____ [p.699] restoration of the climax community. Attempts to reestablish biodiversity in abandoned farmlands as well as other large areas disturbed by agriculture, mining, and other human activities, is known as (15) _____ [p.699] restoration.

Choice

For questions 16–24, choose from the following:

a. primary succession   b. secondary succession

16. ___ Following a disturbance, a patch of habitat or a community moves once again toward the climax state. [p.698]

17. ___ Successional changes begin when a pioneer population colonizes a barren habitat. [p.698]

18. ___ A disturbed area within a community recovers and moves again toward the climax state. [p.698]

19. ___ Many plants in this succession arise from seeds or seedlings that are already present when the process begins. [p.698]

20. ___ Many types are mutualists with nitrogen-fixing bacteria and initially outcompete other plants in nitrogen-poor habitats. [p.698]

21. ___ Involves typically small plants, with brief life cycles, adapted to grow in exposed areas with intense sunlight, swings in temperature, and nutrient-deficient soil [p.698]

22. ___ A successional pattern that occurs in abandoned fields, burned forests, and storm-battered intertidal zones [p.698]

23. ___ Might occur on a new volcanic island or on land exposed by the retreat of a glacier [p.698]

24. ___ Once established, the pioneers improve conditions for other species and often set the stage for their own replacement. [p.698]

Matching

Choose the most appropriate answer for each term.

25. ___ community stability [p.700]
26. ___ endangered species [p.700]
27. ___ geographic dispersal [p.700]
28. ___ jump dispersal [p.700]
29. ___ exotic species [p.700]

A. A rapid geographic dispersal mechanism as when an insect might travel in a ship's hold from an island to the mainland
B. An outcome of forces that have come into an uneasy balance
C. Resident of an established community that has moved from its home range and successfully taken up residence elsewhere; Nile perch, European rabbits, and kudzu are examples
D. Species that are now on rare or threatened species lists or have already become extinct
E. Residents of established communities move out from their home range and successfully take up residence elsewhere; the process may be slow or rapid by expansion of home range, jump dispersal, or extremely slow as in continental drift

40.9. PATTERNS OF BIODIVERSITY [pp.702–703]

Boldfaced, Page-Referenced Terms

[p.703] distance effect _____

_____

[p.703] area effect _____

_____

Short Answer

1. List three factors responsible for creating the higher species diversity values as related to the distance of land and sea from the equator. [p.702]

a. _____

_____

_____

_____

b. _____

_____

_____

_____

c. _____

_____

_____

_____

Fill-in-the-Blanks

The number of coexisting species is highest in the (2) _____ [p.702], and it systematically declines toward the poles. In 1965, a volcanic eruption formed a new island southwest of (3) _____ [p.702], which was named Surtsey. The new island served as a laboratory for studying (4) _____ [p.702]. Within six months, bacteria, fungi, seeds, flies and some seabirds became established on the island. After two years one vascular plant appeared and two years after that, a moss plant was observed. These organisms were all colonists from (5) _____ [p.702].

Islands at great distances from potential colonists receive few colonists; the few that arrive are adapted for long distance (6) _____ [p.703]. This is called the (7) _____ [p.703] effect. Larger islands tend to support more species than small islands at equivalent distances from source areas. This is the (8) _____ [p.703] effect. (9) _____ [p.703] islands tend to have more and varied habitats, more complex topography, and extend farther above sea level. Thus, they favor species (10) _____ [p.703]. Also, being bigger (11) _____ [p.703], they may intercept more colonists.

Most importantly, extinctions suppress (12) _____ [p.703] on (13) _____ [p.703] islands. There, the (14) _____ [p.703] populations are far more vulnerable to storms, volcanic eruptions, diseases, and random shifts in birth and death rates. As for any island, the number of species reflects a balance between (15) _____ [p.703] rates for new species and (16) _____ [p.703] rates for established ones. Small islands that are distant from a source of colonists have low (17) _____ [p.703] and high (18) _____ [p.703] rates, so they support few species once the balance has been struck for their populations.

Problem

19. After consideration of the distance effect, the area effect, and species diversity patterns as related to the equator, answer the following question. There are two islands (B and C) of the same size and topography that are equidistant from the African coast (A), as shown in the illustration below. Which will have the higher species diversity values?

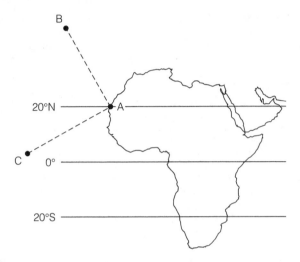

Self-Quiz

____ 1. All the populations of different species that occupy and are adapted to a given habitat are referred to as a(n) _____. [p.690]
   a. biosphere
   b. community
   c. ecosystem
   d. niche

____ 2. The range of all factors that influence whether a species can obtain resources essential for survival and reproduction is called the _____ of a species. [p.690]
   a. habitat
   b. niche
   c. carrying capacity
   d. ecosystem

____ 3. A one-way relationship in which one species benefits and the other is directly harmed is called _____. [p.690]
   a. commensalism
   b. competitive exclusion
   c. parasitism
   d. mutualism

____ 4. A lopsided interaction that directly benefits one species but does not harm or help the other much, if at all, is _____. [p.690]
   a. commensalism
   b. competitive exclusion
   c. predation
   d. mutualism

___ 5. An interaction in which both species benefit is best described as _____. [p.690]
   a. commensalism
   b. mutualism
   c. predation
   d. parasitism

___ 6. A striped skunk being pursued by a predator suddenly turns and releases its foul-smelling odor. This is an example of _____. [p.696]
   a. warning coloration
   b. mimicry
   c. camouflage
   d. moment-of-truth defense

___ 7. When an inexperienced predator attacks a yellow-banded wasp, the predator receives the pain of a stinger and will not attack again. This is an example of _____. [p.696]
   a. mimicry
   b. camouflage
   c. a prey defense
   d. warning coloration
   e. both c and d

___ 8. _____ is represented by foxtail grass, mallow plants, and smartweed because their root systems exploit different areas of the soil in a field. [p.693]
   a. Succession
   b. Resource partitioning
   c. A climax community
   d. A disturbance

___ 9. During the process of community succession, _____. [p.698]
   a. pioneer populations adapt to growing in habitats that cannot support most species
   b. pioneers set the stage for their own replacement
   c. species composition eventually is stable in the form of the climax community
   d. all of the above

___ 10. G. Gause utilized two species of *Paramecium* in a study that described _____. [p.692]
   a. interspecific competition and competitive exclusion
   b. resource partitioning
   c. the establishment of territories
   d. coevolved mutualism

___ 11. The most striking patterns of species diversity on land and in the seas relate to _____. [p.702]
   a. distance effect
   b. area effect
   c. immigration rate for new species
   d. distance from the equator

___ 12. The relationship between the yucca plant and the yucca moth that pollinates it is best described as _____. [p.691]
   a. camouflage
   b. commensalism
   c. competitive exclusion
   d. obligatory mutualism

Chapter Objectives/Review Questions

1. The type of place where you normally find a maple is its _____. [p.690]
2. List five factors that shape the structure of a biological community. [p.690]
3. An organism's _____ is the sum of activities and relationships in which it engages to secure and use the resources necessary for its survival and reproduction. [p.690]
4. Contrast the terms *fundamental niche* and *realized niche*. [p.690]
5. The interaction of a bird's nest and a tree is known as _____. [p.690]
6. In forms of _____, each of the participating species reaps benefits from the interaction. [p.690]
7. In _____ competition, disadvantages flow both ways between species. [p.690]
8. _____ and _____ are interactions that directly benefit one species and directly hurt the other. [p.690]
9. Commensalism, mutualism, and parasitism are all forms of _____, which means "living together." [p.690]
10. Define and cite an example of *obligatory mutualism*. [p.691]
11. Define *intraspecific competition* and *interspecific competition*. [p.692]
12. Describe a study that demonstrates laboratory evidence supporting the competitive exclusion concept. [p.692]

13. Define a *keystone species*. [p.692]
14. Cite one example of resource partitioning. [p.692]
15. A predator gets food from other living organisms, its _____. [p.694]
16. _____ take up residence in or on other living organisms—their hosts—and feed on specific host tissues for part of the life cycle. [p.694]
17. Many of the adaptations of predators (or parasites) and their victims arose through _____. [p.694]
18. What are the general body locations of ectoparasites and endoparasites? [p.695]
19. List the types of organisms classified as microparasites and those classified as macroparasites. [p.695]
20. Cowbirds are classified as _____ parasites. [p.695]
21. Be able to discuss advantages and disadvantages of using parasites as biological controls. [p.695]
22. Be able to completely define and give examples of the following prey defenses: warning coloration, mimicry, moment-of-truth defenses, and camouflage. [p.695]
23. Describe two examples of how predators counter prey defenses with their own marvelous adaptations. [p.697]
24. Define *ecological succession*. [p.698]
25. A(n) _____ community is a stable, self-perpetuating array of species in equilibrium with one another and their habitat. [p.698]
26. Distinguish between primary and secondary succession. [p.698]
27. By a(n) _____-_____ model, a community is adapted to a total pattern of environmental factors. [p.698]
28. Describe how fire disturbances positively affect a community of giant sequoias. [p.699]
29. Distinguish between *natural* and *active* restoration. [p.699]
30. Community _____ is an outcome of forces that have come into uneasy balance. [p.700]
31. Explain how the introduction of exotic species can be disastrous. List five specific examples of species introductions into the United States that have had adverse results. [pp.700–701]
32. List three factors that underlie the existing patterns of biodiversity. [p.702]
33. Describe the distance effect and the area effect. [p.703]
34. Estimate qualitatively the differences in species diversity and abundance of organisms likely to exist on two islands with the following characteristics: Island A has an area of 6,000 square miles, and Island B has an area of 60 square miles; both islands lie at 10° N latitude and are equidistant from the same source area of colonizers. [pp.702–703]

Integrating and Applying Key Concepts

If you were Ruler of All People on Earth, how would you organize industry and human populations in an effort to solve our most pressing pollution problems?

Is there a *fundamental niche* that is occupied by humans? If you think so, describe the minimal abiotic and biotic conditions required by populations of humans in order to live and reproduce. (Note that *"thrive and be happy"* are not criteria.) If you do not think so, state why.

These minimal niche conditions can be viewed as resource categories that must be protected by populations if they are to survive. Do you believe that the cold war between the United States and the Soviet Union primarily involved protection of minimal niche conditions, or do you believe that the cold war was based on other, more (or less) important factors?

a. If the former, how do you think *minimal* niche conditions might have been guaranteed for all humans willing and able to accept certain responsibilities as their contribution toward enabling this guarantee to be met?
b. If the latter, identify what you think those factors are, and explain why you consider them more (or less) important than minimal niche conditions.

41

ECOSYSTEMS

Interactive Exercises

Crêpes for Breakfast, Pancake Ice for Dessert [pp.706–707]

41.1. THE NATURE OF ECOSYSTEMS [pp.708–709]

41.2. HOW DOES ENERGY FLOW THROUGH ECOSYSTEMS? [pp.710–711]

41.3. *Focus on Science:* ENERGY FLOW AT SILVER SPRINGS, FLORIDA [p.712]

Selected Words: herbivores [p.708], *carnivores* [p.708], *omnivores* [p.708], *parasites* [p.708], *energy inputs* [p.708], *nutrient inputs* [p.708], *energy outputs* [p.708], *nutrient outputs* [p.708], *troph* [p.709], *cross-connecting* [p.709], *gross* primary productivity [p.710], *net* amount [p.710]

Boldfaced, Page-Referenced Terms

[p.708] primary producers _____

_____

[p.708] consumers _____

_____

[p.708] decomposers _____

_____

[p.708] detritivores _____

_____

[p.708] ecosystem _____

_____

[p.709] trophic levels _____

_____

[p.709] food chain _____

_____

[p.709] food webs _____

_____

[p.710] primary productivity _____

_____

[p.710] grazing food webs _____

_____

[p.710] detrital food webs _____

_____

[p.710] ecological pyramid _____

_____

[p.711] energy pyramid _____

_____

Choice

For questions 1–15, choose from the following:

a. primary producer   b. consumer   c. detritivore   d. decomposer

1. ___Mule deer [pp.708–709]
2. ___Earthworm [pp.708–709]
3. ___Parasites [pp.708–709]
4. ___The only category lacking heterotrophs [pp.708–709]
5. ___Omnivores [pp.708–709]
6. ___Tapeworm [pp.708–709]
7. ___Herbivores [pp.708–709]
8. ___Wolf [pp.708–709]
9. ___Fungi and bacteria [pp.708–709]
10. ___Carnivores [pp.708–709]
11. ___Green plants [pp.708–709]
12. ___Homo sapiens [pp.708–709]
13. ___Crabs [pp.708–709]
14. ___Autotrophs [pp.708–709]
15. ___Grasshopper [pp.708–709]

Fill-in-the-Blanks

In an ecosystem, autotrophs known as primary (16) _____ [p.708] are able to secure energy directly from their environment for the entire ecosystem of which it is a part. All other organisms in the system are heterotrophs, not self-feeders, and in general, are known as (17) _____ [p.708]. There are several types of these that extract energy from compounds put together by the primary producers: (18) _____ [p.708] eat plants; (19) _____ [p.708] eat animals; (20) _____ [p.708] eat both plants and animals; (21) _____ [p.708] extract energy from living hosts that they live in or on. Still other heterotrophs, known as (22) _____ [p.708], include fungi and many bacteria that obtain energy by breaking down the remains or products of organisms. Other heterotrophs, the (23) _____ [p.708], obtain nutrients from decomposing particles of organic matter. Ecosystems are (24) _____ [p.708] systems, so they cannot be self-sustaining. They require (25) _____ [p.708] input (the sun) and (26) _____ [p.708] inputs (minerals); they also have outputs of the same components. However, (27) _____ [p.708] cannot be recycled and much is lost to the environment; (28) _____ [p.708] do get cycled, but some are still lost. Within virtually any ecosystem, the members fit into a hierarchy of feeding relationships called (29) _____ [p.709] levels. A straight-line sequence of "who eats whom" in an ecosystem (for example, cow eats grass, man eats cow) is called a(n) (30) _____ _____ [p.709]. In reality, the same food source is most likely part of more than one chain, especially at low trophic levels. It is more accurate to think of different food chains as cross-connecting with one another—that is, as a(n) (31) _____ _____ [p.709].

The rate at which an ecosystem's primary producers capture and store a given amount of energy in a specified time interval is the primary (32) _____ [p.710]. The (33) _____ [p.710] amount is the rate of energy storage in plant tissues in excess of the rate of aerobic respiration by the plants. Other factors such as seasonal patterns and distribution through a given habitat also influence the amount of (34) _____ [p.710] primary production.

Energy from a primary source flows in one direction through two categories of food webs. In (35) _____ [p.710] food webs, the energy flows from plants to herbivores, and then through an array of carnivores. In (36) _____ [p.710] food webs, it flows primarily from photosynthetic species through detritivores and decomposers. These two kinds of food webs often cross-connect.

Ecologists often represent the trophic structure of an ecosystem in the form of an ecological (37) _____ [p.710]. In such forms, the primary (38) _____ [p.710] form a broad base for successive tiers of consumers above them. Some pyramids are based on (39) _____ [p.711] or the weight of all the members at each trophic level. Sometimes, a pyramid of (40) _____ [p.711] can be "upside down," with the (41) [choose one] ❏ smallest ❏ largest [p.711] tier on the bottom. Sometimes a more useful way to depict an ecosystem's trophic structure is with a(n) (42) _____ [p.711] pyramid. These pyramids show energy losses at each transfer to a different (43) _____ [p.711] level in the ecosystem. They have a (44) [choose one] ❏ small ❏ large [p.711] energy base at the bottom and are always "right-side up."

The ecological study of energy flow at Silver Springs, Florida, found that about (45) [choose one] ❏ 1 ❏ 5 [p.712] percent of all incoming solar energy was captured by producers prior to transfer to the next trophic level. It was also reported that only about (46) [choose one] ❏ 6 ❏ 8 [p.712] percent to (47) [choose one] ❏ 10 ❏ 16 [p.712] percent of the energy entering one trophic level becomes available for organisms at the next level. In general, the study showed that the efficiency of the energy transfers is so (48) [choose one] ❏ high ❏ low [p.712], ecosystems have usually no more than (49) [choose one] ❏ 4 ❏ 6 [p.712] consumer trophic levels.

Dichotomous Choice

Circle one of two possible answers given between parentheses in each statement.

50. Ecosystems are (open/closed) systems, and so are not self-sustaining. [p.708]
51. Minerals carried by erosion into a lake represent nutrient (input/output). [p.708]
52. Energy (can/cannot) be recycled. [p.708]
53. Nutrients typically (can/cannot) be cycled. [p.708]
54. Ecosystems have energy and nutrient inputs as well as nutrient output and (have/lack) energy output. [p.708]

Matching

Match each organism in the Antarctic food chain given below with the principal trophic level it occupies. Some letters may not be needed.

55. ___emperor penguins [p.709]

56. ___krill [p.709]

57. ___blue whale [p.709]

58. ___diatoms [p.709]

59. ___leopard seal [p.709]

60. ___fishes, small squids [p.709]

61. ___killer whale [p.709]

A. Chemosynthetic autotrophs
B. Herbivores
C. Photosynthetic autotrophs
D. Secondary consumers
E. Tertiary consumers

41.4. BIOGEOCHEMICAL CYCLES—AN OVERVIEW [p.713]

41.5. HYDROLOGIC CYCLE [p.714]

41.6. SEDIMENTARY CYCLES [p.715]

Selected Words: *hydrologic* cycle [p.713], *atmospheric* cycles [p.713], *sedimentary* cycles [p.713]

Boldfaced, Page-Referenced Terms

[p.713] biogeochemical cycle _____

_____

[p.714] hydrologic cycle _____

_____

[p.714] watershed _____

_____

[p.715] phosphorus cycle _____

_____

[p.715] eutrophication _____

_____

Short Answer

1. In what form(s) are elements used as nutrients usually available to producers? [p.713] _____

_____

2. How is the ecosystem's reserve of nutrients maintained? [p.713] _____

_____

_____

3. How does the amount of a nutrient being cycled through most major ecosystems compare with the amount entering or leaving in a given year? [p.713] _____

_____

_____

_____

4. What are the common input sources for an ecosystem's nutrient reserves? [p.713] _____

_____

_____

5. What are the output sources of nutrient loss for land ecosystems? [p.713] _____

_____

_____

Complete the Table

6. Complete the following table to summarize the functions of the three types of biogeochemical cycles.

Biogeochemical Cycle	General Function(s)
[p.713] a. Hydrologic cycle	
[p.713] b. Atmospheric cycles	
[p.713] c. Sedimentary cycles	

Matching

Choose the most appropriate answer for questions 7–10; 11–14 may have more than one answer

7. ___solar energy [p.714]

8. ___source of most water [p.714]

9. ___watershed [p.714]

10. ___water and plants taking up water [p.714]

11. ___forms of precipitation falling to land [p.714]

12. ___deforestation [p.714]

13. ___have important roles in the global hydrologic cycle [p.714]

14. ___forms of atmospheric water [p.714]

A. Mostly rain and snow
B. Important in moving nutrients in biochemical cycles
C. Water vapor, clouds, and ice crystals
D. Where precipitation of a specified region becomes funneled into a single stream or river
E. May have long-term disruptive effects on nutrient availability for an entire ecosystem
F. Slowly drives water through the atmosphere, on or through land mass surface layers, to oceans, and back again
G. Ocean currents and wind patterns
H. Evaporation from the oceans

Fill-in-the-Blanks

A(n) (15) _____ [p.714] is any region in which precipitation becomes funneled into a single stream or river. Most of the water that enters a watershed seeps into the (16) _____ [p.714] or becomes surface runoff that enters (17) _____ [p.714]. Plants withdraw water and its dissolved minerals from the soil, then they lose it by (18) _____ [p.714]. Watershed studies revealed the importance of (19) _____ [p.714] cover in the movements of (20) _____ [p.714] through the ecosystem phase of biogeochemical cycles.

Measurements of watershed inputs and outputs have many practical applications. In studies of young, undisturbed forests in the Hubbard Brook watersheds, each hectare lost only about 8 kilograms or so of (21) _____ [p.714]. Rainfall and the weathering of rocks brought in replacements of this element. Tree roots were also "mining" the soil, so (22) _____ [p.714] was being stored in a growing (23) _____ [p.714] of tree tissues. In some experimental watersheds, (24) _____ [p.714] caused a shift in nutrient ouputs. Calcium and other nutrients (25) _____ [p.714] so slowly that deforestation may disrupt nutrient availability for entire (26) _____ [p.714].

Fill-in-the-Blanks

The Earth's crust is the largest (27) _____ [p.715] for phosphorus and other minerals that move through ecosystems as part of (28) _____ [p.715] cycles. In rock formations on land, phosphorus is typically in the form of (29) _____ [p.715]. By natural weathering and erosion of soil, phosphates enter rivers and streams that transport them to (30) _____ [p.715] sediments. Phosphorus slowly accumulates mainly on submerged (31) "_____" [p.715] of continents. Millions of years go by. Where great movements of (32) _____ [p.715] plates uplift part of the seafloor, phosphates become exposed on drained land surfaces. In time, weathering releases the phosphates from the exposed rocks, and the cycle's (33) _____ [p.715] phase begins again.

The (34) _____ [p.715] phase of the cycle is more rapid than the long-term geochemical phase. All (35) _____ [p.715] require phosphorus for synthesizing phospholipids, NADPH, ATP, nucleic acids, and other compounds. Plants take up dissolved, (36) _____ [p.715] forms of phosphate very rapidly. Herbivores obtain phosphorus by dining on (37) _____ [p.715]; carnivores obtain it by dining on the (38) _____ [p.715]. Both types of animals excrete phosphates in urine and feces. (39) _____ [p.715] activities of soil organisms releases phosphates. Plants take up the mineral and help to rapidly recycle it within the (40) _____ [p.715].

(41) _____ [p.715] use phosphates as key ingredients. Where these compounds are heavily applied, phosphorus that becomes concentrated in runoff from fields alters living conditions in (42) _____ [p.715] and other aquatic ecosystems. Nitrogen, potassium, and phosphorus are three important nutrients that algae and aquatic plants require for their (43) _____ [p.715]. Bacteria fix enough (44) _____ [p.715]. Freshwater ecosystems usually have excess amounts of (45) _____ [p.715]. Most of the phosphorus is locked in (46) _____ [p.715], so it tends to be a limiting factor on plant growth. If the water is enriched with (47) _____ [p.715]-containing fertilizers, the outcome will be dense (48) _____ [p.715] blooms. Activities that increase the concentrations of dissolved nutrients can lead to (49) _____ [p.715], a term referring to nutrient enrichment of any aquatic ecosystem.

41.7. CARBON CYCLE [pp.716–717]

41.8. *Focus on the Environment:* FROM GREENHOUSE GASES TO A WARMER PLANET? [pp.718–719]

41.9. NITROGEN CYCLE [pp.720–721]

41.10. *Focus on Science:* ECOSYSTEM MODELING [p.721]

Selected Words: *Anabaena* [p.720], *Nostoc* [p.720], *Rhizobium* [p.720], *Azotobacter* [p.720]

Boldfaced, Page-Referenced Terms

[p.716] carbon cycle _____

_____

[p.718] greenhouse effect _____

_____

[p.718] global warming _____

_____

[p.720] nitrogen cycle _____

_____

[p.720] nitrogen fixation _____

_____

[p.720] decomposition _____

_____

[p.720] ammonification _____

_____

[p.720] nitrification _____

_____

[p.721] denitrification _____

_____

[p.721] ecosystem modeling _____

_____

[p.721] biological magnification _____

_____

Matching

Choose the most appropriate answer for each term.

1. ___greenhouse gases [p.718]
2. ___carbon dioxide fixation [p.716]
3. ___carbon cycle [pp.716–717]
4. ___ways carbon enters the atmosphere [p.716]
5. ___greenhouse effect [p.718]
6. ___carbon dioxide (CO_2) [p.716]
7. ___oceans [p.716]

A. Form of most of the atmospheric carbon
B. Aerobic respiration, fossil fuel burning, and volcanic eruptions
C. CO_2, CFCs, CH_4, and N_2O
D. Photosynthesizers incorporate carbon atoms into organic compounds
E. Holds most of the carbon in dissolved form
F. Carbon reservoirs ⟶ atmosphere and oceans ⟶ through organisms ⟶ carbon reservoirs
G. Warming of Earth's lower atmosphere due to accumulation of certain gases

Matching

Choose the most appropriate answer for each term.

8. ___nitrogen cycle description [p.720]
9. ___nitrogen fixation [p.720]
10. ___decomposition and ammonification [p.720]
11. ___nitrification [p.720]
12. ___denitrification [p.721]

A. Ammonia or ammonium in soil is stripped of electrons, and nitrite (NO_2^-) is the result; other bacteria convert nitrite to nitrate (NO_3^-).
B. Bacteria convert nitrate or nitrite to N_2 and a bit of nitrous oxide (N_2O).
C. Bacteria and fungi break down nitrogen-containing wastes and plant and animal remains; released amino acids and proteins are used for growth with the excess given up as ammonia or ammonium ions that plants can use.
D. Occurs in the atmosphere (largest reservoir); only certain bacteria, volcanic action, and lightning can convert N_2 into forms that can enter food webs.
E. A few kinds of bacteria convert N_2 to ammonia (NH_3), which dissolves quickly in water to form ammonium (NH_4^+).

Short Answer

13. List reasons that an insufficient soil nitrogen supply is a problem for land plants. [p.721] _____

_____

_____

Fill-in-the-Blanks

(14) _____ _____ [p.721] is a method of identifying and combining crucial bits of information

about an ecosystem through computer programs and models in order to predict the outcome of the next

disturbance. (15) _____ [p.721], a relatively stable hydrocarbon, is a synthetic organic pesticide.

Because it is insoluble in water, winds carry DDT in vapor form, transporting fine particles of it. DDT is also

highly soluble in (16) _____ [p.721], so it can accumulate in the tissues of organisms. Thus, DDT can

show (17) _____ _____ [p.721]. By this occurrence, a nondegradable or slowly degradable

substance becomes more and more (18) _____ [p.721] in the tissues of organisms at the higher trophic

levels of a food (19) _____ [p.721]. Most of the DDT from all organisms that a(n) (20) _____ [p.721]

feeds on during its lifetime ends up in its own tissues. DDT and modified forms of it disrupt (21) _____

[p.721] activities and are often toxic to many aquatic and terrestrial animals.

Self-Quiz

___ 1. An array of organisms and their physical environment, interacting by a one-way flow of energy and a cycling of materials, is a(n) _____. [p.708]
a. population
b. community
c. ecosystem
d. biosphere

___ 2. _____ consume dead or decomposing particles of organic matter. [p.708]
a. Herbivores
b. Parasites
c. Detritivores
d. Carnivores

___ 3. The members of feeding relationships are structured in a hierarchy, the steps of which are called _____. [p.709]
a. organism level
b. energy source level
c. eating level
d. trophic levels

___ 4. In the Antarctic, blue whales feed mainly on _____. [p.709]
a. petrel
b. krill
c. seals
d. fish and small squids

___ 5. Which of the following is a primary consumer? [p.709]
a. cow
b. dog
c. hawk
d. all of the above

___ 6. In a natural community, the primary consumers are _____. [p.708]
a. herbivores
b. carnivores
c. scavengers
d. decomposers

___ 7. A straight-line sequence of who eats whom in an ecosystem is sometimes called a(n) _____. [p.709]
a. trophic level
b. food chain
c. ecological pyramid
d. food web

___ 8. Of the 1,700,000 kilocalories of solar energy that entered an aquatic ecosystem in Silver Springs, Florida, investigators determined that about _____ percent of incoming solar energy was trapped by photosynthetic autotrophs. [p.712]
a. 1
b. 10
c. 25
d. 74

___ 9. A biogeochemical cycle that deals with phosphorus and other nutrients that do not have gaseous forms is the _____ type. [p.713]
a. sedimentary
b. hydrologic
c. nutrient
d. atmospheric

___ 10. _____ is a process in which nitrogenous waste products or organic remains of organisms are decomposed by soil bacteria and fungi that use the amino acids being released for their own growth and release the excess as ammonia or ammonium, which plants take up. [p.720]
a. Nitrification
b. Ammonification
c. Denitrification
d. Nitrogen fixation

___ 11. In the carbon cycle, carbon enters the atmosphere through _____. [p.716]
a. carbon dioxide fixation
b. respiration, burning, and volcanic eruptions
c. oceans and accumulation of plant biomass
d. release of greenhouse gases

___ 12. _____ refers to an increase in concentration of a nondegradable (or slowly degradable) substance in organisms as it is passed along food chains. [p.721]
a. Ecosystem modeling
b. Nutrient input
c. Biogeochemical cycle
d. Biological magnification

Chapter Objectives/Review Questions

1. List the principal trophic levels in an ecosystem of your choice; state the source of energy for each trophic level and give one or two examples of organisms associated with each trophic level. [pp.708–709]
2. Explain why nutrients can be completely recycled but energy cannot. [p.708]
3. A(n) _____ is an array of organisms and their physical environment, all interacting through a flow of energy and a cycling of materials. [p.708]
4. Members of an ecosystem fit somewhere in a hierarchy of energy transfers (feeding relationships) called _____ levels. [p.709]
5. Distinguish between food chains and food webs. [p.709]
6. Define *primary productivity*. [p.710]
7. Compare grazing food webs with detrital food webs. Present an example of each. [p.710]
8. Understand how materials and energy enter, pass through, and exit an ecosystem. [p.710]
9. Ecological pyramids that are based on _____ are determined by the weight of all the members of each trophic level; _____ pyramids reflect the energy losses at each transfer to a different trophic level. [pp.710–711]
10. In _____ cycles, the nutrient is transferred from the environment to organisms, then back to the environment—which serves as a large reservoir for it. [p.713]
11. Be able to discuss water movements through the hydrologic cycle. [p.714]
12. Explain what studies in the Hubbard Brook watershed have taught us about the movement of substances (water, for example) through a forest ecosystem. [p.714]
13. Describe the geochemical and ecosystem phase of the phosphorus cycle. [p.715]
14. The carbon cycle traces carbon movement from reservoirs in the _____ and oceans, through organisms, then back to reservoirs. [p.716]

15. Certain gases cause heat to build up in the lower atmosphere, a warming action known as the _____ effect. [p.718]
16. A major element found in all proteins and nucleic acids moves in an atmospheric cycle called the _____ cycle. [p.720]
17. Define *eutrophication*, and be able to discuss its causes. [p.715]
18. Define the chemical events that occur during nitrogen fixation, decomposition and ammonification, and nitrification. [p.720]
19. Explain why agricultural methods in the United States tend to put more energy into mechanized agriculture in the form of fertilizers, pesticides, food processing, storage, and transport than is obtained from the soil in the form of energy stored in foods. [p.721]
20. Through _____ modeling, crucial bits of information about different ecosystem components are identified and used to build computer models for predicting outcomes of ecosystem disturbances. [p.721]
21. Describe how DDT damages ecosystems; discuss biological magnification. [p.721]

Integrating and Applying Key Concepts

In 1971, *Diet for a Small Planet* was published. Frances Moore Lappé, the author, felt that people in the United States of America wasted protein and ate too much meat. She said, "We have created a national consumption pattern in which the majority, who can pay, overconsume the most inefficient livestock products [cattle] well beyond their biological needs (even to the point of jeopardizing their health), while the minority, who cannot pay, are inadequately fed, even to the point of malnutrition." Cases of marasmus (a nutritional disease caused by prolonged lack of food calories) and kwashiorkor (caused by severe, long-term protein deficiency) have been found in Nashville, Tennessee, and on an Indian reservation in Arizona, respectively. Lappé's partial solution to the problem was to encourage people to get as much of their protein as possible directly from plants and to supplement that with less meat from the more efficient converters of grain to protein (chickens, turkeys, and hogs) and with seafood and dairy products.

Most of us realize that feeding the hungry people of the world is not just a matter of distributing the abundance that exists—it is also a matter of political, economic, and cultural factors. Yet it is still valuable to consider applying Lappés idea to our everyday living. Devise two full days of breakfasts, lunches, and dinners that would enable you to exploit the lowest acceptable trophic levels to sustain yourself healthfully.

42

THE BIOSPHERE

Interactive Exercises

Does a Cactus Grow in Brooklyn? [pp.724–725]

42.1. AIR CIRCULATION PATTERNS AND REGIONAL CLIMATES [pp.726–727]

42.2. OCEANS, LANDFORMS, AND REGIONAL CLIMATES [pp.728–729]

Selected Words: hydrosphere [p.725], lithosphere [p.725], "ozone layer" [p.726], *leeward* [p.729], *windward* [p.729], "mini-monsoons" [p.729]

Boldfaced, Page-Referenced Terms

[p.724] biogeography _____

_____

[p.725] biosphere _____

_____

[p.725] atmosphere _____

_____

[p.725] climate _____

_____

[p.726] temperature zones _____

_____

[p.728] oceans _____

_____

[p.729] rain shadow _____

_____

[p.729] monsoons _____

_____

Matching

Choose the most appropriate answer for each term.

1. ___biogeography [p.724]
2. ___biosphere [p.725]
3. ___hydrosphere [p.725]
4. ___lithosphere [p.725]
5. ___atmosphere [p.725]
6. ___climate [p.725]
7. ___ozone layer [p.726]
8. ___temperature zones [p.726]
9. ___oceans [p.728]
10. ___rain shadow [p.729]
11. ___leeward [p.729]
12. ___windward [p.729]
13. ___monsoons [p.729]
14. ___mini-monsoons [p.729]

A. One continuous body of water that covers 71 percent of the Earth's surface
B. Made up of gases and airborne particles that envelop the Earth
C. Of the world, defined by differences in solar heating at different latitudes and the modified air circulation patterns
D. The sum total of all places in which organisms live
E. Refers to the direction not facing a wind
F. An atmospheric region between 17 and 25 kilometers above sea level
G. Recurring sea breezes along coastlines; warmed air above land rises and cooler marine air moves in
H. The waters of the Earth, including the ocean, polar ice caps, and other forms of liquid and frozen water
I. Refers to the direction from which the wind is blowing
J. Average weather conditions, such as temperature, humidity, wind speed, cloud cover, and rainfall, over time
K. The outer, rocky layer of the Earth
L. An arid or semiarid region of sparse rainfall on the leeward side of high mountains
M. The study of the distribution of organisms, past and present, and of diverse processes that underlie the distribution patterns
N. Patterns of air circulation that affect conditions on the continents lying poleward of warm oceans; intensely heated land draws in moisture-laden air that forms above ocean water

Fill-in-the-Blanks

The numbered items in the following illustrations represent missing information. Complete the blanks in the following narrative to supply that information.

The sun's rays are more concentrated in (15) _____ [p.727] regions than at the poles. The global pattern of air circulation begins as (16) _____ [p.727] equatorial air (17) _____ [p.727] and spreads northward and southward giving up much (18) _____ [p.727] as precipitation (warm air can hold more moisture than cold air), which supports luxuriant tropical forests. The cooled, drier air then (19) _____ [p.727] at latitudes of about 30°, and becomes warmer and drier in areas where deserts form. Air even farther from the equator picks up some (20) _____ [p.727], and (21) _____ [p.727] to higher altitudes, cools, and then gives up (22) _____ [p.727] at latitudes of about 60° C to create another moist belt. The cooled, dry air then (23) _____ [p.727] at the polar regions, where the low temperatures and almost nonexistent precipitation give rise to the cold, dry, polar deserts. The Earth's rotation then modifies the air circulation by deflecting it into worldwide belts of prevailing (24) _____ [p.727] and (25) _____ [p.727] winds.

The world's temperature zones, beginning at the equator and moving toward the poles, are the (26) _____ [p.727], the (27) _____ [p.727] temperate, the (28) _____ [p.727] temperate, and the (29) _____ [p.727]. Finally, the amount of the (30) _____ [p.727] radiation reaching the surface varies annually, owing to the Earth's (31) _____ [p.727] around the sun. This leads to seasonal changes in daylength, prevailing wind directions, and temperature. These factors influence the locations of different ecosystems.

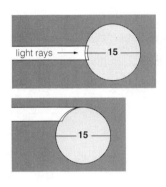

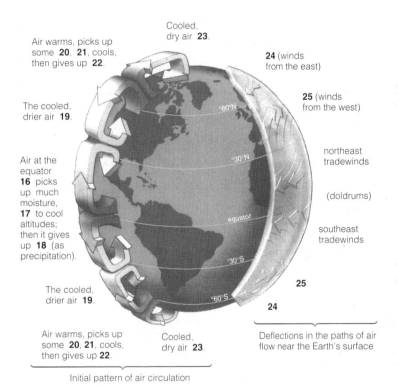

Air warms, picks up some **20**, **21**, cools, then gives up **22**.

Cooled, dry air **23**.

The cooled, drier air **19**.

Air at the equator **16** picks up much moisture, **17** to cool altitudes; then it gives up **18** (as precipitation).

The cooled, drier air **19**.

Air warms, picks up some **20**, **21**, cools, then gives up **22**.

Cooled, dry air **23**.

24 (winds from the east)

25 (winds from the west)

northeast tradewinds

(doldrums)

southeast tradewinds

Deflections in the paths of air flow near the Earth's surface

Initial pattern of air circulation

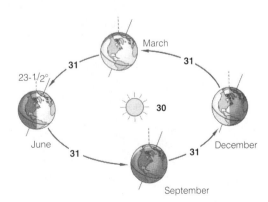

42.3. REALMS OF BIODIVERSITY [pp.730–731]

42.4. SOILS OF MAJOR BIOMES [p.732]

42.5. DESERTS [p.733]

42.6. DRY SHRUBLANDS, DRY WOODLANDS, AND GRASSLANDS [pp.734–735]

Selected Words: soil *profile* [p.732], *shortgrass* prairie [p.734], *tallgrass* prairie [p.734], *monsoon* grasslands [p.735]

Boldfaced, Page-Referenced Terms

[p.731] biogeographic realms _____

_____

[p.731] biome _____

_____

[p.731] ecoregions _____

_____

[p.732] soils _____

_____

[p.733] deserts _____

_____

[p.733] desertification _____

_____

[p.734] dry shrublands _____

_____

[p.734] dry woodlands _____

_____

[p.734] grasslands _____

_____

[p.734] savannas _____

_____

Matching

Choose the most appropriate answer for each term.

1. ___biogeographic realms [p.730]
2. ___biome [p.731]
3. ___soil [p.732]
4. ___deserts [p.733]
5. ___desertification [p.733]
6. ___dry shrublands [p.734]
7. ___soil profile [p.732]
8. ___dry woodlands [p.734]
9. ___grasslands [p.734]
10. ___savannas [p.734]
11. ___monsoon grasslands [p.735]

A. Mixtures of mineral particles and variable amounts of decomposing organic material (humus)
B. Areas receiving less than 25 to 60 centimeters of rain per year; local names include fynbos and chaparral
C. The conversion of grasslands and other productive biomes to dry wastelands
D. Sweep across much of the interior of continents, in the zones between deserts and temperate forests; warm temperatures prevail in summer, winters are extremely cold
E. Areas that dominate when annual rainfall is about 40 to 100 centimeters; dominant trees can be tall but do not form a dense, continuous canopy
F. Subdivision of biogeographic realms; a large region of land characterized by the climax vegetation of the ecosystems within its boundaries
G. Formed in land regions where the potential for evaporation exceeds sparse rainfall
H. Six vast land areas on the Earth, each with distinguishing plants and animals
I. Broad belts of grasslands with a smattering of shrubs or trees; rainfall averages 90 to 150 centimeters a year with prolonged seasonal droughts common
J. The layered structure of soils
K. Form in southern Asia where heavy rains alternate with a dry season; dense stands of tall, coarse grasses form, then die back and often burn in the dry season

Matching

Choose the most appropriate answer for each term.

12. ___tropical rain forest soil [p.732]
13. ___deciduous forest soil [p.732]
14. ___desert soil [p.732]
15. ___grassland soil [p.732]
16. ___coniferous soil [p.732]

A. A horizon: alkaline, deep, rich in humus
B. O horizon: scattered litter; A horizon: rich in organic matter above humus layer unmixed with minerals
C. O horizon: sparse litter; A–E horizons: continually leached
D. O horizon: pebbles, little organic matter; A horizon: shallow, poor soil
E. O horizon: well-defined, compacted mat of organic deposits resulting mainly from activity of fungal decomposers

Choice

For questions 17–26, choose from the following:

a. deserts   b. dry shrublands   c. dry woodlands   d. grasslands   e. savannas

17. ___ Monsoon type that forms dense stands of tall, coarse plants in parts of southern Asia where heavy rains alternate with a dry season [pp.734–735]

18. ___ Biome where the potential for evaporation greatly exceeds rainfall [p.733]

19. ___ Steinbeck's *Grapes of Wrath* speaks eloquently of the disruption of this biome. [p.734]

20. ___ Local names for this biome include fynbos and chaparral; dominant plants often have hardened, tough, evergreen leaves. [p.734]

21. ___ A biome in which dominant trees can be tall but do not form a dense canopy; includes Eucalyptus woodlands of southwestern Australia and oak woodlands of California and Oregon biome [p.734]

22. ___ Home to deep-rooted evergreen shrubs, fleshy-stemmed, shallow-rooted cacti, saguaro, short prickly pear, and ocotillo biome [p.733]

23. ___ Broad belts of grasslands with a smattering of shrubs or trees; prolonged seasonal droughts are common biome [p.734]

24. ___ The dominant animals are grazing and burrowing species; grazing and periodic fires maintain the fringes of this biome. [p.734]

25. ___ Within this biome, dominant fast-growing grasses dominate where rainfall is low, but acacia and other shrubs grow where there is slightly more moisture. [p.734]

26. ___ In summer, lightning-sparked, wind-driven firestorms can sweep through these biomes and swiftly burn shrubs with highly flammable leaves to the ground. [p.734]

42.7. TROPICAL RAIN FORESTS AND OTHER BROADLEAF FORESTS [pp.736–737]

42.8. CONIFEROUS FORESTS [p.738]

42.9. TUNDRA [p.739]

Selected Words: tropical deciduous forests [p.737], *monsoon* forests [p.737], *temperate* deciduous forests [p.737], *taiga* [p.738], *tuntura* [p.739], *arctic* tundra [p.739], *alpine* tundra [p.739]

Boldfaced, Page-Referenced Terms

[p.736] evergreen broadleaf forests _____

_____

[p.736] tropical rain forest _____

_____

[pp.736–737] deciduous broadleaf forests _____

_____

[p.738] coniferous forests _____

_____

[p.738] boreal forests _____

_____

[p.738] southern pine forests _____

_____

[p.739] tundra _____

_____

[p.739] permafrost _____

_____

Choice

For questions 1–16, choose from the following biomes:

a. deciduous broadleaf forests   b. coniferous forests   c. evergreen broadleaf forests   d. tundra

1. ___ Nearly continuous sunlight in summer; short plants grow and flower profusely with rapidly ripening seeds [p.739]

2. ___ Sweep across tropical zones of Africa, the East Indies and Malay Archipelago, southeast Asia, South America, and Central America [p.736]

3. ___ Highly productive forest; decomposition and mineral cycling are rapid in the hot, humid climate biome; tropical rain forests [p.736]

4. ___ Within this biome, complex forests of ash, beech, chestnut, elm, and deciduous oaks once stretched across northeastern North America. [p.737]

5. ___ Boreal forests or taiga; conifers are the primary producers [p.738]

6. ___ Spruce and balsam fir dominate the northern part. [p.738]

7. ___ Biome of the temperate zone; cold winter temperatures; many trees drop all their leaves in winter [p.737]

8. ___ Great treeless plain between the polar ice cap and belts of boreal forests in Europe, Asia, and North America [p.739]

9. ___ Soils are highly weathered, humus-poor, and not good nutrient reservoirs [pp.736,739]

10. ___ Forests that dominate the coastal plains of the south Atlantic and Gulf states; dominant plants are adapted to the dry, sandy, nutrient-poor soil and to natural fires or controlled burns [p.738]

11. ___ Annual rainfall can exceed 200 centimeters and is never less than 130 centimeters. [p.736]

12. ___ Includes tropical deciduous forests, monsoon forests, and temperate deciduous forests [pp.736–737]

13. ___ Cone-bearing trees are the primary producers; most have needle-shaped leaves with a thick cuticle and recessed stomata—adaptations that help the trees conserve water through dry times. [p.738]

14. ___ Pines, scrub oak, and wiregrass grow in New Jersey; palmettos grow below loblolly and other pines in the Deep South. [p.738]

15. ___ Just beneath the surface is a perpetually frozen layer, the permafrost. [p.739]

16. ___ One type is known as alpine; it prevails at high elevations throughout the world, although there is no permafrost beneath the soil. [p.739]

42.10. FRESHWATER PROVINCES [pp.740–741]
42.11. THE OCEAN PROVINCES [pp.742–743]

Selected Words: littoral zone [p.740], limnetic zone [p.740], profundal zone [p.740], phytoplankton [p.740], zooplankton [p.740], *thermocline* [p.740], *oligotrophic* lakes [p.741], *eutrophic* lakes [p.741], *benthic* province [p.742], *pelagic* province [p.742]

Boldfaced, Page-Referenced Terms

[p.740] lake _____

_____

[p.740] spring overturn _____

_____

[p.740] fall overturn _____

_____

[p.741] eutrophication _____

_____

[p.741] streams _____

_____

[p.742] ultraplankton _____

_____

[p.742] marine snow _____

_____

[p.743] hydrothermal vents _____

_____

Fill-in-the-Blanks

A(n) (1) _____ biome [p.740] is a standing body of freshwater with three zones. The shallow, usually well-lit (2) _____ [p.740] zone extends around the shore to the depth at which rooted aquatic plants stop growing. The diversity of organisms is greatest here. The (3) _____ [p.740] zone is the open, sunlit water past the littoral and extends to a depth where photosynthesis is insignificant. Aquatic communities of (4) _____ [p.740] abound here. The (5) _____ [p.740] zone includes all open water below the depth at which wavelengths suitable for photosynthesis can penetrate. Bacterial decomposers in bottom sediments of this zone release nutrients into the water.

Water is densest at 4°C; at this temperature, it sinks to the bottom of its basin, displacing the nutrient-rich bottom water upward and giving rise to spring and fall (6) _____ [p.740]. In spring, ice melts, daylength increases, and the surface waters of a lake slowly warm to (7) _____ °C [p.740]. Surface winds cause a(n) (8) _____ [p.740] overturn in which strong vertical water movements of water carry dissolved oxygen from a lake's surface layer to its depths, and nutrients released by decomposition are brought from the bottom sediments to the surface. The surface layer warms above 4°C by midsummer, becomes less dense, and the lake has developed a middle layer, the (9) _____ [p.740], that prevents vertical mixing. When autumn comes, the upper layer (10) _____ [p.740], becomes denser, then sinks, and the thermocline vanishes. This is termed the (11) _____ [p.740] overturn. Water then mixes vertically, allowing dissolved oxygen to move (12) [choose one] ❏ up ❏ down [p.740] and nutrients to move (13) [choose one] ❏ up ❏ down [p.740]. Primary productivity of a lake corresponds with the seasons. Following a spring overturn, longer daylengths and cycled nutrients support (14) [choose one] ❏ lower ❏ higher [p.740] rates of photosynthesis. By late summer, nutrient shortages are limiting photosynthesis. After the fall overturn, nutrient cycling drives a (15) [choose one] ❏ short ❏ long [p.741] burst of primary activity. Lakes have a trophic nature. (16) _____ [p.741] lakes are deep, nutrient-poor, and low in primary productivity. Lakes that are (17) _____ [p.741] are often shallow, nutrient-enriched, and high in primary productivity. Human activities can determine the trophic condition of lakes. The term (18) _____ [p.741] refers to nutrient enrichment of a lake resulting in reduced water transparency and a community rich in phytoplankton. (19) _____ [p.741] are flowing-water ecosystems that begin as freshwater springs or seeps. Three habitat types are found between headwaters and the river's end: riffles, pools, and (20) _____ [p.741].

Label–Match

Label each numbered item in the accompanying illustration. Complete the exercise by matching and entering the letter of the proper description in the parentheses following each label. [All are from p.742]

21. _____ province ( )

22. _____ province ( )

23. _____ zone ( )

24. _____ zone ( )

A. The entire volume of ocean water
B. All the water above the continental shelves
C. Includes all sediments and rocks of the ocean bottom; begins with continental shelves and extends to deep-sea trenches
D. Water of the ocean basin

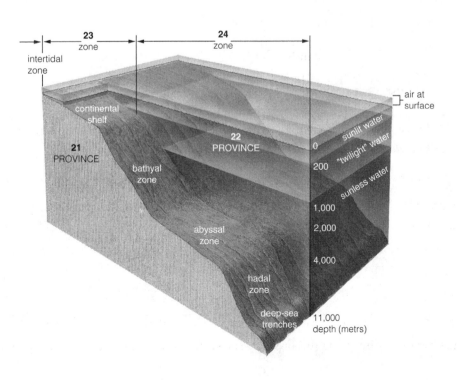

Choice

For questions 25–44, choose from the following:

a. stream ecosystems   b. lake ecosystems   c. ocean

25. ___ Riffles, pools, and runs [p.741]

26. ___ Hydrothermal vent communities of chemosynthetic bacteria in the abyssal zone; possible sites of life's origin [p.742]

27. ___ Ultraplankton contribute 70 percent of the primary productivity. [p.742]

28. ___ Can be oligotrophic or eutrophic [p.741]

29. ___ Submerged mountains, valleys, and plains [p.742]

30. ___ Begin as freshwater springs or seeps [p.741]

31. ___ Sewage and logging on adjacent lands can contribute to eutrophication of this water. [p.741]

32. ___ Average flow volume and temperature depend on rainfall, snowmelt, geography, altitude, and even shade cast by plants. [p.741]

33. ___ Spring and fall overturns [p.740]

34. ___ Has benthic and pelagic provinces [p.742]

35. ___ Especially in forests, these waters import most of the organic matter that supports food webs. [p.741]

36. ___ Created by geologic processes such as retreating glaciers [p.740]

37. ___ They grow and merge as they flow downslope and then often combine. [p.740]

38. ___ Primary productivity there varies seasonally, just as it does on land. [p.742]

39. ___ The final successional stage is a filled-in basin. [p.741]

40. ___ Vast "pastures" of phytoplankton and zooplankton become the basis of detrital food webs. [p.742]

41. ___ Since cities formed, these waters have been sewers for industrial and municipal wastes. [p.741]

42. ___ Has a thermocline by midsummer [p.740]

43. ___ Littoral, limnetic, and profundal zones [p.740]

44. ___ Bathyal, abyssal, and hadal zones [p.742]

42.12. CORAL REEFS AND CORAL BANKS [pp.744–745]

42.13. LIFE ALONG THE COASTS [pp.746–747]

42.14. *Focus on Science:* RITA IN THE TIME OF CHOLERA [pp.748–749]

Selected Words: *Corallina* [p.744], *atolls* [p.744], *fringing reefs* [p.744], *barrier reefs* [p.744], *rocky* shores [p.746], *sandy* and *muddy* shores [p.746], "downwelling" [p.747], *El Niño Southern Oscillation* (ENSO) [p.748]

Boldfaced, Page-Referenced Terms

[p.744] coral reef _____

_____

[p.744] coral banks _____

_____

[p.746] estuary _____

_____

[p.746] intertidal zone _____

_____

[p.747] upwelling _____

_____

[p.747] El Niño _____

_____

Complete the Table

1. Complete the following table, which describes three types of coral reefs.

Reef Type	Description
[p.744] a. Atolls	
[p.744] b. Fringing reefs	
[p.744] c. Barrier reefs	

Choice

For questions 2–22, choose from the following:

    a. coral reefs and banks   b. estuary   c. intertidal zone   d. coastal upwelling   e. ENSO

2. ___ The fogbanks that form along the California coast are one outcome. [p.747]

3. ___ *Lophelia* has constructed large ones in the cold waters of Norway's fjords. [p.744]

4. ___ Waves batter its resident organisms. [p.746]

5. ___ Massive dislocations in global rainfall patterns characterize this event, which corresponds to changes in sea surface temperatures and air circulation patterns. [p.748]

6. ___ Organisms living there must constantly contend with the tides. [p.746]

7. ___ Each wave-resistant formation began with accumulated remains of countless organisms. [p.744]

8. ___ Primary producers here include phytoplankton, salt-tolerant plants and some algae that grow in mud and on plant surfaces. [p.746]

9. ___ In general, an upward movement of cold, deep, often nutrient-rich ocean waters occurring in equatorial currents as well as along the coasts of both continents [p.747]

10. ___ A partly enclosed coastal region where seawater swirls and mixes with nutrient-rich freshwater from rivers, streams, and land runoff [p.746]

11. ___ Most of the substantial ones are formed in clear, warm waters between latitudes 25° north and south. [p.744]

12. ___ Salt marshes are common. [p.746]

13. ___ El Niño [p.748]

14. ___ Wind friction causes surface waters to begin moving and, under the force of the Earth's rotation, moving water is deflected west, away from a coast. [p.747]

15. ___ Many of these structures are being destroyed by human activities. [p.744]

16. ___ Commercial fishing industries of Peru and Chile depend on it. [p.747]

17. ___ A recurring seesaw in atmospheric pressure in the western equatorial Pacific [p.748]

18. ___ Home to exotic and rare fish sold in pet stores and eaten in Asian restaurants [pp.744–745]

19. ___ Sandy and muddy shores [p.746]

20. ___ Reverses the usual westward flow of air and water; displaces the cold, deep Humboldt current; this stops nutrient upwelling along the western coast of South America [p.748]

21. ___ Found along rocky and sandy coastlines [p.746]

22. ___ All show spectacular biodiversity, and vulnerability [p.744]

Self-Quiz

___ 1. The distribution of different types of ecosystems is influenced by _____. [pp.727, 729]
   a. global air circulation patterns
   b. variation in the amount of solar radiation reaching the Earth through the year
   c. surface ocean currents
   d. all of the above

___ 2. In a(n)_____, nutrient-rich freshwater draining from the land mixes with seawater carried in on tides. [p.746]
   a. pelagic province
   b. rift zone
   c. upwelling
   d. estuary

___ 3. A biome with broad belts of grasslands and scattered trees adapted to prolonged dry spells is known as a _____. [p.734]
   a. warm desert
   b. savanna
   c. tundra
   d. taiga

___ 4. The _____ biome is located at latitudes of about 30° north and south, has limited vegetation, and has rapid surface cooling at night. [p.733]
   a. shrublands
   b. savanna
   c. taiga
   d. desert

___ 5. In tropical rain forests, _____. [p.736]
   a. productivity is high
   b. litter does not accumulate
   c. soils are weathered and humus-poor, and have poor nutrient reservoirs
   d. decomposition and mineral cycling are extremely rapid
   e. all of the above

___ 6. In a lake, the open sunlit water with its suspended phytoplankton is referred to as its _____ zone. [p.740]
   a. epileptic
   b. limnetic
   c. littoral
   d. profundal

___ 7. The lake's upper layer cools, the thermocline vanishes, lake water mixes vertically, and once again dissolved oxygen moves down and nutrients move up. This describes the _____. [p.740]
   a. spring overturn
   b. summer overturn
   c. fall overturn
   d. winter overturn

___ 8. The _____ is a permanently frozen, water-impermeable layer just beneath the surface of the _____ biome. [p.739]
   a. permafrost; alpine tundra
   b. hydrosphere; alpine tundra
   c. permafrost; arctic tundra
   d. taiga; arctic tundra

___ 9. _____ are air circulation patterns that influence the continents north or south of warm oceans; low pressure causes moisture-laden air above the neighboring ocean to move inland, resulting in heavy rains. [p.729]
  a. Geothermal ecosystems
  b. Upwellings
  c. Taigas
  d. Monsoons

___10. All of the water above the continental shelves is in the _____. [p.742]
  a. neritic zone of the benthic province
  b. oceanic zone of the pelagic province
  c. neritic zone of the pelagic province
  d. oceanic zone of the benthic province

___11. Complex forests of ash, beech, birch, chestnut, elm, and oaks are found in the _____. [p.737]
  a. tropical deciduous forest
  b. monsoon forest
  c. temperate deciduous forest
  d. evergreen broadleaf forest

___12. Chemoautotrophic bacteria are the primary producers for _____. [p.743]
  a. hydrothermal vent communities
  b. desert communities
  c. lake communities
  d. coniferous forest communities
  e. coral reef communities

Chapter Objectives/Review Questions

1. _____ is the study of the distribution of organisms, past and present, and of diverse processes that underlie the distribution patterns. [p.724]
2. The _____ is the sum total of all places in which organisms live. [p.725]
3. _____ refers to average weather conditions, such as temperature, humidity, wind speed, cloud cover, and rainfall. [p.725]
4. State the reason that most forms of life depend on the ozone layer. [p.726]
5. _____ energy drives Earth's weather systems. [p.726]
6. Be able to describe the causes of global air circulation patterns. [pp.726–727]
7. Describe how the tilt of the Earth's axis affects annual variation in the amount of incoming solar radiation. [p.727]
8. Atmospheric _____ patterns, ocean _____, and landforms influence the distribution and dominant features of different types of ecosystems. [p.729]
9. Mountains, valleys, and other land formations influence _____ climates. [p.729]
10. Describe the cause of the rain shadow effect. [p.729]
11. Broadly, there are six distinct land realms, the _____ realms that were named by W. Sclater and Alfred Wallace. [pp.730–731]
12. Realms are divided into _____. [p.731]
13. _____ are mixtures of mineral particles and variable amounts of decomposing organic material. [p.732]
14. Be able to list the major biomes and briefly characterize them in terms of climate, topography, and organisms. [pp.733–739]
15. The wholesale conversion of grasslands and other productive biomes to desertlike wastelands is known as _____. [p.733]
16. A _____ is a standing body of freshwater with littoral, limnetic, and profundal zones. [p.740]
17. Define *plankton*, *phytoplankton*, and *zooplankton*. [p.740]
18. Describe the spring and fall overturn in a lake in terms of causal conditions and physical outcomes. [pp.740–741]
19. _____ refers to nutrient enrichment of a lake or some other body of water. [p.741]
20. _____ lakes are often deep, poor in nutrients, and low in primary productivity; _____ lakes are often shallow, rich in nutrients, and high in primary productivity. [p.741]
21. Describe a stream ecosystem. [p.741]
22. Be able to fully describe the benthic and pelagic provinces of the ocean. [p.742]

23. Within the pelagic province, all the water above the continental shelves is the _____ zone; the _____ zone is the water of the ocean basin. [p.742]
24. As much as 70 percent of the ocean's primary productivity may be the contribution of _____. [p.742]
25. Describe the unusual hydrothermal vent ecosystems. [pp.742–743]
26. List the three types of coral reefs, and be able to describe their formation. [p.744]
27. List reasons that reefs are being ecologically threatened. [pp.744–745]
28. Be able to descriptively distinguish between estuaries and the intertidal zones. [p.746]
29. State the significance of ocean upwelling. [p.747]
30. Describe conditions of ENSO occurrence and how this phenomenon interrelates ocean surface temperatures, the atmosphere, and the land. [pp.748–749]

Integrating and Applying Key Concepts

One species, *Homo sapiens*, uses about 40 percent of all of Earth's productivity, and its representatives have invaded every biome, either by living there or by dumping waste products there. Many of Earth's residents are being denied the minimal resources they need to survive, while human populations continue to increase exponentially. Can you suggest a better way of keeping Earth's biomes healthy while providing at least the minimal needs of all Earth's residents (not just humans)? If so, outline the requirements of such a system and devise a way in which it could be established.

43

HUMAN IMPACT ON THE BIOSPHERE

Interactive Exercises

An Indifference of Mythic Proportions [pp.752–753]

43.1. AIR POLLUTION—PRIME EXAMPLES [pp.754–755]

43.2. OZONE THINNING—GLOBAL LEGACY OF AIR POLLUTION [p.756]

Boldfaced, Page-Referenced Terms

[p.754] pollutants _____

_____

[p.754] thermal inversion _____

_____

[p.754] industrial smog _____

_____

[p.754] photochemical smog _____

_____

[p.754] PANs _____

_____

[p.754] dry acid deposition _____

_____

[p.754] acid rain _____

_____

[p.756] ozone thinning _____

_____

[p.756] chlorofluorocarbons (CFCs) _____

_____

Fill-in-the-Blanks

(1) _____ [p.754] are substances with which ecosystems have had no prior evolutionary experience. Adaptive mechanisms are not in place to deal with them. When weather conditions trap a layer of dense, cool air beneath a layer of warm air, the situation is known as a(n) (2) _____ _____ [p.754]; this has been a key factor in some of the worst air pollution disasters. Where (3) _____ [p.754] are cold and wet, (4) _____ _____ [p.754] develops as a gray haze over industrialized cities that burn coal and other fossil fuels for manufacturing, heating, and generating electric power. In warm climates, (5) _____ _____ [p.754] develops as a brown haze over large cities located in natural basins. The key culprit is nitric oxide. After release from vehicles, it reacts with (6) _____ [p.754] in the air to form (7) _____ _____ [p.754]. When (7) is exposed to sunlight, it reacts with hydrocarbons, and (8) _____ [p.754] oxidants result. Most hydrocarbons come from spilled or partially burned (9) _____ [p.754]. The main oxidants are ozone and (10) _____ (peroxyacyl nitrates) [p.754]. Even traces can sting eyes, irritate lungs, and damage crops.

Oxides of (11) _____ [p.754] and (12) _____ [p.754] are among the worst pollutants. Coal-burning power plants, metal smelters, and factories emit most (13) _____ [p.754] dioxides. Motor vehicles, gas- and oil-burning power plants, and (14) _____ -rich [p.754] fertilizers produce (15) _____ [p.754] oxides. During dry weather, fine particles of oxides may be briefly airborne and then fall to Earth as dry (16) _____ _____ [p.754]. When the oxides dissolve in atmospheric water, they form weak solutions of (17) _____ [p.754] and (18) _____ [p.754] acids. Strong winds may distribute them over great distances; when they fall to Earth in rain and snow, it is called wet acid deposition, or (19) _____ _____ [p.754]. (19) can be 10 to 100 times more acidic than normal rainwater that has a pH of about (20) _____ [p.754]. The deposited acids eat away at marble buildings, metals, rubber, plastics, and even nylon stockings. They also have the potential to disrupt the physiology of organisms and the chemistry of ecosystems. (21) _____ (CFCs) [p.756] are compounds of chlorine, fluorine, and carbon that are odorless and invisible. They are major factors in reduction of the ozone layer in the atmosphere.

Choice

For questions 22–43, choose from the following aspects of atmospheric pollution:

a. thermal inversion   b. industrial smog   c. photochemical smog   d. acid deposition
e. chlorofluorocarbons   f. ozone layer

22. ___ Develops as a brown, smelly haze over large cities [p.754]

23. ___ Includes the "dry" and "wet" types [p.754]

24. ___ Contributes to ozone reduction more than any other factor [p.756]

25. ___ Where winters are cold and wet, this develops as a gray haze over industrialized cities that burn coal and other fossil fuels. [p.755]

26. ___ Weather conditions trap a layer of cool, dense air under a layer of warm air. [p.754]

27. ___ The cause of London's 1952 air pollution disaster, in which 4,000 died [p.754]

28. ___ Each year, from September through mid-October, it thins down at higher altitudes. [p.756]

29. ___ Methyl bromide, a fungicide, will account for about 15 percent of its thinning if production does not stop. [p.756]

30. ___ Intensifies a phenomenon called smog [p.754]

31. ___ Today most of this forms in cities of China, India, and other developing countries, as well as in coal-dependent countries of eastern Europe. [p.754]

32. ___ Contains airborne pollutants, including dust, smoke, soot, ashes, asbestos, oil, bits of lead, and other heavy metals, and sulfur oxides [p.754]

33. ___ Depending on soils and vegetation cover, some regions are more sensitive than others to this. [p.755]

34. ___ Have contributed to some of the worst local air pollution disasters [p.754]

35. ___ Chemically attack(s) marble buildings, metals, mortar, rubber, plastic, and even nylon stockings [pp.754–755]

36. ___ Its reduction allows more ultraviolet radiation to reach the Earth's surface. [p.756]

37. ___ Reaches harmful levels where the surrounding land forms a natural basin, as it does around Los Angeles and Mexico City [p.754]

38. ___ Tall smokestacks were added to power plants and smelters in an unsuccessful attempt to solve this problem. [p.755]

39. ___ As it thins, it lets more ultraviolet radiation reach the Earth. [p.756]

40. ___ A dramatic rise in skin cancers, eye cataracts, immune system weakening, and harm to photosynthesizers is related to its reduction. [p.756]

41. ___ Found in refrigerators and air conditioners (as the coolants), solvents, and plastic foams [p.756]

42. ___ The main culprit is nitric oxide, produced mainly by cars and other vehicles with internal combustion engines. [p.754]

43. ___ Oxides of sulfur and nitrogen dissolve in water to form weak solutions of sulfuric acid and nitric acid that may fall with rain or snow. [p.754]

43.3. WHERE TO PUT SOLID WASTES? WHERE TO PRODUCE FOOD? [p.757]

43.4. DEFORESTATION—AN ASSAULT ON FINITE RESOURCES [pp.758–759]

43.5. *Focus on Bioethics:* YOU AND THE TROPICAL RAIN FOREST [p.760]

Selected Words: *subsistence* agriculture [p.757], *animal-assisted* agriculture [p.757], *mechanized* agriculture [p.757]

Boldfaced, Page-Referenced Terms

[p.757] green revolution _____

_____

[p.758] deforestation _____

_____

[p.758] shifting cultivation _____

_____

Matching

Choose the most appropriate answer for each term.

1. ___throwaway mentality [p.757]
2. ___green revolution [p.757]
3. ___shifting cultivation [pp.758–759]
4. ___animal-assisted agriculture [p.757]
5. ___deforestation [p.758]
6. ___recycling [p.757]
7. ___watersheds of forested regions [p.758]
8. ___subsistence agriculture [p.757]
9. ___new genetic resources [p.760]
10. ___mechanized agriculture [p.757]

A. Agriculture in developing countries runs on energy inputs from sunlight and human labor
B. An affordable, technologically feasible alternative to "throwaway technology"
C. Act like giant sponges that absorb, hold, and gradually release water
D. Potential benefits to be obtained by genetic engineering and tissue culture methods in the rainforests
E. Runs on energy inputs from oxen and other draft animals
F. An attitude prevailing in the United States and other developed countries that greatly adds to solid waste accumulation
G. Requires massive inputs of fertilizers, pesticides, fossil fuel energy, and ample irrigation to sustain high-yield crops
H. Research directed toward improving crop plants for higher yields and exporting modern agricultural practices and equipment to developing countries
I. Trees are cut and burned, then ashes tilled into the soil; crops are grown for one to several seasons on quickly leached soils that become infertile
J. Removal of all trees from large land tracts; leads to loss of fragile soils and disrupts watersheds; greatest today in Brazil, Indonesia, Columbia, and Mexico

43.6. WHO TRADES GRASSLANDS FOR DESERTS? [p.761]

43.7. A GLOBAL WATER CRISIS [pp.762–763]

Selected Words: primary, secondary, and *tertiary* wastewater treatment [p.763]

Boldfaced, Page-Referenced Terms

[p.761] desertification _____

_____

[p.762] desalinization _____

_____

[p.762] salinization _____

_____

[p.762] water table _____

_____

[p.763] wastewater treatment _____

_____

Dichotomous Choice

Circle one of two possible answers given between parentheses in each statement.

1. Conversion of large tracts of grasslands, or rain-fed or irrigated croplands to a more barren state is known as (subsistence agriculture/desertification). [p.761]
2. Presently, (too many cattle in the wrong places/overgrazing on marginal lands) is the main cause of large-scale desertification. [p.761]
3. In Africa, (domestic cattle/native wild herbivores) trample grasses and compact the soil surfaces as they wander about looking for water. [p.761]
4. A 1978 study by biologist David Holpcraft demonstrated that African range conditions improved in land areas where (domestic cattle/native wild herbivores) were ranched. [p.761]
5. Without irrigation and conservation practices, grasslands that were converted for agriculture often end up as (deserts/forested watersheds). [p.761]

Fill-in-the-Blanks

The Earth has a tremendous supply of water, but most is too (6) _____ [p.762] for human consumption or for agriculture. The removal of salt from seawater is called (7) _____ [p.762]. For most countries this process is not practical due to the costly fuel (8) _____ [p.762] necessary to drive it. Large-scale (9) _____ [p.762] accounts for nearly two-thirds of the human population's use of freshwater. Irrigation of otherwise useless soil can cause a salt buildup, or (10) _____ [p.762], due to evaporation in areas of poor soil drainage. Land that drains poorly also becomes waterlogged and raises the (11) _____ _____ [p.762]. When the (11) is too close to the ground's surface, soil becomes saturated with (12) _____ [p.762] water, which can damage plant roots. A large problem is the fact that water tables are subsiding. For example, overdrafts have depleted half of the Ogallala aquifer that supplies irrigation water for (13) [choose one] ❏ 40 ❏ 20 percent [p.762] of the croplands in the United States. Human sewage, **animal wastes**, toxic chemicals, agricultural runoff, sediments, pesticides, and plant nutrients are all sources of water (14) _____ [p.763] that amplifies the problem of water scarcity. There are three levels of (15) _____ [p.763] treatment. They are primary, secondary, and (16) _____ [p.763] treatments. Most of the (14) is not treated adequately. If the current rates of population growth and water depletion hold, the amount of freshwater available for each person on the planet will be (17) [choose one] ❑ 45–56 ❑ 55–66 ❑ 65–76 percent [p.763] less than it was in 1976. Water, not (18) _____ [p.763], may become the most important fluid of the twenty-first century. National, regional, and global (19) _____ [p.763] for water usage and water rights have yet to be developed.

Complete the Table

20. Complete the following table, which summarizes three levels of treatment methods for maintaining the water quality of polluted wastewater.

Treatment Method	Description
[p.763] a. Primary treatment	
[p.763] b. Secondary treatment	
[p.763] c. Tertiary treatment	

43.8. A QUESTION OF ENERGY INPUTS [pp.764–765]

43.9. ALTERNATIVE ENERGY SOURCES [p.766]

43.10. *Focus on Bioethics:* BIOLOGICAL PRINCIPLES AND THE HUMAN IMPERATIVE [p.767]

Selected Words: total energy [p.764], *net* energy [p.764], supertanker *Valdez* [p.764]

Boldfaced, Page-Referenced Terms

[p.764] fossil fuels _____

_____

[p.764] meltdown _____

_____

[p.766] solar-hydrogen energy _____

_____

[p.766] wind farms _____

_____

[p.766] fusion power _____

_____

Dichotomous Choice

Circle one of two possible answers given between parentheses in each statement.

1. Paralleling the (S-shaped/J-shaped) curve of human population growth is a dramatic rise in total and per capita energy consumption. [p.764]
2. The increase in per capita energy consumption is due to (increased numbers of energy users and to extravagant consumption and waste/energy used to locate, extract, transport, store, and deliver energy to consumers). [p.764]
3. (Total energy/Net energy) is that left over after subtracting the energy used to locate, extract, transport, store, and deliver energy to consumers. [p.764]
4. Fossil fuels are the carbon-containing remains of (plants/plants and animals) that lived hundreds of millions of years ago. [p.764]
5. Even with strict conservation efforts, known petroleum and natural gas reserves may be used up during the (current/next) century. [p.764]
6. The net energy (decreases/increases) as costs of extraction and transportation to and from remote areas increase. [p.764]
7. World coal reserves can meet human energy needs for several centuries, but burning releases sulfur dioxides into the atmosphere and adds to the global problem of (global photochemical smog/global acid deposition). [p.764]
8. By 1990 in the United States, it cost slightly more to generate electricity by nuclear energy than by using coal, but today it costs (more/less). [p.764]
9. The danger in the use of radioactivity as an energy supply during normal operation is with potential (radioactivity escape/meltdown). [p.764]
10. After nearly fifty years of research, scientists (have/have not) agreed on the best way to store high-level radioactive wastes. [p.765]
11. When electrodes in photovoltaic cells exposed to sunlight produce an electric current to split water molecules into oxygen and hydrogen gas (potential fuels), it is known as (fusion power/solar-hydrogen energy). [p.766]
12. California gets 1 percent of its electricity from (fusion power/wind farms). [p.766]
13. (Solar-hydrogen energy/Fusion power) involves a mimic of a process occurring in the sun's environment and may provide a good energy source in about fifty years. [p.766]

Sequence–Classify

Arrange the consumption of world resources in correct hierarchical order. Enter the letter of the energy source of highest consumption next to 14, the letter of the next highest next to 15, and so on. Enter an (N) in the parentheses following the letter of the resource if it is nonrenewable and an (R) if the resource is renewable.

14. ___ ( )   A. Hydropower, geothermal, solar [p.764]

15. ___ ( )   B. Natural gas [p.764]

16. ___ ( )   C. Oil [p.764]

17. ___ ( )   D. Nuclear power [p.764]

18. ___ ( )   E. Biomass [p.764]

19. ___ ( )   F. Coal [p.764]

Self-Quiz

____ 1. Which of the following processes is not generally considered a component of secondary wastewater treatment? [p.763]
   a. screens and settling tanks remove sludge
   b. microbial populations are used to break down organic matter
   c. removal of all nitrogen, phosphorus, and toxic substances
   d. chlorine is often used to kill pathogens in the water

____ 2. When fossil-fuel burning gives off dust, smoke, soot, ashes, asbestos, oil, bits of lead, other heavy metals, and sulfur oxides, it forms _____. [p.754]
   a. photochemical smog
   b. industrial smog
   c. a thermal inversion
   d. both a and c

____ 3. _____ result(s) when nitrogen dioxide and hydrocarbons react in the presence of sunlight. [p.754]
   a. Photochemical smog
   b. Industrial smog
   c. A thermal inversion
   d. Both a and c

____ 4. When weather conditions trap a layer of cool, dense air under a layer of warm air, _____ occurs. [p.754]
   a. photochemical smog
   b. a thermal inversion
   c. industrial smog
   d. acid deposition

____ 5. A major concern of the use of nuclear energy is _____. [pp.764–765]
   a. excessive temperature
   b. meltdown
   c. waste disposal
   d. all of the above

____ 6. Nitrogen oxides dissolve in atmospheric water to form a weak solution of sulfuric acid and nitric acid; this describes _____. [p.754]
   a. photochemical smog
   b. industrial smog
   c. ozone and PANs
   d. acid rain

____ 7. Which of the following statements is false? [p.762]
   a. Ozone reduction allows more ultraviolet radiation to reach the Earth's surface.
   b. CFCs enter the atmosphere and resist breakdown.
   c. Salinization of soils aids plant growth and increases yields.
   d. CFCs already in the air will be there for over a century.

____ 8. "Adequately reduces pollution but is largely experimental and expensive," describes _____ wastewater treatment. [p.763]
   a. quaternary
   b. secondary
   c. tertiary
   d. primary

____ 9. The statement "photovoltaic cells exposed to sunlight produce an electric current that splits water molecules into oxygen and hydrogen gas" refers to _____. [p.766]
   a. fusion power
   b. wind energy
   c. water power
   d. solar-hydrogen energy

____10. Energy inputs from sunlight and human labor is the basis of _____. [p.757]
   a. animal-assisted agriculture
   b. subsistence agriculture
   c. the green revolution
   d. mechanized agriculture

Chapter Objectives/Review Questions

1. Identify the principal air pollutants, their sources, their effects, and the possible methods for controlling each pollutant. [pp.754–755]
2. During a(n) _____ _____, weather conditions trap a layer of cool, dense air under a layer of warm air; trapped pollutants may reach dangerous levels. [p.754]
3. Distinguish photochemical smog from industrial smog. [p.754]
4. Explain what acid rain does to an ecosystem. Contrast those effects with the action of CFCs. [pp.754–755]
5. Discuss the significance of the effects of the ozone layer's thinning to life on Earth. [p.756]
6. List the key sources of air pollutants as related to ozone layer thinning. [p.756]
7. Under the banner of the _____ _____, research has been directed towards improving the genetics of crop plants for higher yields and exporting modern agricultural practices and equipment to the developing countries. [p.757]
8. _____ agriculture runs on energy inputs from sunlight and human labor; _____-_____ agriculture runs on energy inputs from oxen and otherdraft animals; _____ agriculture requires massive inputs of fertilizers, pesticides, and ample irrigation to sustain high-yield crops. [p.757]
9. Explain the repercussions of deforestation that are evident in soils, water quality, and genetic diversity in general. [pp.758–759]
10. _____ _____ involves cutting and burning trees, tilling ashes with soil, planting crops from one to several seasons, and then abandoning the clear plots. [pp.758–759]
11. Examine the effects that modern agriculture has wrought on desert and grassland ecosystems. [p.761]
12. _____ is the conversion of large tracts of natural grasslands to a more desertlike state. [p.761]
13. Explain why desalination is not a practical solution for the shortage of freshwater. [p.762]
14. Define *primary*, *secondary*, and *tertiary* wastewater treatment, and list some of the methods used in each of the three types of treatment. [p.763]
15. Explain the meaning of "the coming water wars." [p.763]
16. _____ energy refers to the amount left over after subtracting the energy that is used to locate, extract, transport, store, and deliver energy to consumers. [p.764]
17. Be able to describe the dangers accompanying a meltdown. [pp.764–765]
18. Describe how our use of fossil fuels, solar energy, and nuclear energy affects ecosystems. [pp.764–765]
19. Briefly characterize solar-hydrogen energy, wind energy, and fusion power as alternative sources of energy. [p.766]
20. List five ways in which you could become personally involved in ensuring that institutions serve the public interest in a long-term, ecologically sound way. [p.767 and the chapter]

Integrating and Applying Key Concepts

1. If you were Ruler of All People on Earth, how would you encourage people to depopulate the cities and adopt a way of life by which they could supply their own resources from the land and dispose of their own waste products safely on their own land?
2. Explain why some biologists believe that the endangered species list now includes all species.

44

AN EVOLUTIONARY VIEW OF BEHAVIOR

Interactive Exercises

Deck the Nest With Sprigs of Green Stuff [pp.770–771]

44.1. BEHAVIOR'S HERITABLE BASIS [pp.772–773]

44.2. LEARNED BEHAVIOR [p.774]

Selected Words: Sturnus [p.770], *Ornithonyssus* [p.771], *intermediate* response [p.772]

Boldfaced, Page-Referenced Terms

[p.771] animal behavior _____

_____

[p.773] song system _____

_____

[p.773] instinctive behavior _____

_____

[p.773] sign stimuli _____

_____

[p.773] fixed action pattern _____

[p.774] learned behavior _____

[p.774] imprinting _____

Matching

Choose the most appropriate answer for each term.

1. ___ intermediate response [p.772]
2. ___ song system [p.773]
3. ___ instinctive behavior [p.773]
4. ___ sign stimuli [p.773]
5. ___ fixed action pattern [p.773]
6. ___ learned behavior [p.773]
7. ___ imprinting [p.773]

A. Animals process and integrate information gained from experiences, then use it to vary or change responses to stimuli
B. Well-defined environmental cues that trigger suitable responses
C. Term applied to genetically based behavioral reactions of hybrid offspring
D. Time-dependent form of learning; triggered by exposure to sign stimuli and usually occurring during sensitive periods of young animals
E. A behavior performed without having been learned by actual environmental experience
F. Consists of several brain structures that will govern activity of a vocal organ's muscles
G. A program of coordinated muscle activity that runs to completion independently of feedback from the environment

Dichotomous Choice

Circle one of two possible answers given between parentheses in each statement.

8. For garter snake populations living along the California coast, the food of choice is (the banana slug/tadpoles and small fishes). [p.772]
9. In Stevan Arnold's experiments, newborn garter snakes that were offspring of coastal parents usually (ate/ignored) a chunk of slug as the first meal. [p.772]
10. Newborn garter snake offspring of (coastal/inland) parents ignored cotton swabs drenched in essence of slug and only rarely ate the slug meat. [p.772]
11. The differences in the behavioral eating responses of coastal and inland snakes (were/were not) learned. [p.772]
12. Hybrid garter snakes with coastal and inland parents exhibited a feeding response that indicated a(n) (environmental/genetic) basis for this behavior. [p.772]
13. In zebra finches and some other songbirds, singing behavior is an outcome of seasonal differences in the secretion of melatonin, a hormone secreted by the (gonads/pineal gland). [p.772]
14. In songbirds in spring, melatonin secretion is suppressed and gonads are released from hormonal suppression; they (increase/decrease) in size and step up their secretions of estrogen and testosterone. [p.773]
15. Even before a male bird hatches, a high (estrogen/testosterone) level stimulates development of a masculinized brain. [p.773]
16. Later, at the start of the breeding season, a male's enlarged gonads secrete even more (estrogen/testosterone) that binds to cells and induces metabolic changes in the sound system to prepare the bird to sing when suitably stimulated. [p.773]
17. (Hormones/Genes) underlie animal behavior—coordinated responses to stimuli. [p.773]

Complete the Table

18. Complete the following table to consider examples of instinctive and learned behavior.

Category	Examples
[p.773] a. Instinctive behavior	
[p.774] b. Learned behavior	

Matching

Choose the most appropriate answer for each category of learned behavior.

19. ___imprinting [p.774]

20. ___classical conditioning [p.774]

21. ___operant conditioning [p.774]

22. ___habituation [p.774]

23. ___spatial or latent learning [p.774]

24. ___insight learning [p.774]

A. Birds living in cities learn not to flee from humans or cars, which pose no threat to them.
B. Chimpanzees abruptly stack several boxes and use a stick to reach suspended bananas out of reach.
C. In response to a bell, dogs salivate even in the absence of food.
D. Bluejays store information about dozens or hundreds of places where they have stashed food.
E. Baby geese formed an attachment to Konrad Lorenz if separated from the mother shortly after hatching.
F. A toad learns to avoid stinging or bad-tasting insects after attempts to eat them.

44.3. THE ADAPTIVE VALUE OF BEHAVIOR [p.775]

Boldfaced, Page-Referenced Terms

[p.775] natural selection _____

_____

[p.775] reproductive success _____

_____

[p.775] adaptive behavior _____

_____

[p.775] social behavior _____

_____

[p.775] selfish behavior _____

_____

[p.775] altruistic behavior _____

_____

[p.775] territory _____

_____

Fill-in-the-Blanks

(1) _____ _____ [p.775] is the result of differences in survival and reproduction among individuals of a population that differ from one another in their heritable traits. Some versions of a trait may be better than others at helping the individual leave offspring. Thus alleles for those versions (2) _____ [p.775] in a population. Using the theory of (3) _____ [p.775] by natural selection as a starting point, one should be able to identify (4) _____ [p.775] forms of behavior—and to discern how they bestow (5) _____ [p.775] benefits that offset reproductive costs (disadvantages) or disadvantages that might be associated with them. If a behavior is (6) _____ [p.775], it must promote the (7) _____ [p.775] production of offspring.

Complete the Table

8. Complete the following table of terms describing forms of individual adaptive behavior.

Behavior	Description
[p.775] a. Reproductive success	
[p.775] b. Adaptive behavior	
[p.775] c. Social behavior	
[p.775] d. Selfish behavior	
[p.775] e. Altruistic behavior	

44.4. COMMUNICATION SIGNALS [pp.776–777]

Selected Words: *signaling* pheromones [p.776], *priming* pheromones [p.776]

Boldfaced, Page-Referenced Terms

[p.776] communication signals _____

_____

[p.776] signaler _____

_____

[p.776] signal receivers _____

_____

[p.776] pheromones _____

_____

[p.776] composite signal _____

_____

[p.776] communication display _____

_____

[p.776] threat display _____

_____

[p.777] courtship displays _____

_____

[p.777] tactile displays _____

_____

[p.777] illegitimate receiver _____

_____

[p.777] illegitimate signalers _____

_____

Choice

For questions 1–12, choose from the following:

a. chemical signal, signaling pheromones
b. chemical signal, priming pheromones
c. acoustical signal
d. composite signal
e. communication display, social signal
f. communication display, threat display
g. communication display, courtship
h. communication display, tactile signal
i. illegitimate signaler
j. illegitimate receiver

1. ___ Assassin bugs hook dead termite bodies on their dorsal surfaces and acquire termite scent; this deception allows assassin bugs to hunt termite victims more easily. [p.777]

2. ___ Male songbirds sing to secure territory and attract a female. [p.776]

3. ___ The play bow of dogs and wolves [p.776]

4. ___ Bombykol molecules released by female silk moths serve as sex attractants. [p.776]

5. ___ A male bird might emit calls while bowing low, as if to peck the ground for food. [p.777]

6. ___ Termites act defensively when detecting scents from invading ants whose scent signals are meant to elicit cooperation from other ants. [p.777]

7. ___ A dominant male baboon's "yawn" that exposes large canines [pp.776–777]

8. ___ A volatile odor in the urine of certain male mice triggers and enhances estrus in female mice. [p.776]

9. ___ After finding a source of pollen or nectar, a foraging honeybee returns to its colony (a hive) and performs a complex dance. [p.777]

10. ___ Tungara frogs issue nighttime calls to females and rival males; the call is a "whine" followed by a "chuck." [p.776]

11. ___ A male firefly has a light-generating organ that emits a bright, flashing signal. [p.777]

12. ___ Ears laid back against the head of a zebra convey hostility, but ears pointing up convey its absence; a zebra with laid-back ears isn't too riled up when its mouth is open just a bit, but when the mouth is gaping, watch out. [p.776]

44.5. MATES, PARENTS, AND INDIVIDUAL REPRODUCTIVE SUCCESS [pp.778–779]

Selected Words: *Harpobittacus* [p.778], *Centrocercus* [p.779]

Boldfaced, Page-Referenced Terms

[p.778] sexual selection _____

_____

[p.779] lek _____

_____

Complete the Table

1. Complete the following table to supply the common names of the animals that fit the text examples of sexual selection.

Animals	Descriptions of Sexual Selection
[pp.778–779] a.	Females select the males that offer them superior material goods; females permit mating only after they have eaten the "nuptial gift" for about five minutes.
[p.779] b.	Males congregate in a lek or communal display ground; each male stakes out a few square meters as his territory; females are attracted to the lek to observe male displays and usually select and mate with only one male.
[p.779] c.	Females of a species cluster in defendable groups at a time they are sexually receptive; males compete for access to the clusters; combative males are favored.
[p.779] d.	Extended parental care improves the likelihood that the current generation of offspring will survive; this behavior comes at a reproductive cost to the adults.

44.6. BENEFITS OF LIVING IN SOCIAL GROUPS [pp.780–781]

44.7. COSTS OF LIVING IN SOCIAL GROUPS [p.782]

44.8. EVOLUTION OF ALTRUISM [p.783]

44.9. *Focus on Science:* WHY SACRIFICE YOURSELF? [pp.784–785]

Selected Words: *Ovibos* [p.780], *Parus* [p.780], caring for one's *relatives* [p.783], *indirect* genetic contribution [p.783], "self-sacrifice" genes [p.783], *Heterocephalus* [p.785], *DNA fingerprinting* [p.785]

Boldfaced, Page-Referenced Terms

[p.780] selfish-herd _____

_____

[p.781] dominance hierarchies _____

_____

[p.783] theory of indirect selection _____

_____

Matching

Choose the most appropriate answer for each term.

1. ___cost-benefit approach [p.780]
2. ___disadvantages to sociality [p.782]
3. ___dominance hierarchy [p.781]
4. ___cooperative predator avoidance [p.780]
5. ___the selfish herd [pp.780–781]

A. Competition for resources, rapid depletion of food resources, cannibalism, and greater vulnerability to disease

B. A simple society brought together by reproductive self-interest; larger, more powerful male bluegills tend to claim the central locations

C. An attempt to identify the costs and benefits of sociality in terms of reproductive success of the individual

D. Adult musk oxen form a circle around their young while they face outward, and the "ring of horns" successfully deters the wolves; writhing, regurgitating reaction of Australian sawfly caterpillers to a disturbance

E. Some individuals of a baboon troop adopt a subordinate status with respect to the other members

Complete the Table

6. Complete the following table to supply examples of each listed category of the costs of living in social groups.

Costs of Living in Social Groups	Examples
[p.782] a. An increase in the competition for food	
[p.782] b. Encourages the spread of contagious diseases and parasites	
[p.782] c. Risks of being killed or exploited by others in the group	

Fill-in-the-Blanks

A(n) (7) _____ [p.783] animal that gives way to a dominant one is acting in its own self-interest. Such (8) _____ [p.783] animals are found among many vertebrate groups. A female wolf's (9) _____ [p.783] success increases when she monopolizes the benefits that males offer. Within a wolf pack, there is usually a(n) (10) _____ [p.783] breeding female and male. The other wolves are (11) _____ [p.783] sisters, aunts, brothers, and uncles. Their (12) _____ [p.783] behavior is to hunt and bring back food to members that remain in the den and guard the pups. Altruistic behavior is most extreme in certain (13) _____ [p.783] societies. When a(n) (14) _____ [p.783] bee plunges its stinger into an invader of the hive, it commits suicide. Altruistic individuals of a social group do not contribute their (15) _____ [p.783] to the next generation, but yet seem to perpetuate the genetic basis for their altruistic behavior over evolutionary time. According to William Hamilton's theory of (16) _____ _____ [p.783], those genes associated with caring for relatives—not one's direct descendants—tend to be favored in certain situations. This form of altruism can be thought of as an extension of (17) _____ [p.783]. For example, if an uncle helps his niece survive long enough to reproduce, he has made a(n) (18) _____ [p.783] genetic contribution to the next generation, as measured in terms of the genes that he and his relatives share. Altruism costs him; he may lose his own opportunities to (19) _____ [p.783]. Similarly, nonbreeding workers in insect societies indirectly promote their "self-sacrifice" genes through (20) _____ [p.783] behavior directed toward relatives. Thus, when a guard bee drives her stinger into a raccoon, she inevitably dies—but her siblings in the hive will perpetuate some of her (21) _____ [p.783]. Sterility and extreme self-sacrifice are rare among social groups of (22) _____ [p.783].

Unlike any other known vertebrate, the highly social naked mole-rat individuals live out their lives as (23) _____ [p.785] helpers in their social group. In each mole-rat clan, there is a single reproducing female, and she mates with one to three males. All other members of the clan care for the "queen" and "king" (or kings) and their offspring. A self-sacrificing naked mole-rat helps to (24) _____ [p.785] a very high proportion of the forms of genes that it carries. Following studies utilizing a technique called DNA fingerprinting, it turns out that the (25) _____ [p.785] of helpers and the helped might be as much as 90 percent identical.

44.10. AN EVOLUTIONARY VIEW OF HUMAN SOCIAL BEHAVIOR [p.786]

Selected Words: *redirected* adaptive behaviors [p.786]

Boldfaced, Page-Referenced Terms

[p.786] adoption _____

_____

Dichotomous Choice

Circle one of two possible answers given between parentheses in each statement.

1. Many people seem to believe that attempts to identify the adaptive value of a particular (animal/human) trait is an attempt to define its moral or social advantage. [p.786]
2. "Adaptive" refers to (a trait with moral value/a trait valuable in gene transmission). [p.786]
3. In many species, adults that have lost their offspring will, if presented with a substitute, (adopt/reject) it. [p.786]
4. John Alcock suggests that husbands and wives who have lost an only child or who fail to produce children themselves should be especially prone to adopt (strangers/relatives). [p.786]
5. The human adoption process (can/cannot) be considered adaptive when indirect selection favors adults who direct parenting assistance to relatives. [p.786]
6. Joan Silk showed that in some traditional societies, people (will not/will) adopt related children far more often than nonrelated ones. [p.786]
7. In large, industrialized societies in which agencies and other means of adoption exist, individuals become parents of (related/nonrelated) children. [p.786]
8. It (is/is not) possible to test evolutionary hypotheses about the adaptive value of human behaviors. [p.786]
9. *Adaptive* behavior and *socially desirable* behavior (are not/are) separate issues. [p.786]
10. Strong parenting mechanisms evolved in the past, and it may be that their redirection toward a (relative/nonrelative) says more about human evolutionary history than it does about the transmission of one's genes. [p.786]

Self-Quiz

___ 1. The observable, coordinated responses that animals make to stimuli are what we call _____. [p.771]
  a. imprinting
  b. instinct
  c. behavior
  d. learning

___ 2. In _____, a particular behavior is performed without having been learned by actual experience in the environment. [p.773]
  a. natural selection
  b. altruistic behavior
  c. sexual selection
  d. instinctive behavior

___ 3. Newly hatched goslings follow any large moving objects to which they are exposed shortly after hatching; this is an example of _____. [p.774]
  a. homing behavior
  b. imprinting
  c. piloting
  d. migration

___ 4. A young toad flips its sticky-tipped tongue and captures a bumblebee that stings its tongue; in the future, the toad leaves bumblebees alone. This is _____. [p.774]
  a. instinctive behavior
  b. a fixed reaction pattern
  c. altruistic
  d. learned behavior

___ 5. Pavlov's dog experiments represent an example of _____. [p.774]
  a. classical conditioning
  b. latent learning
  c. selfish behavior
  d. habituation

___ 6. _____ provides an example of an illegitimate signaler. [p.777]
  a. A soldier termite killing an ant on cue
  b. An assassin bug with acquired termite odor
  c. A termite pheromone alarm signal
  d. The "yawn" of a dominant male baboon

_____ 7. The claiming of the more protected central locations of the bluegill colony by the largest, most powerful males suggests _____. [pp.780–781]
  a. cooperative predator avoidance
  b. the selfish herd
  c. a huge parent cost
  d. self-sacrificing behavior

_____ 8. A chemical odor in the urine of male mice triggers and enhances estrus in female mice. The source of stimulus for this response is a _____. [p.776]
  a. generic mouse pheromone
  b. signaling pheromone
  c. priming pheromone
  d. cue from male mice

_____ 9. Female insects often attract mates by releasing sex pheromones. This is an example of a(n) _____ signal. [p.776]
  a. chemical
  b. visual
  c. acoustical
  d. tactile

_____10. Male birds sing to stake out territories, attract females, and discourage males. This is an example of a(n) _____ signal. [p.776]
  a. chemical
  b. visual
  c. acoustical
  d. tactile

_____11. An example of dominance hierarchy and self-sacrificing behavior is _____. [p.783]
  a. cannibalistic behavior of a breeding pair of herring gulls in a huge nesting colony
  b. members of wolf packs helping others by sharing food or fending off predators even though they do not breed

  c. clumps of regurgitating Australian sawfly caterpillars
  d. a huge colony of prairie dogs being ravaged by a parasite

_____12. When musk oxen form a "ring of horns" against predators, it is _____. [p.780]
  a. a selfish herd
  b. cooperative predator avoidance
  c. self-sacrificing behavior
  d. dominance hierarchy

_____13. Caring for nondescendant relatives favors the genes associated with helpful behavior and is classified as _____. [p.781]
  a. dominance hierarchy
  b. indirect selection
  c. altruism
  d. both b and c

_____14. "_____" means only that a given trait has proved valuable in the transmission of an individual's genes. [p.786]
  a. Dominance hierarchy
  b. Indirect selection
  c. Adaptive
  d. Altruism

_____15. _____ also favors adults who direct parenting behavior toward relatives and so indirectly perpetuate their shared genes. [p.783]
  a. Indirect selection
  b. Moral selection
  c. Redirected selection
  d. Perpetuated selection

Chapter Objectives/Review Questions

1. What explains the fact that coastal and inland garter snakes of the same species have different food preferences? [p.772]
2. Describe the intermediate response obtained in Arnold's experiment with coastal and inland garter snakes. [p.772]
3. Tongue-flicking, body orientation, and strikes at prey by newborn garter snakes are good examples of _____ behavior. [p.772]
4. Describe the origin and formation of a song system. [pp.772–773]
5. Define *sign stimuli*. [p.773]
6. Describe and cite an example of a fixed action pattern. [p.773]

7. When animals incorporate and process information gained from specific experiences and then use the information to vary or change responses to stimuli, it is _____ behavior. [p.774]

8. Define each of the following categories of learned behavior and give one example of each: imprinting, classical conditioning, operant conditioning, habituation, spatial or latent learning, and insight learning. [p.774]

9. What is meant by reproductive success? [p.775]

10. _____ behavior is any behavior that promotes the propagation of an individual's genes and tends to occur at increased frequency in successive generations. [p.775]

11. _____ behavior refers to the cooperative, interdependent relationships among individuals of the species. [p.775]

12. Distinguish between selfish behavior and altruistic behavior. [p.775]

13. A(n) _____ is an area that one or more individuals defend against competitors. [p.775]

14. Examples of _____ signals are chemical, visual, acoustical, and tactile. [p.776]

15. Define the roles of signalers and signal receivers. [p.776]

16. Distinguish between signaling and priming pheromones, and cite an example of each. [p.776]

17. A(n) _____ signal is illustrated by a zebra with laid-back ears and a gaping mouth. [p.776]

18. Describe one example of a threat display. [pp.776–777]

19. Ritualization is often developed to an amazing degree in _____ displays between potential mates. [p.777]

20. An example of a(n) _____ signal is the physical contact of bees in a hive maintaining physical contact during the dance of a returning foraging bee. [p.777]

21. When soldier termites detect ant scents meant for other ants and kill ants on cue, the termites are said to be _____ _____ of a signal meant for individuals of a different species. [p.777]

22. Assassin bugs covered with termite scent are able to use deception to hunt termite victims more easily and as such are acting as _____ signalers. [p.777]

23. Natural selection tends to favor communication signals that promote _____ success. [p.777]

24. Competition among members of one sex for access to mates and selection of mates are the result of a microevolutionary process called _____ _____. [p.778]

25. Be able to discuss mate selection processes by female hangingflies and the female sage grouse. [pp.778–779]

26. List the costs and benefits of parenting in the example of adult Caspian terns. [p.779]

27. Explain the "cost-benefit approach" that evolutionary biologists utilize to find answers to the questions about social life. [p.780]

28. Studies of Australian sawfly caterpillars indicate _____ predator avoidance. [p.780]

29. Define *selfish herd*; cite an example. [pp.780–781]

30. Members of baboon troops recognize a _____ _____ in which some individuals have adopted a subordinate status with respect to the others. [p.781]

31. List disadvantages to sociality. [p.782]

32. Hamilton's theory of _____ selection relates to caring for nondescendant relatives and how this favors genes associated with helpful behavior. [p.783]

33. With _____ behavior, an individual behaves in a self-sacrificing way that helps others but decreases its own chance of reproductive success. [p.783]

34. Be able to discuss the self-sacrificing behavior of honeybees and termites in a eucalyptus forest in Australia. [pp.783,784–785]

35. Explain how DNA fingerprinting was used to establish that self-sacrificing mole-rats help to perpetuate a very high proportion of genes (alleles) that they carry—even though they are not the reproducing mole-rats. [p.785]

36. It is possible to test evolutionary hypotheses about the _____ value of human behaviors; discuss human adoption practices in this context. [p.786]

37. "_____" means only that a specified trait has proved beneficial in the transmission of the genes of an individual that are responsible for that trait. [p.786]

Integrating and Applying Key Concepts

Think about communication signals that humans use and list them. Do you believe a dominance hierarchy exists in human society? Think of examples.

ANSWERS

Chapter 1   Concepts and Methods in Biology

Biology Revisited [pp.2–3]

1.1. DNA, ENERGY, AND LIFE [pp.4–5]
1. Cells; 2. DNA; 3. proteins; 4. amino; 5. Enzymes;
6. RNAs; 7. RNAs; 8. proteins; 9. reproduction;
10. development; 11. Energy; 12. transfer; 13. Metabolism; 14. photosynthesis; 15. ATP; 16. aerobic respiration; 17. responses; 18. receptors; 19. stimuli;
20. stimulus; 21. internal; 22. homeostasis.

1.2. ENERGY AND LIFE'S ORGANIZATION [pp.6–7]
1. F; 2. E; 3. H; 4. J; 5. G; 6. C; 7. L; 8. B; 9. M;
10. D; 11. K; 12. A; 13. I; 14. D; 15. G; 16. B; 17. J;
18. L; 19. E; 20. M; 21. A; 22. I; 23. F; 24. C; 25. H;
26. K; 27. producers; 28. consumers; 29. Decomposers;
30. energy; 31. cycling.

1.3. IF SO MUCH UNITY, WHY SO MANY SPECIES?
  [pp.8–9]
1. species; 2. genus; 3. genus; 4. species; 5. family;
6. order; 7. class; 8. phylum; 9. prokaryotic;
10. eukaryotic; 11. a. Plantae; b. Eubacteria; c. Fungi;
d. Archaebacteria; e. Protista; f. Animalia; 12. D;
13. F; 14. A; 15. E; 16. B; 17. C; 18. G.

1.4. AN EVOLUTIONARY VIEW OF DIVERSITY
  [pp.10–11]
1. a; 2. b; 3. b; 4. a; 5. b; 6. b; 7. a; 8. a; 9. b; 10. b.

1.5. THE NATURE OF BIOLOGICAL INQUIRY
  [pp.12–13]
**1.6. *Focus on Science:* THE POWER OF
  EXPERIMENTAL TESTS** [pp.14–15]
1.7. THE LIMITS OF SCIENCE [p.15]
1. G; 2. A; 3. D; 4. B; 5. E; 6. C; 7. F; 8. O; 9. O;
10. C; 11. O; 12. C; 13. a. Hypothesis; b. Prediction;
c. Theory; d. Scientific experiment; e. Control group;
f. Variable; g. Inductive logic; h. Deductive logic;
14. experiment; 15. control; 16. hypothesis; 17. prediction; 18. belief; 19. inductive logic; 20. deductive logic;
21. variable; 22. test predictions; 23. cause and effect;
24. sampling error; 25. quantitative; 26. subjective;
27. supernatural; 28. conviction.

Self-Quiz
1. d; 2. a; 3. b; 4. c; 5. d; 6. b; 7. d; 8. a; 9. c; 10. e.

Chapter 2   Chemical Foundations for Cells

Checking Out Leafy Clean-Up Crews [pp.20–21]

2.1. REGARDING THE ATOMS [p.22]
**2.2. *Focus on Science:* USING RADIOISOTOPES TO
  TRACK CHEMICALS AND SAVE LIVES** [p.23]
1. M; 2. C; 3. H; 4. D; 5. N; 6. B; 7. F; 8. K; 9. J;
10. L; 11. E; 12. G; 13. A; 14. I.

**2.3. WHAT HAPPENS WHEN ATOM BONDS WITH
  ATOM?** [pp.24–25]
1. C; 2. E; 3. I; 4. H; 5. A; 6. D; 7. F; 8. B; 9. G;
10. a. Calcium, Ca; b. Carbon, C; c. Chlorine, Cl;
d. Hydrogen, H; e. Sodium, Na; f. Nitrogen, N;
g. Oxygen, O; 11. equation; 12. formula; 13. yields;
14. reactants; 15. products; 16. 12; 17. 98 grams.

18.

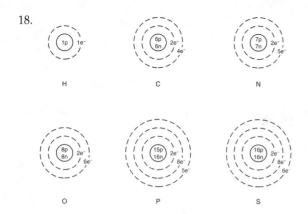

2.4. IMPORTANT BONDS IN BIOLOGICAL MOLECULES [pp.26–27]

1.

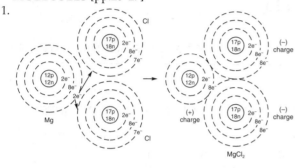

2.

3. In a covalent bond, atoms share electrons to fill their outermost shells. In a nonpolar covalent bond, atoms attract shared electrons equally. An example is the H_2 molecule. In a polar covalent bond, atoms do not share electrons equally, and the bond is positive at one end, negative at the other (for example, the water molecule). 4. In the linear DNA molecule, the two nucleotide chains are held together by hydrogen bonds.

2.5. PROPERTIES OF WATER [pp.28–29]
1. polarity; 2. hydrophilic; 3. hydrophobic; 4. Temperature; 5. hydrogen; 6. evaporation; 7. ice; 8. cohesion; 9. solvent; 10. solutes; 11. dissolved; 12. hydration.

2.6. ACIDS, BASES, AND BUFFERS [pp.30–31]
1. N; 2. K; 3. F; 4. J; 5. H; 6. D; 7. I; 8. A; 9. B; 10. C; 11. O; 12. L; 13. E; 14. G; 15. M; 16. a. 7.3–7.5, slightly basic; b. 6.2–7.4, variable or slightly basic or slightly acid; c. 5.0–7.0, slightly acid to neutral; d. 1.0–3.0, acid; e. 7.8–8.3, basic; f. 3.0, acid.

Self-Quiz
1. c; 2. a; 3. d; 4. d; 5. c; 6. c; 7. a; 8. c; 9. a; 10. d; 11. e.

Chapter 3   Carbon Compounds in Cells

Carbon, Carbon, in the Sky—Are You Swinging Low and High? [pp.34–35]

3.1. PROPERTIES OF ORGANIC COMPOUNDS [pp.36–37]
1. methyl; 2. hydroxyl; 3. ketone; 4. amino; 5. phosphate; 6. carboxyl; 7. aldehyde; 8. Enzymes represent a special class of proteins that speed up specific metabolic reactions. Enzymes mediate five categories of reactions by which most of the biological molecules are assembled, rearranged, and broken apart; 9. a. A juggling of internal bonds converts one type of organic compound into another; b. A molecule splits into two smaller ones; c. Through covalent bonding, two molecules combine to form a larger molecule; d. One or more electrons stripped from one molecule are donated to another molecule; e. One molecule gives up a functional group, which another molecule accepts; 10.

amino acid   amino acid   dipeptide

11. Hydrolysis reactions reverse the chemistry of condensation reactions; in the presence of water, large molecules are split into their component smaller molecules. Both condensation and hydrolysis require the presence of enzymes specific to the particular molecules involved.

3.2. CARBOHYDRATES [pp.38–39]
1.

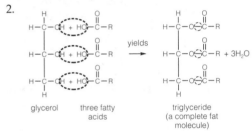

glucose   glucose   maltose   water
(a monosaccharide) (a monosaccharide)   (a disaccharide)

2. a. Sucrose, Oligosaccharide (disaccharide); b. Deoxyribose, Monosaccharide; c. Glucose, Monosaccharide; d. Cellulose, Polysaccharide; e. Ribose, Monosaccharide; f. Lactose, Oligosaccharide (disaccharide); g. Glycogen, Polysaccharide; h. Chitin, Polysaccharide; i. Glycogen, Polysaccharide; j. Starch, Polysaccharide.

3.3. LIPIDS [pp.40–41]
1. a. unsaturated; b. saturated
2.

glycerol   three fatty   triglyceride
        acids   (a complete fat
                     molecule)

3. Phospholipid molecules have two fatty acid tails attached to a glycerol backbone; they have hydrophilic heads that dissolve in water. Phospholipids are the main structural materials of cell membranes; 4. B; 5. D; 6. E; 7. A; 8. B; 9. E; 10. E; 11. E; 12. D; 13. E; 14. A; 15. B; 16. B; 17. D; 18. B.

3.4. AMINO ACIDS AND THE PRIMARY STRUCTURE OF PROTEINS [pp.42–43]
3.5. HOW DOES A PROTEIN'S THREE-DIMENSIONAL STRUCTURE EMERGE? [pp.44–45]

3.6. *Focus on the Environment:* FOOD PRODUCTION AND A CHEMICAL ARMS RACE [p.46]
1. A; 2. C; 3. B;
4.

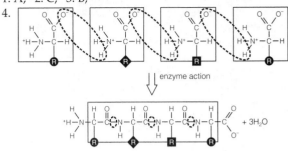

5. K; 6. F; 7. B; 8. J; 9. L; 10. A; 11. D; 12. I; 13. G; 14. E; 15. C; 16. H.

3.7. NUCLEOTIDES AND THE NUCLEIC ACIDS [p.47]
1. B; 2. A; 3. C;
4. Three as shown:

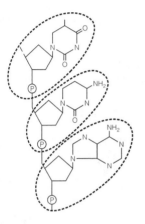

5. B; 6. A; 7. C.

Self-Quiz
1. a. Lipids; b. Nucleic acids; c. Proteins; d. Proteins; e. Nucleic acids; f. Carbohydrates; g. Nucleic Acids; h. Lipids; i. Nucleic acids; j. Lipids; k. Lipids; l. Carbohydrates; 2. d; 3. a; 4. c; 5. c; 6. c; 7. b; 8. d; 9. c; 10. b; 11. b.

Chapter 4   Cell Structure and Function

Animalcules and Cells Fill'd With Juices [pp.50–51]

4.1. BASIC ASPECTS OF CELL STRUCTURE AND FUNCTION [pp.52–53]
4.2. CELL SIZE AND SHAPE [p.54]
4.3. *Focus on Science:* MICROSCOPES—GATEWAYS TO CELLS [pp.54–55]
1. plasma membrane (B); 2. cytoplasm (A); 3. DNA-containing region (C); 4. adhesion; 5. transport;

6. transport; 7. transport; 8. transport; 9. receptor; 10. recognition; 11. J; 12. I; 13. D; 14. C; 15. A; 16. B; 17. H; 18. G; 19. E; 20. F; 21. Cell size is constrained by the surface-to-volume ratio. Past a certain point of growth, the surface area will not be sufficient to admit enough nutrients and to allow enough wastes to exit the cell. As a cell grows, the surface area increases with the square and the volume increases with the cube. 22. F; 23. E; 24. G; 25. C; 26. B; 27. A; 28. D.

4.4. THE DEFINING FEATURES OF EUKARYOTIC CELLS [pp.56–59]
1. a. Nucleus; b. Ribosomes; c. Endoplasmic reticulum; d. Golgi body; e. Various vesicles; f. Mitochondria; g. Cytoskeleton.

4.5. THE NUCLEUS [pp.60–61]
1. First, the nucleus physically separates the cell's DNA from the metabolic machinery of the cytoplasm; this permits easier copying of DNA and the distribution of new DNA molecules into new cells following division. Second, the nuclear envelope helps control the exchange of substances and signals between the nucleus and the cytoplasm. 2. a. Nucleolus; b. Nuclear envelope; c. Chromatin; d. Nucleoplasm; e. Chromosome; 3. ribosomes; 4. cytoplasm; 5. cytomembrane; 6. endoplasmic; 7. Golgi; 8. DNA's; 9. proteins; 10. Lipids; 11. enzymes; 12. Vesicles.

4.6. THE CYTOMEMBRANE SYSTEM [pp.62–63]
1. I; 2. A; 3. B; 4. H; 5. C; 6. C; 7. F; 8. L; 9. K; 10. E; 11. D; 12. G; 13. J.

4.7. MITOCHONDRIA [p.64]
4.8. SPECIALIZED PLANT ORGANELLES [p.65]
1. b; 2. a; 3. c; 4. a; 5. e; 6. a; 7. d; 8. b; 9. e; 10. a; 11. d; 12. b; 13. b; 14. a; 15. c; 16. a; 17. b; 18. d; 19. a; 20. b.

4.9. THE CYTOSKELETON [pp.66–67]

4.10. CELL SURFACE SPECIALIZATIONS [pp.68–69]
1. eukaryotic; 2. Protein; 3. motility; 4. tubulin; 5. actin; 6. Intermediate filaments; 7. assembly; 8. shortens; 9. cytoplasmic streaming; 10. flagella; 11. cilia; 12. cilia; 13. centriole; 14. dynein; 15. C; 16. E; 17. B; 18. A; 19. F; 20. D.

4.11. PROKARYOTIC CELLS—THE BACTERIA [p.70]
1. bacteria; 2. nucleus; 3. wall; 4. cytoplasm; 5. flagella; 6. ribosomes; 7. nucleoid.

Self-Quiz
1. Golgi body (J); 2. vesicle (G); 3. microfilaments (L); 4. mitochondrion (O); 5. chloroplast (M); 6. microtubules (H); 7. central vacuole (C); 8. rough endoplasmic reticulum (P); 9. ribosomes (R); 10. smooth endoplasmic reticulum (D); 11. DNA + nucleoplasm (Q); 12. nucleolus (K); 13. nuclear envelope (A); 14. nucleus (I); 15. plasma membrane (N); 16. cell wall (B); 17. microfilaments (L); 18. microtubules (H); 19. plasma membrane (N); 20. mitochondrion (O); 21. nuclear envelope (A); 22. nucleolus (K); 23. DNA + nucleoplasm (Q); 24. nucleus (I); 25. vesicle (G); 26. lysosome (E); 27. rough endoplasmic reticulum (P); 28. ribosomes (R); 29. smooth endoplasmic reticulum (D); 30. vesicle (G); 31. Golgi body (J); 32. centrioles (F); 33. d; 34. c; 35. d; 36. d; 37. b; 38. a; 39. c; 40. a; 41. b; 42. c; 43. b; 44. d; 45. c; 46. b, c, d, e; 47. a, b, c, d; 48. b, c, d, e; 49. b, c, d, e; 50. c, d; 51. a; 52. a, b, d; 53. b, d; 54. b, c, d, e; 55. b, c, d, e; 56. b, c, d, e; 57. b, c, d, e.

Chapter 5   Ground Rules of Metabolism

You Light Up My Life [pp.74–75]

5.1. ENERGY AND THE UNDERLYING ORGANIZATION OF LIFE [pp.76–77]

5.2. DOING CELLULAR WORK [pp.78–79]
1. The world of life maintains a high degree of organization only because it is being resupplied with energy lost from someplace else. 2. second; 3. T; 4. T; 5. increasing; 6. I; 7. II; 8. I; 9. I; 10. II; 11. Exergonic reactions show a net loss of energy (energy out); an example is the breakdown of food molecules in the human body as the reactants become products. 12. Endergonic reactions show a net gain in energy (energy in); an example is the construction of starch and other large molecules from smaller, energy-poor molecules. 13. metabolic pathway; 14. enzyme; 15. endergonic; 16. exergonic; 17. exergonic; 18. B; 19. A; 20. A; 21. B; 22. intermediate; 23. water; 24. energy; 25. work; 26. phosphate; 27. ATP; 28. adenine; 29. ribose; 30. phosphate; 31. ADP; 32. phosphate; 33. energy; 34. energy; 35. metabolic; 36. ATP/ADP; 37. phosphorylation;

38. energy; 39. reaction; 40. ATP/ADP; 41. three phosphate groups; 42. ribose; 43. adenine; 44. adenosine triphosphate; 45. electron transport; 46. enzymes; 47. oxidized; 48. reduced; 49. energy; 50. work; 51. phosphate.

5.3. ENZYME STRUCTURE AND FUNCTION [pp.80–81]
5.4. FACTORS INFLUENCING ENZYME ACTIVITY [pp.82–83]
5.5. REACTANTS, PRODUCTS, AND CELL MEMBRANES [p.83]
1. Enzymes are usually <u>protein molecules</u> with <u>enormous catalytic power</u> (a few RNA forms have been found to act as enzymes). Enzymes do not cause reactions that would not happen on their own; enzymes are not changed in reactions—they can be used over and over; each enzyme is <u>highly selective about its substrates</u>; an enzyme can recognize both the reactants and the products of a given reaction as its substrate. Enzymes work most effectively at <u>specific temperatures</u> and <u>pH values</u>. Enzymes <u>lower the energy of activation</u> of the reactions

they catalyze. 2. D; 3. F; 4. C; 5. B; 6. A; 7. E; 8. Enzymes; 9. equilibrium; 10. substrate; 11. active site; 12. induced-fit; 13. activation energy; 14. Temperature (pH); 15. pH (temperature); 16. metabolism; 17. Allosteric; 18. feedback inhibition; 19. Shutdown occurs when too much of a particular molecule exists and is not being used by the cell; 20. Production of excess tryptophan molecules could interfere with other cellular activities and waste energy by producing a molecule not needed; 21. Some of the excess end product (tryptophan) molecules bind to the allosteric sites of the first enzyme in the pathway, alter its shape, stop the first, and all subsequent reactions to shut down the pathway; 22. When end-product (tryptophan) molecules are scarce because they are in demand, any of them linked to the first enzyme in the production pathway disengage, freeing up the enzyme to carry out the first reaction, and the subsequent reactions follow producing more end product, tryptophan.

5.6. WORKING WITH AND AGAINST CONCENTRATION GRADIENTS [pp.84–85]
5.7. MOVEMENT OF WATER ACROSS MEMBRANES [pp.86–87]
5.8. EXOCYTOSIS AND ENDOCYTOSIS [p.88]
1. Concentration; 2. gradient; 3. diffusion; 4. osmosis; 5. membrane pumps; 6. transport protein; 7. lipid

bilayer; 8. T; 9. hypotonic; 10. isotonic; 11. T; 12. hypertonic; 13. T; LB: 14, 15, 18, 23; ITP: 16, 17, 19, 20, 21, 22; 24. Their inherent kinetic energy (energy of motion), which increases as temperature increases; 25. Particles collide randomly with each other and the boundaries of their space. Where concentrations of particles are greater, collisions are more frequent and the paths between collisions are shorter. As they diffuse down their concentration gradient, collisions become less frequent and the movement path becomes longer. In this way, any kind of gradient can determine and drive the directional movement of a substance through its surrounding medium and across membranes; 26. Transporting proteins with a pump mechanism driven by a source of energy (generally ATP) to overcome the force of the flow caused by the gradient; 27. cell division, synthesis of cellular structures, movement of contractile proteins, information flow in nervous systems; 28. bulk flow; 29. hydrostatic pressure is a force directed outward against any barrier that encloses a fluid; osmotic pressure is the amount of force opposing any further increase in the volume of a solution; 30. endocytosis.

Self-Quiz
1. c; 2. d; 3. d; 4. d; 5. c; 6. c; 7. e; 8. d; 9. a; 10. d.

Chapter 6   How Cells Acquire Energy

Sunlight and Survival [pp.92–93]

6.1. PHOTOSYNTHESIS—AN OVERVIEW [pp.94–95]
6.2. SUNLIGHT AS AN ENERGY SOURCE [pp.96–97]
1. Autotrophs; 2. carbon dioxide; 3. Photosynthetic; 4. Heterotrophs; 5. animals; 6. aerobic respiration; 7. $12 H_2O + 6CO_2 \rightarrow 6O_2 + C_6H_{12}O_6 + 6H_2O$; 8. (a) Twelve (b) carbon dioxide (c) oxygen (d) glucose (e) six.; 9. light-dependent (light-independent); 10. light-independent (light-dependent); 11. Carbon dioxide; 12. water; 13. glucose; 14. thylakoid; 15. grana; 16. hydrogen; 17. stroma; 18. thylakoid; 19. photon; 20. pigments; 21. chlorophylls; 22. red (blue); 23. blue (red); 24. Carotenoids; 25. G; 26. F; 27. D; 28. E; 29. B; 30. A; 31. C; 32. H; 33. I.

6.3. THE LIGHT-DEPENDENT REACTIONS [pp.98–99]
6.4. A CLOSER LOOK AT ATP FORMATION IN CHLOROPLASTS [p.100]
1. photosystem; 2. light (photon); 3. electron; 4. acceptor; 5. Phosphorylation; 6. P700; 7. channel proteins; 8. ADP; 9. chemiosmotic; 10. a. Type I photosystem: a pigment cluster dominated by P700 that absorbs light energy; b. Electrons: electrons representing energy are ejected from P700 to an electron acceptor but move over the electron transport system, where some of the energy

is used to produce ATP; c. P700: a special chlorophyll molecule that absorbs wavelengths of 700 nanometers and then ejects electrons; d. Electron acceptor: a molecule that accepts electrons ejected from chlorophyll P700 and then passes electrons down the electron transport system; e. Electron transport system: electrons flow through this system, which is composed of a series of molecules bound in the thylakoid membrane that drive the phosphorylation of ADP to produce ATP; f. ADP: ADP undergoes phosphorylation in cyclic pathway; to become ATP; 11. electron acceptor; 12. electron transport system; 13. Type II photosystem; 14. Type I photosystem; 15. photolysis; 16. NADPH; 17. ATP; 18–53. The following numbers should have a check mark (√): 20, 22, 28, 30, 32, 33, 34, 35, 37, 40, 41, 42, 45, 50, and 51. All others should be blank.

6.5. THE LIGHT-INDEPENDENT REACTIONS [p.101]
6.6. FIXING CARBON—SO NEAR, YET SO FAR [pp.102–103]
6.7. *Focus on Science:* LIGHT IN THE DEEP DARK SEA? [p.103]
6.8. *Focus on the Environment:* AUTOTROPHS, HUMANS, AND THE BIOSPHERE [p.104]
1. carbon dioxide (D); 2. carbon fixation (E); 3. PGA (phosphoglycerate) (F); 4. adenosine triphosphate (H); 5. NADPH (G); 6. PGAL phosphoglyceraldehyde (A);

7. sugar phosphates (B); 8. Calvin-Benson cycle (I);
9. ribulose bisphosphate (C); 10. ATP (NADPH);
11. NADPH (ATP); 12. carbon dioxide; 13. ribulose
bisphosphate; 14. PGA; 15. fixation; 16. PGA;
17. PGAL; 18. six; 19. RuBP; 20. carbon; 21. PGALs;
22. sugar phosphate; 23. fixation; 24. light-dependent;
25. ATP (NADPH); 26. NADPH (ATP); 27. Sugar (glucose) phosphate; 28. photorespiration; 29. C3;
30. food; 31. oxygen; 32. oxaloacetate; 33. CAM;

34. carbon dioxide; 35. aerobic respiration; 36. chemo-
(protons); 37. organic (food); 38. H$^+$ (electrons);
39. electrons (H$^+$); 40–83. The following numbers should
have a check mark ($\sqrt{}$): 40, 45, 50, 52, 53, 54, 55, 57, 59, 63,
64, 69, 70, 80, and 81. All others should be blank.

Self-Quiz
1. c; 2. a; 3. d; 4. b; 5. b; 6. a; 7. a; 8. d; 9. c; 10. c.

Chapter 7   How Cells Release Stored Energy

The Killers Are Coming! The Killers Are Coming!
  [pp.108–109]

7.1. HOW DO CELLS MAKE ATP? [pp.110–111]
1. Adenosine triphosphate (ATP); 2. Oxygen withdraws
electrons from the electron transport system and joins
with H$^+$ to form water; 3. Glycolysis followed by some
end reactions (called fermentation), and anaerobic electron transport. Some organisms (including humans) use
fermentation pathways when oxygen supplies are low;
many microbes rely exclusively on anaerobic pathways;
4. ATP; 5. photosynthesis; 6. aerobic respiration;
7. glycolysis; 8. pyruvate; 9. Krebs; 10. Water;
11. ATP; 12. electrons; 13. electron transport;
14. phosphorylation; 15. ATP; 16. Oxygen; 17. anaerobic; 18. Fermentation; 19. electron transport;
20. $C_6H_{12}O_6 + 6O_2 \rightarrow 6CO_2 + 6H_2O + 36$ (38); 21. One
molecule of glucose plus six molecules of **oxygen** (in the
presence of appropriate enzymes) yield **six** molecules of
carbon dioxide plus **six** molecules of water and thirty-six
molecules of ATP.

7.2. GLYCOLYSIS: FIRST STAGE OF ENERGY-
  RELEASING PATHWAYS [pp. 112–113]
1. autotrophic; 2. Glucose; 3. pyruvate; 4. ATP
5. NADH; 6. glucose; 7. ATP; 8. PGAL; 9. phosphate;
10. hydrogen; 11. ATP; 12. water; 13. phosphate;
14. ATP; 15. substrate-level; 16. glucose; 17. pyruvate;
18. three; 19. D; 20. F; 21. B; 22. H; 23. G; 24. A;
25. E; 26. C.

7.3. SECOND STAGE OF THE AEROBIC PATHWAY
  [pp. 114–115]
1. acetyl CoA; 2. Krebs; 3. oxalocetate; 4. citrate;
5. coenzyme A; 6. carbon dioxide; 7. electrons;
8. NAD$^+$ (FAD); 9. FAD (NAD$^+$); 10. inner compartment; 11. inner membrane; 12. outer compartment;
13. outer membrane; 14. cytoplasm; 15. ATP; 16. oxygen (O$_2$); 17. FADH$_2$; 18. NADH; 19. electron transport system.

7.4. THIRD STAGE OF THE AEROBIC PATHWAY
  [pp. 116–117]
1. three; 2. two; 3. transport; 4. chemiosmotic; 5. synthases; 6. ATP; 7. oxygen; 8. mitochondria; 9. thirty-

six (thirty-eight); 10. thirty-eight (thirty-six); 11. c;
12. a, b; 13. c; 14. a; 15. b; 16. a; 17. b; 18. b; 19. a,
b, c; 20. b; 21. c; 22. a, b; 23. b; 24. c; 25. a; 26. c;
27. c; 28. a; 29. b; 30. a; 31. c; 32. c.

7.5. ANAEROBIC ROUTES OF ATP FORMATION
  [pp.118–119]
1. oxygen (O$_2$); 2. fermentation; 3. lactate; 4. ethanol;
5. carbon dioxide; 6. Anaerobic; 7. electron; 8. Glycolysis; 9. pyruvate; 10. NADH; 11. pyruvate; 12. lactate; 13. acetaldehyde; 14. carbon dioxide; 15. ethanol;
16. glycolysis; 17. NAD$^+$; 18. bacteria; 19. ATP;
20. sulfate; 21. hydrothermal vents; 22. food;
23–74. With a check ($\sqrt{}$): 23, 25, 27, 31, 32, 34, 35, 37, 42,
47, 50, 54, 56, 59, 61, 63, 65, [71]; all others lack a check.

7.6. ALTERNATIVE ENERGY SOURCES IN THE
  HUMAN BODY [pp.120–121]
7.7. *Commentary*: PERSPECTIVE ON LIFE [p.122]
1. Figure 6.1 shows how any complex carbohydrate or fat
can be broken down and at least part of those molecules
can be fed into the glycolytic pathway; 2. T; 3. T;
4. Page 122 tells us that energy flows from organized to
less organized forms; thus energy cannot be completely
recycled; 5. Page 122 tells us that Earth's first organisms
were anaerobic fermenters or electron transporters;
6. fatty acids; 7. glycerol; 8. glycolysis; 9. amino acids;
10. Krebs cycle; 11. acetyl-CoA; 12. pyruvate; 13–102.
With a check ($\sqrt{}$): 13, 15, 16, 17, 19, 22, 23, 25, 29, 31, 32,
39, 46, 47, 48, 49, 53, 54, 65, 66, 70, 71, 72, 80, 84, 87, 95,
102; all others lack a check; 103. e; 104. b(c); 105. c,d,e;
106. a; 107. c; 108. h; 109. e; 110. c; 111. b; 112. h;
113. i; 114. e; 115. c; 116. h; 117 g.

Self-Quiz
1. c; 2. c; 3. d; 4. b; 5. d; 6. d; 7. a; 8. d; 9. d;
10. d; 11. C; 12. A, B, D; 13. B, (C), D; 14. C, E; 15. B,
C; 16. A, D; 17. A; 18. B, D; 19. E; 20. A, E.

Chapter 8   Cell Division and Mitosis

Silver in the Stream of Time [pp.126–127]

8.1. DIVIDING CELLS: THE BRIDGE BETWEEN GENERATIONS [p.128]
8.2. THE CELL CYCLE [p.129]
1. G; 2. J; 3. C; 4. E; 5. B; 6. A; 7. I; 8. H; 9. F;
10. D; 11. interphase; 12. mitosis; 13. G1; 14. S;
15. G2; 16. prophase; 17. metaphase; 18. anaphase;
19. telophase; 20. cytoplasm divided or cytokinesis;
21. 15; 22. 12; 23. 14; 24. 13; 25. 11; 26. 20; 27. 11.

8.3. THE STAGES OF MITOSIS—AN OVERVIEW [pp.130–131]
1. interphase-daughter cells (F); 2. anaphase (A); 3. late
prophase (G); 4. metaphase (D); 5. cells at interphase
(E); 6. early prophase (C); 7. transition to metaphase
(B); 8. telophase (H).

8.4. DIVISION OF THE CYTOPLASM [pp.132–133]
1. a; 2. b; 3. a; 4. a; 5. b; 6. a; 7. b; 8. b; 9. a; 10. a

8.5. A CLOSER LOOK AT THE CELL CYCLE [pp.134–135]

8.6. *Focus on Science:* HENRIETTA'S IMMORTAL CELLS [p.135]
1. F; 2. I; 3. H; 4. A; 5. D; 6. J; 7. B; 8. G; 9. E;
10. C; 11. HeLa cells are tumor cells taken from and
named for a cancer patient, Henrietta Lacks, in 1951.
Henrietta Lacks died (at age thirty-one) two months
following her diagnosis of cancer. HeLa cells have continued to divide in culture and are used for cancer research in laboratories all over the world. The legacy of
Henrietta Lacks continues to benefit humans
everywhere.

Self-Quiz
1. a; 2. c; 3. d; 4. a; 5. d; 6. c; 7. e; 8. c; 9. c; 10. d;
11. d.

Chapter 9   Meiosis

Octopus Sex and Other Stories [pp.138–139]

9.1. COMPARISON OF ASEXUAL AND SEXUAL REPRODUCTION [p.140]
9.2. HOW MEIOSIS HALVES THE CHROMOSOME NUMBER [pp.140–141]
1. b; 2. a; 3. a; 4. b; 5. a; 6. b; 7. a; 8. b; 9. a; 10. b;
11. Meiosis; 12. gamete; 13. Diploid; 14. Diploid;
15. Meiosis; 16. sister chromatids; 17. sister
chromatids; 18. one; 19. four; 20. two; 21. interphase
preceding meiosis I; 22. chromosome; 23. haploid;
24. meiosis II; 25. twenty-three 26. E($2n = 2$); 27. D
($2n = 2$); 28. B ($2n = 2$); 29. A ($n = 1$); 30. C ($n = 1$).

9.3. A VISUAL TOUR OF THE STAGES OF MEIOSIS [pp.142–143]
9.4. A CLOSER LOOK AT KEY EVENTS OF MEIOSIS I [pp.144–145]
1. anaphase II (H); 2. metaphase II (F); 3. metaphase I
(A); 4. prophase II (B); 5. telophase II (C); 6. telophase
I (G); 7. prophase I (E); 8. anaphase I (D); 9. F; 10. I;
11. G; 12. D; 13. B; 14. J; 15. A; 16. H; 17. C; 18. E.

9.5. FROM GAMETES TO OFFSPRING [pp.146–147]
1. b; 2. c; 3. a; 4. c; 5. b; 6. b; 7. b; 8. b; 9. c; 10. b;
11. 2 ($2n$); 12. 5(n); 13. 4(n); 14. 1($2n$); 15. 3(n); 16. A;
17. B; 18. E; 19. D; 20. C; 21. During prophase I of
meiosis, crossing over and genetic recombination occur.
During metaphase I of meiosis, the two members of each
homologous chromosome assort independently of the
other pairs. Fertilization is a chance mix of different
combinations of alleles from two different gametes.

9.6. MEIOSIS AND MITOSIS COMPARED [pp.148–149]
1. a. Mitosis; b. Mitosis; c. Meiosis; d. Meiosis;
e. Meiosis; f. Meiosis; g. Mitosis; h. Mitosis; i. Meiosis; 2. C; 3. F; 4. D; 5. A; 6. B; 7. E; 8. 4; 9. 8;
10. 4; 11. 8; 12. 2.

Self-Quiz
1. a; 2. a; 3. d; 4. b; 5. c; 6. b; 7. d; 8. a; 9. b; 10. c.

Chapter 10   Observable Patterns of Inheritance

10.1. MENDEL'S INSIGHTS INTO PATTERNS OF INHERITANCE [pp.152–153]

10.2. MENDEL'S THEORY OF SEGREGATION [pp.156–157]

10.3. INDEPENDENT ASSORTMENT [pp.158–159]
1. I;   2. B;   3. D;   4. J;   5. H;   6. M;   7. O;   8. E;   9. F;   10. K;   11. A;   12. G;   13. N;   14. C;   15. L;   16. monohybrid;   17. probability;   18. fertilization;   19. segregation;   20. testcross;   21. 1:1;   22. dihybrid;   23. independent assortment;   24. genotype: 1/2 *Tt*; 1/2 *tt*, phenotype: 1/2 tall; 1/2 short;   25. a. 1 tall : 1 short; 1 heterozygous tall : 1 homozygous short; b. all tall; 1 homozygous tall : 1 heterozygous tall; c. all short; all homozygous short; d. 3 tall : 1 short; 1 homozygous tall : 2 heterozygous tall : 1 homozygous short; e. 1 tall : 1 short; 1 homozygous short : 1 heterozygous tall; f. all tall; all heterozygous tall; g. all tall; all homozygous tall; h. all tall; 1 heterozygous tall : 1 homozygous tall;   26. a. 9/16 pigmented eyes, right-handed; b. 3/16 pigmented eyes, left-handed; c. 3/16 blue-eyed, right-handed; d. 1/16 blue-eyed, left-handed (note Punnett square below). 27. Albino = *aa*, normal pigmentation = *AA* or *Aa*. The woman of normal pigmentation with an albino mother is genotype *Aa*; the woman received her recessive gene (*a*) from her mother and her dominant gene (*A*) from her father. It is likely that half of the couple's children will be albinos (*aa*) and half will have normal pigmentation but be heterozygous (*Aa*). 28. a. *F*₁: black trotter; *F*₂: nine black trotters, three black pacers, three chestnut trotters, one chestnut pacer; b. black pacer; c. *BbTt*; d. *bbtt*, chestnut pacers and *BBTT*, black trotters.

	BR	Br	bR	br
BR	BBRR	BBRr	BbRR	BbRr
Br	BBRr	BBrr	BbRr	Bbrr
bR	BbRR	BbRr	bbRR	bbRr
br	BbRr	Bbrr	bbRr	bbrr

10.4. DOMINANCE RELATIONS [p.160]

10.5. MULTIPLE EFFECTS OF SINGLE GENES [p.161]

10.6. INTERACTIONS BETWEEN GENE PAIRS [pp.162–163]
1. a. Incomplete dominance; b. Codominance; c. Multiple allele system; d. Epistasis; e. Pleiotropy.
2. a. phenotype: all pink, genotype: all *RR'*; b. phenotype: all white, genotype: all *R'R'*; c. phenotype: 1/2 red; 1/2 pink, genotype: 1/2 *RR*; 1/2 *RR'*; d. phenotype: all red, genotype: all *RR*.
3. The man must have sickle-cell trait with the genotype Hb^AHb^S, and the woman he married would have a normal genotype, Hb^AHb^A. The couple could be told that the probability is 1/2 that any child would have sickle-cell trait and 1/2 that any child would have the normal genotype.
4. Both the man and the woman have the genotype Hb^SHb^S. The probability of children from this marriage is: 1/4 normal, Hb^AHb^A; 1/2 sickle-cell trait, Hb^AHb^S; 1/4 sickle-cell anemia, Hb^SHb^S.   5. Genotypes: 1/4 I^AI^A; 1/4 I^AI^B; 1/4 I^Ai; 1/4 I^Bi, phenotypes: 1/2 A; 1/4 AB; 1/4 B.
6. Genotypes: 1/4 I^AI^B; 1/4 I^Bi; 1/4 I^Ai; 1/4 ii, phenotypes: 1/4 AB; 1/4 B; 1/4 A; 1/4 O.   7. Genotypes: all I^Ai, phenotypes: all A.   8. Genotypes: all ii, phenotypes: all O.
9. Genotypes: 1/4 I^AI^A; 1/2 I^AI^B; 1/4 I^BI^B, phenotypes: 1/4 A; 1/2 AB; 1/4 B.   10. 1/4 color; 3/4 white.   11. 3/4 color; 1/4 white.   12. 1/4 color; 3/4 white.   13. 3/8 black; 1/2 yellow; 1/8 brown.   14. The genotype of the male parent is *RrPp* and the genotype of the female parent is *rrpp*. The offspring are 1/4 walnut comb, *RrPp*; 1/4 rose comb, *Rrpp*; 1/4 pea comb, *rrPp*; 1/4 single comb, rrpp.   15. The genotype of the walnut-combed male is *RRpp* and the genotype of the single-combed female is *rrpp*. All offspring are rose comb with the genotype *Rrpp*.

10.7. HOW CAN WE EXPLAIN LESS PREDICTABLE VARIATIONS? [pp.164–165]

10.8. EXAMPLES OF ENVIRONMENTAL EFFECTS ON PHENOTYPE [p.166]
1. b;   2. b;   3. a;   4. b;   5. a.

Self-Quiz
1. d;   2. b;   3. a;   4. c;   5. d;   6. b;   7. e;   8. a;   9. a;   10. c;   11. d.

Integrating and Applying Key Concepts
DdPp × *Ddpp*

Chapter 11   Chromosomes and Human Genetics

The Philadelphia Story [pp.170–171]

11.1. THE CHROMOSOMAL BASIS OF INHERITANCE—AN OVERVIEW [p.172]
11.2. *Focus on Science:* KARYOTYPING MADE EASY [p.173]

1. Philadelphia;   2. Leukemias;   3. karyotype;   4. spectral;   5. genes;   6. homologous;   7. Alleles;   8. wild;   9. crossing over;   10. recombination;   11. sex;   12. autosomes;   13. karyotype;   14. in vitro;   15. colchicine;   16. metaphase;   17. centrifugation;   18. centromeres.

11.3. SEX DETERMINATION IN HUMANS [pp.174–175]
11.4. EARLY QUESTIONS ABOUT GENE LOCATIONS [pp.176–177]

1. E;   2. C;   3. A;   4. D;   5. B;   6. Two blocks of the Punnett square should be XX and two blocks should be XY;   7. sons;   8. mothers;   9. daughters;   10. a. F_1 flies all have red eyes: 1/2 heterozygous red females : 1/2 red-eyed males; b. F_2 Phenotypes: females all have red eyes; males 1/2 red eyes, 1/2 white eyes; Genotypes: females are $X^W X^W$;   11. a.

11.5. HUMAN GENETIC ANALYSIS [pp.178–179]

1. A;   2. E;   3. F;   4. B;   5. C;   6. G;   7. D;   8. A genetic abnormality is nothing more than a rare, uncommon version of a trait that society may judge as abnormal or merely interesting; a genetic disorder is an inherited condition that sooner or later causes mild to severe medical problems; a syndrome is a recognized set of symptoms that characterize a given disorder; a genetic disease is illness caused by a person's genes increasing susceptibility to infection or weakening the response to it;   9. a. Autosomal recessive;   b. Autosomal dominant;   c. X-linked recessive;   d. X-linked recessive;   e. Autosomal dominant;   f. X-linked dominant;   g. Changes in chromosome number;   h. Autosomal dominant;   i. Changes in chromosome structure;   j. Changes in chromosome number;   k. X-linked recessive;   l. X-linked recessive; m. Autosomal dominant;   n. Changes in chromosome number;   o. X-linked recessive;   p. Autosomal recessive.

11.6. INHERITANCE PATTERNS [pp.180–181]
11.7. *Focus on Health:* TOO YOUNG TO BE OLD [p.182]

1. b;   2. a and c;   3. b;   4. c;   5. a;   6. b;   7. b;   8. b;   9. a;   10. c;   11. c;   12. b;   13. b;   14. a;   15. b;   16. c;   17. a;   18. The woman's mother is heterozygous normal, *Aa*, the woman is also heterozygous normal, *Aa*. The albino man, *aa*, has two heterozygous normal parents, *Aa*. The two normal children are heterozygous normal, *Aa*; the albino child is *aa*;   19. Assuming the father is heterozygous with Huntington's disorder and the mother normal, the chances are 1/2 that the son will develop the disease;   20. If only male offspring are considered, the probability is 1/2 that the couple will have a color-blind son;   21. The probability is that 1/2 of the sons will have hemophilia; the probability is 0 that a daughter will express hemophilia; the probability is that 1/2 of the daughters will be carriers;   22. If the woman marries a normal male, the chance that her son would be color-blind is 1/2. If she marries a color-blind male, the chance that her son would be color blind is also 1/2.

11.8. CHANGES IN CHROMOSOME STRUCTURE [pp.182–183]
11.9. CHANGES IN CHROMOSOME NUMBER [pp.184–185]

1. duplication (B);   2. inversion (C);   3. deletion (A);   4. translocation (D);   5. a. With aneuploidy, individuals have one extra or one less chromosome; a major cause of human reproductive failure;   b. With polyploidy, individuals have three or more of each type of chromosome; common in flowering plants and some animals but lethal in humans;   c. Nondisjunction is due to a failure of one or more pairs of chromosomes to separate in mitosis or meiosis; some or all forthcoming cells will have too many or too few chromosomes;   6. All gametes will be abnormal;   7. One-half of the gametes will be abnormal;   8. About half of all flowering plant species are polyploids but this condition is lethal for humans;   9. A tetraploid cell has four of each type of chromosome; a trisomic ($2n + 1$) individual will have three of one type of chromosome and two of every other type; a monosomic ($2n - 1$) individual will have only one of one type of chromosome but two of every other type;   10. c;   11. b;   12. c;   13. d;   14. a;   15. b;   16. d;   17. c;   18. a;   19. d.

11.10. *Focus on Bioethics:* PROSPECTS IN HUMAN GENETICS [pp.186–187]

1. a. Prenatal diagnosis;   b. Abortion;   c. Genetic counseling;   d. Preimplantation diagnosis;   e. Phenotypic treatments;   f. Genetic screening.

Self-Quiz

1. d;   2. d;   3. b;   4. c;   5. b;   6. a;   7. c;   8. b;   9. b;   10. c.

Chapter 12   DNA Structure and Function

Cardboard Atoms and Bent-Wire Bonds
[pp.190–191]

12.1. DISCOVERY OF DNA FUNCTION [pp.192–193]
1. a. Miescher: identified "nuclein" from nuclei of pus cells and fish sperm; discovered DNA; b. Griffith: discovered the transforming principle in *Streptococcus pneumoniae*; live, harmless R cells were mixed with dead S cells; R cells became S cells; c. Avery: reported that the transforming substance in Griffith's bacteria experiments was probably DNA, the substance of heredity; d. Hershey and Chase: worked with radioactive sulfur (protein) and phosphorus (DNA) labels; T4 bacteriophage and *E. coli* demonstrated that labeled phosphorus was in bacteriophage DNA and contained hereditary instructions for new bacteriophages. 2. virus; 3. bacterial; 4. viruses (bacteriophages); 5. proteins; 6. ^{32}P; 7. bacteriophage (viral); 8. ^{35}S; 9. ^{32}P; 10. genetic material; 11. DNA; 12. proteins.

12.2. DNA STRUCTURE [pp.194–195]
12.3. *Focus on Bioethics:* ROSALIND'S STORY [p.196]
1. A five-carbon sugar called deoxyribose, a phosphate group, and one of the four nitrogen-containing bases; 2. guanine (pu); 3. cytosine (py); 4. adenine (pu); 5. thymine (py) 6. deoxyribose (B); 7. phosphate group (G); 8. purine (C); 9. pyrimidine (A); 10. purine (E); 11. pyrimidine (D); 12. nucleotide (F); 13. T; 14. T; 15. F, sugar; 16. T; 17. T; 18. pairing; 19. constant; 20. sequence; 21. different; 22. Living organisms have so many diverse body structures and behave in different ways because the many different habitats of Earth have selected those genotypes most able to survive in those habitats. The remaining genotypes have perished. The directions that code for the building of those body structures and which enable the specific successful behaviors reside in DNA or in a few cases, RNA. All living organisms follow the same rules for base pairing between the two nucleotide strands in DNA; adenine always pairs with thymine in undamaged DNA and cytosine always pairs with guanine. All living organisms must extract energy from food molecules and the reactions of glycolysis occur in virtually all of Earth's species. That means that similar enzyme sequences enable similar metabolic pathways to occur. While virtually all living organisms on Earth use the same code and the same enzymes during replication, transcription and translation, the particular array of proteins being formed differs from individual to individual even of the same species according to the sequences of nitrogenous bases that make up an individual's chromosome(s), and therein lies the key to the enormous diversity of life on Earth: no two individuals have the exact same array of proteins in their phenotypes. 23. If one observes the sugar-phosphate sides of the DNA "ladder," one sees that one strand runs from 5' to 3' and the other runs from 3' to 5'. The numerals 3' and 5' are used to identify specific carbon atoms in each deoxyribose molecule.

12.4. DNA REPLICATION AND REPAIR [pp.196–197]
12.5. *Focus on Science:* DOLLY, DAISIES, AND DNA [p.198]

1.

T–	A	T	–A
G–	C	G	–C
A–	T	A	–T
C–	G	C	–G
C–	G	C	–G
C–	G	C	–G
old	new	new	old

2. F, adenine bonds to thymine (during replication) or uracil (during transcription); 3. F, it is a conserving process because each "new" DNA molecule contains one "old" strand from the parent cell attached to a strand of "new" complementary nucleotides that were assembled from stockpiles in the cell; 4. T; 5. T.

Self-Quiz
1. d; 2. d; 3. a; 4. d; 5. b; 6. d; 7. c; 8. a; 9. d; 10. d.

Chapter 13   From DNA to Proteins

Beyond Byssus [pp.200–201]

13.1. HOW IS DNA TRANSCRIBED INTO RNA? [pp.202–203]
1. sequence; 2. gene; 3. transcription (translation); 4. translation (transcription); 5. transcription; 6. translation; 7. protein; 8. folded; 9. structural (functional); 10. functional (structural); 11. a. ribosomal RNA; rRNA; RNA molecule that associates with certain proteins to form the ribosome, the "workbench" on which polypeptide chains are assembled; b. messenger RNA; mRNA; RNA molecule that moves to the cytoplasm, complexes with the ribosome where translation will result in polypeptide chains; c. transfer RNA; tRNA; RNA molecule that moves into the cytoplasm, picks up a specific

amino acid, and moves it to the ribosome where tRNA pairs with a specific mRNA code word for that amino acid; 12. RNA molecules are single-stranded, while DNA has two strands; uracil substitutes in RNA molecules for thymine in DNA molecules; ribose sugar is found in RNA, while DNA has deoxyribose sugar; 13. Both DNA replication and transcription follow base-pairing rules; nucleotides are added to a growing RNA strand one at a time as in DNA replication; 14. Only one region of a DNA strand serves as a template for transcription; transcription requires different enzymes (three types of RNA polymerase); the results of transcription are single-stranded RNA molecules, but replication results in DNA, a double-stranded molecule; 15. C; 16. B; 17. E; 18. A; 19. D; 20. A-U-G-U-U-C-U-A-U-U-G-U-A-A-U-A-A-A-G-G-A-U-G-G-C-A-G-U-A-G; 21. DNA (E); 22. introns (B); 23. cap (F); 24. exons (A); 25. tail (D); 26. mature mRNA transcript (C).

13.2. DECIPHERING THE mRNA TRANSCRIPTS [pp.204–205]

13.3 HOW IS mRNA TRANSLATED? [pp.206–207]
1. F; 2. B; 3. G; 4. H; 5. C; 6. A; 7. E; 8. D; 9. a. initiation; b. chain elongation; c. chain termination; 10. mRNA transcript: AUG UUC UAU UGU AAU AAA GGA UGG CAG UAG; 11. tRNA anticodons: UAC AAG AUA ACA UUA UUU CCU ACC GUC AUC; 12. amino acids: (start) met phe tyr cys asn lys gly try gln stop; 13. amino acids; 14. three; 15. one; 16. mRNA; 17. codon; 18. mRNA; 19. assembly (synthesis); 20. Transfer; 21. amino acid; 22. protein (polypeptide); 23. codon; 24. anticodon.

13.4. DO MUTATIONS AFFECT PROTEIN SYNTHESIS? [pp.208–209]
1. Viruses, ultraviolet radiation, and certain chemicals are examples; 2. mutagens; 3. substitution; 4. amino acid; 5. hemoglobin; 6. transposable; 7. DNA (H); 8. transcription (J); 9. intron (E); 10. exon (A); 11. mature mRNA transcript (L); 12. tRNAs (C); 13. rRNA subunits (G); 14. mRNA (B); 15. anticodon (K); 16. amino acids (D); 17. tRNA (F); 18. ribosome-mRNA complex (I); 19. polypeptide (M).

Self-Quiz
1. c; 2. b; 3. c; 4. a; 5. c; 6. a; 7. a; 8. d; 9. d.

Chapter 14   Controls Over Genes

When DNA Can't Be Fixed [pp.212–213]

14.1. OVERVIEW OF GENE CONTROLS [p.214]
14.2. CONTROLS IN BACTERIAL CELLS [pp.214–215]
1. a. prevents transcription enzymes (RNA polymerases) from binding to DNA; this is negative transcription control; b. Hormones; c. specific base sequences on DNA that serve as binding sites for control agents; before RNA assembly can occur on DNA, the enzymes must bind with the promoter site; d. Operators; e. Activator protein; 2. regulator gene (K); 3. gene that codes for an enzyme (G); 4. repressor protein (E); 5. promoter (J); 6. operator (B); 7. lactose operon (A); 8. RNA polymerase (D); 9. repressor protein bound to lactose molecule (F); 10. lactose (C); 11. mRNA transcript (I); 12. lactose-degrading enzymes (H); 13. *Escherichia coli*; 14. operon; 15. controls; 16. regulator; 17. promoter; 18. negative control; 19. low; 20. RNA polymerase (mRNA transcription); 21. blocks; 22. repressor protein; 23. operator; 24. needed (required).

14.3. CONTROLS IN EUKARYOTIC CELLS [pp.216–217]
14.4. MANY LEVELS OF CONTROL [p.218]
14.5. *Focus on Science:* **LOST CONTROLS AND CANCER** [pp.218–219]
1. All cells in the body descend from the same zygote; as cells divide to form the body, they become specialized in composition, structure, and function—they differentiate through selective gene expression; 2. DNA (genes); 3. cell differentiation; 4. selective; 5. controls; 6. regulatory; 7. activators; 8. Barr; 9. mosaic; 10. anhidrotic ectodermal dysplasia; 11. X chromosome; 12. increase, high, start; 13. T; 14. T; 15. bring about; 16. T; 17. abnormal.

Self-Quiz
1. d; 2. b; 3. b; 4. b; 5. c; 6. c; 7. d; 8. b; 9. e; 10. a.

Chapter 15   Recombinant DNA and Genetic Engineering

Mom, Dad, and Clogged Arteries [pp.222–223]

15.1. A TOOLKIT FOR MAKING RECOMBINANT DNA [pp.224–225]

15.2. PCR—A FASTER WAY TO AMPLIFY DNA [p.226]

1. "Good" alleles that code for producing normal LDL receptors on cell surfaces exist in DNA libraries. Through PCR, they can be copied over and over again. Through recombinant DNA techniques, the good alleles can be spliced into harmless viruses that are allowed to infect liver cells surgically removed from the patient and cultured in the laboratory. Later about a billion of the modified cells can be infused into the patient's hepatic portal vein, which leads directly to the liver. Some modified cells with the substituted "good" alleles may take up residence, produce the receptors and begin removing cholesterol from the bloodstream; 2. The bacterial chromosome, a circular DNA molecule, contains all the genes necessary for normal growth and development. Plasmids, small, circular molecules of "extra" DNA, carry only a few genes and are self-replicating; 3. small circles of DNA in bacteria; 4. T; 5. mutations; 6. recombinant DNA; 7. species; 8. amplify; 9. protein; 10. research; 11. Genetic engineering; 12. B; 13. D; 14. G; 15. C; 16. E; 17. F; 18. A; 19. expression (translation); 20. engineered; 21. introns; 22. transcriptase; 23. mRNA; 24. cDNA; 25. Enzyme; 26. DNA; 27. cDNA; 28. template (transcript); 29. splicing enzymes; 30. B; 31. F; 32. E; 33. C; 34. A; 35. D; 36. a. after a DNA library is inserted into a host cell's cloning vector (often a plasmid), repeated replications and divisions of the host cells produce multiple, identical copies of DNA fragments, or cloned DNA; b. a DNA strand "copied" from a mature mRNA transcript; c. a replication enzyme that joins the short fragments of DNA; d. PCR is a method of creating many copies of chromosomal DNA or cDNA inside test tubes; e. bacterial enzymes that cut apart DNA molecules injected into the cell by viruses; several hundred have been identified; f. a process that uses a viral enzyme to transcribe mRNA into DNA, which can then be inserted into a plasmid for amplification; the same cDNA can be inserted into bacteria, which they will command to make a specific protein;

15.3. *Focus on Science:* **DNA FINGERPRINTS** [p.227]

15.4. HOW IS DNA SEQUENCED? [p.228]

15.5. FROM HAYSTACKS TO NEEDLES— ISOLATING GENES OF INTEREST [p.229]

1. a. a collection of DNA fragments produced by restriction enzymes and incorporated into plasmids; b. a unique pattern of DNA fragments inherited from each parent that can be used to map the human genome, apprehend criminals, and resolve cases of disputed paternity and maternity; c. a nucleic acid hybridization technique is used to identify bacterial colonies harboring the DNA (gene) of interest; a short nucleotide sequence is assembled from radioactively labeled subunits (part of the sequence must be complementary to that of the desired gene); d. several procedures allow researchers to determine the nucleotide sequence of a DNA fragment (see text Figure 15.6); e. base-pairing between nucleotide sequences from different sources can indicate that the two different sources carry the same gene(s); f. RFLPs are variations in the banding patterns of DNA fragments from different individuals; the variations occur because no two individuals (except identical twins) have identical base sequences in their entire DNA set. Restriction enzymes cut different DNA molecules at different sites and into different numbers of DNA fragments; 2. fingerprint; 3. genomes; 4. tandem repeats; 5. gel electrophoresis; 6. sequencing; 7. amplified; 8. bacterium; 9. probes; 10. contain the gene of interest that is to be located; 11. bacterial cells adhere to the filter and mirror the distribution of the colonies on the culture plate; 12. bursts open the bacterial cells; the released DNA sticks to the filter and can be unwound to single strands; 13. the probes hybridize by base-pairing only with complementary strands from the colony containing the interesting gene; 14. will be exposed by the radioactive tags and the pattern formed will allow the colony to be located and isolated; 15. TCCATGGACCA.

15.6. USING THE GENETIC SCRIPTS [p.230]

15.7. DESIGNER PLANTS [pp.230–231]

15.8. GENE TRANSFERS IN AMIMALS [pp.232–233]

15.9. *Focus on Bioethics:* **WHO GETS ENHANCED?** [p.233]

15.10. SAFETY ISSUES [p.234]

1. insulin; 2. heavy metals; 3. evolutionary; 4. Researchers are working to sequence the estimated 3 billion nucleotides present in human chromosomes. 5. If human collagen can be produced, it may be used to correct various skin, cartilage, and bone disorders. If human serum albumin can be produced, it can help control blood pressure. 6. The plasmid of *Agrobacterium* can be used as a vector to introduce desired genes into cultured plant cells; *Agrobacterium* was used to deliver a firefly gene (bioluminescent) into cultured tobacco plant cells. 7. Certain cotton plants have been genetically engineered for resistance to herbicides that kill surrounding weeds, but leave cotton plants alone. 8. In separate experiments the rat and human somatotropin genes became integrated into the mouse DNA. The mice grew much larger than their normal littermates. 9. Bacteria that have specific proteins on their cell surfaces facilitate ice crystal formation on whatever substrate the bacteria are located; bacteria without the ability to synthesize those proteins ("ice-minus") have been genetically engineered and were sprayed upon

strawberry plants. Nothing bad happened.   10. Genetically engineered bacteria, harmless to begin with, have had specific genes spliced into them that allow them to degrade oil. If "fail-safe" genes are spliced in also, specific cues in the environment will cause the fail-safe gene to trip into action, membrane function will be destroyed and the genetically engineered bacteria will die;   11. ice-minus;   12. body cells (genetic material);   13. gene therapy;   14. eugenic engineering.

Self-Quiz
1. a; 2. a; 3. b; 4. c; 5. d; 6. a; 7. c; 8. b; 9. a; 10. d.

Chapter 16   Microevolution

Designer Dogs [pp.238–239]

16.1. EARLY BELIEFS, CONFOUNDING DISCOVERIES [pp.240–241]
16.2. A FLURRY OF NEW THEORIES [pp.242–243]
16.3. DARWIN'S THEORY TAKES FORM [pp.244–245]
1. G;   2. H;   3. J;   4. I;   5. D;   6. E;   7. F;   8. C;   9. B;   10. A;   11. b;   12. a;   13. b;   14. a;   15. a;   16. b;   17. a;   18. b;   19. a;   20. b;   21. b;   22. b;   23. a. John Henslow;   b. Cambridge University;   c. H.M.S. *Beagle*;   d. Charles Lyell;   e. Thomas Malthus;   f. Galápagos Islands;   g. Natural selection;   h. Alfred Wallace;   i. *Archaeopteryx*.

16.4. INDIVIDUALS DON'T EVOLVE— POPULATIONS DO [pp.246–247]
16.5. *Focus on Science:* WHEN IS A POPULATION *NOT* EVOLVING? [pp.248–249]
1. M;   2. P;   3. M;   4. B;   5. P;   6. M;   7. B;   8. P;   9. B;   10. M;   11. D;   12. C;   13. E;   14. B;   15. A;   16. Phenotypic variation is due to the effects of genes and the environment. The environment can mediate how the genes governing traits are expressed in an individual. For example, ivy plants grown from cuttings of the same parent all have the same genes, but the amount of sunlight affects the genes governing leaf growth. Ivy plants growing in full sun have smaller leaves than those growing in full shade. Offspring inherit genes, not phenotype;   17. There are five conditions: no genes are undergoing mutation; a very, very large population; the population is isolated from other populations of the species; all members of the population survive and reproduce; and mating is random;   18. a. 0.64 *BB*, 0.16 *Bb*, 0.16 *Bb*, and 0.04 *bb*;   b. genotypes: 0.64 *BB*, 0.32 *Bb*, and 0.04 *bb*; phenotypes: 96% black, 4% gray;   c.

Parents F$_1$	B sperm	b sperm
0.64 *BB*	0.64	0
0.32 *Bb*	0.16	0.16
0.04 *bb*	0	0.04
Totals =	0.80	0.20

19. Find (b) first, then (c), and finally (a).   a. $2pq = 2 \times (0.9) \times (0.1) = 2 \times (0.09) = 0.18 = 18\%$, which is the percentage of heterozygotes;   b. $p^2 = 0.81$, $p = \sqrt{0.81} = 0.9 =$ the frequency of the dominant allele;   c. $p + q = 1$, $q = 1.00 - 0.9 = 0.1 =$ the frequency of the recessive allele;   20. a. homozygous dominant $= p^2 \times 200 = (0.8)^2 \times 200 =$ $0.64 \times 200 = 128$ individuals;   b. $q = (1.00 - p) = 0.20$; homozygous recessive $= q^2 \times 200 = (0.2)^2 \times 200 = (0.04) \times (200) = 8$ individuals;   c. heterozygotes $= 2pq \times 200 = 2 \times 0.8 \times 0.2 \times 200 = 0.32 \times 200 = 64$ individuals. Check: $128 + 8 + 64 = 200$;   21. If $p = 0.70$, since $p + q = 1$, $0.70 + q = 1$; then $q = 0.30$, or 30 percent;   22. If $p = 0.60$, since $p + q = 1$, $0.60 + q = 1$; then $q = 0.40$; thus, $2pq = 0.48$, or 48 percent;   23. D;   24. H;   25. I;   26. J;   27. G;   28. C;   29. A;   30. F;   31. B;   32. E;   33. population;   34. Hardy-Weinberg;   35. allele frequencies;   36. genetic equilibrium;   37. mutations.

16.6. NATURAL SELECTION REVISITED [p.249]
16.7. DIRECTIONAL CHANGE IN THE RANGE OF VARIATION [pp.250–251]
16.8. SELECTION AGAINST OR IN FAVOR OF EXTREME PHENOTYPES [pp.252–253]
16.9. SPECIAL TYPES OF SELECTION [pp.254–255]
1. a. the most common phenotypes are favored; average human birth weight of 7 pounds;   b. allele frequencies shift in a steady, consistent direction in response to a new environment or a directional change in an old one; light to dark forms of peppered moths;   c. forms at both ends of the phenotypic range are favored, and intermediate forms are selected against; finches on the Galápagos Islands;   2. a. directional;   b. disruptive;   c. stabilizing;   3. natural;   4. Stabilizing;   5. Directional;   6. Disruptive;   7. Gall-making flies;   8. Sickle-cell anemia;   9. stabilizing;   10. allele;   11. balanced polymorphism;   12. sexual dimorphism;   13. selection;   14. Sexual;   15. sexual;   16. e;   17. c;   18. b;   19. d;   20. b;   21. a;   22. b;   23. b;   24. c, d;   25. a;   26. e.

16.10. GENE FLOW [p.255]
16.11. GENETIC DRIFT [pp.256–257]
1. a;   2. b;   3. a;   4. b;   5. a;   6. a;   7. a;   8. b;   9. a;   10. b;   11. In the founder effect, a few individuals leave a population and establish a new one; by chance, allele frequencies will differ from the original population. In bottlenecks, disease, starvation, or some other stressful situation nearly eliminates a population; relative allele frequencies are randomly changed.

Self-Quiz
1. c;   2. d;   3. d;   4. b;   5. d;   6. c;   7. b;   8. a;   9. c;   10. e.

Chapter 17   Speciation

The Case of the Road-Killed Snails [pp.260–261]

17.1. ON THE ROAD TO SPECIATION [pp.262–263]
17.2. SPECIATION IN GEOGRAPHICALLY ISOLATED POPULATIONS [pp.264–265]
1. c;   2. a;   3. b;   4. a;   5. b;   6. a;   7. c;   8. c;   9. a;   10. c;
11. C (pre);   12. A (pre);   13. B (post);   14. G (pre);   15. H (post);   16. E (pre);   17. D (pre);   18. F (post);   19. c;
20. a;   21. e;   22. d;   23. b;   24. f;   25. a. one;   b. two;
c. considerable divergence begins between B and C;   d. D (C);   e. B and C;   26. Speciation;   27. species;   28. Divergence;   29. reproductive isolating;   30. geographic.

17.3. MODELS FOR OTHER SPECIATION ROUTES [pp.266–267]
17.4. PATTERNS OF SPECIATION [pp.268–269]
1. a. Sympatric;   b. Allopatric;   c. Parapatric;   2. sympatric;   3. parapatric;   4. allopatric;   5. allopatric;   6. allopatric;   7. a. Seven pairs;   b. AA;   c. Seven pairs;
d. BB;   e. Fertilization of nonreduced gametes produced a fertile tetraploid;   f. Fertilization of nonreduced gametes resulted in a fertile hexaploid;   g. *Triticum monococcum*; the unknown wild wheat; *T. tauschii*;
h. polyploidy (hybridization);   i. hybridization (polyploidy);   8. a. Horizontal branching;   b. Many branchings of the same lineage at or near the same point in geologic time;   c. A branch that ends before the present;   d. Softly angled branching;   e. Vertical continuation of a branch;   f. A dashed line;   9. I;   10. H;   11. E;   12. C;
13. J;   14. D;   15. A;   16. F;   17. B;   18. G.

Self-Quiz
1. c;   2. d;   3. d;   4. a;   5. c;   6. c;   7. b;   8. a;   9. c;   10. d;
11. e;   12. e;   13. c;   14. b.

Chapter 18   The Macroevolutionary Puzzle

Of Floods and Fossils [pp.272–273]

18.1. FOSSILS—EVIDENCE OF ANCIENT LIFE [pp.274–275]
1. F;   2. J(D);   3. I;   4. A;   5. E;   6. H;   7. C;   8. D(J);   9. B;
10. G;   11. b;   12. c;   13. a;   14. d;   15. e.

18.2. EVIDENCE FROM COMPARATIVE MORPHOLOGY [pp.276–277]
18.3. EVIDENCE FROM PATTERNS OF DEVELOPMENT [pp.278–279]
18.4. EVIDENCE FROM COMPARATIVE BIOCHEMISTRY [pp.280–281]
1. T;   2. low;   3. convergence;   4. convergence;   5. differences;   6. d (a);   7. c;   8. a (d);   9. e;   10. b;   11. D;   12. D;
13. D;   14. B;   15. B;   16. B;   17. D;   18. B;   19. M;   20. M;
21. Several disciplines contribute to understanding evolutionary relationships among the huge diversity of organisms on Earth: (1) understanding the fossil record and the geologic time scale, (2) understanding homologous and analogous structures, (3) knowledge of similarities in patterns of development, and (4) comparisons of DNA, RNA, proteins, and similar metabolic pathways associated with particular species.

18.5. IDENTIFYING SPECIES, PAST AND PRESENT [pp.282–283]
1. B;   2. E;   3. A;   4. D;   5. C.;   6. kingdom; phylum (or division); class; order; family; genus; species; 7. kingdom: Plantae; division: Anthophyta; class: Dicotyledonae; order: Asterales; family: Asteraceae; genus: *Archibaccharis*; species: *linearilobas*;   8. G;   9. C;
10. J;   11. H;   12. B;   13. A;   14. E;   15. D;   16. I;   17. F;
18. C. Woese and C. Bult have determined that populations of archaebacteria and eubacteria differ substantially in the makeup of their cell walls, plasma membranes, chemical composition, and gene arrays.

18.6. EVIDENCE OF A CHANGING EARTH [pp.284–285]
18.7. *Focus on Science:* DATING PIECES OF THE MACROEVOLUTIONARY PUZZLE [p.286]
1. uniformity;   2. continents;   3. supercontinent;
4. mantle;   5. iron;   6. polarity;   7. accept;   8. seafloor spreading;   9. plumes;   10. crust;   11. Missouri;
12. plates;   13. Andes;   14. radioactive decay;
15. layer(s);   16. geologic time scale;   17. radiometric dating;   18. radioisotope;   19. and 20 use either a combination of (a) older, . . . less or (b) more recent, . . . more.

Self-Quiz
1. c;   2. b;   3. b;   4. a;   5. d;   6. b;   7. a;   8. b;   9. a;   10. e;
11. d;   12; c.

Chapter 19   The Origin and Evolution of Life

In the Beginning . . . [pp.290–291]

19.1. CONDITIONS ON THE EARLY EARTH
[pp.292–293]
19.2. EMERGENCE OF THE FIRST LIVING CELLS
[pp.294–295]

1. f; 2. e; 3. a; 4. c; 5. d; 6. b; 7. f; 8. c; 9. b; 10. a;
11. d; 12. f; 13. e; 14. c (b); 15. a; 16. b; 17. e; 18. c;
19. a; 20. b.

19.3. ORIGIN OF PROKARYOTIC AND
EUKARYOTIC CELLS [pp.296–297]
19.4. *Focus on Science:* **WHERE DID ORGANELLES**
COME FROM? [pp.298–299]
19.5. LIFE IN THE PALEOZOIC ERA [pp.300–301]

1. b; 2. b; 3. a; 4. b; 5. a; 6. a; 7. b; 8. a; 9. a; 10. b;
11. a; 12. b; 13. a; 14. a; 15. b; 16. organelles;
17. gene; 18. plasma; 19. channel; 20. ER; 21. genes;
22. plasmids; 23. prokaryotic; 24. endosymbiosis;
25. eukaryotes; 26. Electron; 27. oxygen; 28. bacteria;
29. ATP; 30. host; 31. incapable; 32. mitochondria;
33. ATP; 34. DNA; 35. chloroplasts; 36. host; 37. eubacteria; 38. protistans; 39. F; 40. A; 41. G; 42. C;
43. E; 44. D; 45. B; 46. F; 47. A, B, C, D, E, G.

19.6. LIFE IN THE MESOZOIC ERA [pp.302–303]
19.7. *Focus on Science:* **HORRENDOUS END TO**
   DOMINANCE [p.304]
19.8. LIFE IN THE CENOZOIC ERA [p.305]
19.9. SUMMARY OF EARTH AND LIFE HISTORY
   [pp.306–307]

1. D; 2. C; 3. E; 4. A; 5. B; 6. D; 7. E; 8. B; 9. F;
10. A; 11. G; 12. C; 13. D, E; 14. B; 15. A, C, F, G;
16. a. Mesozoic, 240–205; b. Archean, 4,600–3,800;
c. Paleozoic, 435–410; d. Paleozoic, 550–505;
e. Archean, 4,600–3,800; f. Mesozoic, 135–65; g. Paleozoic, 360–280; h. Proterozoic, 2,500–570; i. Archean,
3,800–2,500; j. Paleozoic, 290–240; k. Proterozoic,
600–550; l. Mesozoic, 181–65; m. Proterozoic,
2,500–570; n. Cenozoic, 38–1.65; o. Cenozoic, 1.65–
present; p. Paleozoic, 440–435; q. Paleozoic, 435–360.

Self-Quiz

1. b; 2. c; 3. c; 4. b; 5. e; 6. d; 7. a; 8. d; 9. c; 10. b.

Chapter 20   Bacteria, Viruses, and Protistans

The Unseen Multitudes [pp.310–311]

20.1. CHARACTERISTICS OF BACTERIA [pp.312–313]
20.2. BACTERIAL DIVERSITY [pp.314–315]
20.3. BACTERIAL GROWTH AND REPRODUCTION
   [p.316]

1. b; 2. d; 3. c; 4. e; 5. a; 6. prokaryotic; 7. plasmids;
8. wall; 9. capsule; 10. micrometers; 11. prokaryotic
fission; 12. cocci; 13. bacilli; 14. spirilla; 15. positive;
16. Archaebacteria; 17. thermophiles; 18. eubacteria;
19. cyanobacteria; 20. nitrogen-fixation; 21. sulfur;
22. nitrogen; 23. *Rhizobium*; 24. endospores;
25. *Clostridium botulinum* (*C. tetani*); 26. *Clostridium tetani*
(*C. botulinum*); 27. Lyme disease; 28. spirochete;
29. membrane receptors; 30. sunlight; 31. decomposers;
32. K; 33. d, H; 34. c, C; 35. b, B; 36. a, F; 37. c, I;
38. a, G; 39. c, E; 40. a, A; 41. c, D.

20.4. CHARACTERISTICS OF VIRUSES [p.317]
20.5. VIRAL MULTIPLICATION CYCLES [pp.318–319]
20.6. *Focus on Health:* **INFECTIOUS PARTICLES**
   TINIER THAN VIRUSES [p.319]

1. a. Nonliving, infectious agents, smaller than the smallest cells; require living cells to act as hosts for their replication; not acted upon by antibiotics.   b. The core can be
DNA or RNA; the capsid can be protein and/or lipid.
c. Bacteriophage viruses may use the lytic pathway, in
which the virus quickly subdues the host cells and replicates itself and descendants are released as the cell undergoes lysis; or they may use a temperate pathway, in
which viral genes remain inactive inside the host cell
during a period of latency, which may be a long time,
before activation and lysis.   2. a. Possible answers include Herpes simplex (a Herpesvirus), HIV (a Retrovirus), or influenza virus.   b. Herpes simplex: DNA
virus. Initial infection is a lytic cycle that causes Herpes
(sores) on mucous membranes on mouth or genitals.
Recurrent infections are temperate. Most cells are in
nerves and skin. No immunity. No cure. HIV: RNA
virus. Host cells are specific white blood cells. There is a
latency period that may last longer than a year before
host tests positive for HIV. As white blood cells are destroyed, the host's immune system is progressively destroyed (AIDS). No cure exists.   3. virus; 4. nucleic acid;
5. protein coat (viral capsid); 6. Viruses; 7. Lysis;
8. viroids; 9. Retroviruses (HIV); 10. lysogenic;
11. Nanometers; 12. micrometers; 13. 86,000; 14. latency; 15. prions; 16. BSE prion, prion; 17. HIV, RNA;
18. Herpesviruses, DNA; 19. a; 20. c; 21. e; 22. b;
23. d.

20.7. KINGDOM AT THE CROSSROADS [p.320]
20.8. PARASITIC OR PREDATORY MOLDS
[pp.320–321]
20.9. THE ANIMAL-LIKE PROTISTANS [pp.322–323]
20.10. THE NOTORIOUS SPOROZOANS [p.324]
20.11. *Focus on Science:* **THE NATURE OF INFECTIOUS DISEASES** [p.325]
1. a. E; b. E; c. E; d. E; e. P; f. E; g. E; 2. Chytrids; 3. water molds ; 4. slime molds; 5. saprobes; 6. slime molds; 7. reproductive structures; 8. spores; 9. plasmodial; 10. meters; 11. late blight; 12. fish; 13. downy mildew; 14. chytrids; 15. enzymes; 16. pseudopodia; 17. Foraminiferans; 18. radiolarians; 19. plankton; 20. freshwater; 21. contractile vacuoles; 22. gullet; 23. enzyme-filled vesicles; 24. trypanosomes; 25. *Plasmodium*; 26. mosquito; 27. gametes; 28. C, E, I; 29. C, E, K; 30. B, F; 31. A; 32. D, H; 33. B, F; 34. B, F, G; 35. D; 36. A; 37. B; 38. C; 39. E; 40. C; 41. E; 42. D; 43. A; 44. B.

20.12. A SAMPLING OF THE (MOSTLY) SINGLE-CELLED ALGAE [pp.326–327]
20.13. RED ALGAE [p.327]
20.14. BROWN ALGAE [p.328]
20.15. GREEN ALGAE [pp.328–329]
1. chloroplasts; 2. eyespot; 3. heterotrophic; 4. algae; 5. Chrysophytes; 6. diatoms; 7. fucoxanthin; 8. silica (glass); 9. diatomaceous deposits; 10. filters; 11. phytoplankton; 12. dinoflagellates; 13. agar; 14. marine; 15. nori; 16. brown; 17. algin; 18. photosynthetic; 19. cellulose; 20. starch; 21. a. Chlorophyll $\underline{a}$, phycobilins; b. Rhodophyta; c. agar, used as a moisture-preserving agent and culture medium; carrageenan is a stabilizer of emulsions; d. *Porphyra, Eucheuma*; e. Chlorophylls $\underline{a}$ and $\underline{c}_1$ and $\underline{c}_2$ + various carotenoids such as fucoxanthin; f. Phaeophyta; g. Algin, used as a thickener, emulsifier, and stabilizer of foods, cosmetics, medicines, paper, and floor polish; also are sources of mineral salts and fertilizer; h. *Macrocystis* (sea palm), *Ectocarpus, Laminaria*; i. Chlorophylls $\underline{a}$ and $\underline{b}$; j. Chlorophyta; k. Chlorophytes form much of the phytoplankton base of many food webs that support humans; l. *Volvox, Ulva, Spirogyra, Chlamydomonas*; 22. zygote, F; 23. resistant zygote, B; 24. meiosis and germination, H; 25. asexual reproduction, C; 26. asexual production, E; 27. gametes meet, G; 28. cytoplasmic fusion, A; 29. fertilization, D; 30. (+): a, c, d, e, g; (-) b, f, h, i, j.

Self-Quiz
1. a; 2. b; 3. c; 4. b; 5. a; 6. c; 7. a; 8. a; 9. b; 10. d; 11. A, E; 12. A, I; 13. A, B; 14. A, E, J; 15. A, D; 16. A, H; 17. A, F, G; 18. A, C; 19. F; 20. H; 21. C; 22. G; 23. D; 24. B; 25. A; 26. E; 27. A, C; 28. A; 29. A; 30. B, E; 31. B, F; 32. A; 33. D; 34. A.

Chapter 21   Fungi

Ode to the Fungus Among Us [pp.332–333]

21.1. CHARACTERISTICS OF FUNGI [p.334]
1. Symbiosis; 2. mutualism; 3. lichen; 4. mycorrhizae; 5. decomposers; 6. D; 7. E; 8. C; 9. H; 10. G; 11. B; 12. A; 13. F.

21.2. CONSIDER THE CLUB FUNGI [pp.334–335]
1. C; 2; D; 3. B; 4. A; 5. E; 6. reproductive; 7. mycelium; 8. club; 9. Nuclear; 10. diploid; 11. Meiosis; 12. spores; 13. cytoplasmic; 14. dikaryotic.

21.3. SPORES AND MORE SPORES [pp.336–337]
21.4. *Focus on Science:* **A LOOK AT THE UNLOVED FEW** [p.337]
1. hyphae; 2. gametangia; 3. zygospore; 4. spores; 5. mycelium; 6. E; 7. C; 8. D; 9. A; 10. F; 11. B; 12.a. *Cryphonectria parasitica;* b. *Allomyces capsulatus;* c. *Epidermophyton floccosum;* d. *Venturia inequalis;* e. *Claviceps purpurea.*

21.5. THE SYMBIONTS REVISITED [p.338]
1. Symbiosis; 2. mutualism; 3. lichen; 4. mycobiont; 5. photobiont; 6. sac; 7. hostile; 8. fungus; 9. photobiont; 10. photobiont's; 11. shelter; 12. parasite; 13. hypha; 14. cytoplasm; 15. mycobiont (photobiont); 16. photobiont (mycobiont); 17. layers; 18. mycorrhizae; 19. efficiently; 20. ectomycorrhiza; 21. temperate; 22. club; 23. endomycorrhizae; 24. penetrate; 25. zygomycetes; 26. absorptive; 27. soil.

Self-Quiz
1. *Polyporus* (c); 2. *Pilobolus* (b); 3. fly agaric mushroom (c); 4. cup fungus (a); 5. algae and fungi—lichen (e); 6. algae and fungi—lichen (e); 7. big laughing mushroom (c); 8. *Penicillium* (a); 9. b; 10. d; 11. d; 12. c; 13. a; 14. a; 15. c; 16. c; 17. d; 18. a; 19. a and c; 20. b.

Chapter 22 Plants

Pioneers in a New World [pp.340–341]

22.1. EVOLUTIONARY TRENDS AMONG PLANTS
[pp.342–343]
1. C; 2. D; 3. E; 4. A; 5. B; 6. a. Well-developed root systems; b. Well-developed shoot systems; c. Lignin production; d. Xylem and phloem; e. Cuticle production; f. Stomata; g. Gametophytes; h. Sporophytes; i. Spores; j. Heterospory; k. Pollen grains; l. Seeds.

22.2. BRYOPHYTES [pp.344–345]
1. air; 2. T; 3. rhizoids; 4. T; 5. mosses; 6. sporophytes; 7. T; 8. Peat; 9. water; 10. nonvascular; 11. gametophytes; 12. gametophytes; 13. sperms; 14. Fertilization; 15. zygote; 16. sporophyte; 17. Meiosis; 18. spores; 19. gametophytes.

22.3. *Focus on the Environment:* **ANCIENT CARBON TREASURES** [p.345]
22.4. EXISTING SEEDLESS VASCULAR PLANTS
[pp.346–347]
1. c; 2. b; 3. e; 4. a; 5. d; 6. d; 7. e; 8. c; 9. a; 10. d; 11. e; 12. d; 13. c; 14. c; 15. e; 16. b; 17. d; 18. d; 19. b; 20. e; 21. rhizome; 22. sorus; 23. Meiosis; 24. spores; 25. spores; 26. gametophyte; 27. gametophyte; 28. sperm; 29. egg; 30. fertilization; 31. zygote; 32. embryo.

22.5. THE RISE OF THE SEED-BEARING PLANTS
[pp.348–349]

22.6. *Focus on the Environment:* **GOOD-BYE, FORESTS**
[p.349]
1. C; 2. E; 3. A; 4. B; 5. D; 6. sporophyte; 7. cones; 8. cones; 9. ovule; 10. meiosis; 11. gametophyte; 12. eggs; 13. meiosis; 14. pollen; 15. Pollination; 16. tube; 17. Sperm; 18. Fertilization; 19. embryo.

22.7. GYMNOSPERM DIVERSITY [p.350]
1. b; 2. d; 3. a; 4. b; 5. c; 6. a; 7. e; 8. b; 9. e; 10. c; 11. d; 12. e.

22.8. ANGIOSPERMS—THE FLOWERING, SEED-BEARING PLANTS [p.351]
22.9. VISUAL OVERVIEW OF FLOWERING PLANT LIFE CYCLES [p.352]
22.10. *Commentary:* **SEED PLANTS AND PEOPLE**
[p.353]
1. B; 2. C; 3. E; 4. F; 5. D; 6. A; 7. H; 8. G.

Self-Quiz
1. a. gametophyte, none or simple vascular tissue, no; b. sporophyte, yes, no; c. sporophyte, yes, no; d. sporophyte, yes, no; e. sporophyte, yes, yes; f. sporophyte, yes, yes; 2. b; 3. c; 4. a; 5. c; 6. d; 7. b; 8. a; 9. d; 10. c; 11. c.

Chapter 23 Animals: The Invertebrates

Madeleine's Limbs [pp.356–357]

23.1. OVERVIEW OF THE ANIMAL KINGDOM
[pp.358–359]
23.2. PUZZLES ABOUT ORIGINS [p.360]
23.3. SPONGES—SUCCESS IN SIMPLICITY
[pp.360–361]
1. O; 2. I; 3. H; 4. E; 5. P; 6. A; 7. F; 8. K; 9. C; 10. J; 11. D; 12. G; 13. N; 14. M; 15. L; 16. B; 17. a. Placozoa; b. Sponges; c. Cnidaria; d. Turbellarians, flukes, tapeworms; e. Nematoda; f. Snails, slugs, clams, squids, octopuses; g. Annelida; h. Crustaceans, spiders, insects; i. Echinodermata; 18. c; 19. a; 20. c; 21. d; 22. c; 23. b; 24. c; 25. c; 26. H; 27. K; 28. D; 29. J; 30. B; 31. C; 32. I; 33. E; 34. A; 35. G; 36. F.

23.4. CNIDARIANS—TISSUES EMERGE [pp.362–363]
1. Cnidaria; 2. nematocysts; 3. medusa; 4. polyp; 5. gastrodermis; 6. epidermis; 7. epithelium; 8. Nerve; 9. receptor; 10. contractile; 11. mesoglea; 12. hydrosta-

tic; 13. corals; 14. gonads; 15. planulas; 16. Dinoflagellates; 17. feeding polyp; 18. reproductive polyp; 19. female medusa; 20. planula larva.

23.5. ACOELOMATE ANIMALS—AND THE SIMPLEST ORGAN SYSTEMS [pp.364–365]
1. b, c; 2. a; 3. c; 4. a; 5. c; 6. c; 7. a; 8. a; 9. b; 10. a; 11. c; 12. a; 13. c; 14. c; 15. c; 16. b; 17. c; 18. b; 19. branching gut; 20. pharynx (protruding); 21. brain; 22. nerve cord; 23. ovary; 24. testis; 25. planarian (genus name = *Dugesia*); 26. no; 27. yes; 28. no coelom or acoelomate.

23.6. ROUNDWORMS [p.365]
23.7. *Focus on Health:* **A ROGUE'S GALLERY OF WORMS** [pp.366–367]
1. roundworm; 2. no; 3. pseudocoelom; 4. pseudocoelomate; 5. abundant; 6. bilateral; 7. cylindrical; 8. digestive; 9. coelom; 10. nutrients; 11. Parasitic; 12. living; 13. experimental; 14. schistomiasis; 15. pin-

worm; 16. juvenile; 17. pigs; 18. elephantiasis; 19. sexually; 20. larvae; 21. snail; 22. larvae; 23. human; 24. Larvae; 25. intermediate; 26. human; 27. intestine; 28. proglottids; 29. organs; 30. proglottids; 31. feces; 32. larval; 33. intermediate.

23.8. TWO MAJOR DIVERGENCES [p.367]
23.9. A SAMPLING OF MOLLUSKS [pp.368–369]
1. b; 2. b; 3. a; 4. b; 5. a; 6. b; 7. a; 8. a; 9. b; 10. a; 11. III; 12. I; 13. II; 14. mouth; 15. anus; 16. gill; 17. heart; 18. radula; 19. foot; 20. shell; 21. stomach; 22. mouth; 23. gill; 24. mantle; 25. muscle; 26. foot; 27. stomach; 28. internal shell; 29. mantle; 30. reproductive organ; 31. gill; 32. ink sac; 33. tentacle; 34. mollusk; 35. shell; 36. mantle; 37. gills; 38. foot; 39. radula; 40. eyes (tentacles); 41. tentacles (eyes); 42. d; 43. b; 44. b; 45. a; 46. d; 47. b; 48. c; 49. d; 50. b; 51. d; 52. d; 53. b; 54. d; 55. c; 56. b; 57. d; 58. c; 59. d; 60. b; 61. d; 62. b; 63. d; 64. c; 65. b.

23.10. ANNELIDS—SEGMENTS GALORE [pp.370–371]
1. F; 2. L; 3. B; 4. I; 5. D; 6. G; 7. N; 8. C; 9. K; 10. J; 11. A; 12. H; 13. E; 14. M; 15. coelom; 16. cuticle; 17. nerve cord; 18. seta; 19. nephridium; 20. body wall; 21. hearts; 22. vessels; 23. mouth; 24. brain; 25. nerve cord; 26. earthworm; 27. Annelida; 28. segmentation and a closed circulatory system; 29. Protostome; 30. yes; 31. bilateral; 32. yes.

23.11. ARTHROPODS—THE MOST SUCCESSFUL ORGANISMS ON EARTH [p.372]
1. e; 2. f; 3. a; 4. b; 5. f; 6. a; 7. f; 8. d; 9. b; 10. c; 11. e; 12. d; 13. a; 14. Arthropods, as a group, have the highest number of species, occupy the most habitats, and have very efficient defenses against predators and competitors, and the capacity to exploit the greatest amounts and kinds of foods.

23.12. A LOOK AT SPIDERS AND THEIR KIN [p.373]
23.13. A LOOK AT THE CRUSTACEANS [pp.374–375]
23.14. *HOW MANY LEGS?* [p.375]
1. E; 2. G; 3. F; 4. D; 5. C; 6. B; 7. A; 8. an exoskeleton; 9. lobsters and crabs; 10. similar; 11. Lobsters and crabs; 12. Barnacles; 13. Copepods; 14. barnacles; 15. barnacles; 16. molts; 17. millipedes; 18. centipedes; 19. Millipedes; 20. Centipedes; 21. cephalothorax; 22. abdomen; 23. swimmerets; 24. legs; 25. cheliped; 26. antennae; 27. lobster; 28. crustaceans; 29. exoskeleton and jointed legs; 30. yes; 31. bilateral; 32. yes; 33. poison gland; 34. brain; 35. heart; 36. spinnerets; 37. book lung; 38. chelicerates.

23.15. A LOOK AT INSECT DIVERSITY [pp.376–377]
1. J; 2. F; 3. E; 4. B; 5. I; 6. A; 7. D; 8. G; 9. H; 10. C.

23.16. THE PUZZLING ECHINODERMS [pp.378–379]
1. deuterostomes; 2. echinoderms; 3. calcium carbonate; 4. skeleton; 5. radial; 6. bilateral; 7. brain; 8. nervous; 9. arm; 10. tube; 11. water; 12. ampulla; 13. muscle; 14. whole; 15. digesting; 16. anus; 17. lower stomach; 18. upper stomach; 19. anus; 20. gonad; 21. coelom; 22. digestive gland; 23. eyespot; 24. tube feet; 25. starfish; 26. deuterostome; 27. water vascular; 28. a. brittle star; b. sea urchin; c. feather star (crinoid); d. sea cucumber; 29. Echinodermata; 30. A water vascular system and a body wall with spines, spicules, or plates; 31. radial.

Self-Quiz
1. a; 2. c; 3. a; 4. d; 5. b; 6. c; 7. d; 8. c; 9. d; 10. a; 11. a; 12. b; 13. g, F; 14. b, CGI; 15. f, AD; 16. h, E; 17. d, J; 18. i, L; 19. e, BK; 20. a, H.

Chapter 24   Animals: The Vertebrates

Making Do (Rather Well) With What You've Got [pp.382–383]

24.1. THE CHORDATE HERITAGE [p.384]
24.2. INVERTEBRATE CHORDATES [pp.384–385]
24.3. EVOLUTIONARY TRENDS AMONG THE VERTEBRATES [pp.386–387]
24.4. EXISTING JAWLESS FISHES [p.387]
24.5. EXISTING JAWED FISHES [pp.388–389]
1. nerve cord; 2. pharynx; 3. notochord; 4. invertebrate; 5. vertebrates 6. lancelets; 7. filter-feeding; 8. gill slits; 9. Tunicates (sea squirts); 10. tadpoles; 11. notochord; 12. torsion bar; 13. larva; 14. mutation; 15. sex organs; 16. predators; 17. cephalochordates (lancelets); 18. vertebrae; 19. predators; 20. jaws; 21. brain; 22. paired (fleshy); 23. fleshy; 24. lungs; 25. circulatory; 26. adult tunicate (sea squirt); 27. lancelet; 28. notochord; 29. pharynx with gill slits; 30. mouth; 31. gut; 32. pharynx; 33. intestine; 34. atrial opening (exit); 35. oral opening; 36. midgut; 37. pharyngeal gill slits; 38. anus; 39. notochord; 40. dorsal, tubular nerve cord; 41. early jawless fish (agnathan); 42. Supporting structures; 43. gill slit; 44. placoderm; 45. jaws; 46. hemichordates; 47. tunicates; 48. lancelets; 49. lampreys, hagfishes; 50. jawed, armored fishes; 51. cartilaginous fishes; 52. bony fishes; 53. Amphibia; 54. Reptilia; 55. Aves; 56. Mammalia; 57. sharks; 58. gill slits; 59. bony; 60. ray-finned; 61. C; 62. G; 63. J; 64. F; 65. M; 66. H; 67. N; 68. I; 69. B; 70. O; 71. D; 72. E; 73. E; 74. K; 75. L; 76. A; 77. all jawless; 78. filter-feeders (also rasping or sucking); 79. jaws; 80. cartilage; 81. bone; 82. ray-finned fishes; 83. lobe-finned fishes

such as the coelacanth; 84. branch ⑤; 85. branch ④; 86. Silurian; 87. about 375 million years ago.

24.6. AMPHIBIANS [pp.390–391]
24.7. THE RISE OF REPTILES [pp.392–393]
24.8. BIRDS [pp.394–395]
24.9. THE RISE OF MAMMALS [pp.396–397]
1. lungs; 2. fins; 3. Skeletal elements; 4. brains; 5. balance; 6. circulatory; 7. blood; 8. insects; 9. salamanders; 10. water; 11. reproduce; 12. toxins; 13. insects; 14. reptilian; 15. limb; 16. amniote; 17. turtles (lizards); 18. lizards (turtles); 19. internal; 20. Carboniferous; 21. turtles; 22. Carboniferous; 23. Triassic; 24. Jurassic; 25. Birds; 26. four; 27. synapsid; 28. Carboniferous; 29. G; 30. D; 31. E; 32. B; 33. C; 34. A; 35. F; 36. A; 37. A; 38. A; 39. C; 40. C; 41. D; 42. D; 43. a. dry, scaly skin; b. four-chambered heart; c. feather development; d. hair development; e. loss of limbs. 44. embryo (notochord); 45. albumin; 46. yolk sac; 47. reptiles; 48. feathers; 49. sternum (breastbone); 50. air cavities; 51. four-chambered heart; 52. migratory; 53. amniote; 54. platypus; 55. pouched (marsupials); 56. placental; 57. four; 58. hair.

24.10. THE PRIMATES [pp.398–399]
24.11. EMERGENCE OF EARLY HUMANS
  [pp.400–401]

24.12. *Focus on Science:* **OUT OF AFRICA—ONCE , TWICE, OR . . .** [p.402]
1. mammals; 2. hair (mammary glands); 3. orders; 4. Primates; 5. depth; 6. prehensile; 7. grasping; 8. smell; 9. brains; 10. culture; 11. handbones; 12. teeth; 13. 60; 14. rodents; 15. insects; 16. Miocene; 17. drier; 18. grasslands; 19. 10; 20. 5; 21. apes; 22. 10; 23. 5; 24. 4.4; 25. australopiths; 26. bipedal; 27. 2.5; 28. *Homo erectus*; 29. fire; 30. tool; 31. Neandertals; 32. *Homo erectus*; 33. biochemical; 34. immunological; 35. 40,000; 36. *Homo sapiens*; 37. b; 38. c; 39. c; 40. c; 41. b; 42. blank [not a primate]; 43. a; 44. a; 45. c; 46. e; 47. c; 48. c; 49. d; 50. a; 51. e; 52. e; 53. b; 54. b; 55. B, G, H, J; 56. A, G, J; 57. E, H, I, K; 58. B, G, H, J; 59. C, D, L.

Self-Quiz
1. d; 2. a; 3. c; 4. c; 5. a; 6. a; 7. a; 8. b; 9. d; 10. d, H; 11. b, B; 12. i, F; 13. g, D; 14. h, A; 15. c, E; 16. a, I; 17. e, G; 18. f, C; 19. early amphibian, Amphibia; 20. Arctic fox, Mammalia; 21. soldier fish, Osteichthyes; 22. Ostracoderm, Agnatha; 23. owl, Aves; 24. marine turtle, Reptilia; 25. shark, Chondrichthyes; 26. coelacanth, Osteichthyes (lobe-finned fish); 27. reef ray, Chondrichthyes; 28. tunicate (sea squirt), Urochordata; 29. lancelet, Cephalochordata; 30. B; 31. H; 32. C; 33. D; 34. A; 35. E; 36. G; 37. F

Chapter 25   Plant Tissues

Plants Versus the Volcano [pp.406–407]

25.1. OVERVIEW OF THE PLANT BODY [pp.408–409]
1. ground tissue (C); 2. vascular tissues (E); 3. dermal tissues (D); 4. shoot system (A); 5. root system (B); 6. E; 7. D; 8. B; 9. F; 10. A; 11. C; 12. radial; 13. tangential; 14. transverse or cross; 15. shoot apical meristem (D); 16. shoot transitional meristems (B); 17. root transitional meristem (E); 18. root apical meristem (A); 19. lateral meristems (C); 20. a. Epidermis; primary; b. Ground tissues; primary; c. Vascular tissues; primary; d. Vascular tissues; secondary; e. Periderm; secondary.

25.2. TYPES OF PLANT TISSUES [pp.410–411]
1. sclerenchyma; 2. sclerenchyma; 3. collenchyma; 4. sclerenchyma; 5. parenchyma; 6. sclerenchyma; 7. b; 8. c; 9. a; 10. a; 11. b; 12. b; 13. a; 14. c; 15. a; 16. a; 17. c; 18. pits (xylem); 19. cytoplasm (xylem); 20. tracheids (xylem); 21. vessel (xylem); 22. vessel (xylem); 23. sieve (phloem); 24. companion (phloem); 25. sieve (phloem); 26. sieve (phloem); 27. a. Vessel members and tracheids; no; conduct water and dissolved minerals absorbed from soil, mechanical support. b. Sieve-tube members and companion cells; yes; transport sugar and other solutes. 28. a. Primary plant body;

cutin in the cuticle layer over epidermal cells restricts water loss and resists microbial attack; openings (stomata) permit water vapor and gases to enter and leave the plant. b. Secondary plant body; replaces epidermis to cover roots and stems. 29. a. One; in threes or multiples thereof; usually parallel; one pore or furrow; distributed throughout ground stem tissue. b. Two; in fours or fives or multiples thereof; usually netlike; three pores or pores with furrows; positioned in a ring in the stem.

25.3. PRIMARY STRUCTURE OF SHOOTS
  [pp.412–413]
1. primordium; 2. apical meristem; 3. bud; 4. shoot; 5. leaf; 6. procambium; 7. protoderm; 8. procambium; 9. ground meristem; 10. epidermis; 11. cortex; 12. procambium; 13. pith; 14. primary xylem; 15. primary phloem; 16. epidermis; 17. ground tissue; 18. vascular bundle; 19. sclerenchyma cells; 20. air space; 21. xylem vessel; 22. sieve tube; 23. companion cell; 24. epidermis; 25. cortex; 26. vascular bundle; 27. pith; 28. vessels; 29. meristematic cells; 30. sieve tubes; 31. fibers.

25.4. A CLOSER LOOK AT LEAVES [pp.414–415]
1. blade; 2. petiole (leaf stalk); 3. axillary bud; 4. node; 5. stem; 6. sheath; 7. dicot; 8. monocot; 9. palisade

mesophyll (D); 10. spongy mesophyll (B); 11. lower epidermis (A); 12. stoma (C); 13. leaf vein (E); 14. C; 15. A; 16. B; 17. E; 18. D.

25.5. PRIMARY STRUCTURE OF ROOTS [pp.416–417]
1. root hair (H); 2. endodermis (G); 3. pericycle (B); 4. epidermis (E); 5. cortex (D); 6. root apical meristem (F); 7. root cap (A); 8. endodermis (G); 9. pericycle (B); 10. primary phloem (C); 11. primary; 12. lateral; 13. taproot; 14. lateral; 15. fibrous; 16. hairs; 17. vascular; 18. pericycle; 19. cortex; 20. pith; 21. air; 22. oxygen; 23. endodermis; 24. control; 25. pericycle; 26. lateral; 27. cortex.

25.6. ACCUMULATED SECONDARY GROWTH—THE WOODY PLANTS [pp.418–419]
1. C; 2. K; 3. M; 4. F; 5. E; 6. N; 7. G; 8. L; 9. B; 10. A; 11. J; 12. D; 13. H; 14. I; 15. vessel (E); 16. early wood (F); 17. late wood (B); 18. bark (D); 19. one (C); 20. and 21. two and three (A); 22. vascular cambium (G).

Self-Quiz
1. b; 2. a; 3. b; 4. c; 5. a; 6. d; 7. d; 8. d; 9. c; 10. c; 11. b; 12. b.

Chapter 26   Plant Nutrition and Transport

Flies for Dinner [pp.422–423]

26.1. SOIL AND ITS NUTRIENTS [pp.424–425]
1. F; 2. C; 3. I; 4. B; 5. H; 6. A; 7. J; 8. E; 9. G; 10. D; 11. hydrogen; 12. thirteen; 13. ionic; 14. hydrogen; 15. macronutrients; 16. micronutrients; 17. a. iron, micronutrient; b. potassium, macronutrient; c. magnesium, macronutrient; d. chlorine, micronutrient; e. manganese, micronutrient; f. molybdenum, micronutrient; g. nitrogen, macronutrient; h. sulfur, macronutrient; i. boron, micronutrient; j. zinc, micronutrient; k. calcium, macronutrient; l. copper, micronutrient; m. phosphorus, macronutrient.

26.2. HOW DO ROOTS ABSORB WATER AND MINERAL IONS? [pp.426–427]
1. exodermis (B); 2. root hair (E); 3. epidermis (H); 4. vascular cylinder (J); 5. cortex (C); 6. endodermis (I); 7. cytoplasm (A); 8. water movement (G); 9. Casparian strip (D); 10. endodermal cell wall (F); 11. C; 12. D; 13. A; 14. B; 15. E; 16. mutualism; 17. Nitrogen "fixed" by bacteria; 18. root nodules; 19. scarce minerals that the fungus is better able to absorb; 20. obtaining sugars and nitrogen-containing compounds.

26.3. HOW IS WATER TRANSPORTED THROUGH PLANTS? [pp.428–429]
1. water; 2. xylem; 3. tracheids (vessels); 4. vessels (tracheids); 5. transpiration; 6. hydrogen; 7. cohesion; 8. tension; 9. roots; 10. xylem.

26.4. HOW DO STEMS AND LEAVES CONSERVE WATER? [pp.430–431]
1. 90 percent; 2. T; 3. T; 4. stomata; 5. opens; 6. T; 7. T; 8. decrease; 9. opens; 10. T; 11. night; 12. day.

26.5. HOW ARE ORGANIC COMPOUNDS DISTRIBUTED THROUGH PLANTS? [pp.432–433]
1. photosynthesis; 2. starch; 3. proteins; 4. seeds; 5. fats; 6. Protein; 7. starches; 8. solutes; 9. sucrose; 10. sucrose; 11. D; 12. F; 13. B; 14. G; 15. A; 16. C; 17. E.

Self-Quiz
1. c; 2. e; 3. c; 4. d; 5. b; 6. d; 7. b; 8. a; 9. b; 10. c.

Chapter 27   Plant Reproduction and Development

A Coevolutionary Tale [pp.436–437]

27.1. REPRODUCTIVE STRUCTURES OF FLOWERING PLANTS [pp.438–439]
27.2. *Focus on Environment:* POLLEN SETS ME SNEEZING [p.439]
1. sporophyte (C); 2. flower (B); 3. meiosis (D); 4. gametophyte (F); 5. gametophyte (E); 6. fertilization (A); 7. sepal; 8. petal; 9. stamen; 10. filament; 11. anther; 12. carpel; 13. stigma; 14. style; 15. ovary; 16. ovule; 17. receptacle; 18. I; 19. C; 20. G; 21. F; 22. B; 23. E; 24. H; 25. A; 26. D.

27.3. A NEW GENERATION BEGINS [pp.440–441]
1. anther; 2. pollen sac; 3. microspore mother; 4. meiosis; 5. microspores; 6. pollen tube; 7. sperm-producing; 8. pollen; 9. stigma; 10. male gametophyte; 11. ovule; 12. integuments; 13. meiosis; 14. megaspores; 15. megaspore; 16. megaspore; 17. mitosis; 18. eight; 19. embryo sac; 20. female gametophyte; 21. two; 22. endosperm; 23. egg; 24. double fertilization; 25. endosperm; 26. embryo; 27. seed; 28. coat; 29. Chemical and molecular cues guide a pollen tube's growth through tissues of the style and the ovary, toward the egg chamber and sexual destiny; 30. The em-

bryo sac is the site of double fertilization; 31. a. Fusion of one egg nucleus (*n*) with one sperm nucleus (*n*); the plant embryo (2*n*); eventually develops into a new sporophyte plant. b. Fusion of one sperm nucleus (*n*) with the endosperm mother cell (2*n*); endosperm tissues (3*n*); nourishes the embryo within the seed.

27.4. FROM ZYGOTE TO SEEDS AND FRUITS
[pp.442–443]
1. a. "Seed leaves" that develop from two lobes of meristematic tissue of the embryo; b. Seeds are mature ovules; c. Integuments of the ovule harden into the seed coat; d. A fruit is a mature ovary; 2. nucleus; 3. vacuole; 4. zygote; 5. embryo; 6. embryo; 7. seed coat; 8. shoot tip; 9. cotyledons; 10. sporophyte; 11. endosperm; 12. root tip; 13. mature; 14. fruit; 15. mitotic; 16. sporophyte; 17. ovule; 18. fruit; 19. cotyledons; 20. two; 21. one; 22. endosperm; 23. germinates; 24. thin; 25. enzymes; 26. seedling; 27. ovule; 28. endosperm; 29. ovary; 30. coat; 31. seed; 32. ovule; 33. fruits; 34. dry; 35. ovaries; 36. receptacle; 37. c; 38. c; 39. b; 40. a; 41. a; 42. c; 43. b; 44. b; 45. a; 46. b; 47. e; 48. b; 49. d; 50. c; 51. f; 52. b; 53. c; 54. c.

27.5. ASEXUAL REPRODUCTION OF FLOWERING PLANTS [pp.444–445]
1. E; 2. G; 3. I; 4. F; 5. H; 6. D; 7. B; 8. C; 9. A; 10. D; 11. E; 12. A; 13. B; 14. C.

27.6. PATTERNS OF EARLY GROWTH AND DEVELOPMENT—AN OVERVIEW [pp.446–447]
1. embryo; 2. germination; 3. environmental; 4. imbibition; 5. ruptures; 6. aerobic; 7. meristematic;

8. root; 9. primary root; 10. germination; 11. heritable (genetic); 12. genes; 13. zygote; 14. genes; 15. cytoplasmic; 16. metabolic; 17. selective; 18. hormones; 19. Interactions; 20. cotyledons; 21. hypocotyl; 22. primary; 23. cotyledons; 24. foliage; 25. primary; 26. cotyledons; 27. coleoptile; 28. branch; 29. primary; 30. foliage; 31. stem; 32. adventitious; 33. branch; 34. primary; 35. prop; 36. foliage; 37. coleoptile.

27.7. EFFECTS OF PLANT HORMONES [pp.448–449]
27.8. *Focus on Science:* FOOLISH SEEDLINGS! [p.449]
1. e; 2. c; 3. a; 4. d; 5. a; 6. e; 7. a; 8. b; 9. a; 10. c; 11. e; 12. c; 13. e; 14. d; 15. F; 16. H; 17. E; 18. J; 19. G; 20. C; 21. I; 22. B; 23. A; 24. K; 25. D.

27.9. ADJUSTMENTS IN THE RATE AND DIRECTION OF GROWTH [pp.450–451]
1. a; 2. c; 3. d; 4. b; 5. d; 6. b; 7. a; 8. a; 9. c; 10. b.

27.10. BIOLOGICAL CLOCKS AND THEIR EFFECTS [pp.452–453]
1. D; 2. I; 3. H; 4. J; 5. B; 6. A; 7. G; 8. C; 9. F; 10. E; 11. Pr; 12. Pfr; 13. Pfr; 14. Pr; 15. response; 16. Photoperiodism; 17. Pfr; 18. enzymes; 19. longer; 20. shorter; 21. mature; 22. long-day; 23. short-day.

27.11. LIFE CYCLES END, AND TURN AGAIN [p.454]
1. b; 2. d; 3. c; 4. b; 5. a; 6. a; 7. c; 8. a; 9. a; 10. a.

Self-Quiz
1. c; 2. d; 3. c; 4. b; 5. c; 6. c; 7. d; 8. d; 9. c; 10. a; 11. a; 12. d; 13. d; 14. c.

Chapter 28   Tissues, Organ Systems, and Homeostatis

Meerkats, Humans, It's All the Same [pp.458–459]

28.1. EPITHELIAL TISSUE [pp.460–461]
1. Anatomy; 2. unlike tissues; 3. T; 4. T; 5. impermeable; 6. Exocrine; 7. Exocrine; 8. tissue; 9. organ; 10. metabolic; 11. internal environment; 12. organ; 13. organ system (urinary system, excretory system); 14. organism; 15. homeostasis; 16. tissue; 17. epithelial; 18. connective (epithelial); 19. Epithelial; 20. free; 21. basement membrane; 22. lining; 23. protective; 24. junctions; 25. Gland; 26. exocrine; 27. hormones; 28. endocrine; 29. b. air sacs; e. columnar; c. cuboidal; f. gut (small intestine); d. nephrons; a. squamous.

28.2 CONNECTIVE TISSUE [pp.462–463]
28.3. MUSCLE TISSUE [p.464]
28.4. NERVOUS TISSUE [p.465]
28.5. *Focus on Science*: FRONTIERS IN TISSUE RESEARCH [p.465]

1. groups of; 2. T; 3. synchronously with; 4. T; 5. T; 6. T; 7. ground substance; 8. Loose; 9. Tendons (Ligaments); 10. ligaments (tendons); 11. Laboratory-grown epidermis; 12. designer organ; 13. F; 14. E; 15. C; 16. A, H; 17. B; 18. D; 19. G; 20. H; 21. A; 22. connective, D, 9, 11; 23. epithelial, G, 1, (6), 12; 24. muscle, I, 8, (11); 25. muscle, J, 5, (11); 26. connective, E, 6, 7, 11, (14); 27. gametes, 2; 28. connective, B, 10, 16; 29. epithelial, H, (1), 6, (12); 30. connective, (1), 6, 15; 31. nervous, 3; 32. muscle, C, 13; 33. epithelial, F, 1, 6, 12; 34. connective, A, 4, 7, 14.

28.6. ORGAN SYSTEMS [pp.466–467]
1. germ cells; 2. gametes; 3. meiosis; 4. somatic; 5. mitosis; 6. tissues; 7. ectoderm; 8. endoderm; 9. mesoderm; 10. mesoderm; 11. endoderm; 12. ectoderm; 13. circulatory; 14. respiratory; 15. urinary (= excretory); 16. skeletal; 17. endocrine; 18. immune lymphatic; 19. reproductive; 20. digestive; 21. muscular; 22. nervous; 23. integumentary; 24. G; 25. I;

26. H; 27. B; 28. F; 29. E; 30. C; 31. J; 32. A; 33. D;
34. K; 35. thoracic; 36. cranial; 37. pelvic; 38. midsagittal; 39. frontal; 40. anterior; 41. transverse;
42. inferior; 43. fluid; 44. chemical/hormonal; 45. nervous.

28.7. HOMEOSTASIS AND SYSTEMS CONTROL
[pp.468–469]

1. positive; 2. T; 3. an effector; 4. T; 5. positive;
6. negative; 7. integrator; 8. glands; 9. receptors;
10. hypothalamus; 11. effectors; 12. sweat.

Self-Quiz

1. d; 2. b; 3. c; 4. c; 5. d; 6. c; 7. c; 8. d; 9. a; 10. d;
11. d; 12. a; 13. A; 14. H; 15. E; 16. F; 17. K; 18. I;
19. J; 20. G; 21. B; 22. D; 23. C

Chapter 29   Integration and Control: Nervous Systems

Why Crack the System? [pp.472–473]

29.1. NEURONS—THE COMMUNICATION
SPECIALISTS [pp.474–475]
29.2. A CLOSER LOOK AT ACTION POTENTIALS
[pp.476–477]

1. neurons; 2. Neuroglial; 3. Sensory; 4. interneurons;
5. motor; 6. cell body; 7. Dendrites; 8. signals (stimuli); 9. axon; 10. cell body; 11. axon; 12. endings;
13. voltage differential; 14. resting membrane potential;
15. action potential (nerve impulse); 16. disturbance;
17. potassium; 18. Na; 19. Channel (transport);
20. Transport; 21. potassium; 22. sodium; 23. 30;
24. 10; 25. gates; 26. sodium–potassium pumps; 27. localized; 28. Graded; 29. duration; 30. trigger zone;
31. action potential; 32. all-or-nothing; 33. threshold;
34. negative; 35. positive; 36. resting; 37. trigger zone
(axonal membrane); 38. channel proteins; 39. gates;
40. sodium-potassium pump; 41. lipid bilayer; 42. action potential; 43. threshold; 44. resting membrane
potential; 45. milliseconds; 46. millivolts

29.3. CHEMICAL SYNAPSES [pp.478–479]
29.4. PATHS OF INFORMATION FLOW [pp.480–481]
29.5. *Focus on Health:* SKEWED INFORMATION
FLOW [p.482]

1. chemical synapse; 2. neurotransmitters; 3. Acetylcholine (ACh); 4. excitatory; 5. inhibitory; 6. postsy-
naptic; 7. Endorphins; 8. Synaptic integration;
9. summed; 10. myelin; 11. exposed node; 12. reflex;
13. stretch reflex; 14. inhibitory; 15. axon; 16. myelin
sheath; 17. blood vessels; 18. axons; 19. exposed node;
20. Schwann cell (myelin sheath); 21. sensory neuron;
22. inter; 23. motor neuron; 24. receptor; 25. cell body;
26. axon; 27. axon endings; 28. dendrites; 29. H, F;
30. E; 31. K; 32. I; 33. J; 34. B; 35. C; 36. G; 37. A, D

29.6. INVERTEBRATE NERVOUS SYSTEMS
[pp.482–483]

29.7. VERTEBRATE NERVOUS SYSTEMS—AN
OVERVIEW [pp.484–485]
29.8. THE MAJOR EXPRESSWAYS [pp.486–487]

1. radial; 2. nerve net; 3. danger; 4. Flatworms;
5. cephalization; 6. somatic; 7. autonomic; 8. sympathetic; 9. parasympathetic; 10. central nervous system;
11. vertebral column; 12. cranial; 13. spinal cord;
14. cervical nerves; 15. thoracic nerves; 16. lumbar
nerves; 17. sacral nerves; 18. coccygeal nerves; 19. autonomic; 20. sympathetic; 21. parasympathetic; 22. increases; 23. decreases; 24. spinal cord; 25. white
matter; 26. gray matter; 27. reflexes; 28. spinal cord;
29. ganglion; 30. vertebra; 31. spinal nerve;
32. meninges; 33. gray matter; 34. white matter

29.9. THE VERTEBRATE BRAIN [pp.488–489]
29.10. MEMORY [p.490]
29.11. *Focus on Health:* DRUGGING THE BRAIN
[pp.490–491]

1. medulla oblongata; 2. cerebellum; 3. pons; 4. midbrain; 5. reticular formation; 6. forebrain; 7. hypothalamus; 8. sexual; 9. thalamus; 10. cerebrospinal fluid;
11. memory; 12. short-term memory; 13. long-term
memory; 14. Endorphins (Enkephalins); 15. pain;
16. neurotransmitters; 17. I; 18. H; 19. A; 20. L;
21. K; 22. J; 23. F; 24. B; 25. D; 26. G; 27. C; 28. E;
29. cerebellum; 30. medulla oblongata; 31. pons;
32. hypothalamus; 33. thalamus; 34. corpus callosum;
35. D; 36. A; 37. D; 38. D; 39. A; 40. B; 41. C; 42. C;
43. D

Self-Quiz

1. a; 2. a; 3. d; 4. b; 5. a; 6. M; 7. G; 8. D; 9. F;
10. K; 11. C; 12. I; 13. J; 14. B; 15. A; 16. L; 17. H;
18. E.

Chapter 30   Sensory Reception

Different Strokes for Different Folks [pp.494–495]

30.1. OVERVIEW OF SENSORY PATHWAYS [p.496]
30.2. SOMATIC SENSATIONS [p.497]
30.3. SENSES OF HEARING AND BALANCE
[pp.498–499]
1. receptors; 2. stimulus; 3. Chemoreceptors;
4. mechanoreceptors; 5. photoreceptors; 6. thermoreceptors; 7. sensation; 8. perception; 9. nerve pathways; 10. brain regions; 11. D; 12. C; 13. B; 14. A;
15. E; 16. E; 17. B; 18. B; 19. B; 20. C; 21. B;
22. touch; 23. cold; 24. skin; 25. skeletal;
26. mechanoreceptors; 27. Free; 28. Pain;
29. Mechanoreceptors; 30. skin; 31. vestibular apparatus; 32. free nerve endings (C); 33. Ruffini endings (A);
34. Meissner corpuscle (D); 35. epidermis; 36. dermis;
37. Pacinian corpuscle (B); 38. brain; 39. frequency;
40. number; 41. hearing; 42. amplitude; 43. frequency;
44. higher; 45. mechanoreceptors; 46. middle;
47. cochlea; 48. organ of Corti; 49. vestibular apparatus;

50. middle earbones (malleus, incus, stapes); 51. cochlea;
52. auditory nerve; 53. tympanic membrane/eardrum;
54. oval window; 55. basilar membrane; 56. tectorial
membrane

30.4. SENSE OF VISION [pp.500–501]
30.5. VISUAL PERCEPTION [p.502]
30.6. SENSES OF TASTE AND SMELL [p.503]
1. Light; 2. photoreceptor; 3. Vision; 4. compound
eyes; 5. Eyes; 6. cornea; 7. retina; 8. focal point;
9. Visual accommodation; 10. Farsighted; 11. Cone;
12. fovea; 13. vitreous body; 14. cornea; 15. iris;
16. lens; 17. aqueous humor; 18. ciliary muscle;
19. retina; 20. fovea; 21. optic nerve; 22. blind
spot/optic disk; 23. sclera

Self-Quiz

1. a; 2. c; 3. d; 4. b; 5. a; 6. e; 7. G; 8. E; 9. F;
10. B; 11. C(F); 12. H; 13. D; 14. A.

Chapter 31   Endocrine Control

Hormone Jamboree [pp.506–507]

31.1. THE ENDOCRINE SYSTEM [pp.508–509]
1. E; 2. A; 3. B; 4. F; 5. C; 6. D; 7. a. 1, six releasing
and inhibiting hormones; synthesizes ADH, oxytocin;
b. 2, ACTH, TSH, FSH, LH, GSH; c. 2, stores and secretes two hypothalamic hormones, ADH and oxytocin;
d. 3, sex hormones of opposite sex, cortisol, aldosterone;
e. 3, epinephrine, norepinephrine; f. 4, estrogens, progesterone; g. 5, testosterone; h. 6, melatonin; i. 7, thyroxine and triiodothyronine; j. 8, parathyroid hormone
(PTH); k. 9, thymosins; l. 10, insulin, glucagon, somatostatin.

31.2. SIGNALING MECHANISMS [pp.510–511]
1. a; 2. b; 3. a; 4. b; 5. b; 6. a; 7. b; 8. b; 9. b; 10. b.

**31.3. THE HYPOTHALAMUS AND PITUITARY
GLAND** [pp.512–513]
1. A(H); 2. P(G); 3. A(A); 4. A(I); 5. A(D); 6. I(B);
7. P(E); 8. A(C); 9. A(F); 10. hypothalamus; 11. posterior; 12. anterior; 13. intermediate; 14. releasers;
15. inhibitors.

**31.4. EXAMPLES OF ABNORMAL PITUITARY
OUTPUT** [p.514]
1. a. Gigantism; b. Pituitary dwarfism; c. Diabetes
insipidus; d. Acromegaly.

**31.5. SOURCES AND EFFECTS OF OTHER
HORMONES** [p.515]
1. a. D (c); b. I (e); c. F (h); d. A (g); e. H (k); f. E (d);
g. C (a); h. K (j); i. B (i); j. J (f); k. G (b); l. F (l); m. A
(m); n. B (o); o. C (n).

**31.6. FEEDBACK CONTROL OF HORMONAL
SECRETIONS** [pp.516–517]
1. E; 2. F; 3. D; 4. H; 5. I; 6. J; 7. A; 8. G; 9. B;
10. C; 11. thyroid; 12. metabolic (metabolism); 13. iodine; 14. iodide; 15. TSH; 16. goiter; 17. Hypothyroidism; 18. hypothyroid; 19. Hyperthyroidism;
20. gonads; 21. hormones; 22. testes; 23. ovaries;
24. testosterone; 25. secondary; 26. gametes.

31.7. RESPONSES TO LOCAL CHEMICAL CHANGES
[pp.518–519]
1. parathyroid; 2. PTH; 3. calcium; 4. calcium; 5. reabsorption; 6. D_3; 7. intestinal; 8. rickets; 9. a.
Glucagon, Causes glycogen (a storage polysaccharide)
and amino acids to be converted to glucose in the liver
(glucagon raises the glucose level); b. Insulin, Stimulates glucose uptake by liver, muscle, and adipose cells;
promotes synthesis of proteins and fats, and inhibits
protein conversion to glucose (lowers the glucose level);
c. Somatostatin, Helps control digestion; can block secretion of insulin and glucagon; 10. rises; 11. excessive;
12. energy; 13. ketones; 14. insulin; 15. Glucagon;
16. type 1 diabetes; 17. type 1 diabetes; 18. type 2 diabetes; 19. Type 2 diabetes.

Chapter 32   Protection, Support, and Movement

Of Men, Women, and Polar Huskies [pp.524–525]

32.1. VERTEBRATE SKIN [p.526]
32.2. *Focus on Health:* **SUNLIGHT AND SKIN** [p.527]
1. epidermis; 2. dermis; 3. hypodermis; 4. adipose;
5. hair; 6. sensory nerve ending; 7. sebaceous gland;
8. smooth muscle; 9. hair follicle; 10. sweat gland;
11. blood vessels; 12. epidermis; 13. dermis; 14. hypodermis; 15. a; 16. c; 17. d; 18. b; 19. c; 20. c; 21. e.

32.3. SKELETAL SYSTEMS [pp.528–529]
32.4. CHARACTERISTICS OF BONE [pp.530–531]
1. contract (relax); 2. relax (contract); 3. contractile;
4. radial; 5. hydrostatic; 6. exoskeleton; 7. Bones;
8. muscles; 9. Haversian; 10. Red marrow; 11. calcium
(phosphate); 12. phosphate (calcium); 13. osteoclasts;
14. sex; 15. osteoporosis; 16. axial; 17. appendicular;
18. Synovial; 19. cartilage; 20. synovial; 21. osteoarthritis; 22. rheumatoid arthritis; 23. nutrient canal; 24. yellow marrow; 25. compact bone; 26. spongy bone;
27. connective tissue covering (periosteum); 28. Haversian system; 29. Haversian canal (blood vessel);
30. mineral deposits (calcium phosphate); 31. osteocyte
(bone cell); 32. cranium; 33. clavicle; 34. sternum;
35. scapula; 36. radius; 37. carpal bones; 38. femur;
39. tibia; 40. tarsal bones; 41. metatarsal bones.

32.5. SKELETAL-MUSCULAR SYSTEMS [pp.532–533]
32.6. A CLOSER LOOK AT MUSCLES [pp.534–535]
1. Skeletal; 2. bones; 3. tendons; 4. Skeletal; 5. bones;
6. joints; 7. biceps; 8. triceps; 9. bones; 10. Tendons;
11. D; 12. C; 13. A; 14. E; 15. B; 16. F; 17. triceps
brachii; 18. pectoralis major; 19. external oblique;
20. rectus abdominis; 21. adductor longus; 22. quadriceps femoris; 23. tibialis anterior; 24. gastrocnemius;
25. deltoid; 26. biceps brachii; 27. biceps contracted;
28. triceps relaxed; 29. biceps relaxed; 30. triceps contracted; 31. muscle; 32. muscle cell (muscle fiber);
33. myofibril; 34. sarcomere; 35. myosin filament;
36. actin filament; 37. myofibrils; 38. sarcomeres;
39. myosin; 40. actin; 41. sliding-filament; 42. myosin;
43. sarcomere; 44. dephosphorylation of creatine phosphate; 45. glycolysis alone; 46. aerobic respiration;
47. ATP; 48. Its oxygen-requiring reactions provide
most of the ATP needed for muscle contraction during
prolonged, moderate exercise; 49. It provides the en-

ergy to make myosin filaments slide along actin filaments; 50. Supplies phosphate to ADP → ATP, which
powers muscle contraction for a short time because creatine phosphate supplies are limited; 51. Stored in muscles and in the liver, glucose is stored by the animal body
in this form of starch; 52. An anaerobic pathway in
which glucose is broken down to lactate with a small
yield of ATP; this pathway operates during intense exercise; 53. Projections from the thick filaments of myosin
that bind to actin sites and form temporary cross-bridges. Making and breaking these cross-bridges cause
myosin filaments to be pulled to the center of a sarcomere; 54. The repetitive unit of muscle contraction. Many
sarcomeres constitute a myofibril. Many myofibrils constitute a muscle cell.

32.7. CONTROL OF MUSCLE CONTRACTION [p.536]
32.8. PROPERTIES OF WHOLE MUSCLES [p.537]
1. excitability; 2. neuron; 3. action potential; 4. motor;
5. ATP; 6. creatine phosphate; 7. Contraction; 8. bone;
9. calcium; 10. sarcoplasmic reticulum; 11. actin;
12. active transport; 13. action potentials; 14. sarcoplasmic reticulum; 15. diameter; 16. motor; 17. motor
unit; 18. muscle twitch; 19. twitch; 20. Tetanus;
21. weak; 22. frequency (rate); 23. axon of motor neuron serving one motor unit; 24. neuromuscular junction;
25. axon of another motor neuron serving another motor
unit; 26. individual muscle cells; 27. time that stimulus
is applied; 28. contraction phase; 29. relaxation phase;
30. time (seconds); 31. force; 32. Used to clear the actin
binding sites of any obstacles to cross-bridge formation
with myosin heads; 33. Supplies commands (signals) to
muscle cells to contract and relax; 34. The endoplasmic
reticulum of a muscle cell. Stores calcium ions and releases them in response to incoming signals from motor
neurons; uses active transport to bring the calcium ions
back inside; 35. Anabolic steroids; 36. muscle;
37. acne; 38. testes; 39. aggression; 40. facial;
41. menstrual periods; 42. voice.

Self-Quiz
1. e; 2. c; 3. c; 4. d; 5. b; 6. d; 7. a; 8. c; 9. d; 10. b;
11. e; 12. d; 13. a; 14. e; 15. b; 16. c; 17. b; 18. d;
19. a; 20. e.

Chapter 33   Circulation

Heartworks [pp.540–541]

33.1. CIRCULATORY SYSTEMS—AN OVERVIEW [pp.542–543]

1. nutrients (food); 2. wastes; 3. closed; 4. Blood;
5. heart; 6. interstitial fluid; 7. heart; 8. rapidly; 9. capillary; 10. solutes; 11. lymphatic; 12. digestive (respiratory); 13. respiratory (digestive); 14. respiratory;
15. urinary; 16. pulmonary; 17. systemic; 18. oxygen;
19. oxygen; 20. heart; 21. one; 22. heart(s); 23. blood vessels; 24. hearts; 25. open circulatory system; blood is pumped into short tubes that open into spaces in the body's tissues, mingles with tissue fluids, then is reclaimed by open-ended tubes that lead back to the heart; 26. closed circulatory system; blood flow is confined within blood vessels that have continuously connected walls and is pumped by 5 pairs of "hearts."

33.2. CHARACTERISTICS OF BLOOD [pp.544–545]
33.3. *Focus on Health:* BLOOD DISORDERS [p.546]
33.4. BLOOD TRANSFUSION AND TYPING [pp.546–547]

1. connective; 2. pH; 3. phagocytic; 4. 4–5; 5. iron;
6. cell count; 7. 50 to 60; 8. bone marrow; 9. Stem cells; 10. Neutrophils; 11. Platelets; 12. nucleus;
13. four; 14. nucleus; 15. nine; 16. AB; 17. O;
18. Women; 19. Rh-; 20. a. Plasma proteins; b. Red blood cells; c. Neutrophils; d. Lymphocytes;
e. Platelets; 21. stem; 22. red blood (E); 23. platelets (C); 24. neutrophils (A); 25. B (D); 26. T (D);
27. monocytes; 28. macrophages (A); 29. vitamin B_{12} deficiency anemia; 30. hemorrhagic anemia; 31. iron deficiency anemia; 32. sickle-cell anemia; 33. polycythemia; 34. infectious mononucleosis; 35. leukemia.

33.5. HUMAN CARDIOVASCULAR SYSTEM [pp.548–549]
33.6. THE HEART IS A LONELY PUMPER [pp.550–551]

1. pulmonary; 2. systemic; 3. atrium; 4. ventricle;
5. systole; 6. diastole; 7. ventricles; 8. atrial; 9. jugular; 10. superior vena cava; 11. pulmonary; 12. hepatic;
13. renal; 14. inferior vena cava; 15. iliac; 16. femoral;
17. femoral; 18. iliac; 19. abdominal; 20. renal;
21. brachial; 22. coronary; 23. pulmonary; 24. ascending aorta (aortic arch); 25. carotid; 26. aorta; 27. left pulmonary veins; 28. semilunar valve; 29. left ventricle;
30. inferior vena cava; 31. atrioventricular valve;
32. right pulmonary artery; 33. superior vena cava.

33.7. BLOOD PRESSURE IN THE CARDIOVASCULAR SYSTEM [pp.552–553]
33.8. FROM CAPILLARY BEDS BACK TO THE HEART [pp.554–555]
33.9. *Focus on Health:* CARDIOVASCULAR DISORDERS [pp.556–557]
33.10. HEMOSTASIS [p.558]

1. vein; 2. artery; 3. arteriole; 4. capillary; 5. smooth muscle, elastic fibers; 6. valve; 7. artery; 8. capillary;
9. endothelial; 10. capillary bed (diffusion zone); 11. interstitial; 12. Valves; 13. veins (venules); 14. venules (veins); 15. Arteries; 16. ventricles; 17. Arterioles;
18. pressure; 19. stroke; 20. coronary occlusion;
21. Atherosclerotic plaque; 22. low; 23. thrombus;
24. hemostasis; 25. platelet plug formation; 26. coagulation; 27. collagen; 28. insoluble; 29. pressure;
30. blood pressure cannot remain constant because it passes through various kinds of vessels that have varied structures; 31. aorta; 32. pressure; 33. increases;
34. elastic; 35. little; 36. does not drop much; 37. arterioles; 38. nervous; 39. medulla oblongata; 40. beat more slowly; 41. relax; 42. vasodilation.

33.11. LYMPHATIC SYSTEM [pp.558–559]

1. tonsils; 2. right lymphatic duct; 3. thymus gland;
4. thoracic duct; 5. spleen; 6. bone marrow; 7. Lymph;
8. fats; 9. small intestine.

Self-Quiz

1. d; 2. e; 3. c; 4. e; 5. e; 6. a; 7. b; 8. a; 9. a; 10. a.

Chapter 34   Immunity

Russian Roulette, Immunological Style [pp.562–563]

34.1. THREE LINES OF DEFENSE [p.564]
34.2. COMPLEMENT PROTEINS [p.565]
34.3. INFLAMMATION [pp.566–567]

1. mucous; 2. Lysozyme; 3. Gastric; 4. bacterial;
5. clotting; 6. Phagocytic (Macrophage); 7. complement system; 8. Antibodies; 9. histamine; 10. capillaries;
11. Basophils secrete histamine and prostaglandins that change permeability of blood vessels in damaged or irritated tissues. 12. Eosinophils attack parasitic worms by secreting corrosive enzymes. 13. Neutrophils are the most abundant white blood cells; they quickly phagocytize bacteria and reduce them to molecules that can be used for other purposes. 14. Monocytes mature into macrophages that slowly go about engulfing foreign agents and cleaning out dead and damaged cells.

34.4. THE IMMUNE SYSTEM [pp.568–569]
34.5. LYMPHOCYTE BATTLEGROUNDS [p.570]
34.6. CELL-MEDIATED RESPONSES [pp.570–571]
1. nonspecific; 2. immune system; 3. MHC marker; 4. nonself; 5. lymphocytes; 6. B cell; 7. T cell; 8. thymus; 9. viruses; 10. cancer; 11. antigen-presenting cells (macrophages); 12. virgin helper T cells; 13. memory T cells; 14. intracellular; 15. antigen-presenting B cells; 16. effector B cells; 17. antibodies; 18. extracellular; 19. a; 20. d; 21. e; 22. b; 23. c; 24. macrophage; 25. antigen-MHC complex; 26. helper T; 27. cytotoxic T; 28. T cell; 29. primary immune response; 30. memory cells; 31. antigens; 32. natural killer; 33. perforins

34.7. ANTIBODY-MEDIATED RESPONSES
[pp.572–573]
34.8. *Focus on Health:* **CANCER AND IMMUNOTHERAPY** [p.573]
34.9. IMMUNE SPECIFICITY AND MEMORY
[pp.574–575]
1. immunoglobulin; 2. binding sites; 3. antigen; 4. phagocytes; 5. extracellular pathogens; 6. Antibodies; 7. helper T; 8. antigen-MHC complex; 9. virgin B cell; 10. antibody; 11. effector B; 12. memory B cell;

13. c; 14. e; 15. a; 16. b; 17. d; 18. antibody; 19. antigen; 20. virgin B; 21. memory cells; 22. effector cells; 23. clonal selection hypothesis; 24. nonself; 25. Recombination; 26. billion; 27. Cancer; 28. monoclonal (pure) antibodies

34.10. DEFENSES ENHANCED, MISDIRECTED, OR COMPROMISED [pp.576–577]
34.11. *Focus on Health:* **AIDS—THE IMMUNE SYSTEM COMPROMISED** [pp.578–579]
1. immunization; 2. active; 3. primary immune response; 4. memory cells; 5. Allergy; 6. autoimmune response; 7. Myasthenia gravis; 8. Rheumatoid arthritis; 9. human immunodeficiency virus; 10. male homosexuals; 11. enveloped retroviruses; 12. reverse transcriptase; 13. 860; 14. 30,600,000; 15. 21; 16. a, b, c, d; 17. a, b; 18. a, b; 19. b, c, (d); 20. b, c, (d); 21. a, b, c; 22. b, c, (d); 23. a, b, c, d; 24. a, b, c.

Self-Quiz
1. d; 2. b; 3. b; 4. a; 5. a; 6. e; 7. e; 8. e; 9. a; 10. a; 11. H; 12. D; 13. E; 14. J; 15. B; 16. G; 17. C; 18. I; 19. F; 20. A

Chapter 35   Respiration

Conquering Chomolungma [pp.582–583]

35.1. THE NATURE OF RESPIRATION [p.584]
35.2. INVERTEBRATE RESPIRATION [p.585]
35.3. VERTEBRATE RESPIRATION [pp.586–587]
1. gill; 2. countercurrent flow; 3. tracheas; 4. 21; 5. aerobic metabolism; 6. O_2 (oxygen); 7. carbon dioxide (CO_2); 8. respiration; 9. pressure gradient; 10. high; 11. lowest; 12. high; 13. lower; 14. surface area; 15. partial pressure; 16. Hemoglobin; 17. lung; 18. airways; 19. blood; 20. Hypoxia; 21. water out; 22. blood vessel (gill filament); 23. oxygen-poor blood; 24. oxygen-rich blood; 25. water; 26. blood

35.4. HUMAN RESPIRATORY SYSTEM [pp.588–589]
1. ventilating; 2. alveoli; 3. capillaries; 4. carbon dioxide; 5. pleural membrane; 6. larynx; 7. glottis; 8. bronchi; 9. bronchioles; 10. alveoli; 11. vocalizations; 12. intercostal muscles; 13. diaphragm; 14. pharynx; 15. epiglottis; 16. vocal cords; 17. trachea; 18. bronchus; 19. bronchioles; 20. chest cavity; 21. abdominal cavity; 22. alveolar duct; 23. bronchiole; 24. alveolus, alveoli; 25. capillary; 26. a. Alveoli; b. Bronchial tree; c. Diaphragm; d. Larynx; e. Pharynx

35.5. BREATHING—CYCLIC REVERSALS IN AIR PRESSURE GRADIENTS [pp.590–591]
35.6. GAS EXCHANGE AND TRANSPORT
[pp.592–593]
35.7. *Focus on Health:* **WHEN THE LUNGS BREAK DOWN** [pp.594–595]
1. diaphragm; 2. rib cage; 3. increases; 4. drops; 5. intrapleural; 6. diaphragm; 7. intercostal; 8. rises; 9. greater; 10. passive; 11. Erythropoietin; 12. stem cells; 13. oxygen-delivery; 14. partial pressure; 15. Diffusion; 16. carbon dioxide (CO_2); 17. Hemoglobin; 18. bicarbonate (HCO_3^-); 19. systemic (low-pressure); 20. carbonic anhydrase; 21. partial pressure; 22. medulla oblongata; 23. Emphysema; 24. lung cancer.

Self-Quiz
1. c; 2. e; 3. a; 4. c; 5. d; 6. a; 7. a; 8. a; 9. b; 10. d; 11. H; 12. E; 13. J; 14. K; 15. G; 16. N; 17. F; 18. L; 19. B; 20. D; 21. A; 22. C; 23. M; 24. I

Chapter 36 Digestion and Human Nutrition

Lose It—And It Finds Its Way Back [pp.598–599]

36.1. THE NATURE OF DIGESTIVE SYSTEMS [pp.600–601]
36.2. OVERVIEW OF THE HUMAN DIGESTIVE SYSTEM [p.602]

1. Nutrition; 2. carbohydrates; 3. particles; 4. molecules; 5. absorbed; 6. incomplete; 7. circulatory; 8. complete; 9. opening; 10. Motility; 11. secretion; 12. stomach; 13. small intestine; 14. anus; 15. accessory; 16. pancreas; 17. circulatory; 18. respiratory; 19. carbon dioxide; 20. urinary; 21. salivary glands; 22. liver; 23. gall-bladder; 24. pancreas; 25. anus; 26. large intestine (colon); 27. small intestine; 28. stomach; 29. esophagus; 30. pharynx; 31. mouth (oral cavity)

36.3. INTO THE MOUTH, DOWN THE TUBE [p.603]
36.4. DIGESTION IN THE STOMACH AND SMALL INTESTINE [pp.604–605]

1. a. Mouth; b. Salivary glands; c. Stomach; d. Small intestine; e. Pancreas; f. Liver; g. Gallbladder; h. Large intestine; i. Rectum; 2. salivary amylase; 3. epiglottis; 4. esophagus; 5. peristalsis; 6. Pepsin; 7. starches; 8. salivary amylase; 9. disaccharide; 10. stomach; 11. small intestine; 12. amylase; 13. small intestine; 14. disaccharidases; 15. stomach; 16. pepsins; 17. small intestine; 18. pancreas; 19. amino acids; 20. Lipase; 21. small intestine; 22. fatty acid; 23. Bile; 24. gallbladder; 25. lipase; 26. small intestine; 27. small intestine; 28. Carboxypeptidase; 29. Pancreatic nucleases; 30. T; 31. cellular respiration

36.5. ABSORPTION IN THE SMALL INTESTINE [pp.606–607]
36.6. DISPOSITION OF ABSORBED ORGANIC COMPOUNDS [p.608]
36.7. THE LARGE INTESTINE [p.609]

1. Constructing hormones, nucleotides, proteins, and enzymes; 2. Monosaccharides, free fatty acids, and glycerol; 3. The three uses are (a) to construct components of cells and storage forms (such as glycogen) and specialized derivatives such as steroids and acetylcholine; (b) to convert to amino acids as needed; and (c) to serve as a source of energy; 4. glucose; 5. fats (lipids); 6. glycogen; 7. liver; 8. proteins; 9. hormones; 10. nucleotides; 11. fat; 12. ammonia; 13. urea; 14. T; 15. T; 16. small intestine

36.8. HUMAN NUTRITIONAL REQUIREMENTS [pp.610–611]
36.9. VITAMINS AND MINERALS [pp.612–613]
36.10. *Focus on Science:* WEIGHTY QUESTIONS, TANTALIZING ANSWERS [pp.614–615]

1. food pyramid; 2. carbohydrates; 3. bread; 4. 6–11; 5. vegetable; 6. 3–5; 7. fruit; 8. 2–4; 9. apples; 10. berries; 11. meat; 12. proteins; 13. 2–3; 14. amino acids; 15. milk; 16. 2–3; 17. 0; 18. a. 2,070; b. 2,900; c. 1,230; 19. a. 22.2, +8 lb.; b. 20.71, –21 lb.; c 21.65, +3 lb.; 20. Complex carbohydrates; 21. 55 to 60; 22. Phospholipids; 23. energy reserves; 24. 30; 25. essential fatty acids; 26. Proteins; 27. essential; 28. milk (eggs); 29. eggs (milk); 30. Vitamins; 31. Minerals; 32. a. Consult Figure 36.16 (men's column, 6′1″). 178 + 6 = 184 lb. If he is 195 lb., he is 11 lb. overweight; 33. To maintain that weight, he can ingest 195 × 10 = 1,950 kilocalories each day. To reach the desired weight (184 lb.) he must increase his exercise level and eat less of the required food groups until he reaches 184 lb. Thereafter, to maintain that weight, he should ingest 1,840 kilocalories each day. The excess 11 pounds should be lost gradually by adopting an everyday exercise program that over many months would gradually eliminate the excess kilocalories that are stored mostly in the form of fat; the smallest range of serving sizes shown in Figure 36.14 will help keep the total caloric intake to about 1600 kcal. 34. a. 6 servings; b. bread, cereal, rice, pasta; 35. a. 2 servings; b. fruits; 36. a. 3 servings; b. vegetables; 37. a. 2 servings; b. milk, yogurt, or cheese; 38. a. 2 servings; b. legumes, nuts, poultry, fish, or meats; 39. a. Scarcely any; b. added fats and simple sugars.

Self-Quiz

1. b; 2. b; 3. a; 4. e; 5. d; 6. c; 7. d; 8. b; 9. a; 10. b; 11. c; 12. D; 13. C; 14. H; 15. J; 16. B; 17. A; 18. I; 19. F; 20. E; 21. G

Chapter 37 The Internal Environment

Tale of the Desert Rat [pp.618–619]

37.1. URINARY SYSTEM [pp.620–621]

1. metabolism; 2. urine; 3. lungs; 4. sweating; 5. Thirst; 6. metabolism; 7. ammonia; 8. urea; 9. uric acid; 10. kidney; 11. ureter; 12. urinary bladder; 13. urethra; 14. cortex; 15. medulla; 16. ureter; 17. glomerular capillaries; 18. proximal tubule; 19. Bowman's capsule; 20. distal tubule; 21. collecting duct; 22. loop of Henle

37.2. URINE FORMATION [pp.622–623]
37.3. *Focus on Health:* WHEN KIDNEYS BREAK DOWN [p.624]

1. T; 2. T; 3. kidneys; 4. nephrons; 5. bloodstream;
6. arteriole; 7. Bowman's capsule; 8. glomerulus;
9. blood pressure; 10. solutes; 11. blood capillaries
(peritubular capillaries); 12. tubular secretion;
13. Ureters; 14. urinary bladder; 15. urethra; 16. aldosterone; 17. sodium; 18. less; 19. blood pressure;
20. renal failure; 21. infectious agents 22. ADH; 23. inhibited; 24. volume

37.4. THE ACID–BASE BALANCE [p.624]
37.5. ON FISH, FROGS, AND KANGAROO RATS
   [p.625]

37.6. MAINTAINING BODY TEMPERATURE
   [pp.626–627]
1. kidneys; 2. H^+; 3. 7.43; 4. Acids; 5. bases; 6. lowered; 7. H^+; 8. bicarbonate (HCO_3^-); 9. urinary;
10. loops of Henle; 11. water; 12. water; 13. solutes;
14. very dilute; 15. hypothalamus; 16. central; 17. pilomotor response; 18. Peripheral vasoconstriction;
19. hypothermia; 20. T; 21. T; 22. T; 23. D; 24. B;
25. A; 26. C; 27. G; 28. F; 29. E

Self-Quiz
1. d; 2. c; 3. e; 4. a; 5. d; 6. d; 7. d; 8. b; 9. e; 10. c;
11. b

Chapter 38   Reproduction and Development

From Frog to Frog and Other Mysteries [p.630–631]

38.1. THE BEGINNING: REPRODUCTIVE MODES
   [pp.632–633]
38.2. STAGES OF DEVELOPMENT—AN OVERVIEW
   [pp.634–635]
1. asexual; 2. environmental; 3. variation; 4. courtship;
5. gamete; 6. reproductive; 7. potential; 8. offspring;
9. watery; 10. gametes; 11. internal; 12. reproductive;
13. nourish; 14. yolk; 15. tiny; 16. birds; 17. embryo;
18. F; 19. B; 20. C; 21. A; 22. E; 23. D; 24. a. mesoderm; b. ectoderm; c. endoderm; d. mesoderm; e.
ectoderm; f. mesoderm; g. endoderm; h. mesoderm;
i. mesoderm.

38.3. EARLY MARCHING ORDERS [pp.636–637]
38.4. HOW DO SPECIALIZED TISSUES AND
   ORGANS FORM? [pp.638–639]
1. J; 2. E; 3. F; 4. H; 5. M; 6. C; 7. I; 8. D; 9. K;
10. G; 11. A; 12. B; 13. L; 14. b; 15. b; 16. a; 17. b;
18. b; 19. a; 20. b; 21. b; 22. a; 23. a; 24. b; 25. a;
26. b; 27. b.

38.5. PATTERN FORMATION [pp.640–641]
1. mutations; 2. pattern formation; 3. master; 4. regulatory; 5. embryo; 6. suppressed; 7. selective; 8. gradients; 9. Homeotic; 10. master; 11. D; 12. E; 13. A;
14. B; 15. C; 16. induction; 17. mesoderm (ectoderm);
18. ectoderm (mesoderm); 19. mesoderm; 20. apical
ectodermal; 21. ceases; 22. more; 23. developing;
24. morphogens; 25. distance; 26. selective expression;
27. sequence; 28. homeotic; 29. regulatory; 30. induc-
tions; 31. development; 32. organ; 33. phyletic;
34. organizer.

38.6. REPRODUCTIVE SYSTEM OF HUMAN MALES
   [pp.642–643]
38.7. MALE REPRODUCTIVE FUNCTION
   [pp.644–645]

1. mitotic (mitosis); 2. seminiferous; 3. meiosis;
4. sperm; 5. epididymis; 6. vas deferens; 7. urethra;
8. Seminal vesicles; 9. Prostate gland; 10. Bulbourethral;
11. Leydig; 12. Testosterone; 13. testosterone; 14. anterior; 15. hypothalamus; 16. decrease; 17. LH; 18. Sertoli; 19. increased; 20. depressed; 21. Sertoli;
22. hypothalamus; 23. anterior pituitary; 24. Sertoli
cells; 25. Leydig cells.

38.8. REPRODUCTIVE SYSTEM OF HUMAN
   FEMALES [pp.646–647]
38.9. FEMALE REPRODUCTIVE FUNCTION
   [pp.648–649]
38.10. VISUAL SUMMARY OF THE MENSTRUAL
   CYCLE [p.650]
1. ovary; 2. oviduct; 3. uterus; 4. cervix;
5. myometrium; 6. endometrium; 7. vagina; 8. labia
majora; 9. labia minora; 10. clitoris; 11. urethra;
12. Meiosis; 13. I; 14. 300,000; 15. follicle; 16. hypothalamus; 17. anterior; 18. follicle; 19. oocyte;
20. Glycoprotein; 21. zona pellucida; 22. FSH (LH);
23. LH (FSH); 24. estrogen; 25. blood; 26. oocyte;
27. cytoplasm; 28. secondary oocyte; 29. polar bodies;
30. haploid; 31. gametes; 32. sexual; 33. pituitary;
34. LH; 35. follicle; 36. enzymes; 37. secondary oocyte;
38. ovulation; 39. pregnancy; 40. endometrium;
41. progesterone; 42. endometrium; 43. ovulation;
44. cervical; 45. mucus; 46. corpus luteum; 47. granulosa; 48. Progesterone; 49. blastocyst; 50. mucus;
51. vagina; 52. endometrium; 53. twelve; 54. hypothalamus; 55. follicles; 56. blastocyst; 57. prostaglandins;
58. progesterone (estrogen); 59. estrogen (progesterone);
60. blastocyst; 61. menstrual; 62. menopause.

38.11. PREGNANCY HAPPENS [p.651]
38.12. FORMATION OF THE EARLY EMBRYO
   [pp.652–653]
38.13. EMERGENCE OF THE VERTEBRATE BODY
   PLAN [p.654]

38.14. WHY IS THE PLACENTA SO IMPORTANT?
[p.655]

38.15. EMERGENCE OF DISTINCTLY HUMAN FEATURES [pp.656–657]

38.16. *Focus on Health:* **MOTHER AS PROVIDER, PROTECTOR, POTENTIAL THREAT** [pp.658–659]

1. sperm; 2. oviduct; 3. enzymes; 4. II; 5. ovum; 6. diploid; 7. D; 8. G; 9. H; 10. F; 11. B; 12. A; 13. E; 14. C; 15. C; 16. B; 17. F; 18. H; 19. A; 20. I; 21. G; 22. E; 23. D; 24. D; 25. B; 26. A; 27. C; 28. C; 29. A; 30. B; 31. maternal; 32. blood vessels; 33. maternal; 34. uterus; 35. fetal; 36. embryonic; 37. umbilical cord; 38. intervillus; 39. chorionic villus; 40. amniotic; 41. C; 42. E; 43. A; 44. D; 45. B; 46. yolk sac; 47. embryonic disk; 48. amniotic cavity; 49. chorionic cavity; 50. primitive streak; 51. neural groove; 52. future brain; 53. somites; 54. pharyngeal (gill) slits; 55. four; 56. pharyngeal (gill) slits; 57. somites; 58. five; 59. forelimb (upper); 60. d; 61. b; 62. f; 63. a; 64. e; 65. g; 66. c; 67. a; 68. d; 69. a; 70. b.

38.17. FROM BIRTH ONWARD [pp.660–661]

38.18. CONTROL OF HUMAN FERTILITY [pp.662–663]

1. E; 2. C; 3. I; 4. F; 5. J; 6. A; 7. G; 8. D; 9. H; 10. B; 11. J; 12. F; 13. K; 14. D; 15. H; 16. C; 17. L; 18. A; 19. E; 20. I; 21. B; 22. G.

38.19. *Focus on Health:* **SEXUALLY TRANSMITTED DISEASES** [pp.664–665]

38.20. *Focus on Bioethics:* **TO SEEK OR END PREGNANCY** [p.666]

1. A, F; 2. A, C; 3. D, E; 4. F; 5. F; 6. D; 7. C; 8. F; 9. C; 10. C, F; 11. C; 12. F; 13. D; 14. B; 15. A, B, C, D, E, F; 16. In vitro; 17. Abortion.

Self-Quiz

1. b; 2. a; 3. b; 4. a; 5. d; 6. e; 7. c; 8. e; 9. c; 10. a; 11. b; 12. d; 13. e; 14. d; 15. a; 16. d; 17. e; 18. a; 19. b; 20. c.

Chapter 39   Population Ecology

Tales of Nightmare Numbers [pp.670–671]

39.1. CHARACTERISTICS OF POPULATIONS [p.672]

39.2. *Focus on Science:* **ELUSIVE HEADS TO COUNT** [p.673]

39.3. POPULATION SIZE AND EXPONENTIAL GROWTH [pp.674–675]

1. K; 2. H; 3. D; 4. I; 5. B; 6. F; 7. A; 8. G; 9. J; 10. L; 11. C; 12. E; 13. immigration; 14. emigration; 15. migrations; 16. zero population; 17. per capita; 18. heads; 19. 0.4; 20. 0.1; 21. reproduction; 22. time; 23. 0.3; 24. $G = rN$; 25. J; 26. exponential; 27. doubling; 28. biotic potential; 29. a. It increases; b. It decreases; c. It must increase; 30. population growth rate; 31. It must decrease; 32. 100,000; 33. a. 100.000; b. 300,000.

39.4. LIMITS ON THE GROWTH OF POPULATIONS [pp.676–677]

39.5. LIFE HISTORY PATTERNS [pp.678–679]

39.6. *Focus on Science:* **NATURAL SELECTION AND THE GUPPIES OF TRINIDAD** [p.679]

1. limiting; 2. Carrying capacity; 3. logistic; 4. carrying capacity; 5. increases; 6. decreases; 7. density-dependent; 8. density-independent; 9. dependent; 10. carrying capacity; 11. independent; 12. life history; 13. insurance; 14. cohort; 15. life; 16. "survivorship"; 17. Survivorship; 18. III; 19. I; 20. II; 21. I; 22. F; 23. A, E, H; 24. B, C, D; 25. G.

39.7. HUMAN POPULATION GROWTH [pp.680–681]

39.8. CONTROL THROUGH FAMILY PLANNING [pp.682–683]

39.9. POPULATION GROWTH AND ECONOMIC DEVELOPMENT [pp.684–685]

39.10. SOCIAL IMPACT OF NO GROWTH [p.685]

1. a. 1962–1963; b. 2025 or sooner; c. Depends on the age and optimism of the reader; 2. 6 billion; 3. T; 4. short; 5. T; 6. sidestepped; 7. cannot; 8. 9; 9. resources; 10. pollution; 11. Family planning; 12. birth; 13. two; 14. female; 15. fertility; 16. 3; 17. 6.5; 18. baby-boomers; 19. one-third; 20. thirties; 21. China; 22. stabilize; 23. reproductive; 24. D; 25. C; 26. B; 27. A; 28. B; 29. D; 30. A; 31. C; 32. industrial; 33. decreasing; 34. smaller; 35. transitional; 36. transitional; 37. economic; 38. immigration; 39. 16; 40. 4.7; 41. 21; 42. 25; 43. fifty; 44. 25; 45. 1; 46. 3; 47. 3; 48. 12.9; 49. 258; 50. growth; 51. social; 52. older; 53. economic; 54. postponed; 55. cultural; 56. carrying capacity.

Self-Quiz

1. d; 2. a; 3. b; 4. a; 5. d; 6. d; 7. a; 8. b; 9. a; 10. a; 11. c; 12. c; 13. b; 14. d; 15. b; 16. a; 17. a.

Chapter 40   Community Interactions

No Pigeon Is an Island [pp.688–689]

40.1. WHICH FACTORS SHAPE COMMUNITY STRUCTURE? [p.690]
40.2. MUTUALISM [p.691]
1. habitat; 2. community; 3. niche; 4. fundamental;
5. realized; 6. neutral; 7. directly; 8. commensalistic;
9. mutualism; 10. mutualistic; 11. interspecific;
12. Predation (Parasitism); 13. parasitism (predation);
14. symbiosis; 15. a. It cannot complete its life cycle in
any other plant, and its larvae eat only yucca seeds.
b. The yucca moth is the plant's only pollinator.

40.3. COMPETITIVE INTERACTIONS [pp.692–693]
40.4. PREDATION AND PARASITISM [pp.694–695]
40.5. *Focus on the Environment:* **THE COEVOLUTIONARY ARMS RACE** [pp.696–697]
1. Intraspecific; 2. Interspecific; 3. Interspecific;
4. competitive exclusion; 5. keystone; 6. keystone;
7. mussels; 8. resource petitioning; 9. pigeons;
10. size; 11. root; 12. E; 13. C; 14. G; 15. B; 16. F;
17. D; 18. A; 19. camouflage; 20. warning coloration;
21. Mimicry; 22. Moment of truth; 23. adaptations;
24. F; 25. G; 26. H; 27. B; 28. I; 29. C; 30. E; 31. H;
32. E; 33. G; 34. F; 35. C.

40.6. FORCES CONTRIBUTING TO COMMUNITY STABILITY [pp.698–699]
40.7. COMMUNITY INSTABILITY [p.700]
40.8. *Focus on the Environment:* **EXOTIC AND ENDANGERED SPECIES** [pp.700–701]

1. succession; 2. Pioneer; 3. pioneers; 4. climax;
5. primary; 6. replacement; 7. secondary; 8. facilitate;
9. succession; 10. climax-pattern; 11. community;
12. pioneers; 13. fires; 14. natural; 15. active; 16. b;
17. a; 18. b; 19. b; 20. a; 21. a; 22. b; 23. a; 24. a;
25. B; 26. D; 27. E; 28. A; 29. C.

40.9. PATTERNS OF BIODIVERSITY [pp.702–703]
1. a. Resource availability tends to be higher and more
reliable. Tropical latitudes have more sunlight of greater
intensity, rainfall amount is higher, and the growing
season is longer. Vegetation grows all year long to support diverse herbivores, etc. b. Species diversity might
be self-reinforcing. When a greater number of plant
species compete and coexist, a greater number of herbivore species evolve because no herbivore can overcome
the chemical defenses of all kinds of plants. Then more
predators and parasites evolve in response to the diversity of prey and hosts. c. The rates of speciation in the
tropics have exceeded those of background extinction. At
higher latitudes, biodiversity has been suppressed during times of mass extinction. 2. tropics; 3. Iceland;
4. biodiversity; 5. Iceland; 6. dispersal; 7. distance;
8. area; 9. Larger; 10. diversity; 11. targets; 12. biodiversity; 13. small; 14. small; 15. immigration; 16. extinction; 17. immigration; 18. extinction; 19. Island C.

Self-Quiz

1. b; 2. b; 3. c; 4. a; 5. b; 6. d; 7. e; 8. b; 9. d; 10. a;
11. d; 12. d.

Chapter 41   Ecosystems

Crêpes for Breakfast, Pancake Ice for Dessert
[pp.706–707]

41.1. THE NATURE OF ECOSYSTEMS [pp.708–709]
41.2. HOW DOES ENERGY FLOW THROUGH ECOSYSTEMS? [pp.710–711]
41.3. *Focus on Science:* **ENERGY FLOW AT SILVER SPRINGS, FLORIDA** [p.712]
1. b; 2. c; 3. b; 4. a; 5. b; 6. b; 7. b; 8. b; 9. b; 10. b;
11. a; 12. b; 13. c; 14. a; 15. b; 16. producers; 17. heterotrophs; 18. herbivores; 19. carnivores; 20. omnivores; 21. parasites; 22. decomposers; 23. detritivores;
24. open; 25. energy; 26. nutrient; 27. energy; 28. nutrients; 29. trophic; 30. food chain; 31. food web;
32. productivity; 33. net; 34. net; 35. grazing; 36. detrital; 37. pyramid; 38. producers; 39. biomass;
40. biomass; 41. smallest; 42. energy; 43. trophic;
44. large; 45. 1; 46. 6; 47. 10; 48. low; 49. 4;

50. open; 51. input; 52. cannot; 53. can; 54. have;
55. E; 56. B; 57. D; 58. C; 59. E; 60. D; 61. E.

41.4. BIOGEOCHEMICAL CYCLES—AN OVERVIEW [p.713]
41.5. HYDROLOGIC CYCLE [p.714]
41.6. SEDIMENTARY CYCLES [p.715]
1. Usually as mineral ions such as ammonium (NH_4^+);
2. Inputs from the physical environment and the cycling
activities of decomposers and detritivores; 3. The
amount of a nutrient being cycled through the ecosystem
is greater; 4. Common sources are rainfall or snowfall,
metabolism (such as nitrogen fixation), and weathering
of rocks; 5. Losses of mineral ions occurs by runoff;
6. a. Oxygen and hydrogen move in the form of water
molecules; b. A large portion of the nutrient is in the
form of atmospheric gas such as carbon and nitrogen
(mainly CO_2); c. Nutrients are not in gaseous forms;

nutrients move from land to the seafloor and only "return" to land through geological uplifting of long duration; phosphorus is an example; 7. F; 8. H; 9. D; 10. B; 11. A (C); 12. E; 13. G (H); 14. C; 15. watershed; 16. soil; 17. streams; 18. transpiration; 19. vegetation; 20. nutrients; 21. calcium; 22. calcium; 23. biomass; 24. deforestation; 25. cycle; 26. ecosystems; 27. reservoir; 28. biogeochemical; 29. phosphates; 30. ocean; 31. shelves; 32. crustal; 33. geochemical; 34. ecosystem; 35. organisms; 36. ionized; 37. plants; 38. herbivores; 39. Decomposition; 40. ecosystem; 41. Fertilizers; 42. lakes; 43. growth; 44. nitrogen; 45. potassium; 46. sediments; 47. phosphorus; 48. algal; 49. eutrophication.

41.7. CARBON CYCLE [pp.716–717]
41.8. *Focus on the Environment:* FROM GREENHOUSE GASES TO A WARMER PLANET? [pp.718–719]

41.9. NITROGEN CYCLE [pp.720–721]
41.10. *Focus on Science:* ECOSYSTEM MODELING [p.721]
1. C; 2. D; 3. F; 4. B; 5. G; 6. A; 7. E; 8. D; 9. E; 10. C; 11. A; 12. B; 13. Soil nitrogen compounds are vulnerable to being leached and lost from the soil; some fixed nitrogen is lost to air by denitrification; nitrogen fixation comes at high metabolic cost to plants that are symbionts of nitrogen-fixing bacteria; losses of nitrogen are enormous in agricultural regions through the tissues of harvested plants, soil erosion, and leaching processes; 14. Ecosystem modeling; 15. DDT; 16. water; 17. biological magnification; 18. concentrated; 19. chain; 20. consumer; 21. metabolic.

Self-Quiz
1. c; 2. c; 3. d; 4. b; 5. a; 6. a; 7. b; 8. a; 9. a; 10. b; 11. b; 12. d.

Chapter 42   The Biosphere

Does a Cactus Grow in Brooklyn? [pp.724–725]

42.1. AIR CIRCULATION PATTERNS AND REGIONAL CLIMATES [pp.726–727]
42.2. OCEANS, LANDFORMS, AND REGIONAL CLIMATES [pp.728–729]
1. M; 2. D; 3. H; 4. K; 5. B; 6. J; 7. F; 8. C; 9. A; 10. L; 11. E; 12. I; 13. N; 14. G; 15. equatorial; 16. warm; 17. rises or ascends; 18. moisture; 19. descends; 20. moisture; 21. ascends; 22. moisture; 23. descends; 24. east (easterlies); 25. west (westerlies); 26. tropical; 27. warm; 28. cool; 29. cold; 30. solar (sun's); 31. rotation.

42.3. REALMS OF BIODIVERSITY [pp.730–731]
42.4. SOILS OF MAJOR BIOMES [p.732]
42.5. DESERTS [p.733]
42.6. DRY SHRUBLANDS, DRY WOODLANDS, AND GRASSLANDS [pp.734–735]
1. H; 2. F; 3. A; 4. G; 5. C; 6. B; 7. J; 8. E; 9. D; 10. I; 11. K; 12. C; 13. B; 14. D; 15. A; 16. E; 17. d; 18. a; 19. d; 20. b; 21. c; 22. a; 23. e; 24. d; 25. e; 26. b.

42.7. TROPICAL RAIN FORESTS AND OTHER BROADLEAF FORESTS [pp.736–737]
42.8. CONIFEROUS FORESTS [p.738]
42.9. TUNDRA [p.739]
1. d; 2. c; 3. c; 4. a; 5. b; 6. b; 7. a; 8. d; 9. c, d; 10. b; 11. c; 12. a; 13. b; 14. b; 15. d; 16. d.

42.10. FRESHWATER PROVINCES [pp.740–741]
42.11. THE OCEAN PROVINCES [pp.742–743]
1. lake; 2. littoral; 3. limnetic; 4. plankton; 5. profundal; 6. overturns; 7. 4; 8. spring; 9. thermocline; 10. cools; 11. fall; 12. down; 13. up; 14. higher; 15. short; 16. Oligotrophic; 17. eutrophic; 18. eutrophication; 19. Streams; 20. runs; 21. benthic (C); 22. pelagic (A); 23. neritic (B); 24. oceanic (D); 25. a; 26. c; 27. c; 28. b; 29. c; 30. a; 31. b; 32. a; 33. b; 34. c; 35. a; 36. b; 37. a; 38. c; 39. b; 40. c; 41. a; 42. b; 43. b; 44. c.

42.12. CORAL REEFS AND CORAL BANKS [pp.744–745]
42.13. LIFE ALONG THE COASTS [pp.746–747]
42.14. *Focus on Science:* RITA IN THE TIME OF CHOLERA [pp.748–749]
1. a. Atolls are ring-shaped coral reefs that enclose or almost enclose a shallow lagoon; b. Fringing reefs form next to the land's edge in regions of limited rainfall, as on the leeward side of tropical islands; c. Barrier reefs form around islands or parallel with the shore of a continent. A calm lagoon forms behind them; 2. d; 3. a; 4. c; 5. e; 6. c; 7. a; 8. b; 9. d; 10. b; 11. a; 12. b; 13. e; 14. d; 15. a; 16. d; 17. e; 18. a; 19. c; 20. e; 21. c; 22. a.

Self-Quiz
1. d; 2. d; 3. b; 4. d; 5. e; 6. b; 7. c; 8. c; 9. d; 10. c; 11. c; 12. a.

Chapter 43   Human Impact on the Biosphere

An Indifference of Mythic Proportions [pp.752–753]

43.1. AIR POLLUTION—PRIME EXAMPLES
[pp.754–755]
**43.2. OZONE THINNING—GLOBAL LEGACY OF
AIR POLLUTION** [p.756]
1. Pollutants; 2. thermal inversion; 3. winters; 4. industrial smog; 5. photochemical smog; 6. oxygen; 7. nitrogen dioxide; 8. photochemical; 9. gasoline; 10. PANs; 11. sulfur (nitrogen); 12. nitrogen (sulfur); 13. sulfur; 14. nitrogen; 15. nitrogen; 16. acid deposition; 17. sulfuric (nitric); 18. nitric (sulfuric); 19. acid rain; 20. 5; 21. Chlorofluorocarbons; 22. c; 23. d; 24. e; 25. b; 26. a; 27. b; 28. f; 29. f; 30. a; 31. b; 32. b; 33. d; 34. a (b); 35. d; 36. f; 37. c; 38. d; 39. e; 40. f; 41. e; 42. c; 43. d.

**43.3. WHERE TO PUT SOLID WASTES? WHERE TO
PRODUCE FOOD?** [p.757]
**43.4. DEFORESTATION—AN ASSAULT ON FINITE
RESOURCES** [pp.758–759]
43.5. *Focus on Bioethics:* **YOU AND THE TROPICAL
RAIN FOREST** [p.760]
1. F; 2. H (G); 3. I; 4. E; 5. J; 6. B; 7. C; 8. A; 9. D; 10. G.

43.6. WHO TRADES GRASSLANDS FOR DESERTS?
[p.761]
43.7. A GLOBAL WATER CRISIS [pp.762–763]

1. desertification; 2. overgrazing on marginal lands; 3. domestic cattle; 4. native wild herbivores; 5. deserts; 6. salty; 7. desalinization; 8. energy; 9. agriculture; 10. salinization; 11. water table; 12. saline; 13. 20; 14. pollution; 15. wastewater; 16. tertiary; 17. 55–66; 18. oil; 19. policies; 20. a. Screens and settling tanks remove sludge, which is dried, burned, dumped in landfills, or treated further; chlorine is often used to kill pathogens in water, but does not kill them all. b. Microbial populations are used to break down organic matter after primary treatment but before chlorination. c. It removes nitrogen, phosphorus, and toxic substances, including heavy metals, pesticides, and industrial chemicals; it is largely experimental and expensive.

43.8. A QUESTION OF ENERGY INPUTS [pp.764–765]
43.9. ALTERNATIVE ENERGY SOURCES [p.766]
43.10. *Focus on Bioethics:* **BIOLOGICAL PRINCIPLES
AND THE HUMAN IMPERATIVE** [p.767]
1. J-shaped; 2. increased numbers of energy users and to extravagant consumption and waste. 3. Net energy; 4. plants; 5. next; 6. decreases; 7. global acid deposition; 8. less; 9. meltdown; 10. have not; 11. solar-hydrogen energy; 12. wind farms; 13. Fusion power; 14. C (N); 15. B (N); 16. F (N); 17. A (R); 18. D (N); 19. E (R).

Self-Test
1. c; 2. b; 3. a; 4. b; 5. d; 6. d; 7. c; 8. c; 9. d; 10. b.

Chapter 44   An Evolutionary View of Behavior

Deck the Nest With Sprigs of Green Stuff
[pp.770–771]

44.1. BEHAVIOR'S HERITABLE BASIS [pp.772–773]
44.2. LEARNED BEHAVIOR [p.774]
1. C; 2. F; 3. E; 4. B; 5. G; 6. A; 7. D; 8. the banana slug; 9. ate; 10. inland; 11. were not; 12. genetic; 13. pineal gland; 14. increase; 15. estrogen; 16. testosterone; 17. Genes; 18. a. Cuckoo birds are social parasites in that adult females lay eggs in the nests of other bird species; young cuckoos instinctively eliminate the natural-born offspring (eggs are maneuvered onto their backs and pushed out of the nest) and then receive the undivided attention of their unsuspecting foster parents. b. Young toads instinctively capture edible insects with sticky tongues; if a bumblebee is captured and then stings the tongue, the toad learns to leave bumblebees alone; 19. E; 20. C; 21; F; 22. A; 23. D; 24. B

44.3. THE ADAPTIVE VALUE OF BEHAVIOR [p.775]
1. Natural selection; 2. increase; 3. evolution; 4. adaptive; 5. reproductive; 6. adaptive; 7. individual's; 8. a. Consideration of individual survival and production of offspring. b. Any behavior that promotes propagation of an individual's genes and tends to occur at increased frequency in future generations. c. Cooperative, interdependent relationships among individuals of the same species. d. Within a population, any behavior that increases an individual's chances to produce or protect offspring of its own, regardless of the consequences for the population. e. Within a population, a self-sacrificing behavior. The individual behaves in a way that helps others but decreases its own chances to produce offspring.

44.4. COMMUNICATION SIGNALS [pp.776–777]
1. i; 2. c; 3. e; 4. a; 5. g; 6. j; 7. f; 8. b; 9. h; 10. c; 11. g; 12. d.

44.5. MATES, PARENTS, AND INDIVIDUAL REPRODUCTIVE SUCCESS [pp.778–779]

1. a. Hangingflies; b. Sage grouse; c. Lions, sheep, elk, elephant seals, and bison; d. Caspian terns.

44.6. BENEFITS OF LIVING IN SOCIAL GROUPS [pp.780–781]
44.7. COSTS OF LIVING IN SOCIAL GROUPS [p.782]
44.8. EVOLUTION OF ALTRUISM [p.783]
44.9. *Focus on the Science:* WHY SACRIFICE YOURSELF? [pp.784–785]

1. C; 2. A; 3. E; 4. D; 5. B; 6. a. Royal penguins, herring gulls, cliff swallows, and prairie dogs live in huge colonies and must compete for a share of the same food resources. b. Under crowded living conditions, the individual and its offspring are more likely to be weakened by pathogens and parasites that are more readily transmitted from host to host in crowded groups. Plagues spread like wildfire through densely crowded human populations; this is especially the case in settlements and cities with chronic infestations of rats and fleas, and with inadequate or nonexistent sewage treatment and medical care. c. Breeding pairs of herring gulls will quickly cannibalize their neighbors' eggs or young chicks; long-lived lions compete for permanent hunting territories even though they can live for an extended time between kills; a lion pride of three or more actually eat less well than one or a pair of lions; male lions intent on taking over a pride will show infanticidal behavior and will kill the cubs; group living is costly for lionesses in terms of food intake and reproductive success; they stick together to defend territories against smaller groups of rivals; aggressive males almost always kill the cubs of a single lioness, but occasionally a group of two or more lionesses can save some of the cubs.
7. subordinate; 8. altruistic; 9. reproductive; 10. dominant; 11. nonbreeding; 12. altruistic; 13. insect; 14. worker; 15. genes; 16. indirect selection; 17. parenting; 18. indirect; 19. reproduce; 20. altruistic; 21. genes; 22. vertebrates; 23. nonbreeding; 24. perpetuate; 25. genes (alleles).

44.10. AN EVOLUTIONARY VIEW OF HUMAN SOCIAL BEHAVIOR [p.786]

1. human; 2. a trait valuable in gene transmission; 3. adopt; 4. strangers; 5. can; 6. will; 7. nonrelated; 8. is; 9. are; 10. nonrelative.

Self-Quiz

1. c; 2. d; 3. b; 4. d; 5. a; 6. b; 7. b; 8. c; 9. a; 10. c; 11. b; 12. b; 13. d; 14. c; 15. a.